U0223079

影印版说明

本书总结了大量关于金属塑性成形过程中微观组织演变的建模和控制的最新研究结果，深入系统地讨论了钢变形过程中的微观组织演化规律，综述了钢相变的建模模拟、统一本构方程和微合金钢的加工硬化，分析了包括铝材成形中的时效行为等在内的其他材料成形加工中的微观组织演化现象。

金属成形过程中的组织控制是人们长期致力研究的课题。本书适合冶金、材料加工等行业的工程技术人员使用，也可供高等院校相关专业的师生参考。

Jianguo Lin　英国皇家工程院院士、伦敦帝国理工学院机械系教授。

Daniel Balint　英国伦敦帝国理工学院机械系讲师。

Maciej Pietrzyk　波兰矿业冶金学院冶金和材料科学系教授。

材料科学与工程图书工作室

联系电话 0451-86412421

0451-86414559

邮　　箱 yh_bj@aliyun.com

xuyaying81823@gmail.com

zhxh6414559@aliyun.com

WOODHEAD PUBLISHING IN MATERIALS

影印版

金属成形过程中的组织演化

Microstructure evolution in metal forming processes

Edited by Jianguo Lin, Daniel Balint
and Maciej Pietrzyk

哈爾濱工業大學出版社
HARBIN INSTITUTE OF TECHNOLOGY PRESS

黑版贸审字08-2017-080号

Microstructure evolution in metal forming processes
Jianguo Lin, Daniel Balint, Maciej Pietrzyk
ISBN: 978-0-85709-074-4

Elsevier (Singapore) Pte Ltd.
3 Killiney Road
#08-01 Winsland House I
Singapore 239519
Tel: (65) 6349-0200
Fax: (65) 6733-1817

First Published 2017

图书在版编目（CIP）数据

金属成形过程中的组织演化=Microstructure evolution in metal forming processes：英文 /（英）林建国（Jianguo Lin），（英）丹尼尔・巴林（Daniel Balint），（波）麦克杰・P（Maciej Pietrzyk）主编.—影印本.—哈尔滨：哈尔滨工业大学出版社，2017.10
ISBN 978-7-5603-6392-9
Ⅰ.①金… Ⅱ.①林… ②丹… ③麦… Ⅲ.①金属材料－成形－研究－英文 Ⅳ.①TG14

中国版本图书馆CIP数据核字（2017）第001922号

责任编辑 许雅莹 杨 桦 张秀华
出版发行 哈尔滨工业大学出版社
社 址 哈尔滨市南岗区复华四道街10号 邮编 150006
传 真 0451-86414749
网 址 http://hitpress.hit.edu.cn
印 刷 哈尔滨市石桥印务有限公司
开 本 660mm × 980mm 1/16 印张 25.5
版 次 2017 年 10 月第 1 版 2017 年 10 月第 1 次印刷
书 号 ISBN 978-7-5603-6392-9
定 价 260.00元

Microstructure evolution in metal forming processes

Edited by
Jianguo Lin, Daniel Balint and Maciej Pietrzyk

Oxford Cambridge Philadelphia New Delhi

Contents

Contributor contact details *xi*

Part I General principles 1

1 Understanding and controlling microstructural evolution in metal forming: an overview 3
T. Ishikawa, Nagoya University, Japan

1.1 Introduction 3
1.2 How microstructure evolves in metal forming 4
1.3 Models for predicting the microstructural evolution of carbon steels 6
1.4 Strengthening mechanisms and relation between microstructure and mechanical properties 10
1.5 Emerging techniques to control microstructure evolution in metal forming 12
1.6 Advanced high-strength steels (AHSS) 13
1.7 Conclusion and future trends 14
1.8 References 15

2 Techniques for modelling microstructure in metal forming processes 17
Y. Chastel, Renault, France and R. Logé and M. Bernacki, MINES ParisTech, France

2.1 Introduction: importance of microstructure prediction in metal forming 17
2.2 General features of models based on state variables 18
2.3 Coupling between homogeneous microstructure description and constitutive laws 20
2.4 Mean field approach: an example of discontinuous dynamic recrystallization 24
2.5 Recrystallization modelling at the microscopic scale: overview and future trends 28

2.6 Future trends 32
2.7 References 32

3 Modelling techniques for optimizing metal forming processes 35
J. Kusiak, D. Szeliga and Ł. Sztangret, AGH – University of Science and Technology, Poland

3.1 Introduction 35
3.2 Optimization strategies 36
3.3 Nature-inspired optimization techniques: genetic algorithms, evolutionary algorithms, particle swarm optimization and simulated annealing 42
3.4 Application of metamodelling and optimization strategies in metal forming – case studies 52
3.5 Conclusions and future trends 62
3.6 Acknowledgements 64
3.7 References 64

4 Recrystallisation and grain growth in hot working of steels 67
B. López and J. M. Rodriguez-Ibabe, CEIT and Tecnun (University of Navarra), Spain

4.1 Introduction 67
4.2 Grain refinement due to recrystallisation 68
4.3 Grain growth after recrystallisation 76
4.4 Recrystallisation–precipitation interactions 77
4.5 Modelling methods 88
4.6 Case studies in metal forming 104
4.7 Sources of further information and advice 109
4.8 References 109

5 Severe plastic deformation for grain refinement and enhancement of properties 114
A. Rosochowski, University of Strathclyde, UK and L. Olejnik, Warsaw University of Technology, Poland

5.1 Introduction 114
5.2 Principles of severe thermo-mechanical treatment 117
5.3 Severe plastic deformation (SPD) processes 122
5.4 Properties of ultrafine-grained (UFG) metals produced by SPD 129
5.5 Applications of UFG metals 131
5.6 Sources of further information and advice 134
5.7 References 135

Part II Microstructure evolution in the processing of steel 143

6 Modelling phase transformations in steel 145
M. Pietrzyk, AGH – University of Science and Technology, Poland and R. Kuziak, Institute for Ferrous Metallurgy, Poland

6.1 Introduction 145
6.2 Phase transformation in steels 145
6.3 Experimental techniques 146
6.4 Modelling methods 153
6.5 Application in rolling and annealing of dual-phase steels 171
6.6 Discussion and future trends 175
6.7 Sources of further information and advice 177
6.8 References 177

7 Determining unified constitutive equations for modelling hot forming of steel 180
J. Lin, Imperial College London, UK, J. Cao, RTC Innovation Ltd, UK and D. Balint, Imperial College London, UK

7.1 Introduction 180
7.2 The form of unified constitutive equations for hot metal forming 181
7.3 Methods for integrating constitutive equations 185
7.4 Objective functions for optimisation 191
7.5 Optimisation methods for determining the material constants in constitutive equations 198
7.6 Case studies 201
7.7 Conclusion 207
7.8 References 207

8 Modelling phase transformations in hot stamping and cold die quenching of steels 210
J. Cai and J. Lin, Imperial College London, UK and J. Wilsius, ArcelorMittal, France

8.1 Introduction 210
8.2 Phase transformations on heating: experimentation and modelling 214
8.3 Phase transformations on cooling: experimentation and modelling 222
8.4 Conclusion and future trends 234
8.5 References 235

9 Modelling microstructure evolution and work hardening in conventional and ultrafine-grained microalloyed steels 237
J. Majta and K. Muszka, AGH – University of Science and Technology, Poland

9.1 Introduction 237
9.2 Thermomechanical and severe plastic deformation processing of ultrafine-grained microalloyed (MA) steels 239
9.3 The principles of deformation-induced grain refinement 240
9.4 Effects of microstructure evolution on mechanical properties of ultrafine-grained microalloyed steel 243
9.5 Application, results and discussion 245
9.6 Multiscale modelling of the flow stress of conventional and ultrafine-grained microalloyed steels 249
9.7 Conclusion and future trends 258
9.8 References 259

Part III Microstructure evolution in the processing of other metals 265

10 Aging behavior and microstructure evolution in the processing of aluminum alloys 267
D. Shan and L. Zhen, Harbin Institute of Technology, China

10.1 Introduction 267
10.2 Microstructure evolution during plastic processing: the effects of hot working on microstructure and properties 269
10.3 Microstructure evolution during plastic processing: the effects of cold working on microstructure and properties 275
10.4 Aging behavior and age hardening 276
10.5 Characterization and test methods 284
10.6 Case studies and applications 286
10.7 Conclusion and future trends 293
10.8 Acknowledgments 294
10.9 References 294

11 Microstructure control in creep–age forming of aluminium panels 298
L. Zhan, Central South University, China and J. Lin and D. Balint, Imperial College London, UK

11.1 Introduction to the creep–age forming (CAF) process and its importance 298
11.2 The importance of precipitation control in CAF 301
11.3 Testing methods for stress/strain ageing 307
11.4 Modelling of precipitation hardening 313

11.5 Applications and future trends 331
11.6 References 333

12 Microstructure control in processing nickel, titanium and other special alloys 337
C. Sommitsch, R. Radis and A. Krumphals, Graz University of Technology, Austria and M. Stockinger and D. Huber, Böhler Schmiedetechnik GmbH & Co KG, Austria

12.1 Introduction 337
12.2 Application of special alloys such as nickel-based alloys, titanium alloys and titanium aluminides 339
12.3 Production processes 344
12.4 Microstructures and mechanical properties 350
12.5 Materials modelling and process simulation 361
12.6 Process and materials optimization: case study 365
12.7 Future trends 376
12.8 Sources of further information and advice 377
12.9 References 378

Index *384*

Contributor contact details

(* = main contact)

Editors

Professor Jianguo Lin* and Dr Daniel Balint
Department of Mechanical Engineering
Imperial College London
Exhibition Road
London SW7 2AZ
UK

E-mail: Jianguo.Lin@imperial.ac.uk; D.Balint@imperial.ac.uk

Maciej Pietrzyk
AGH – University of Science and Technology
Krakow
Poland

E-mail: Maciej.Pietrzyk@agh.edu.pl

Chapter 1

Professor Takashi Ishikawa
Department of Materials Science and Engineering
Nagoya University
Furo-cho, Chikusa-ku
Nagoya 464-8603
Japan

E-mail: ishikawa@numse.nagoya-u.ac.jp

Chapter 2

Professor Yvan Chastel*
RENAULT
DIMat – Materials Engineering
Director
1 avenue du Golf 78288 Guyancourt
France

E-mail: yvan.chastel@renault.com

Dr Roland Logé and Dr Marc Bernacki
CEMEF (Centre de Mise en Forme des Matériaux)
MINES ParisTech
UMR CNRS 7635
BP 207, 06904 Sophia Antipolis
France

Chapter 3

Jan Kusiak, Danuta Szeliga and Łukasz Sztangret
Department of Applied Computational Science and Modelling
AGH – University of Science and Technology
al. Mickiewicza 30, 30-059 Krakow
Poland

E-mail: kusiak@agh.edu.pl

Chapter 4

B. López* and J. M. Rodriguez-Ibabe
Materials Department
CEIT and Tecnun
University of Navarra
P° de Manuel Lardizabal, 15
20018 Donostia-San Sebastian
Basque Country
Spain

E-mail: blopez@ceit.es; jmribabe@ceit.es

Chapter 5

Dr A. Rosochowski*
Department of Design, Manufacture and Engineering Management
University of Strathclyde
James Weir Building
75 Montrose Street
Glasgow G1 1XJ
UK

E-mail: a.rosochowski@strath.ac.uk

Dr L. Olejnik
Institute of Manufacturing Technologies
Warsaw University of Technology
85 Narbutta Street
Warsaw 02-524
Poland

E-mail: l.olejnik@wip.pw.edu.pl

Chapter 6

Maciej Pietrzyk*
AGH – University of Science and Technology
Krakow
Poland

E-mail: Maciej.Pietrzyk@agh.edu.pl

Roman Kuziak
Institute for Ferrous Metallurgy
Gliwice
Poland

Chapter 7

Professor Jianguo Lin*
Department of Mechanical Engineering
Imperial College London
Exhibition Road
London SW7 2AZ
UK

E-mail: Jianguo.Lin@imperial.ac.uk

Dr Jian Cao
RTC Innovation Ltd
Unit 201F, Argent Centre
60 Frederick Street
Birmingham B1 3HS
UK

E-mail: j.cao@rtcinnovation.com

Dr Daniel Balint
Department of Mechanical Engineering
Imperial College London
Exhibition Road
London SW7 2AZ
UK

E-mail: D.Balint@imperial.ac.uk

Chapter 8

Dr Jingqi Cai* and Professor Jianguo Lin
Department of Mechanical Engineering
Imperial College London
Exhibition Road
London SW7 2AZ
UK

E-mail: Jingqi.cai@imperial.ac.uk; Jianguo.Lin@imperial.ac.uk

Dr Joel Wilsius
ArcelorMittal
Arcelor Atlantique et Lorraine
Arcelor Research Automotive Applications
BP 30109, 1 route de Saint-Leu
60761 Montataire Cedex
France

E-mail: Joel.wilsius@arcelormittal.com

Chapter 9

Professor Janusz Majta* and Dr Krzysztof Muszka
Department of Metals Engineering and Industrial Computer Science
AGH – University of Science and Technology
Al. Mickiewicza 30
Krakow 30-059
Poland

E-mail: majta@metal.agh.edu.pl

Chapter 10

Professor Debin Shan* and Professor Liang Zhen
School of Materials Science and Engineering
Harbin Institute of Technology
No.92 West Dazhi Street, Nangang District
Harbin 150001
P. R. China

E-mail: d.b.shan@gmail.com

Chapter 11

Dr Lihua Zhan*
Institute of Metallurgical Machinery
School of Mechanical & Electrical Engineering
Central South University
Changsha
Hunan, 410083
P. R. China

E-mail: yjs-cast@mail.csu.edu.cn

Professor Jianguo Lin and Dr Daniel Balint
Department of Mechanical Engineering
Imperial College London
Exhibition Road
London SW7 2AZ
UK

E-mail: Jianguo.Lin@imperial.ac.uk; D.Balint@imperial.ac.uk

Chapter 12

Professor Christof Sommitsch*, Dr Rene Radis and Alfred Krumphals
Institute for Materials Science and Welding
Graz University of Technology
Kopernikusgasse 24
8010 Graz
Austria

E-mail: christof.sommitsch@tugraz.at

Dr Martin Stockinger and Daniel Huber
Böhler Schmiedetechnik GmbH & Co KG
Mariazellerstrasse 25
8605 Kapfenberg
Austria

Part I

General principles

1

Understanding and controlling microstructural evolution in metal forming: an overview

T. ISHIKAWA, Nagoya University, Japan

Abstract: This chapter describes the importance of microstructure control in metal forming. The physical metallurgy of the thermo-mechanical treatment used is dependent on the various metallurgical mechanisms involved in processing. The development of the structure by thermo-mechanical treatment is a result of the interrelation of recrystallization, grain growth, precipitation and transformation. A basic overview of these phenomena and their modeling is provided, along with an explanation of the strengthening mechanisms and the relation between microstructure and mechanical properties in metallic materials, especially steels. Basic techniques for the control of microstructure evolution are described and, finally, future trends are outlined.

Key words: thermo-mechanical control processing TMCP, recrystallization, recovery, grain growth, precipitation, transformation, controlled rolling.

1.1 Introduction

Thermo-mechanical control processing (TMCP) is now a widely used method for controlling microstructure and the resultant (principally mechanical) properties in both the ferrous and the nonferrous industries.[1–5] Correct control of the microstructure, in addition to the selection of appropriate alloying elements and a suitable composition, is therefore extremely important. There are many potential benefits offered by appropriate TMCP, particularly if the same properties can be achieved by optimizing the microstructure without the addition of alloying elements, especially rare earth elements and minor metals. Use of common metals rather than rare elements also contributes substantially to environmental conservation.

Previously, the prediction and control of microstructural evolution and mechanical properties relied on the knowledge and experience of the individual engineer. In steelmaking, where processing is complex and a dynamic microstructure evolves, these individual predictions are time-consuming, requiring a great deal of effort. Furthermore, consistent control is almost impossible. However, recent advancements in physical metallurgy, in rolling and metal forming technology, in thermo-mechanical processing, and in computer engineering have allowed microstructures and mechanical properties during production to be predicted. Computer-integrated manufacturing is leading to increased productivity, reduced manufacturing costs, savings in materials and improvements in product quality (Fig. 1.1).[6] Changes in microstructures and

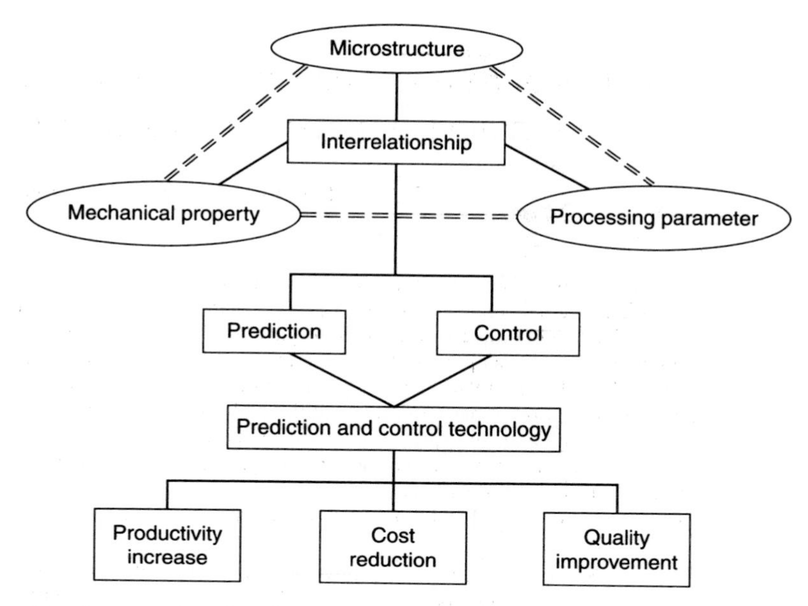

1.1 Diagram showing the concept and goal of prediction and control technology.[6]

mechanical properties can be predicted accurately using mathematical models based on physical metallurgy.

1.2 How microstructure evolves in metal forming

A wide range of mechanisms are responsible for the evolution of microstructures in materials. Microstructural evolution, which also alters the macroscopic properties of materials, is driven by mechanical and thermal loading of the material, chemical transformations, and other conditions such as energetic-particle bombardment in nuclear reactors.[7] The physical metallurgy of a thermo-mechanical treatment is dependent on the various metallurgical mechanisms that take place during processing. The interrelation of recrystallization, recovery, grain growth, precipitation and transformation, and so on, leads to the development of microstructure through thermo-mechanical treatment.

1.2.1 Recovery

Recovery is a process by which deformed grains can reduce their stored energy through the removal of strains in their crystals. These strains, principally due to dislocations, are introduced by plastic deformation and work to increase the yield strength of the material. Since recovery reduces the dislocation density, the process is normally accompanied by a reduction in the strength of the material and an immediate increase in its ductility.

1.2.2 Recrystallization

Recrystallization is a process by which deformed grains are replaced by a new set of nondeformed grains that nucleate and grow until the original grains have been entirely consumed. Recrystallization is usually accompanied by a reduction in the strength and hardness of the material and a simultaneous increase in its ductility. Recrystallization that occurs during hot forming is called dynamic recrystallization, while that which takes place in the interpass periods or after the forming passes, is known as static recrystallization.

1.2.3 Grain growth

Grain growth occurs at higher temperatures, when some of the recrystallized fine grains start to grow rapidly. Grain growth is inhibited by second-phase particles that pin the grain boundaries.

1.2.4 Transformation

The polymorphic transformation from high-temperature face-centered cubic iron (austenite, or γ) to low-temperature body-centered cubic iron (ferrite, or α) is a basic technology used in the thermal treatment of steels. Fe–C alloys with a concentration of carbon under about 2 wt% are defined as carbon steels. The carbon in a steel may be completely dissolved in austenite at high temperatures. Although the content of Si, Mn or other elements may be higher than that of carbon, the carbon can still have the greatest impact on the nature of the iron. Even a small amount of carbon may significantly affect the structure of the iron. The microstructure changes depending on the cooling rate are shown in Fig. 1.2.

1.2.5 Precipitation

The importance of small microalloying additions of carbonitride-forming elements such as niobium, vanadium and titanium in increasing the strength of low-carbon, low alloy steels is now well established.[8–15] The main role of these minor alloying additions is to form fine dispersions of carbonitrides, which, firstly, can control the austenite grain size, if out of solution during austenitization, and, secondly, can precipitate in both austenite and ferrite during cooling from the solution treatment temperature. Control of these precipitation processes during the thermo-mechanical treatment of steel products can bring about high strength levels while maintaining acceptable ductility.

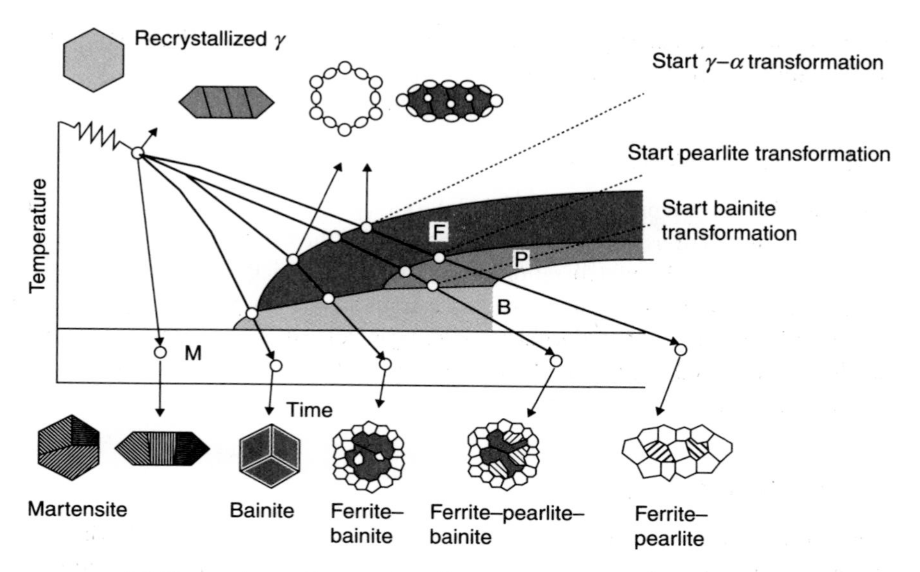

1.2 Dynamic continuous-cooling transformation (CCT) diagrams.

1.2.6 Preferred orientation (texture)

Severe deformation causes a reorientation of the grains into a preferred orientation, known as texture. Certain crystallographic planes tend to orient themselves in a preferred manner with respect to the direction of maximum strain. The preferred orientation resulting from plastic deformation is strongly dependent on the available slip and twinning systems, but is not affected by processing variables such as roll diameter and roll speed.

1.2.7 Mechanical fibering (fibrous texture)

A fibrous texture is produced along the direction of maximum stress acting on the material. The geometry of the flow and the amount of deformation are the most important variables. Mechanical fibering improves the mechanical properties along the fiber direction, with the transverse direction having inferior properties.

1.3 Models for predicting the microstructural evolution of carbon steels

A number of mathematical models for predicting the microstructural evolution of carbon steels have been proposed by researchers in several different countries. From a historical viewpoint, two studies have contributed substantially to the promotion of the development of mathematical models for predicting microstructural evolution and final mechanical properties.[16] One is Irvine and Pickering's study,[17] which showed that the tensile strength had a linear relationship to the 50% transformation temperature, regardless of chemical composition, for all kinds of microstructure. Their result indicates that the tensile strength can be calculated if the transformation behavior can be predicted. The other major study was carried out by Sellars and Whiteman.[18, 19] These authors proposed the first mathematical model for predicting microstructural evolution during multi-pass hot rolling. This model allowed the extensive potential applications of computer metallurgy to be envisaged.

Fig. 1.3[20] shows the conceptual scheme of an integrated model for predicting the evolution of microstructures. As this model is intended to be applied to hot rolling processes, the three relevant processes involved are the slab reheating process, the hot rolling process and the cooling process. During the first process, austenite grains grow; in the second, the austenitic microstructure is refined owing to recrystallization; and in the cooling process, the $\gamma \rightarrow \alpha$ transformation occurs. Equations for recrystallization and grain growth are shown in Table 1.1,[20] and can be applied to low C–Mn steels. In order to predict the microstructure of other steels, such as high carbon steels, special steels or microalloyed steels, basic tests must be used to determine each parameter.

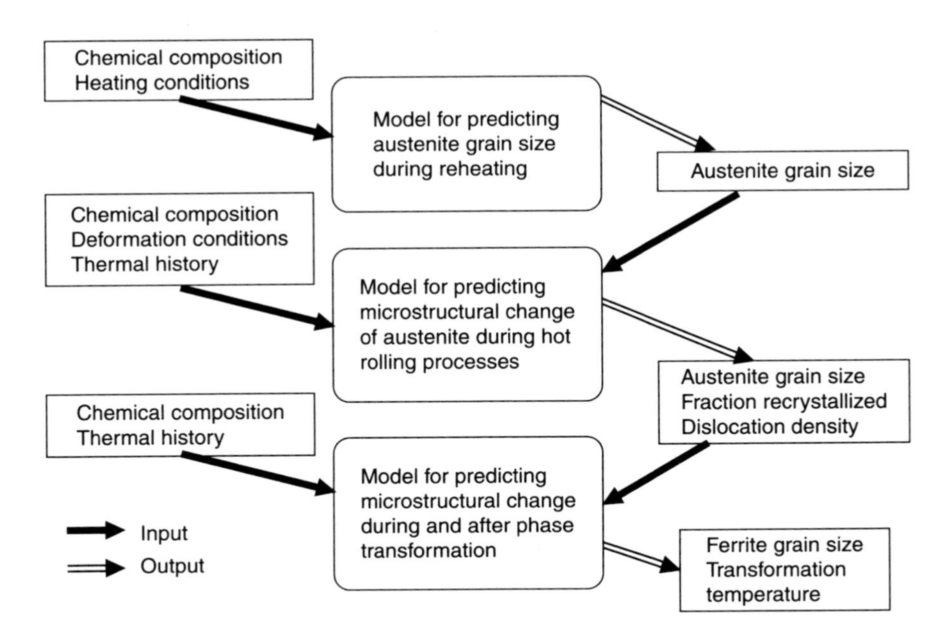

1.3 Conceptual scheme of an integrated model for predicting the evolution of microstructure.[20]

Table 1.1 Equations for prediction of recrystallization and grain growth[20]

Phenomena	Calculation model
Critical strain for dynamic recrystallization	$\varepsilon_c = 4.76 \times 10^{-4} \exp(8000/T)$
Grain size of dynamically recrystallized grain	$d_{dyn} = 22600\left[\dot{\varepsilon}\exp\left(Q/RT\right)\right]^{-0.27} = Z^{-0.27}$, $Q = 63800$ cal/mol
Fraction dynamically recrystallized	$X_{dyn} = 1 - \exp\left[-0.693\left(\frac{\varepsilon - \varepsilon_c}{\varepsilon_{0.5}}\right)^2\right]$ $\varepsilon_{0.5} = 1.144 \times 10^{-5} d_0^{0.28} \dot{\varepsilon}^{0.05} \times \exp\left(6420/T\right)$
Dislocation density in dynamically recrystallized grain	$\rho_s = 87300\left[\dot{\varepsilon} \times \exp\left(Q/RT\right)\right]^{0.248} = 87300Z^{0.248}$ $\rho_s = \rho_{s0}\exp\left[-90\exp\left(-8000/T\right) \times t^{0.7}\right]$
Dislocation density	$\rho_e = \frac{C}{b}\left(1 - e^{-b\varepsilon}\right) + \rho_0 e^{-b\varepsilon}$
Grain growth of dynamically recrystallized grain	$d_p = d_{dyn} + 1.1 \times \left(d_{pd} - d_{dyn}\right) \times y$ $d_{pd} = 5380 \times \exp\left(-6840/T\right)$ $y = 1 - \exp\left[-295\,\dot{\varepsilon}^{0.1}\exp\left(-8000/T\right) \times t\right]$
Grain size of statically recrystallized grain	$d_{st} = 5/\left(Sv \times \varepsilon\right)^{0.6}$ $Sv = \frac{24}{\pi d_0}\left(0.491e^{\varepsilon} + 0.155e^{-\varepsilon} + 0.1433e^{-3\varepsilon}\right)$
Fraction statically recrystallized	$X_{st} = 1 - \exp\left[-0.693\left(\frac{t - t_0}{t_{0.5}}\right)2\right]$ $t_{0.5} = 2.2 \times 10^{-12} Sv^{-0.5} \times \dot{\varepsilon}^{-0.2} \times \varepsilon^{-2} \times \exp\left(30000/T\right)$
Change in dislocation density due to recovery	$\rho_r = \rho_e \exp\left[-90 \times \exp\left(-8000/T\right) \times t^{0.7}\right]$
Grain growth	$d^2 = d_{st}^2 + 1.44 \times 10^{12} \times \exp\left(-Q/RT\right) \times t$

Johnson–Mehl-type models based on the theory of nucleation and growth are used to model the progress of transformations.[21] The transformation progress models used for calculating the distribution of the transformation ratio X in a cross section at a prescribed position in a steel strip are shown in Table 1.2.[21] The ferritic transformation ratio X_F, the pearlitic transformation ratio X_P and the bainitic transformation ratio X_B are calculated using transformation progress models for the ferritic, pearlitic and bainitic transformations. The sum of these transformation ratios is used as the transformation ratio X of the steel strip.

Table 1.2 Johnson–Mehl-type transformation models[21]

Transformation	Transformation speed	Nucleation and growth rate	Coefficient
Ferrite	Nucleation–growth	$1 = T^{-1/2} D \cdot \exp\left(-\frac{k_3}{RT\Delta G_V^2}\right)$	$k_1 = 1.7476 \times 10^6$
	$\frac{dx}{dt} = 4.046\left(k_1 \frac{6}{d_\gamma^4} \tau I G^3\right)^{1/4}$		$k_2 = 8.933 \times 10^{-12} \exp\left(\frac{21100}{T}\right)$
	$\left(\ln\frac{1}{1-x}\right)^{3/4}(1-x)$	$G = \frac{1}{2r} D \frac{C_{\gamma\alpha} - C_\gamma}{C_\gamma - C_\alpha}$	$k_3 = (\mathrm{cal}^3/\mathrm{mol}^3) = 0.957 \times 10^9$
Pearlite	Saturation–growth	$G = \Delta T \cdot D \cdot (C_{\gamma\alpha} - C_{\gamma\beta})$	$k_2 = 6.72 \times 10^6$
Bainite	$\frac{dx}{dt} = k_2 \frac{6}{d_\gamma} G(1-x)$	$G = \frac{1}{2r} D \cdot \left(\frac{C_{\gamma\alpha} - C_\gamma}{C_\gamma - C_\alpha}\right)$	$k_2 = 6.816 \times 10^{-4} \exp\left(\frac{3431.5}{T}\right)$

Note: d_γ, γ particle size; D, diffusion coefficient of C in γ (m²/s); C_γ, C mole fraction concentration in γ; C_α, C mole fraction concentration in ferrite; $C_{\gamma\alpha}$, C molarity in γ of the γ/α interface; $C_{\gamma\beta}$, C molarity in γ of the γ/cementite interface; ΔT, supercooling from Ae_1 (K); r, curvature radius of growth interface.

1.4 Strengthening mechanisms and relation between microstructure and mechanical properties

The following are considered to be the main strengthening mechanisms for steels.

1.4.1 Grain refinement

A fine-grained material is harder and stronger than a coarse-grained material, since the larger number of grain boundaries in the fine-grained material impede dislocation motion. The following general relationship between the yield stress σ (tensile strength) and the grain size d was proposed by Hall[22] and Petch:[23]

$$\sigma = \sigma_0 + kd^{-1/2} \tag{1.1}$$

A Japanese national project called the Ferrous Super Metal Project was started in 1997, with the aim of pursuing ultrafine-grained microstructures below 1 μm for plain C–Si–Mn steels.[24] Tests were carried out using heavy deformation as follows: I, in a cooled austenite region at temperatures from 500 to 700°C, by inducing low-temperature diffusional transformation; II, in a ferrite region with second-phase dispersion at around 700°C, by inducing low-temperature recrystallization; and III, in a ferrite region at around 550°C, just below A_{c1}, by spontaneous reverse transformation. These three types of heavy deformation processing are illustrated in Fig. 1.4.[24]

1.4.2 Precipitation hardening

The strength and hardness of some metal alloys may be enhanced by the presence of extremely small, uniformly-dispersed particles in the matrix of the original phase. Precipitation hardening, or age hardening, requires a second phase that is soluble at high temperature but has a limited solubility at lower temperatures. The size, shape, volume fraction and distribution of the particles are key factors in improving precipitation hardening. High-strength alloys seem to consist of fine, strong particles well distributed in a deformed matrix. Fig. 1.5[25] shows an example

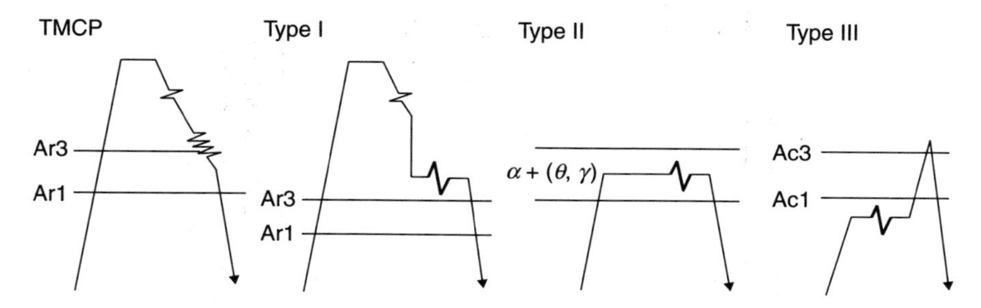

1.4 Schematic illustration showing three types of heavy deformation processing, including conventional TMCP.[24]

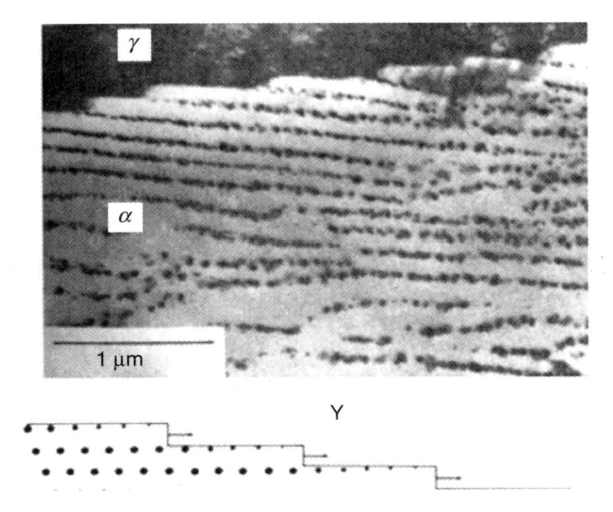

1.5 Nanosized VC produced by interphase boundary precipitation.[25]

of precipitation of VC at an interface in a carbon steel occurring at the point of transformation from austenite to ferrite.

1.4.3 Solid solution hardening

Solid solution hardening due to carbon has a major effect on the strength of martensite, but ductility can be obtained only at low carbon levels. Although alloying elements affect hardenability, they have only a minor effect on hardness, simply reducing it at high carbon levels by causing austenite to be retained. Solute atoms are atoms that have been introduced into the matrix. There are two types of solid solutions: in substitutional solid solutions, the solute and solvent atoms are similar in size, causing the solute atoms to occupy lattice sites; and in interstitial solid solutions, the solute atoms are of smaller size than the solvent atoms, causing the solute atoms to occupy interstitial sites in the solvent lattice.

1.4.4 Work hardening

Work hardening is an important strengthening process in steels, and is particularly important for obtaining high strength levels in rod and wire, in both plain carbon and alloy steels. Indeed, without the addition of special alloying elements, plain carbon steels can be raised to strength levels above 1500 MPa simply by the phenomenon of work hardening.

Increasing the dislocation density increases the yield strength, which in turn results in a higher shear stress being required to move the dislocations. This process is easily observed when a material is worked at room temperature.

1.4.5 Martensite strengthening

Martensitic strengthening is obtained when austenite is transformed into martensite by a diffusionless shear-type process during quenching. Martensitic transformations occur in many alloy systems, but the most pronounced effect has been observed with steels.

1.5 Emerging techniques to control microstructure evolution in metal forming

The thermo-mechanical processing of steel may be classified into three broad categories, depending on whether the deformation process occurs before, during or after the phase transformation.

1.5.1 Deformation completed prior to the transformation of austenite

The processes involved are high-temperature thermo-mechanical treatment (HTMT), controlled rolling and low-temperature thermo-mechanical treatment (LTMT), also called ausforming. Controlled rolling and thermo-mechanical controlled processing have been developed for microalloyed ferrite–pearlite steels.

1.5.2 Deformation during the transformation of austenite

These processes include the isoforming process, where spheroidal carbides form within a ferrite matrix during deformation of metastable austenite, and the treatment of transformation-induced plasticity (TRIP) steels, where metastable austenite is initially deformed above the M_s temperature and finally cold-worked at room temperature to produce strain-induced martensite. The isoforming technique is applied to low alloy, or pearlite-forming, steels, whereas the treatment of TRIP steels is confined to stainless steels.

1.5.3 Deformation after the transformation of austenite

These processes include marstraining, marforming, and strain tempering or warm working. In marforming, the martensite is cold-worked prior to tempering to induce a dislocation substructure that improves the distribution of temper carbides.

1.5.4 Controlled rolling

Controlled rolling is a means by which the properties of steels can be improved to a level equivalent to those of more highly alloyed or heat-treated steels. Controlled rolling consists of three stages: (a) deformation in the recrystallization region at high temperatures, (b) deformation in the nonrecrystallization region in a low-temperature range above A_{r3} and (c) deformation in the austenite–ferrite region. The importance of deformation in the nonrecrystallization region lies in the division of the austenite grains into several blocks as a result of the introduction of deformation bands into the grains. Deformation in the austenite–ferrite region gives a mixed structure consisting of equiaxed grains and subgrains after transformation; this further increases the strength and toughness. The fundamental difference between conventionally hot-rolled and controlled-rolled steels is that the nucleation of ferrite occurs exclusively at austenite grain boundaries in the former, while it occurs in the interior of grains as well as at grain boundaries in the latter, leading to a more refined grain structure. Refinement of the structure is aided by the addition of microalloying elements such as Nb, V and Ti. The improved strength and toughness of controlled-rolled steels have been shown to be a result of the fine grain size.

1.6 Advanced high-strength steels (AHSS)

In general, the elongation decreases as the strength increases (Fig. 1.6) in metallic materials. However, advanced high-strength steels (AHSS) have more elongation at equivalent strength, equating to better formability. Conventional high-strength steels (HSS) are hardened by solid solution hardening, precipitation or grain refining, whereas AHSS are hardened by phase transformations, where the microstructure may include martensite, bainite and retained austenite. AHSS, including dual-phase steels, TRIP steels, complex-phase steels and martensitic steels, are superior in both strength and ductility to conventional HSS: they thus facilitate energy absorption during impact and ensure safety while reducing weight.

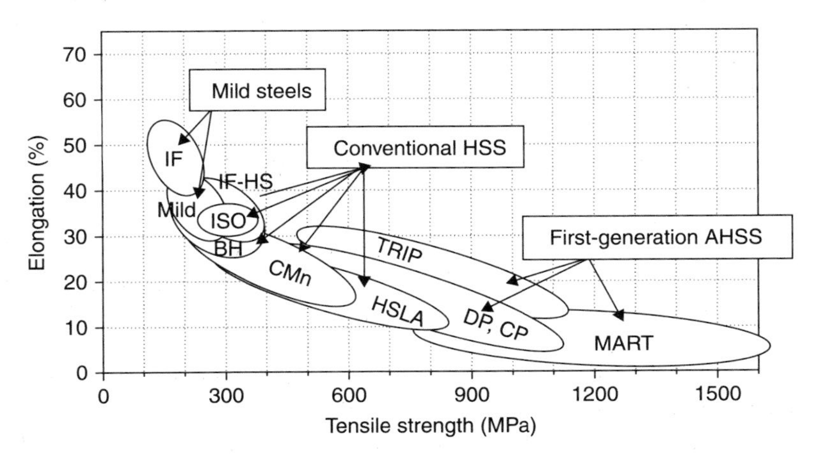

1.6 Relation between elongation and tensile strength of steels.

1.6.1 Dual-phase (DP) steels

Dual-phase (DP) steels are composed of ferrite and 5–20% of martensite, and their strength ranges from 500 to 1200 MPa. DP steels have a microstructure of mainly soft ferrite, with islands of hard martensite dispersed throughout. The strength level of these grades is related to the amount of martensite in the microstructure. DP steels characterized by a low yield ratio and high work-hardening ratio are widely used in automotive components that require high strength, good crashworthiness and good formability.

1.6.2 Transformation-induced plasticity (TRIP) steels

TRIP steels, with a microstructure of ferrite, bainite and 5–15% retained austenite, include hot-rolled, cold-rolled and hot-dipped galvanized products with strengths ranging from 600 to 800 MPa. TRIP steels have high elongation and an excellent, sustainable work-hardening ratio, making them suitable for stretch forming.

1.6.3 Complex-phase (CP) steels

Complex-phase (CP) steels have a microstructure similar to that of TRIP steels, except that CP steels have no retained austenite. Thanks to hard phases such as martensite and bainite, and with some help from precipitation hardening, the strength of CP steels ranges from 800 to 1000 MPa.

1.6.4 Martensitic steels

Martensitic steels, or hot-stamping or die-quenched steels, contain mainly Mn and boron as alloying elements, and thus have excellent hardenability. The hot stamping process consists of heating blanks to austenitization, and then press forming while the blanks are still red hot and soft. Finally, the formed parts are quenched to hard phases such as martensite within the die. The tensile strengths of these steels are typically between 900 and 1500 MPa.

1.7 Conclusion and future trends

The manufacturing of products with excellent quality is the ultimate goal of technological development. To achieve this goal, the properties of manufactured products must be predicted, and both the chemistry and the production process must be carefully designed. Recent advances in physical metallurgy, rolling technology and computer control have made great contributions to the field of structure and property prediction and to the development of a control model allowing the prediction of microstructural evolution and mechanical properties. It is expected that the maximum benefits from manufacturing control will be obtained by using predictions derived from metallurgical models.

Resource conservation, energy reduction, yield improvement, recycling and reduction in the weight of parts are becoming increasingly important. In the future, microstructure control will therefore be expected to assist in the development of new processing technologies that offer improvements in the above areas. One example of this is the manufacture of parts such as those shown in Fig. 1.7[26] with high strength and good machinability. This kind of new technology will be applied to many microalloyed steels. It will prove necessary to develop a precise system for the simulation of microstructural evolution during forming in order to optimize process design for these types of gradient functional parts.

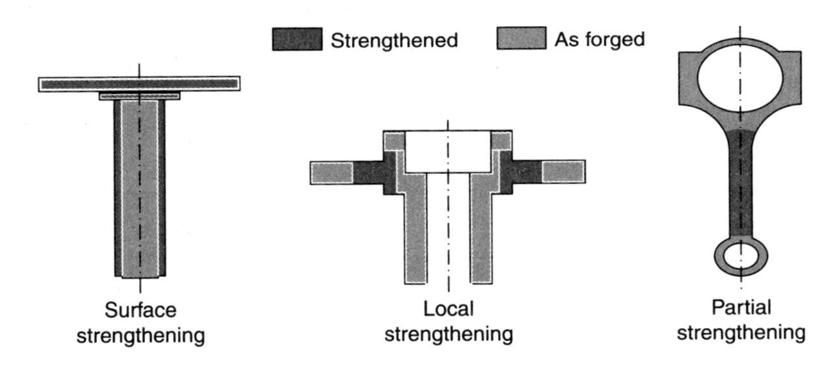

1.7 Gradient functional components.[26]

1.8 References

1. *Proc. Int. Symp. on Accelerated Cooling of Steel*, ed. by Southwick P D, AIME, New York, 1986.
2. *Proc. Int. Symp. on Accelerated Cooling of Steel*, edited by Ruddle G E and Crawley A F, Oxford, 1988.
3. *Proc. 32nd Mechanical Working and Steel Processing Conf.*, AIME ISS, New York, 1991.
4. Yoshie A, Fujioka M, Watanabe Y, Nishioka K, Morikawa H, 'Modelling of microstructural evolution and mechanical properties of steel plates produced by thermo-mechanical control process', *ISIJ Int.*, 1992, 32, 395–403.
5. Watanabe Y, Shimomura S, Funato K, Nishioka K, Yoshie A, Fujioka M, 'Integrated model for microstructural evolution and properties of steel plates manufactured in production line', *ISIJ Int.*, 1992, 32, 405–413.
6. Kwon O, 'Technology for the prediction and control of microstructural changes and mechanical properties in steel', *ISIJ Int.*, 1992, 32, 350–358.
7. El-Azab A, Wegand D, Simmons J, Special Issue on Microstructure Evolution in Materials, *Model. Simul. Mater. Sci. Eng.*, 2009, Vol. 17.

8. Dutta B, Sellars C M, 'Effect of composition and process variables on Nb(C, N) precipitation in niobium microalloyed austenite', *Mater. Sci. Technol.*, 1987, 3, 197–206.
9. Saito Y, Shiga C, Enami T, *Int. Conf. on Physical Metallurgy of Thermomechanical Processing of Steels and Other Metals*, ed. by Tamura I, ISIJ, Tokyo, 1988, 753.
10. Akamatsu S, Matsumura Y, Senuma T, Yada G H, Ishikawa S, 'Modelling of NbC precipitation kinetics in hot deformed austenite on Nb bearing low carbon steels', *Tetsu-to-Hagane*, 1989, 75, 933–941.
11. Liu W J, Jonas J J, 'Nucleation kinetics of Ti carbonitride in microalloyed austenite', *Metall. Trans. A*, 1989, 20, 689–697.
12. Park S H, Jonas J J, 'Mathematical modelling of CCP (continuous-cooling-precipitation) diagrams', *Proc. Int. Symp. on Mathematical Modelling of Hot Rolling of Steel*, ed. by Yue S, CIM, Hamilton, Ontario, 1990, 446–456.
13. Akamatsu S, Senuma T, Yada H, 'A mathematical model of NbC precipitation in austenite of hot deformed niobium bearing low carbon steels', *Proc. Int. Symp. on Mathematical Modelling of Hot Rolling of Steel*, ed. by Yue S, CIM, Hamilton, Ontario, 1990, 467–476.
14. Liu W J, Hawbolt E B, Samarasekera I V, 'Finite difference modelling of the growth and dissolution of carbonitride precipitates in austenite', *Proc. Int. Symp. on Mathematical Modelling of Hot Rolling of Steel*, ed. by Yue S, CIM, Hamilton, Ontario, 1990, 477–487.
15. Zou H, Kirkaldy J S, *Proc. Int. Symp. on Fundamentals and Application of Ternary Diffusion*, ed. by Purdy G R, CIM, Hamilton, Ontario, 1990, 184.
16. Senuma T, Suehiro M, Yada H, 'Mathematical models for predicting microstructural evolution and mechanical properties of hot strips', *ISIJ Int.*, 1992, 32, 423–432.
17. Irvine K J, Pickering F B, 'Low-carbon bainitic steels', *J. Iron Steel Inst.*, 1957, 187, 292–309.
18. Sellars C M, Whiteman J A, 'Recrystallization and grain growth in hot rolling', *Met. Sci*, 1979, 13, 187–194.
19. Sellars C M, *Sheffield Int. Conf. on Working and Forming Processes*, ed. by Sellars C M and Davies G J, Metals Society, London, 1980, 3.
20. Senuma T, Takemoto Y, 'Model for predicting the microstructural evolution of extralow carbon steels', *ISIJ Int.*, 2008, 48, 1635–1639.
21. Ogai H, Ito M, Hirayama R, 'Consistent shape prediction simulator after hot rolling mill', *Nippon Steel Tech. Rep.*, 2004, 89, 43–49.
22. Hall E O, 'The deformation and ageing of mild steel', *Proc. Phys. Soc., Ser. B*, 1951, 64, 747–753.
23. Petch N J, 'The cleavage strength of polycrystals', *J. Iron Steel Inst.*, 1953, 174, 25–28.
24. Niikura M, Fujioka M, Adachi Y, Matsukura A, Yokota T, *et al.*, 'New concepts for ultra refinement of grain size in Super Metal Project', *J. Mater. Process. Technol.*, 2001, 117, 341–364.
25. Kamikawa N, Abe Y, Miyamoto G, Furuhara T, 'Tensile behavior of low carbon steels with nano-sized alloy carbides produced by interphase boundary precipitation', *Proc. 2nd Int. Symp.on Steel Science (ISSS2009)*, 2009, 179–182.
26. Isogawa S, 'Current status and future prospect of controlled forging', *Proc. 11th Asian Symp. on Precision Forging*, 2010, 20–25.

2

Techniques for modelling microstructure in metal forming processes

Y. CHASTEL, RENAULT, France and R. LOGÉ and M. BERNACKI, MINES ParisTech, France

Abstract: This chapter discusses several types of numerical models for metallurgical evolution. First, some basic notions of microstructure representations and microstructure state variables, some features of hardening and recovery and some features of recrystallization are recalled. Then, constitutive models coupled with state variables are introduced and examples of applications are given. Mean field methods are also presented and applied to necklace structures produced in discontinuous dynamic recrystallization. Finally, future trends are described, with an emphasis on digital material models and how they will provide powerful models of recrystallization on the microscopic scale.

Key words: microstructure, state variables, recrystallization, mean field, digital material.

2.1 Introduction: importance of microstructure prediction in metal forming

In the mechanics of metal forming, emphasis is placed on the macroscopic flow of the material and on the stress state it is experiencing. Over recent decades, a vast collection of analytic and numerical methods has been developed to provide full 3D macroscopic analyses of all processes, namely predictions of stress, strain and temperature fields throughout a metal as it is processed. These methods rely on continuum mechanics and integrate process conditions as macroscopic boundary conditions. Depending on the process, various mechanical integral formulations for solids can be used, either Eulerian formulations, particularly for steady-state or quasi-steady-state flow, or updated Lagrangian formulations, for dealing with non-stationary operations.

Metallurgical modelling comes into play when the target properties need to match the material microstructure or, more often, when a final microstructure is required which achieves the desired final material properties. Metallurgical analyses inform our understanding of process–property relationships and of how optimizing the process can optimize the properties. This is today's challenge both in academic research and in industry. Metallurgical simulations also provide a means to imagine and validate processing routes for novel or even virtual grades of metals. In the near future, one will even be able to evaluate how a new alloy should be designed in order to behave in a particular way during process operations to achieve the desired properties. Coupling the metallurgical state to the mechanical behaviour can be performed in several different ways, assuming that the macroscopic calculations of the mechanical and temperature fields are accurate enough.

The focus of this chapter is on modelling techniques for microstructure evolution during forming, that is to say, typically under large strains and under cold, warm or hot forming conditions. Several levels of observation and description are available today when microscopy or chemical analysis techniques are used. As a general feature, one should look at a metal as a polycrystalline ensemble. An appropriate length scale is that of a representative elementary volume (REV) for which a thermo-mechanical loading can be determined using continuum mechanics analysis on the macroscopic scale. This mesoscopic scale is typically the scale of a polycrystal, i.e. a collection of grains including all the different phases, substructures or precipitates that are present. The use of a finer length scale would rely on explicit descriptions of dislocations or atoms. Such models can provide information about interactions between the individual defects in the microstructure, but they are limited to the analysis of small volumes of material. An extremely promising set of approaches rely on explicit modelling of the heterogeneity of the metal at different microscopic and mesoscopic scales – using finite element models, for instance – and on the derivation of more tractable models for large-scale macroscopic simulations. This approach will be discussed as a future trend in the last section of this chapter.

2.2 General features of models based on state variables

2.2.1 Notion of microstructure state variables

Based on observations with devices such as electron microscopes, combined with chemical and crystallographic analyses, a large set of microstructural features can be identified and quantified to describe a metallurgical state. One can then make use of metallurgical state variables describing features such as grains, metallurgical phases and crystallographic texture, and morphological parameters of the grains/phases/inclusions, to name but a few. If a description of the material behaviour is sought, the well-known correlation between the flow stress and the dislocation density or the grain size can provide a first hint about the appropriate state variables S and their integration into a constitutive law:

$$\sigma = f(\sigma_0, \varepsilon, \dot{\varepsilon}, T, S) \qquad [2.1]$$

where σ_0 is a threshold stress, and ε, $\dot{\varepsilon}$, T are the strain, strain rate and temperature, respectively.

However, the way in which microstructures are described is inherently linked to the type of microstructure evolution model being considered. For instance, when dealing with recrystallization processes, one often finds either simplified approaches using analytical models of the Johnson–Mehl–Avrami–Kolmogorov (JMAK) type (Avrami, 1939; Dehghan-Manshadi *et al.*, 2008; Jonas *et al.*, 2009), or more elaborate numerical schemes based on explicit representations of microstructures, meshed in different ways, which will be discussed in the last section of this chapter (Rollett *et al.*, 1992; Yazdipour *et al.*, 2008; Hallberg *et al.*, 2010; Takaki *et al.*, 2009; Logé *et al.*, 2008; Bernacki *et al.*, 2008; Bernacki *et al.*, 2009; Bernacki *et al.*, 2010). In the former case, the microstructure is typically reduced to a scalar value, the average grain size. At the other extreme, the second type of model includes topological aspects of the microstructure, and is associated with a large number of variables.

Intermediate approaches consider a series of state variables which give significant information about the microstructure, but do not require an explicit construction. The choice of the state variables is made such that the evolution equations of the state variables can have a clear physical basis (Montheillet *et al.*, 2009; Roucoules *et al.*, 2003; Sandstrom and Lagneborg, 1975; Estrin, 1998; Thomas *et al.*, 2007).

Since the driving forces for recrystallization are related to stored energies and local grain boundary curvatures, it is meaningful to consider dislocation densities and grain sizes as state variables (Montheillet *et al.*, 2009). Depending on the accuracy sought within the model, a number of grains that are representative of the microstructure can be defined or generated, each one being associated with an average dislocation density and size. The microstructure description is then finalized by fixing the volume fractions associated with the different representative grains.

An example of a general evolution of a scalar microstructure variable S is

$$\dot{S} = S_0 \cdot \left(1 - \frac{S}{S^*}\right) \cdot \dot{\bar{\varepsilon}} \tag{2.2}$$

where S_0 is a constant, and S^* is a limit value which can be defined as a function of the Zener–Hollomon parameter Z:

$$Z = \dot{\bar{\varepsilon}}.\exp\left(\frac{Q}{RT}\right) \tag{2.3}$$

Here, Q is an activation energy for the active metallurgical phenomena, R is the gas constant, and $\dot{S}$ is the particle derivative of S:

$$\dot{S} = \frac{\partial S}{\partial t} + v.\frac{\partial S}{\partial x} \tag{2.4}$$

for a velocity field v. And, indeed, when one is dealing with a coupled problem, the first step consists in calculating a velocity field, before integrating microstructural kinetic laws such as,

$$\dot{S} - f(\dot{\varepsilon}, T, S) = 0 \qquad [2.5]$$

2.2.2 Nucleation and grain boundary migration

A number of nucleation laws have been considered in the literature, and many assume that a critical dislocation density must be reached before nucleation can happen. The probability of activating nucleation usually increases with dislocation density and temperature, and, when dealing with necklace-type nucleation, with the amount of grain boundaries. Under dynamic conditions, an increasing strain rate generally leads to an increased critical dislocation density, but also to an increased probability of activating nucleation beyond the dislocation density threshold.

The driving force ΔE for grain boundary migration is described as the sum of a stored-energy term related to the dislocation content (Humphreys and Hatherly, 2004) and a capillarity term (Hillert, 1965) related to the grain boundary energy and curvature:

$$\Delta E = \tau.\Delta\rho + 2.\gamma_{\mathrm{b}}.\Delta(1/r) - E_{\mathrm{th}} \qquad [2.6]$$

where $\tau \approx \mu b^2/2$ is the average energy per unit dislocation length, γ_{b} is the energy per unit area of the grain boundary, and $r = D/2$ is its radius of curvature. E_{th} is a threshold energy which stands for pinning effects. $\Delta\rho$ and $\Delta(1/r)$ take account of differences in dislocation density and grain size on either side of the boundary. The grain boundary velocity is described by a kinetic relation (Humphreys and Hatherly, 2004),

$$v = m.\Delta E \qquad [2.7]$$

where m is the grain boundary mobility. Physically, $v > 0$ indicates that the grain is growing, and $v < 0$ indicates that the grain is shrinking.

2.3 Coupling between homogeneous microstructure description and constitutive laws

Microstructure coupling proves necessary in constitutive laws when the material behaviour depends strongly on the metallurgical state. In phenomenological approaches, the strain is often used as an internal variable. This is clearly an error, since plasticity is strain-path dependent. The strain rate should be considered as the prime state variable, and the strain path history will then be taken into account quite naturally. As discussed above, the dislocation density and subgrain size can be used as state variables for fine-scale isotropic constitutive models that integrate metallurgical evolution.

2.3.1 Strain hardening, recovery and flow stress

Following Kocks (1976) and Mecking and Kocks (1981) and assuming a constant mean free path for the dislocations (Estrin and Mecking, 1984), the evolution of the dislocation density as a function of the equivalent plastic strain is given by

$$\frac{\partial \rho}{\partial \varepsilon} = K_1 - K_2\rho \qquad [2.8]$$

where K_1 and K_2 represent the strain hardening and the recovery terms, respectively. The flow stress associated with a given representative grain i is determined from its dislocation density using the Taylor equation (Li, 1962),

$$\sigma_i = \sigma^0 + M\alpha\mu b\sqrt{\rho_i} \qquad [2.9]$$

where μ is the shear modulus, b is the Burgers vector, M is the Taylor factor, σ^0 is a 'dislocation-free' yield stress, and α is a constant set equal to 0.2. The material flow stress $\bar{\sigma}$ is then obtained from Eq. 2.9 by considering a volume average:

$$\bar{\sigma} = \langle \sigma_i \rangle \qquad [2.10]$$

2.3.2 Modelling strain-hardening stages and flow stress

As a first example, let us consider the material behaviour of C–Mn steel during two-phase (austenite–ferrite) hot rolling. The behaviour of each phase has to be determined separately so that the phase transformation kinetics can then be used reasonably straightforwardly. We concentrate here on the constitutive laws for the austenite phase.

The influence of the thermo-mechanical path on the material behaviour of austenite is analysed assuming that the stress remains proportional to the square root of the dislocation density ρ. Relying on this postulate, experimental compression results such as (σ, dσ/dε) curves can be transformed into microstructure relationships such as (ρ, dρ/dε) curves or, similarly, (σ^2, $\sigma\cdot$dσ/dε) curves (Figs 2.1–2.4). Note that the Taylor factor is then supposed to remain constant. The tails of the curves for a 1050°C deformation show dynamic recrystallization with a greater softening than for simple recovery–work-hardening combinations.

Three or four regimes can be identified in these graphs. The rate of variation of the dislocation density first increases very fast with the dislocation density. This first regime relates primarily to work hardening. If the variation is parabolic, it corresponds to a hardening function with the form of $\sqrt{\rho}$, and if it is linear, the hardening can be considered as constant. The second regime shows a decrease in the rate of variation with dislocation density. This regime starts after a maximum, which is greater for lower temperatures and higher strain rates. Two variants can be observed for the subsequent regime, depending on the deformation parameters. For slow deformation rates, a second quasi-linear decrease is found. For high strain rates, a threshold or even a slight increase is measured. Comparisons of experimental and best-fit model results obtained with Eqs 2.8 and 2.9 are shown in Fig. 2.5.

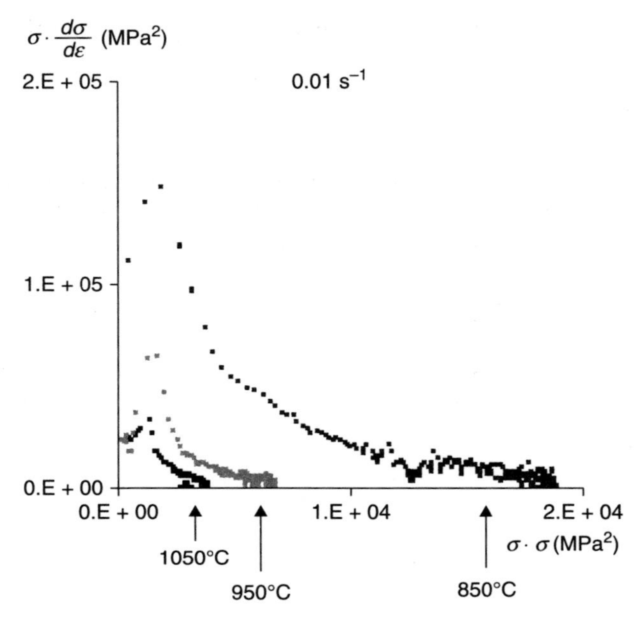

2.1 (σ^2, $\sigma \cdot$ d σ/dε) curves for a strain rate of 0.01 s^{-1}.

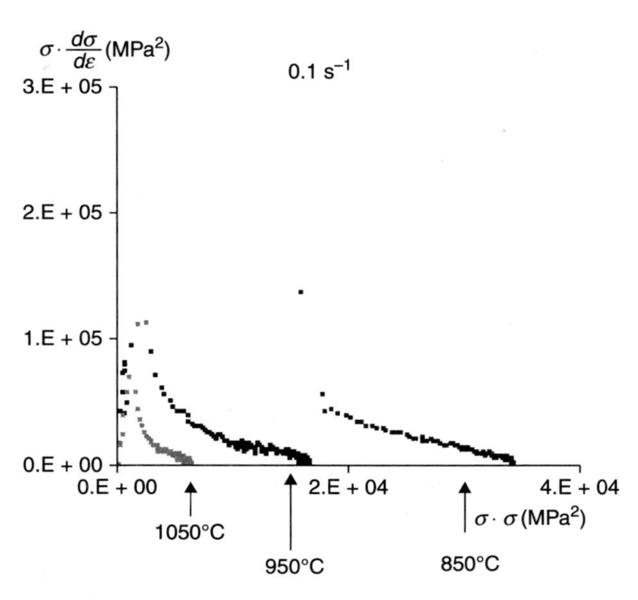

2.2 (σ^2, $\sigma \cdot$ dσ/dε) curves for a strain rate of 0.1 s^{-1}.

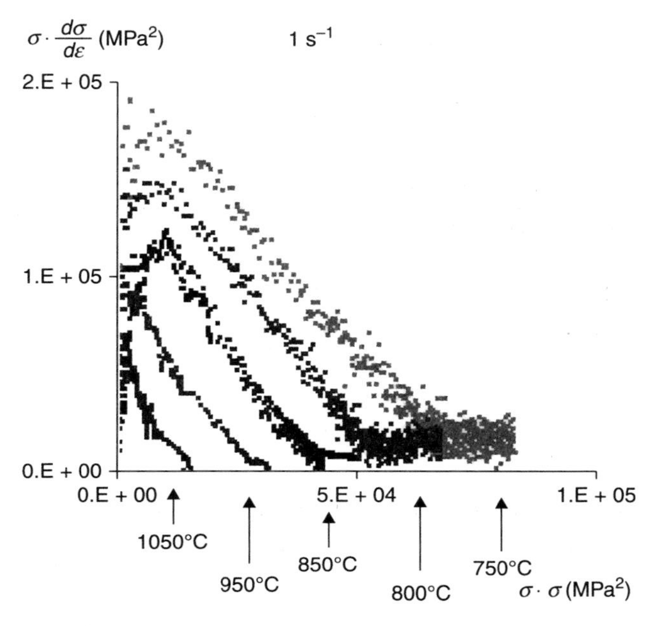

2.3 (σ^2, $\sigma \cdot$ d σ/dε) curves for a strain rate of 1 s^{-1}.

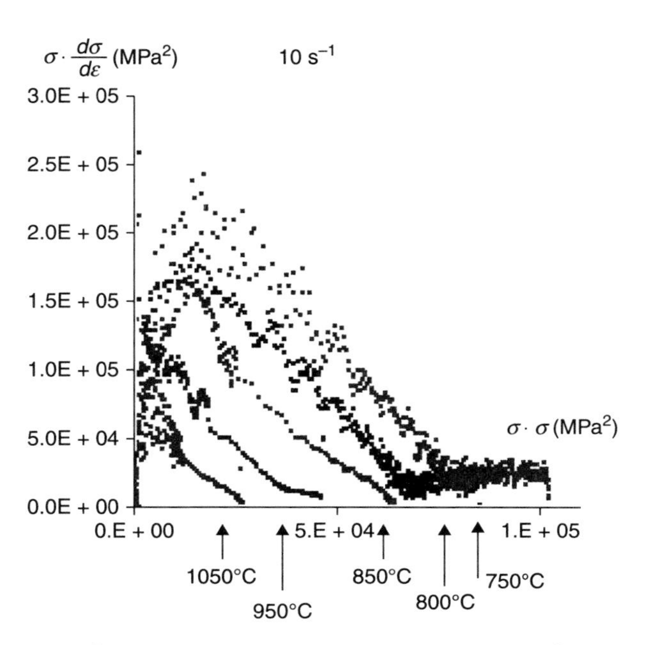

2.4 (σ^2, $\sigma \cdot$ dσ/dε) curves for a strain rate of 10 s^{-1}.

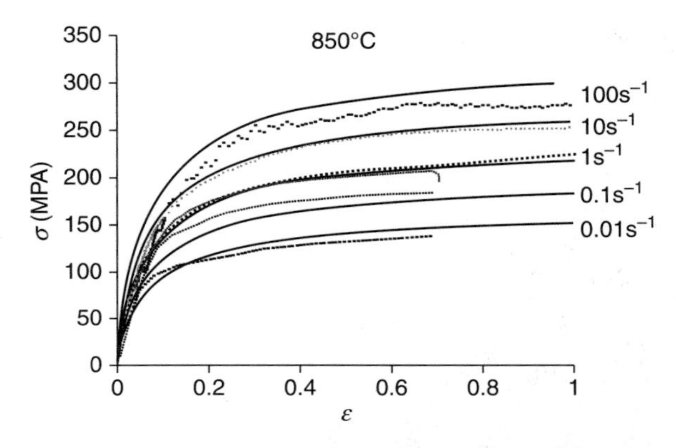

2.5 Experimental compression curves (dashed lines) and model curves.

2.4 Mean field approach: an example of discontinuous dynamic recrystallization

2.4.1 Notion of multiple homogeneous equivalent media

In a metal, each grain or phase is surrounded by neighbours, which are unknown if the topology has been ignored in the data structure of the model. Since the evolution of each representative grain is dictated by its interaction with an unknown neighbourhood, the latter can be conveniently identified as having the average properties of the polycrystalline aggregate (Montheillet *et al.*, 2009). This procedure is equivalent to defining a homogeneous equivalent medium (HEM) interacting with all representative grains and therefore defining their evolution. The state variables defining the HEM themselves evolve as a volume average of the representative state variables.

An interesting extension of the above method for defining unknown neighbourhoods consists in considering two HEMs instead of one (a two-site versus a one-site method). Indeed, in recrystallization processes, a microstructure consisting of a mixture of recrystallized (RX) and non-recrystallized (NR) grains often appears. RX grains typically have a lower dislocation density than NR grains, which means that the boundaries between RX and NR zones will move faster than the others. These boundaries are therefore called 'mobile'. The overall growth or shrinkage rate of a representative grain will depend on the surface fractions of grain boundaries with the various combinations of RX and NR HEMs on either side. These surface fractions themselves depend on the topology of the microstructure and continue to evolve as long as this topology has not reached a steady state.

We concentrate in this section on necklace structures formed in discontinuous dynamic recrystallization. For simplicity, subscripts i and j will be used from now on to designate RX and NR grains, respectively. The grain shape is approximated as circular, and so the curvature term in the capillarity part of the energy gradient used in Eq. 2.6 can be simplified to $(d-1)/r$, where d is the space dimension and r is the radius of the grain considered. In 3D, different driving forces are then distinguished using the notation $\Delta E_k^{\mathrm{HEM}} = E_{\mathrm{HEM}} - E_k - E_{\mathrm{th}}$, with HEM = NR or RX, $k = i$ or j, and $E = \tau\rho + 2\gamma_{\mathrm{b}}/r$.

2.4.2 Migration of RX–NR interfaces

Using Eq. 2.7, the aggregate variations of the recrystallized volume $\Delta V_{\mathrm{RX/NR}}$ and the non-recrystallized volume $\Delta V_{\mathrm{NR/RX}}$ during a time step Δt, considering only the movement of the mobile fractions of the grain boundaries, are given by

$$\Delta V_{\mathrm{RX/NR}} = \sum_{\mathrm{RX}} \gamma^{\mathrm{RX}} N_i \, \Delta V_{i,\mathrm{RX/NR}} = \sum_{\mathrm{RX}} \gamma^{\mathrm{RX}} N_i 4\pi r_i^2 m \, \Delta E_i^{\mathrm{NR}} \, \Delta t \tag{2.11}$$

$$\Delta V_{\mathrm{NR/RX}} = \sum_{\mathrm{NR}} \gamma^{\mathrm{NR}} N_j \, \Delta V_{j,\mathrm{NR/RX}} = \sum_{\mathrm{NR}} \gamma^{\mathrm{NR}} N_j \, 4\pi r_j^2 m \, \Delta E_j^{\mathrm{RX}} \, \Delta t \tag{2.12}$$

respectively, where the meaning of the surface fractions γ^{RX} and γ^{NR} is illustrated in Fig. 2.6, and N_i is the number of grains in the aggregate corresponding to the representative grain i. Volume conservation implies

$$\Delta V_{\mathrm{RX/NR}} = -\Delta V_{\mathrm{NR/RX}} \tag{2.13}$$

Combining Eqs 2.11, 2.12 and 2.13 leads to

$$\gamma^{\mathrm{NR}} = -\gamma^{\mathrm{RX}} \frac{\sum_{\mathrm{RX}} r_i^2 \, \Delta E_i^{\mathrm{NR}}}{\sum_{\mathrm{NR}} r_j^2 \, \Delta E_j^{\mathrm{RX}}} \tag{2.14}$$

or

$$\gamma^{\mathrm{RX}} = -\gamma^{\mathrm{NR}} \frac{\sum_{\mathrm{NR}} r_j^2 \, \Delta E_j^{\mathrm{RX}}}{\sum_{\mathrm{RX}} r_i^2 \, \Delta E_i^{\mathrm{NR}}} \tag{2.15}$$

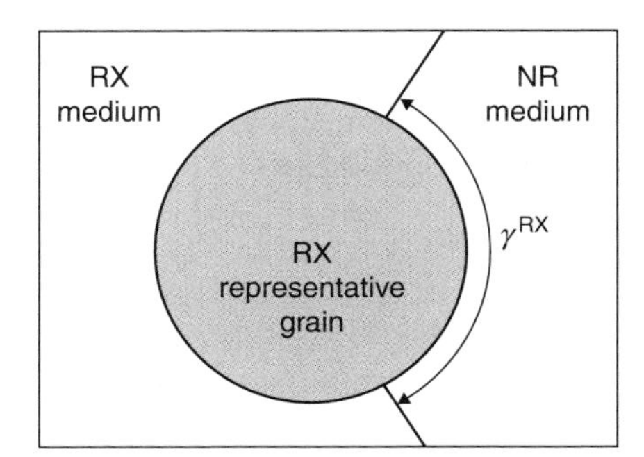

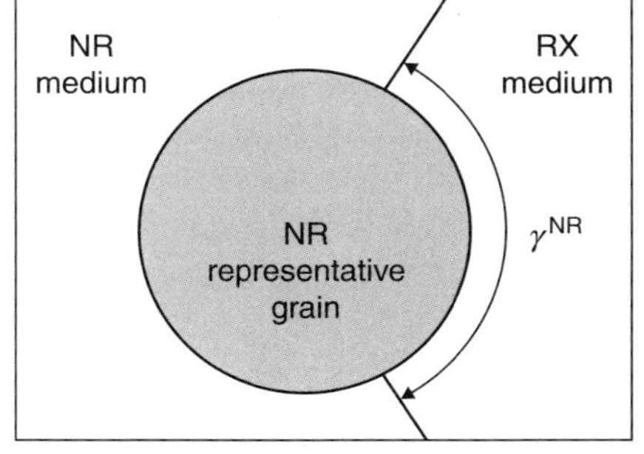

2.6 Schematic representation of mobile surface fractions.

When dealing with a necklace structure, one can set $\gamma^{RX} = 1$ at the beginning of recrystallization, and compute γ^{NR} from Eq. 2.14. When γ^{NR} reaches a value of 1, this means that the first necklace has been formed. Recrystallization then continues when we set $\gamma^{NR} = 1$, and compute γ^{RX} from Eq. 2.15.

2.4.3 Migration of RX–RX and NR–NR interfaces

Although RX–RX and NR–NR interfaces are characterized by smaller magnitudes of driving forces, their motion itself results in a reduction in the dislocation densities and therefore influences both the stress–strain behaviour and the recrystallization kinetics. Based on Eq. 2.7, one can describe the motion of these interfaces by distinguishing growing grains, for which $\Delta E_k^{HEM} > 0$, and shrinking grains, for which $\Delta E_k^{HEM} < 0$. Growing grains can be described by the following equations:

$$\Delta V_{i,\mathrm{RX/RX}} = 4\pi r_i^2 m\, \Delta E_i^{\mathrm{RX}}\, \Delta t,\ \Delta V_{\mathrm{RX/RX}} = \sum_{\Delta E_i^{\mathrm{RX}}>0} \left(1-\gamma^{\mathrm{RX}}\right) N_i\, \Delta V_{i,\mathrm{RX/RX}} \quad [2.16]$$

$$\Delta V_{j,\mathrm{NR/NR}} = 4\pi r_j^2 m\, \Delta E_j^{\mathrm{NR}}\, \Delta t,\ \Delta V_{\mathrm{NR/NR}} = \sum_{\Delta E_j^{\mathrm{NR}}>0} \left(1-\gamma^{\mathrm{NR}}\right) N_j\, \Delta V_{j,\mathrm{NR/NR}} \quad [2.17]$$

The volume changes of shrinking grains, on the other hand, are computed by redistributing the total volume changes of the growing grains given by Eqs 2.16 and 2.17:

$$\Delta V_{i,\mathrm{RX/RX}} = -\Delta V_{\mathrm{RX/RX}} \frac{r_i^2\, \Delta E_i^{\mathrm{RX}}}{\sum_{\Delta E_k^{\mathrm{RX}}<0} N_k r_k^2\, \Delta E_k^{\mathrm{RX}}} \quad [2.18]$$

$$\Delta V_{j,\mathrm{NR/NR}} = -\Delta V_{\mathrm{NR/NR}} \frac{r_j^2\, \Delta E_j^{\mathrm{NR}}}{\sum_{\Delta E_k^{\mathrm{NR}}<0} N_k r_k^2\, \Delta E_k^{\mathrm{NR}}} \quad [2.19]$$

2.4.4 Overall volume changes

Using Eqs 2.11–2.13 and 2.16–2.19, the volume changes of RX and NR grains attributed to grain boundary migration alone can be expressed as

$$\Delta V_i = \gamma^{\mathrm{RX}}\, \Delta V_{i,\mathrm{RX/NR}} + (1-\gamma^{\mathrm{RX}})\, \Delta V_{i,\mathrm{RX/RX}} \quad [2.20]$$

$$\Delta V_j = \gamma^{\mathrm{NR}}\, \Delta V_{j,\mathrm{NR/RX}} + (1-\gamma^{\mathrm{NR}})\, \Delta V_{j,\mathrm{NR/NR}} \quad [2.21]$$

Volume conservation is automatically satisfied when these relationships are used, which means that $\sum_i \Delta V_i + \sum_j \Delta V_j = 0$. If nucleation conditions are met, volume conservation continues to apply, where volume is transferred to a new representative grain.

A zone that has just been swept by a boundary is almost dislocation-free. The total amount of dislocations in the RX grain is not affected by the resulting increase in volume, but the average dislocation density in the grain decreases according to

$$\Delta(\rho V)=0 \qquad [2.22]$$

$$\frac{\Delta\rho}{\rho}=-\frac{\Delta V}{V} \qquad [2.23]$$

2.4.5 Application to 304L stainless steel

In the results shown in Fig. 2.7, a constant strain rate of $0.01\,\mathrm{s}^{-1}$ was chosen, starting with an initial grain size of 35 μm (Dehghan-Manshadi *et al.*, 2008). The effect of temperature on the recrystallization kinetics and flow stress is well predicted, the accuracy being better for the former. The flow stress at 1000°C might, however, have been overestimated owing to self-heating in the experiment, which was not taken into account in the model. Strain rate effects can be captured in a similar way, as illustrated by Bernard *et al.* (2011).

The effect of the initial grain size, and the steady-state behaviour, can be investigated further. Figure 2.8 shows the evolution of X and $\bar{D}_{RX}$ for two initial grain sizes, 200 and 35 μm. The typical evolution of $\bar{D}_{RX}$ includes a fast increase followed by stabilization at a steady-state value. The steady state is reached as

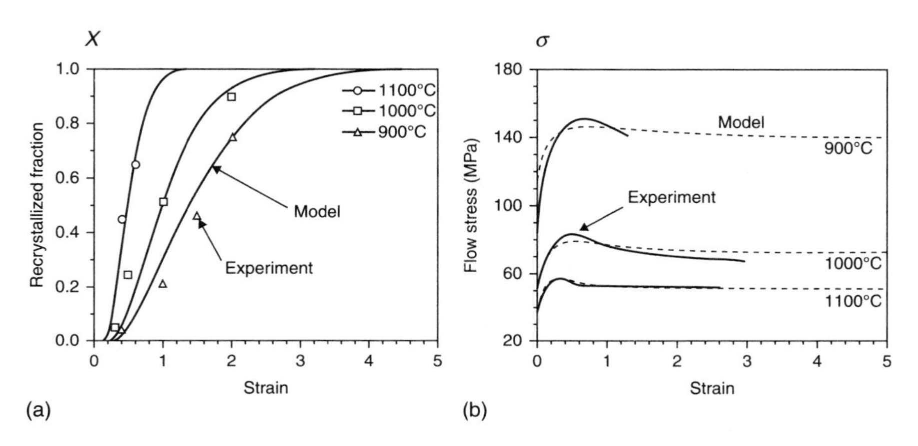

2.7 Predicted effects of temperature on (a) recrystallization kinetics and (b) flow stress for 304L steel (initial grain size = 35 μm, strain rate = $0.01\,\mathrm{s}^{-1}$). Comparison between model predictions and experimental data.

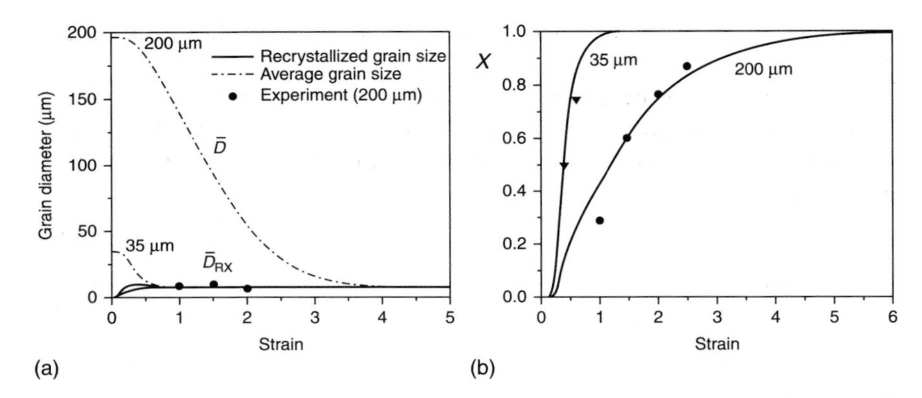

2.8 Effects of initial grain size on (a) grain size evolution and (b) recrystallization kinetics for 304L stainless steel, at 1100°C and $0.01\,s^{-1}$. Comparison between model predictions (lines) and experimental data (symbols).

soon as the material is fully recrystallized, and the deformation required to reach such a steady state is strongly dependent on the initial grain size. The difference in the kinetics of recrystallization is *automatically captured*, i.e. the identified model parameters based on data with one initial grain size, predict accurately the behaviour for the other initial grain size.

It should be noted that the grain size (and, likewise, the flow stress) stabilizes at *the same value*, independent of the initial microstructure, which is a well-known feature of discontinuous dynamic recrystallization. The full distribution of grain sizes actually becomes independent of strain as soon as X reaches 1. The steady state can also be shown to evolve in the expected way as a function of the Zener–Hollomon parameter Z, i.e. towards smaller grain sizes and increased flow stresses for increased values of Z.

2.5 Recrystallization modelling at the microscopic scale: overview and future trends

2.5.1 Digital representations of microstructures

The influence of microstructural heterogeneities on material processing is an issue of prime importance. If it is clear that the mean field approaches described in the previous section are a step forward in the global understanding of recrystallization phenomena, it is also obvious that these models must be calibrated and improved by the use of models which take into account explicitly the topological and physical features of the microstructure, such as the size distribution and the morphology of the grains, the existence of second-phase particles, the heterogeneity of the mobility and of the energy per unit area of the grain boundaries, and the nucleation criteria; an accurate calculation of the stored energy is also necessary. This challenge explains the development, over the last twenty years, of new numerical methods that generate digital materials on a microscopic or mesoscopic scale, which are intended to be statistically or exactly equivalent to the microstructure under consideration in terms of topology and attributes (Dawson, 2000; Dawson *et al.*, 2005; Brahme *et al.*, 2006). It has also proved necessary to connect such digital descriptions to finite difference or finite element calculations. The common approaches to generating statistically equivalent polycrystals involve the so-called Voronoi tessellation method (VTM), which consists in creating N Voronoi nuclei randomly and defining each Voronoi cell as the region containing all points which are closer to one particular nucleus than to any other nucleus. Despite the good geometric correlation between Voronoi tessellations and many cellular structures, the VTM shows some limitations owing to the impossibility of satisfying, for example, the condition of a given statistical distribution of cell volumes. Xu and Li (2009) described the divergence between the statistical properties of the grains classically observed in an equiaxed polycrystal and the results obtained with the VTM; they proposed a constrained Voronoi tessellation method (CVTM). Another approach to improving the classical VTM consists in using the so-called Laguerre–Voronoi tessellation method (LVTM), where one assigns a radius, or weight, to each nucleus and incorporates the distribution of these weights into the Voronoi tessellation (Imai *et al.*, 1985). These approaches allow one to match statistical data concerning the grain volumes in a Voronoi tessellation quite precisely. Other interesting developments have shown the possibility of matching the shape ratios of grains by using sets of ellipsoids coupled with the Voronoi tessellation procedure (Brahme *et al.*, 2006).

The generation of statistical virtual polycrystals remains a fertile research domain, but the transition from the mere generation of a statistical REV made of Voronoi cells to thermo-mechanical and metallurgical calculations performed in a REV is not straightforward. Nevertheless, various multiscale numerical models have been constructed to simulate explicitly, on a microscopic scale, grain growth and primary recrystallization, with or without a description of the nucleation phase (Rollett, 1997; Miodownik, 2002).

2.5.2 Probabilistic methods

Probabilistic methods associated with voxels (or pixels in 2D) can be based on grain structure descriptions; for instance, Monte Carlo (MC) (Rollett and Raabe, 2001) and cellular automaton (CA) approaches have been successfully applied to recrystallization (Rollett and Raabe, 2001; Kugler and Turk, 2004; Kugler and Turk, 2006; Raabe, 2002) and grain growth (Rollett and Raabe, 2001; Wang and Liu, 2003; Holm *et al.*, 2001).

The standard MC method, as derived from the Potts model (a multistate Ising model), applies probabilistic rules to each cell and at each time step of the simulation. In this model, the interfaces between the grains are implicitly defined thanks to the membership of the cells in the various grains. In this context, some probabilistic evolution rules are applied to describe recrystallization phenomena. In the case of grain growth, these rules are enforced by minimizing a Hamiltonian form which sums the interfacial-energy components, and the topological events appear in a natural way (Rollett and Raabe, 2001). Moreover, this formalism can be extended to primary recrystallization (Rollett and Raabe, 2001), as well as to Smith–Zener pinning effects when particles are added to the system (Miodownik, 2002). This model proves quite efficient in 3D, and numerically inexpensive compared with deterministic methods (Hassold and Holm, 1993). However, the comparison between MC results and experiments is not straightforward (Rollett, 1997). Indeed, as discussed in the literature, no real length and timescales are enforced directly. Furthermore, the standard form of the model does not result in a linear relationship between the migration rate and the stored energy.

The CA method uses physically based rules to determine the rate of propagation of a transformation from one cell to a neighbouring cell (Raabe, 2002), and can therefore be readily applied to the kinetics of microstructure change in a real system. In the case of recrystallization, the switching rule is simple: an unrecrystallized cell will switch to a recrystallized state if one of its neighbours is recrystallized. In the standard CA method, the states of all cells are simultaneously updated; this is clearly efficient from a numerical point of view. However, in this framework, it remains difficult to use the curvature as a driving force for grain boundary migration (Rollett and Raabe, 2001), and, as for MC simulations, it is not entirely clear how the computational kinetics relates to the kinetics of recrystallization and grain growth in real materials. Some recent developments have illustrated extended capabilities of this approach with respect to these issues (Janssens, 2010).

2.5.3 Modelling grain interfaces

Three main methods can be found in the literature for finite element calculations on virtual microstructures: the 'vertex' models (VMs), also called 'front-tracking' models; the 'phase field' method (PFM); and the 'level set' method (LSM). The VMs are the only type of method to be based on an explicit description of the grain interfaces using surface elements of the finite element mesh or grid considered.

Historically, VMs were developed to describe the grain growth stage, and not primary recrystallization (Nagai *et al.*, 1990). In these models, the grain boundaries are considered as continuous interfaces transported at a velocity defined by the local curvature of the grain boundaries. The main idea is to model the interfaces by a set of points and to move these points at each time-increment by using the velocity and the normal to the interface. This explains the other name for this approach, 'front tracking'. Complex topological events such as the disappearance of grains and node dissociations are treated by means of a set of rules which is completed by a repositioning of the nodes. More recently, VMs have been extended in order to take into account both recrystallization and grain growth, together with site-saturated nucleation (Piekos *et al.*, 2008a,b). In 2D, results for isotropic grain growth show good agreement with the theory (Piekos *et al.*, 2008b). It remains difficult to deal with the non-natural treatment of topological events, particularly in 3D, where the set of rules becomes very complex and numerically expensive (Weygand *et al.*, 2001; Siha and Weygand, 2010).

The PFM (Chen, 1995; Militzer, 2011) and the LSM (Sethian, 1996; Osher and Sethian, 1988; Zhao *et al.*, 1996) have many points in common. In both approaches, a grain interface is described implicitly by means of an artificial field, so that tracking interfaces as in the VMs is not an issue any more. However, these two approaches require one to work with high-order finite elements or with very fine finite element meshes to obtain an accurate description of the interfaces. The artificial field is built as a continuous approximation to the Heaviside function in the PFM (hence the notion of a diffuse interface), whereas it corresponds to a distance function at the interfaces in the LSM. The initial concept of the PFM was to describe the locations of two phases (Collins and Levine, 1985). This concept has been extended to deal with more complex problems involving more than two phases and for modelling microstructure evolution (Chen, 2002; Karma, 2001). In the case of polycrystalline microstructures, each grain orientation is used as a non-conserved order parameter field and the free energy density of a grain is formulated as a Landau expansion in terms of the structural order parameters. The grain boundary energy is introduced as gradients of the structural order parameters, and the boundaries themselves are represented by an isovalue of the order parameter fields. As in the MC and CA methods, topological events are treated in a natural way as a result of energy minimization. Numerous publications have illustrated the potential of this approach for the modelling of ideal normal grain growth in a 2D context (Chen, 2002) and, more recently, in 3D (Moelans *et al.*, 2009), and for static and dynamic recrystallization (Takaki *et al.*, 2008; Takaki *et al.*, 2009), abnormal grain growth (Ko *et al.*, 2009), and Zener pinning phenomena (Chang *et al.*, 2009).

Discussions in the literature have pointed out the current limitations of the PFM regarding the ability of the free energy density functions to reproduce the physical properties of the material. Another problem arises from the very rapid changes in the phase field across the diffuse interfaces, which can lead to very expensive and intensive calculations, particularly for 3D systems.

The LSM concept is quite old (Sethian, 1996; Osher and Sethian, 1988; Zhao *et al.*, 1996), but its application to grain growth and static recrystallization is recent. The first models for 2D and 3D primary recrystallization included a site-saturated or continuous nucleation stage (Logé *et al.*, 2008; Bernacki *et al.*, 2008; Bernacki *et al.*, 2009). They were applied to very simple microstructures, and then improved to be able to deal with more realistic 2D and 3D microstructures and to make a link with stored energies induced by large plastic deformations (Logé *et al.*, 2008; Bernacki *et al.*, 2009; Resk *et al.*, 2009). Anisotropic meshing and remeshing techniques can be used to accurately describe interfaces, both for modelling plastic deformation using crystal plasticity and for updating the grain boundary network at the recrystallization stage (Logé *et al.*, 2008; Bernacki *et al.*, 2008; Bernacki *et al.*, 2009; Resk *et al.*, 2009). Interestingly, it has also been shown that the distribution of stored energy in a polycrystal resulting from plastic deformation leads to deviations from the JMAK theory (Logé *et al.*, 2008; Bernacki *et al.*, 2008; Bernacki *et al.*, 2009).

Recently, the LSM has been used by Bernacki *et al.* (2011) and Elsey *et al.* (2009) to model 2D and 3D isotropic grain growth. The current main weakness of the LSM remains, as in the PFM, the numerical cost, since meshes of very fine elements have to be used to track the grain interfaces, particularly in 3D.

2.6 Future trends

As illustrated by this global overview, the development of numerical methods to model recrystallization and grain growth on the microscopic scale is a very active research area. Given the number of current publications, the CA method and the PFM and LSM seem very promising strategies and should be able to include increasingly complex ingredients and physical phenomena. Some future trends are clearly the following:

- linking recrystallization models with crystal plasticity in order to better predict the stored energy during deformation and the nucleation of new grains, and also to describe dynamic recrystallization;
- taking account of second-phase particles, which are crucial for predicting grain boundary migration kinetics;
- dealing with heterogeneous grain boundary energy and mobility, heterogeneous distributions of second-phase particles, and heterogeneous stored-energy fields, in order to study abnormal grain growth phenomena.

2.7 References

Avrami M (1939), *J Chem Phys*, 7, 1103–1112.
Bergstrom Y (1970), *Mater Sci Eng*, 5, 193–200.
Bernacki M, Chastel Y, Coupez T, Logé R E (2008), *Scr Mater*, 58, 1129–1132.
Bernacki M, Resk H, Coupez T, Logé R E (2009), *Model Simul Mater Sci Eng*, 17, 064006.
Bernacki M, Coupez T, Logé R E (2011), *Scr Mater*, 64, 525–528.
Bernard P, Bag S, Huang K, Logé R E (2011), 'A two-site mean field model of discontinuous dynamic recrystallization', *Mater Sci Eng A*, 528, 7357–7367.
Blaz L, Sakai T, Jonas J J (1983), *Metal Sci*, 17, 609–616.
Brahme A, Alvi MH, Saylor D *et al.* (2006), *Scr Mater*, 55, 75–80.
Chang K, Feng W, Chen L Q (2009), *Acta Mater*, 57, 5229–5236.
Chen L Q (1995), *Scr Metall Mater*, 32, 115–120.

Chen L Q (2002), *Annu Rev Mater Res*, 32, 113–140.
Collins J B, Levine H (1985), *Phys Rev B*, 31, 6119–6122.
Cram D G, Zurob H S, Brechet Y J M, Hutchinson C R (2009), *Acta Mater*, 57, 5218–5228.
Dawson P R (2000), *Int J Solids Struct*, 37, 115–130.
Dawson P R, Miller M P, Han T-S, *et al.* (2005), *Metall Mater Trans A*, 36, 1627–1641.
Deb K, Pratap A, Agarwal S, Meyarivan T (2002), *IEEE Trans Evol Comput*, 6, 182–197.
Dehghan-Manshadi A, Barnett M R, Hodgson P D (2008), *Metall. Mater. Trans. A*, 39A, 1359–1370.
Doherty R D, Hughes D A, Humphreys F J, Jonas J J, Juul Jensen D, *et al.* (1997), *Mater Sci Eng A*, 238, 219–274.
Elsey M, Esedoglu S, Smereka P (2009), *J Comput Phys*, 228, 8015–8033.
Estrin Y (1998), *J Mater Process Technol*, 80–81, 33–39.
Estrin Y, Mecking H (1984), *Acta Metall*, 32, 57–70.
Fan X G, Yang H, Sun Z C, Zhang D W (2010), *Mater Sci Eng A*, 527, 5368–5377.
Glover G, Sellars C M (1973), *Metall Trans A*, 4, 765–775.
Hallberg H, Wallin M, Ristinmaa M (2010), *Comput Mater Sci*, 49, 25–34.
Hassold G N, Holm E A (1993), *J Comput Phys*, 7, 97–107.
Hillert M (1965), *Acta Metall*, 13, 227–238.
Holm E A, Hassold G N, Miodownik M A (2001), *Acta Mater*, 49, 2981–2991.
Humphreys F J, Hatherly M. (2004), *Recrystallization and Related Annealing Phenomena*, 2nd edn. Oxford: Elsevier.
Imai H, Iri M, Murota K (1985), *SIAM J Comput*, 14, 93–105.
Janssens K G F (2010), *Comput Simul*, 80, 1361–1381.
Jonas J J, Quelennec X, Jiang L, Martin E (2009), *Acta Mater*, 57, 2748–2756.
Karma A (2001), *Phys Rev Lett*, 87, 11570.
Ko K J, Cha P A, Srolovitz D, Hwang N M (2009), *Acta Mater*, 57, 838–845.
Kugler G, Turk R (2004), *Acta Mater*, 52, 4659–4668.
Kugler G, Turk R (2006), *Comput Mater Sci*, 37, 284–291.
Kocks U F (1976), *J Eng Mater Technol*, 98, 76–85.
Li J C M (1962), in: Newkirk J B and Wernick J H (eds), *Direct Observation of Imperfections in Crystals*. New York: Interscience, 234.
Logé R E, Bernacki M, Resk H, Delannay L, Digonnet H, *et al.* (2008), *Philos Mag*, 88, 3691–3712.
Lurdos O (2008), *Lois de comportement et recristallisation dynamique: approche empirique*, PhD thesis, Ecole Nationale Supérieure des Mines de Saint-Etienne, France.
Luton M J, Sellars C M (1969), *Acta Metall*, 17, 1033–1043.
McQueen H J (2004), *Mater Sci Eng A*, 387–389, 203–208.
McQueen H J, Jonas J J (1984), *J Appl Metal Work*, 3, 233–241.
Mecking H, Kocks U F (1981), *Acta Metall*, 29, 1865–1875.
Militzer M (2011), *Curr Opin Solid State Mater Sci*, 15, 106–115.
Miodownik M (2000), *Scr Mater*, 42, 1173–1177.
Miodownik M A (2002), *J Light Met*, 2, 125–135.
Moelans N, Wendler F, Nestler B (2009), *Comput Mater Sci*, 46, 479–490.
Montheillet F, Lurdos O, Damamme G (2009), *Acta Mater*, 57, 1602–1612.
Nagai T, Ohta S, Kawasaki K, Okuzono T (1990), *Phase Transit*, 28, 177–211.
Osher S, Sethian J A (1988), *J Comput Phys*, 79, 12–49.
Piekos K, Tarasiuk J, Wierzbanowski K, Bacroix B (2008a), *Comput Mater Sci*, 42, 36–42.
Piekos K, Tarasiuk J, Wierzbanowski K, Bacroix B (2008b), *Comput Mater Sci*, 42, 584–594.

Poliak E I, Jonas J J (1996), *Acta Mater*, 44, 127–136.
Raabe D (2002), *Annu Rev Mater Res*, 32, 53–76.
Resk H, Delannay L, Bernacki M, Coupez T, Logé R E (2009), *Model Simul Mater Sci Eng*, 17, 075012.
Rollett A D (1997), *Prog Mater Sci*, 42, 79–99.
Rollett A D, Raabe D (2001), *Comput Mater Sci*, 21, 69–78.
Rollett A D, Luton M J, Srolovitz D J (1992), *Acta Metall Mater*, 40, 43–55.
Roucoules C, Pietrzyk M, Hodgson P D (2003), *Mater Sci Eng A*, 339, 1–9.
Ryan N D, McQueen H J (1990), *Can Metall Q*, 29, 147–162.
Sah J P, Richardson G J, Sellars C M (1974), *Metal Sci*, 8, 325–331.
Sakai T, Jonas J J (1984), *Acta Metall*, 32, 189–209.
Sakai T, Akben M G, Jonas J J (1983), *Acta Metall*, 31, 631–642.
Sakui S, Sakai T, Takeishi K (1977), *Trans Iron Steel Inst Japan*, 17, 718–725.
Sandstrom R, Lagneborg R (1975), *Acta Metall*, 23, 387–398.
Sandstrom R, Lagneborg R (1975), *Scr Met*, 9, 59–65.
Sethian J A (1996), *Level Set Methods*. Cambridge: Cambridge University Press.
Siha M, Weygand D (2010), *Model Simul Mater Sci Eng*, 18, 015010.
Stewart G R, Jonas J J, Montheillet F (2004), *ISIJ Int*, 44, 1581–1589.
Takaki T, Yamanaka A, Higa Y, Tomita Y (2008), *J Comput.-Aided Mater Des*, 14, 75–84.
Takaki T, Hisakuni Y, Hirouchi T, Yamanaka A, Tomita Y (2009), *Comput Mater Sci*, 45, 881–888.
Thomas J P, Montheillet F, Semiatin S L (2007), *Metall Mater Trans A*, 38A, 2095–2109.
Underwood E E (1981), *Quantitative Stereology*. Addison-Wesley.
Venugopal S, Sivaprasad P V, Vasudevan M, Mannan S L, Jha S K, *et al.* (1995), *J Mater Process Technol*, 59, 343–350.
Wang C, Liu G (2003), *Mater Lett*, 57, 4424–4428.
Weiss I, Sakai T, Jonas J J (1984), *Metal Sci*, 18, 77–84.
Weygand D, Brechet Y, Lepinoux J (2001), *Adv Eng Mater*, 3, 67–71.
Xu T, Li M (2009), *Philos Mag*, 89, 349–374.
Yazdipour N, Davies C H J, Hodgson P D (2008), *Comput Mater Sci*, 44, 566–576.
Zhao H K, Chan T, Merriman B, Osher S (1996), *J Comput Phys*, 127, 179–195.

3
Modelling techniques for optimizing metal forming processes

J. KUSIAK, D. SZELIGA and Ł. SZTANGRET,
AGH – University of Science and Technology, Poland

Abstract: This chapter presents optimization techniques and strategies, and their applications to solving problems associated with metal forming processes. Most of the classical optimization models for such processes are strongly non-linear and demand long computing times for complex numerical simulations. More robust and time-effective optimization methods have been intensively researched. Probabilistic, nature-inspired optimization techniques belonging to this group of robust methods, as well as metamodel-driven and approximation-based optimization strategies, are discussed here. Some case studies of the application of these methods to particular metal forming problems are presented.

Key words: optimization, optimization strategies, nature-inspired optimization techniques, metamodelling, inverse analysis, die shape design, metal forming processes.

3.1 Introduction

Optimization problems are faced almost everywhere in the control of modern metal forming processes and in the optimization of process parameters. The common classical iterative optimization procedures, which are very efficient for solving mathematical problems, fail when applied to the optimization of materials processing. This is due to the fact that the models of such processes are strongly non-linear and are associated with advanced methods of numerical solution such as finite element (FE) analysis. FE models provide a good approximation to the processes simulated, but are usually time-consuming. Therefore, the optimization of these processes requires long calculation times, which often exceed acceptable limits, and this makes the whole optimization procedure useless from a practical point of view. These problems have prompted the development of effective computer models and new optimization strategies, which are discussed in this chapter.

Another problem in optimization analysis is finding the global optimum, which is difficult in the case of an objective function with a complex form. The classical methods often terminate at a local optimum. Therefore, effective optimization techniques have been developed that are robust in terms of taking account of local optima. Probabilistic, nature-inspired optimization techniques belong to this group of robust methods, and the general principles of these techniques are discussed here. Some results on the optimization of metal forming problems using these methods, are also presented.

3.2 Optimization strategies

The analysis of most materials engineering processes is based on computer simulations built on the basis of complex numerical models. The modelling of real metallurgical processes is based mainly on finite element analysis, which usually requires time-consuming computations. Owing to the excessive computational time needed, the use of such models for optimization (see Fig. 3.1) is problematic. Two different approaches to increasing the efficiency of optimization procedures are presented here. The first is based on a metamodel-driven optimization (MDO) strategy, and the second on an approximation-based optimization (ABO) strategy. These two methods allow rationalization of the computation time for optimization.

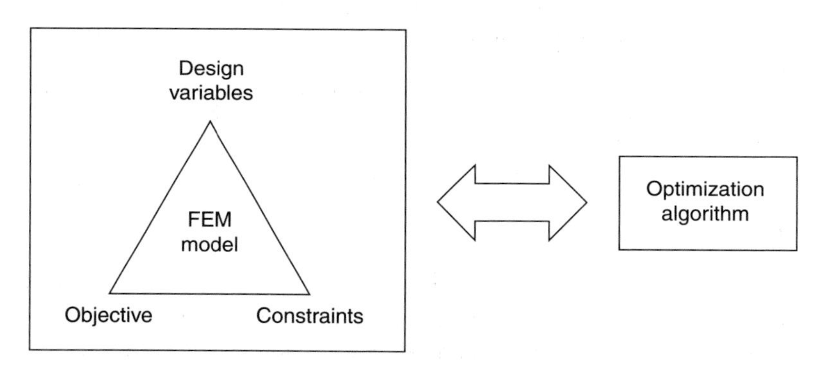

3.1 Flow chart of the classical optimization procedure for real materials processing.

3.2.1 Metamodel-driven optimization strategy

The analysis and optimization of a process require models which precisely describe the reaction of the process to variations in the input signals. Let y be a measured value of the process output and $\mathbf{x}$ the vector of input signals. The relationship between the output and input signals is described by the equation

$$y = f(\mathbf{x}) + \varepsilon \qquad [3.1]$$

where ε is a random measurement error with a normal distribution.

The form of the function f is usually unknown for real processes. Therefore, a model of the process described by a function g is defined, where g is an approximation to the function f. Then, an approximate value of the process output is evaluated according to the following equation:

$$\hat{y} = g(\mathbf{x}) \quad [3.2]$$

For such a function g, the output signal y of the process to be analysed is estimated using the following relationship:

$$y = \hat{y} + h(\varepsilon) \quad [3.3]$$

where h is a function representing the total error of the approximation and measurements.

As has already been mentioned, the modelling of real industrial processes, with implementation of optimizers based on the classical approach (see Fig. 3.2(a)) applied to on-line control systems, is not efficient in practice. This problem has motivated researchers to look for new and more efficient optimization strategies, including the application of metamodels (Persson *et al.*, 2005; Karakasis and Giannakoglou, 2005). In this approach, the industrial process under consideration is replaced by a metamodel (see Fig. 3.2(b)). A metamodel of a process is an

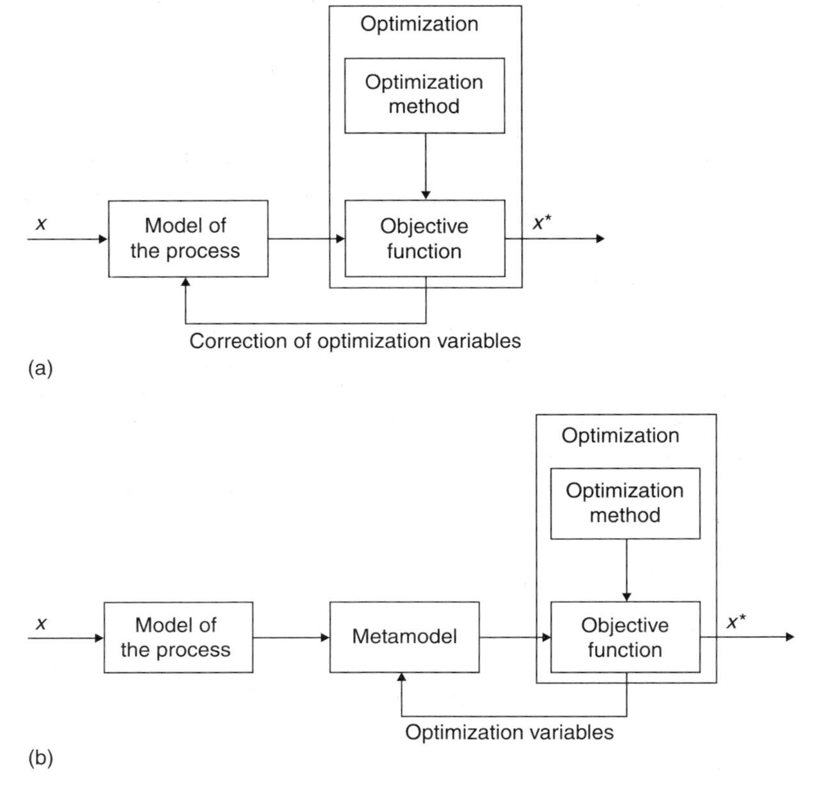

3.2 Flow charts of (a) classical and (b) metamodel-based optimization procedures.

abstraction created on the basis of a lower-level model developed using mathematical techniques. In other words, a metamodel is a 'model of a model'. The metamodel approximates the real process and decreases the computation time of process simulations. It allows the application of classical optimization techniques to search for optimal solutions.

The general idea of a metamodel is the construction of an approximation, based on results from a lower-level model of the process to be analysed. The construction of the metamodel is preceded by collection of a set of values y_i of the process model. The accuracy of the metamodel depends strongly on the number of these points. A higher number of points gives lower values of the approximation error. However, an increase in the number of points also leads to an increase in the computation time, as a result of time-consuming process simulations. Therefore, the number of these points needs to be limited to a necessary minimum, which is usually evaluated by one of the design-of-experiment (DoE) methods (Antony, 2003; Mead, 1990).

The metamodel replaces the model of the process to be analysed and allows:

- examination of the influence of variable input signals on the output signals of the process;
- simple integration of the metamodel with codes that simulate the process;
- a decrease in the computation time and, as a consequence, an increase in the effectiveness of the optimization.

Two commonly used metamodelling techniques are discussed in this chapter: the response surface methodology (RSM) and artificial neural networks (ANNs).

3.2.2 Response surface methodology

The aim of the response surface methodology is to obtain an approximation to the output of a model for given values of the input signals. The approximation problem is formulated as a search for a function g, described by Eq. 3.2, that gives output values close to the modelled function f. One of the simplest methods is to select a number of functions g_k and make the assumption that the function g describing the process can be expressed by the following combination of the g_k:

$$g(\mathbf{x}) = \Sigma_{k=1}^{m} \beta_k g_k(\mathbf{x}), \quad [3.4]$$

where the functions g_k are the known basis functions, and the corresponding parameters β_k are unknown parameters. These parameters can be unambiguously evaluated if the functions g_k are linearly independent and restricted to selected classes. The most popular basis functions are low-order polynomials. For small curvatures of the response surface, polynomials of first order are used:

$$g(\mathbf{x}) = \beta_0 + \Sigma_{i=1}^{n} \beta_i x_i \quad [3.5]$$

where n is the size of the vector of the input signals $\mathbf{x}$. In the case of a significant curvature, the response surface is approximated by second-order polynomials:

$$g(\mathbf{x}) = \beta_0 + \Sigma_{i=1}^{n} \beta_i x_i + \Sigma_{i=1}^{n} \beta_{ii} x_i^2 + \Sigma_{i=1}^{n} \Sigma_{j=1;\, i \neq j}^{n} \beta_{ij} x_i x_j \qquad [3.6]$$

The parameters β of the polynomials in Eqs 3.5 and 3.6 are evaluated using regression analysis.

Another class of basis functions often used in the RSM is that of radial basis functions (RBFs). The most popular RBFs are Gaussian, quadratic and inverse quadratic functions. A detailed description of this method, which is commonly used in metamodelling, has been given by Myers and Montgomery (1995).

The metamodel built with the response surface methodology is applied in further analysis of the process.

3.2.3 Artificial neural networks

An artificial neural network is an information-processing system built from a number of elements called artificial neurons (Fig. 3.3). The input vector of a neuron is composed of n signals $x_1, \ldots, x_n$, and each of these signals is multiplied by a synaptic weight coefficient w_i. These weight parameters allow the impact of the input signal on the output of the neuron to be controlled.

The goal of an artificial neuron is to generate an output signal that depends on the input signals and is close to the observed output of the modelled object. The output signal depends on an activation function Φ and the synaptic weights. Whereas the activation function is selected in the first step of the design of the neuron, the weight parameters undergo variation during the training process. The target of the training process is to estimate the values of the weight parameters that enable the trained neuron to react to the input signals in the same way as the modelled object.

Neurons, during the training process, learn to react to the input signals and to generate the model output. In a supervised training process, a training data set of pairs (X, D) is given, where $X = \{\mathbf{x}^1, \mathbf{x}^2, \ldots, \mathbf{x}^N\}$ is a set of N input vectors $\mathbf{x} = (x_1, x_2, \ldots, x_n)$, and D is a set of corresponding observations of the desired outputs. The goal of training is modification of the weight parameters w_i corresponding to

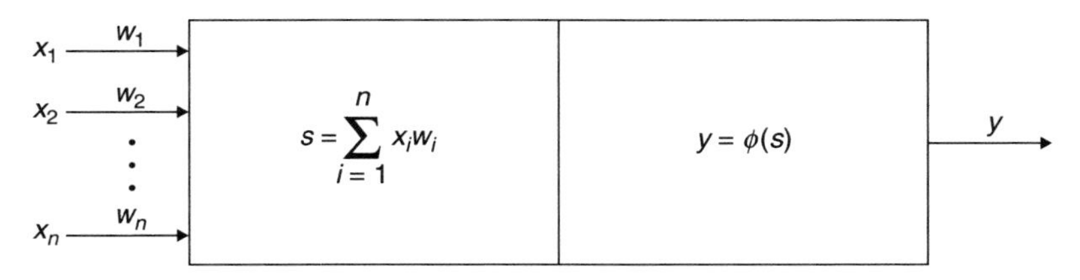

3.3 Diagram of a single artificial neuron ($x_1, \ldots, x_n$, input signals; $w_1, \ldots, w_n$, synaptic weight parameters; Φ, activation function).

every input signal x_i. During the supervised training process, the difference between the values of the desired output signals d_i and the output signals y_i obtained from the neuron is minimized with respect to the weights w_i. A well-trained neuron predicts the correct answer y with the specified accuracy. The initial values of the weights w_i are set randomly.

Artificial neural networks are composed of a defined number of interconnected neurons of the kind described above, arranged in three layers: the input, hidden and output layers (see Fig. 3.4). Feedforward neural networks are commonly used in practice. The number of neurons in the input layer depends on the number of input signals. The number of hidden layers is chosen on the basis of the network designer's experience. In most case studies, there have been one or two layers. Mostly, there is one neuron in the output layer, corresponding to the output signal of a modelled process (Tadeusiewicz, 1993).

The trained ANN is used as a metamodel of an object (the process). The design of a metamodel based on the ANN approach consists of the following three steps:

1. Definition of the ANN architecture (network topology and activation functions).
2. Supervised training of the ANN.
3. Testing of the ANN.

More information about ANNs can be found in Bishop (1995), Haykin (1999) and Hagan *et al.* (1996).

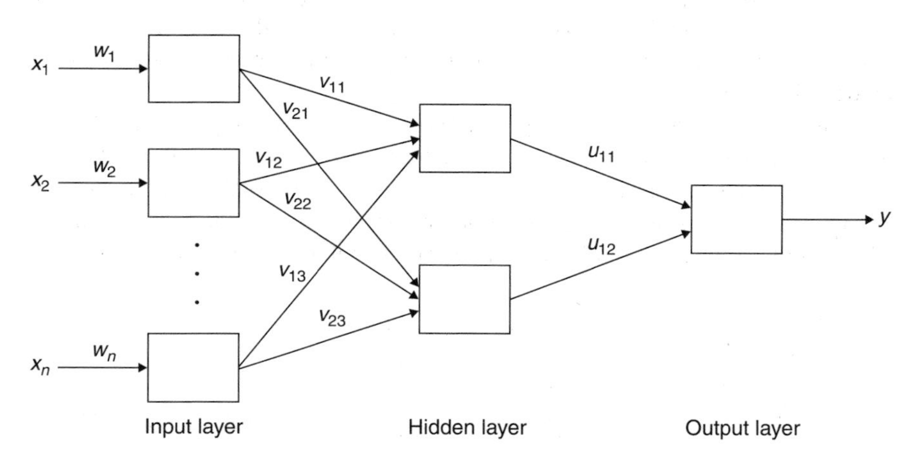

3.4 Feedforward artificial neural network with n–2–1 structure (n neurons in the input layer, two neurons in the hidden layer and one neuron in the output layer): $(x_1, x_2, \ldots, x_n)$, vector of input signals; y, output signal of the network; w_i, v_{ij}, u_{ij}, synaptic weights linked to the network signals.

The main feature of ANNs is their ability to learn and, as a consequence, a good capacity to approximate multivariate nonlinear functions. Because of this, they appear to be useful tools for modelling complex objects and processes. Since ANNs are also very efficient in terms of computational time, they are used in metamodelling and can also be implemented in on-line control systems.

3.2.4 Approximation-based optimization (ABO) strategy

The idea of the approximation-based optimization strategy (Kusiak *et al.*, 2009) is somewhat similar to the response surface methodology. In both methods, the goal function f is replaced by a selected approximation function g. The function g is built based on an initial data set ($X^{(0)}$, $Y^{(0)}$). The elements of the set $Y^{(0)}$ are the values of the goal function calculated at the points of the input data set $X^{(0)}$. Instead of using the goal function f, the method searches for the minimum of the optimization problem for the approximation function g. Once the optimal solution $\mathbf{x}^*$ is found, the value $y^* = f(\mathbf{x}^*)$ of the goal function is calculated and the values of $\mathbf{x}^*$ and y^* are added to the initial data set ($X^{(0)}$, $Y^{(0)}$), giving a new data set ($X^{(1)}$, $Y^{(1)}$). A new approximation $g^{(1)}$ is then built from the modified data set ($X^{(1)}$, $Y^{(1)}$), and the procedure of searching for the minimum solution is repeated until a stop criterion is reached (see Fig. 3.5).

Algorithm of the ABO strategy

The optimization procedure begins with an initial set of design variables $X^{(0)}$ selected according to the DoE method. The selected data set $X^{(0)}$ is composed of M n-dimensional vectors. For each $\mathbf{x}^i \in X^{(0)}$, $i = 1, 2, \ldots, M$, the goal function y^i

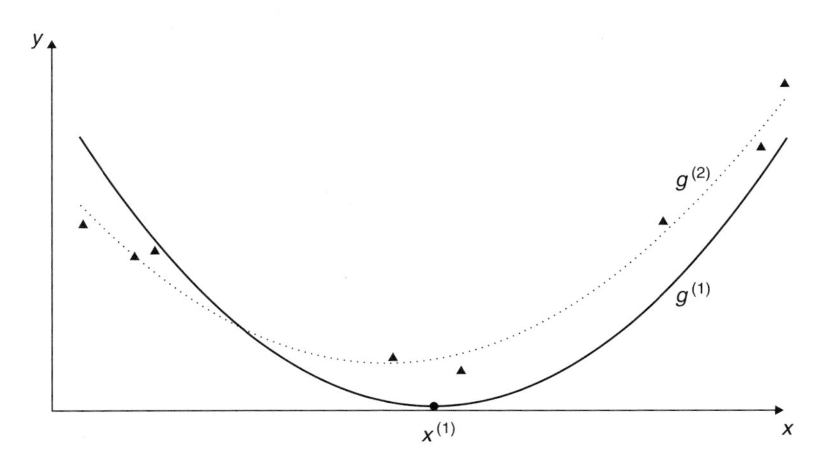

3.5 Idea of the ABO strategy (▲, values of the goal function f; solid line, approximation function $g^{(1)}$; $x^{(1)}$, minimum of the function $g^{(1)}$; dotted line, the approximation $g^{(2)}$ obtained in the next iteration).

$=f(\mathbf{x}^i)$ is calculated, defining the goal function data set $Y^{(0)} = \{y^1, y^2, \ldots, y^M\}$. Let $\mathbf{x}^{(0)}$ be the point of the data set $X^{(0)}$ for which the goal function has the smallest value. We substitute $y^{(0)} = f(\mathbf{x}^{(0)})$ and the value of $y^{(0)}$ becomes the first approximation to the minimum of the goal function f. We set $j = 1$, and then every iteration j has the following four steps:

Step 1. The data set $X^{(j-1)} = \{\mathbf{x}^1, \mathbf{x}^2, \ldots, \mathbf{x}^{M+j-1}\}$ and its corresponding set of goal function values $Y^{(j-1)} = \{y^1, y^2, \ldots, y^{M+j-1}\}$ are defined. On the basis of pairs $(\mathbf{x}^i, y^i)$, $i = 1, 2, \ldots, M + j + 1$, define the function $g^{(j)}$ as the approximation to the goal function f.

Step 2. Search for the minimum of the function $g^{(j)}$ defined in Step 1 using one of the known optimization techniques. Let the point $\mathbf{x}^{(j)}$ be the minimum of the function $g^{(j)}$.

Step 3. Calculate the goal function $y^j = f(\mathbf{x}^{(j)})$ for the point $\mathbf{x}^{(j)}$ determined in Step 2.

Step 4. If $y^{(j)} \leq y^{(j-1)}$ and $|y^{(j)} - y^{(j-1)}| < \varepsilon$ (where ε is the assumed accuracy), stop the calculations. The optimal solution is $\mathbf{x}^* = \mathbf{x}^{(j)}$. Otherwise, define the new data sets $X^{(j)} = X^{(j-1)} \cup \{\mathbf{x}^{(j)}\}$ and $Y^{(j)} = Y^{(j-1)} \cup \{y^{(j)}\}$ and increase the iteration index according to $j = j + 1$. Go to Step 1.

A flow chart of the algorithm of the ABO method is presented in Fig. 3.6. Following the principles of this method, the optimal solution is searched for among the values of the approximations to the goal function. As a consequence, a significant decrease in the time required for the optimization procedure is achieved.

3.3 Nature-inspired optimization techniques: genetic algorithms, evolutionary algorithms, particle swarm optimization and simulated annealing

Solving optimization problems characterized by a multimodal objective function with more than one local minimum appears to be difficult and sometimes fails when the classical, deterministic gradient or non-gradient methods are applied. The optimization problems associated with practical engineering problems often involve such multimodal objective functions. Therefore, effective methods that search for a global minimum not only within a local region but also over the whole domain of the objective function have been developed. The probabilistic methods belong to this group of optimization techniques. Owing to the use of a random search, which is the basis of these methods, the probability of finding the global minimum is increased. The algorithms of the probabilistic methods described in this chapter are results of the observation of nature and its perfect mechanisms: evolution, the behaviour of populations of individuals, etc. The optimization algorithms inspired by nature belong to the group of heuristic algorithms. These algorithms can be applied to any kind of objective function: non-linear, discontinuous or multimodal. The optimization of such functions

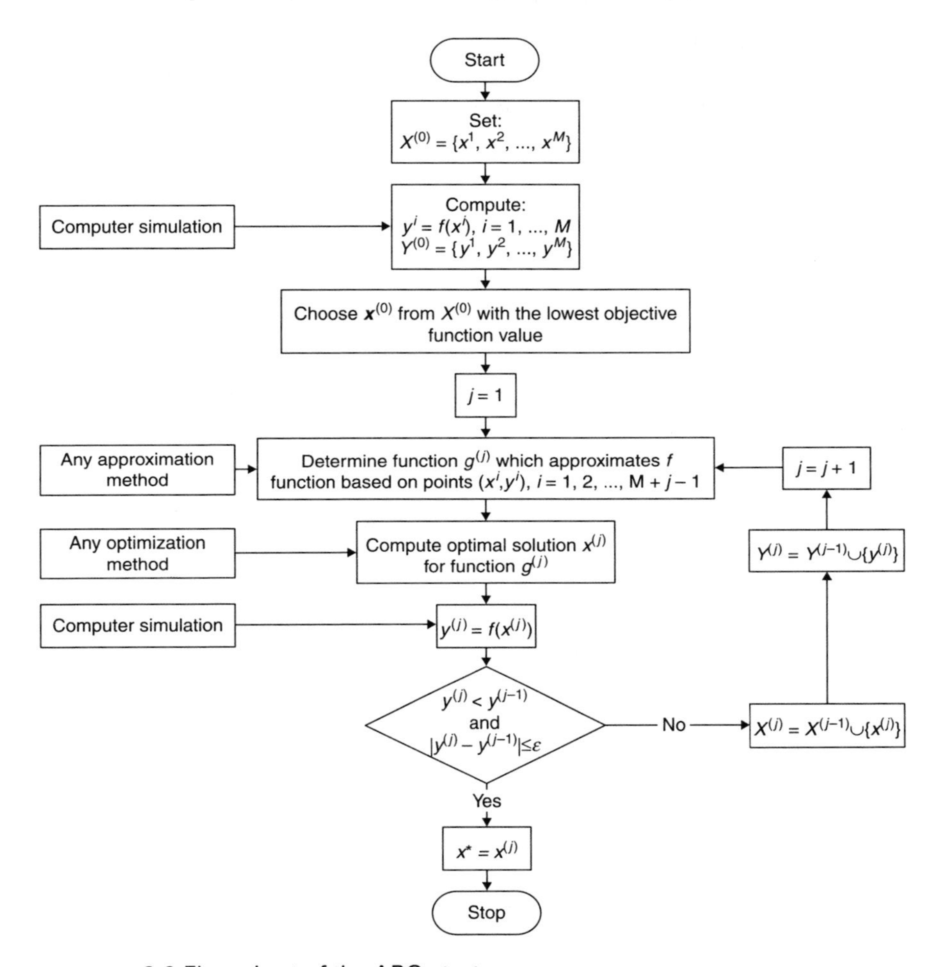

3.6 Flow chart of the ABO strategy.

using classical, conventional algorithms is more difficult and sometimes even impossible. The following nature-inspired algorithms are discussed in this chapter:

- genetic algorithms;
- evolutionary algorithms;
- particle swarm optimization;
- the simulated annealing algorithm method.

3.3.1 Genetic algorithms

Genetic algorithms (GAs) are optimization methods that are inspired by selection, evolution and inheritance in nature (Arabas 2004). The original form was developed by John Holland at the University of Michigan. These methods apply a few simple mechanisms: natural selection, genetic recombination and mutation. Genetic algorithms differ from classical optimization methods primarily by the following key features:

- the encoded form of the solution;
- the fact that search starts from a population of initial solutions;
- the use of probabilistic selection rules.

The greatest advantage of genetic algorithms is the lack of restrictions on the form of the objective function. GAs do not require a knowledge of gradients or higher derivatives of the objective function. GAs should be applied to problems for which classical and other methods are too difficult to implement or excessively time-consuming. Probabilistic selection rules are used in these algorithms, and for this reason, GAs do not always find the optimal solution.

The terminology used in genetic algorithms is rather unique because it has been taken directly from biology. The basic terms, together with their explanation, are given in Table 3.1.

Genetic algorithms process a population of individuals. Each individual is encoded by means of a chain of defined length using a finite alphabet. The *zero–one* alphabet is the most often used, in which the individuals are coded with a natural binary code (NBC). Using an NBC, it is easy to code natural numbers from the interval $[0, 2^n - 1]$, where n is the length of the chain of genes. To encode

Table 3.1 The basic terms used in the description of genetic algorithms

Term	Meaning
Chromosome	Series of genes; encoded form of a potential solution
Gene	Single component of a chromosome
Individual	Potential solution (point in the search space)
Genotype	Set of a specific individual's chromosomes, g
Phenotype	Individual in non-coded form, h
Population	Set of individuals of a specific size, P
Fitness	Numerical value defining the quality of a solution represented by a selected individual, F
Parent	Individual selected for a crossover operation
Descendant	Individual obtained in a crossover operation
Crossover	Process of recombination of genes, leading to the generation of new descendants as a result of the exchange of fragments of the parents' chromosomes
Mutation	Process of change in a single gene in a chromosome
Selection	Selection of individuals, usually best adapted to the parents' population

real numbers from the interval $[u, v]$ with a specified accuracy r, a linear mapping of the interval $[u, v]$ onto the interval $[0, 2^n - 1]$ is performed. Then, for an individual's genotype g and phenotype h, the following relationship is defined:

$$h = u + g \cdot r \qquad [3.7]$$

where r is the accuracy and is given by the equation

$$r = \frac{v - u}{2^n - 1} \qquad [3.8]$$

The pseudocode of a general genetic algorithm is as follows:

$t \leftarrow 0$;
initialization of population $P^{(t)}$;
for each $g \in P^{(t)}$ compute fitness $F(g)$;
do
  $T^{(t)} \leftarrow (P^{(t)})$ selection;
  $O^{(t)} \leftarrow (T^{(t)})$ crossover and mutation;
  for each $g \in O^{(t)}$ compute fitness $F(g)$;
  $P^{(t+1)} \leftarrow O^{(t)}$;
  $t \leftarrow t + 1$;
while (stop condition);

In the first step, the initial population is generated by sampling an appropriate number of individuals. This step is run only once. A new, continually improved initial population is generated in each iteration by applying the following three genetic operators:

- selection (reproduction);
- crossover;
- mutation.

The selection operator is responsible for the selection of individuals creating the population of parents. The selection is random but depends on the values of the objective function for the separate individuals. The higher the value of the objective function, the higher the individual's probability of being selected. In addition to the value of the individual's objective function, the fitness index of the entire population is another factor that influences the probability of selection of a specific individual. This index is equal to the sum of the values of the fitness function for all individuals in the population. There are many ways to execute the selection operation. The method of the roulette wheel is the simplest one. The name of the method refers to a wheel divided into sectors. Each individual has its own sector, the size of which is proportional to the value of the fitness function. A single spin of the wheel selects one individual to be copied to the population of parents. Then, crossover and mutation operators are executed on this population. The crossover operator proceeds in two stages. First, two individuals are randomly selected from the population. Then, the selected pair is subjected to a crossover operation with a probability p_k. A crossover point k ($k < n$) is drawn and the genes in the positions from $k + 1$ to n are exchanged between parent individuals, defining two descendant individuals. The crossover operation is illustrated by the following example.

Let the following two individuals (parents) be selected for crossover:

10101010
01010101

The crossover point $k = 4$ is drawn and the genes in the positions from 5 to 8 are exchanged. The crossover operation results in the following two descendant individuals:

1010|1010 1010|0101
→
0101|0101 0101|1010

The descendant individuals undergo mutation (with a mutation probability p_m) in the next step of the algorithm. A number from the interval [0, 1] is drawn for each gene in the chromosome, and if it is less than or equal to the chosen value of p_m, the gene's value is changed to the opposite value ($0 \leftrightarrow 1$). The mutation operation is shown below for one of the descendant individuals previously obtained. Assuming that the mutation probability p_m is equal to 0.1 and the sequence of numbers 0.82, 0.35, **0.05**, 0.24, 0.74, 0.16, 0.63, 0.47 is drawn for the first descendant, the third gene has to mutate. The individual (10100101) becomes (10000101) after mutation.

After completion of the mutation step, individuals are recorded in the descendant population, which becomes the initial population for the next iteration of the algorithm. The number of iterations is limited to a given maximum value, and the algorithm stops at this value irrespective of the results obtained so far. The calculations also stop if the evaluated solution is sufficiently close to the optimal solution.

3.3.2 Evolutionary algorithms

Evolutionary algorithms (EAs) are based on the same assumptions as genetic algorithms. However, in contrast to GAs, the individuals in the population are not encoded. Evolutionary algorithms have three basic strategies:

- the $(1 + 1)$ strategy;
- the $(\mu + \lambda)$ strategy;
- the (μ, λ) strategy.

These strategies differ in the population size, the operators used and the way individuals are selected for the initial population for the next iteration. Brief descriptions of each strategy are given below.

(1 + 1) strategy

In the (1 + 1) strategy, the population consists of one individual only. The crossover operator does not exist in this case. In each iteration, one new individual is generated by means of a mutation operator. The selection is limited to the selection of an individual with the highest value of the objective function. The algorithm's pseudocode is shown below:

$t \leftarrow 0$;
$\mathbf{x}^{(t)}$ initialization;
do
  $\mathbf{x}'^{(t)} \leftarrow (\mathbf{x}^{(t)})$ mutation;
  if $(f(\mathbf{x}'^{(t)}) > f(\mathbf{x}^{(t)}))$
    $\mathbf{x}^{(t+1)} \leftarrow \mathbf{x}'^{(t)}$;
  else
    $\mathbf{x}^{(t+1)} \leftarrow \mathbf{x}^{(t)}$;
  $t \leftarrow t + 1$;
while (stop condition).

The mutation operator generates a new individual by adding a random number to each gene. The number added is a product of a mutation range and a random number with a normal distribution $N(0, 1)$:

$$x_i'^{(t)} = x_i^{(t)} + \sigma\xi_{N(0,1),i} \qquad [3.9]$$

where σ is the mutation range, and ξ is a random variable with a normal distribution $N(0, 1)$.

The mutation range is a parameter that describes the magnitude of the current changes in the solution. The operator of the variation of mutation range is called the 1/5 *success rule*. It runs as follows:

- if, within k consecutive iterations, the number of successful mutations (i.e. when the descendant individual turns out to be better than its parent) is higher than one fifth of all mutations, then the mutation range is increased: $\sigma' = c_i\sigma$;
- if the number of successful mutations is lower than one fifth of all mutations, then the mutation range is decreased: $\sigma' = c_d\sigma$;
- if the number of successful mutations is equal to one fifth of all mutations, then the mutation range does not change.

The constants c_i and c_d have the following values:

- $c_d = 0.82$;
- $c_i = 1/0.82$.

The advantage of the (1 + 1) strategy is its speed of operation, due to the fact there is only one individual in the population. The disadvantage is low resistivity to local minima.

($\mu + \lambda$) strategy

In the ($\mu + \lambda$) strategy, a population consisting of μ individuals is processed. Each individual consists of two chromosomes. One chromosome contains a vector of independent variables **x** (representing a point in the search space), and the second contains a vector of standard deviation values $\boldsymbol{\sigma}$ that is used by the mutation operator. A crossover operator exists, in addition to the mutation operator. The selection operator, by sampling with replacement from a baseline population of size μ, generates a parent population of individuals with a size of λ (hence the name of the strategy). By analogy to genetic algorithms, the probability of drawing a specific individual is proportional to the value of the objective function. The pseudocode of the ($\mu + \lambda$) strategy is as follows:

$t \leftarrow 0$;
initialization of population $P^{(t)}$;
for each $g \in P^{(t)}$ compute fitness $F(g)$;
do
  $T^{(t)} \leftarrow (P^{(t)})$ selection;
  $O^{(t)} \leftarrow (T^{(t)})$ crossover and mutation;
  for each $g \in O^{(t)}$ compute fitness $F(g)$;
  $P^{(t+1)} \leftarrow \mu$ of the best individuals from $P^{(t)} \cup O^{(t)}$;
  $t \leftarrow t + 1$;
while (stop condition).

The initial population is randomly generated. The process of mutation is divided into three stages. In the first stage, a number with a normal distribution $N(0, 1)$ is drawn. In the second stage, each individual's standard deviation changes according to the relationship

$$\sigma_i' = \sigma_i \cdot e^{(\tau' \xi_{N(0,1)} + \tau \xi_{N(0,1),i})} \qquad [3.10]$$

where τ and τ' are the parameters of the algorithm, expressed as follows:

$$\tau = \frac{K}{\sqrt{2n}} \qquad [3.11]$$

$$\tau' = \frac{K}{\sqrt{2\sqrt{n}}} \qquad [3.12]$$

The value of K is usually taken as equal to 1, and n is equal to the dimension of the decision space. Using updated values of the vector $\boldsymbol{\sigma}'$, the components x_i of the vector $\mathbf{x} \in X$ are modified:

$$x_i' = x_i + \sigma_i' \xi_{N(0,1),i} \qquad [3.13]$$

The advantage of the mutation operator in the ($\mu + \lambda$) strategy compared with the (1 + 1) strategy is the lack of an arbitrary value characterizing the mutation operation. The effect of adaptation of the mutation range in the $(\mu + \lambda)$ strategy is a consequence of the selection mechanism (favouring better individuals).

The crossover operator is executed in a different way. A method which evaluates the mean value of the chromosomes of the parents is commonly used. A random number, denoted by $\xi_{u(0,1)}$, in the interval [0, 1] with a uniform distribution, is used. Descendant individuals are generated according to the following relationships:

$$a = \xi_{U(0,1)} \quad [3.14]$$

$$\mathbf{x}'^{1} = a\mathbf{x}^{1} + (1-a)\mathbf{x}^{2} \quad [3.15]$$

$$\mathbf{x}'^{2} = a\mathbf{x}^{2} + (1-a)\mathbf{x}^{1} \quad [3.16]$$

$$\boldsymbol{\sigma}'^{1} = a\boldsymbol{\sigma}^{1} + (1-a)\boldsymbol{\sigma}^{2} \quad [3.17]$$

$$\boldsymbol{\sigma}'^{2} = a\boldsymbol{\sigma}^{2} + (1-a)\boldsymbol{\sigma}^{1} \quad [3.18]$$

In contrast to genetic algorithms, all individuals are subjected to the mutation and crossover procedures. The selection of μ individuals from the total of the baseline population and the descendant populations defines the population for the next iteration.

(μ, λ) strategy

This strategy is similar to the $(\mu + \lambda)$ strategy. The only difference is that the initial population of μ individuals for the next iteration is selected from the descendant population only.

3.3.3 Particle swarm optimization method

The particle swarm optimization (PSO) method is based on mechanisms observed in nature. However, unlike the GA and EA methods, it is not based on the theory of evolution but on observations of the behaviour of populations of individuals. Particles (identified with the solutions of the problem under consideration) traverse the decision space (the area inhabited by the population) by following the particle representing the best behaviour found so far, while remembering the best position at which they have been so far. Two vectors, a position and a velocity, describe each particle. In each iteration, a new velocity vector $\mathbf{v}$ is evaluated, and this determines the change of the position $\mathbf{x}$ of the particle.

During initialization, a random position $\mathbf{x}_i$ and velocity $\mathbf{v}_i$ are chosen for each particle. If the velocity is too low, the swarm is not able to search the entire permissible area, whereas an excessively high velocity makes the particles 'bump' against the limits of the permissible area. The velocity vector changes according to the relationship:

$$\mathbf{v}_i^{(t+1)} = w\mathbf{v}_i^{(t)} + c_1 r_1 (\mathbf{p}^g - \mathbf{x}_i^{(t)}) + c_2 r_2 (\mathbf{p}_i - \mathbf{x}_i^{(t)}) \quad [3.19]$$

where $\mathbf{x}_i^{(t)}$ and $\mathbf{v}_i^{(t)}$ are the position and velocity, respectively, of the ith particle in the tth iteration; $\mathbf{p}^g$ defines the best position found so far by the swarm; $\mathbf{p}_i$ is the best solution found so far by the ith particle; w is the inertia coefficient; c_1 and c_2 are acceleration coefficients (also called training coefficients); and r_1 and r_2 are random numbers from the interval [0, 1] with a uniform distribution. The new particle's position is defined as follows:

$$\mathbf{x}_i^{(t+1)} = \mathbf{x}_i^{(t)} + \mathbf{v}_i^{(t+1)} \quad [3.20]$$

After displacement of all particles to their new positions, they are subjected to an assessment and the leader of the swarm is selected. The algorithm's pseudocode is presented below:

$t \leftarrow 0$;
initialization of the swarm $P^{(t)}$;
do
  for each $g \in P^{(t)}$ compute fitness $F(g)$;
  update $\mathbf{p}^g$;
  for each $i = 1, \ldots, |P^{(t)}|$ **do**
    update $\mathbf{p}_i$;
    determine velocity vector $\mathbf{v}_i^{(t)}$;
    determine position vector $\mathbf{x}_i^{(t)}$;
  $t \leftarrow t + 1$;
while (stop condition).

The values of the coefficients affect the behaviour of the swarm. The value of the inertia coefficient is usually selected from the interval [0, 1]. A higher value is favourable for global searching of the solution space, and a lower value for local searching. Usually, the value is constant throughout the entire optimization process. However, it may also vary. In this case, at the beginning, a high value is assumed, enabling global searching, and then, as the maximum that is sought is being approached, it gradually decreases. The acceleration coefficients are usually equal and selected from the interval [0, 2]. The values are selected on the assumption of the maximum particle velocities. The stop criterion is defined by a maximum number of iterations or a satisfactory value of the solution.

3.3.4 Simulated annealing algorithm

Unlike the optimization methods described above, the simulated annealing (SA) algorithm is not inspired by live creatures' behaviour, but by the process of annealing of metals or alloys. Annealing consists of heating a material to a defined temperature, maintaining it at this temperature for some time and then cooling. The simulated annealing algorithm searches for a minimal solution of the objective function iteratively. If the newly determined solution in the current iteration is better than the previous solution, it always becomes the current solution, and if it is worse, it becomes the current solution with some probability. This probability depends on the difference between the values of the objective function and on a parameter of the algorithm that is identified with the temperature of an annealing process and it is expressed by the equation

$$p = e^{\Delta f / T} \tag{3.21}$$

where Δf is the difference between the values of the objective function for the current solution and the newly created solution, and T is the current temperature.

If the new solution has a lower value of the objective function (i.e. it is better), then the probability that it becomes the current solution in the next iteration is higher than 1. The algorithm's pseudocode is as follows:

$t \leftarrow 0$;
$\mathbf{x}^{(t)}$ initialization;
$T^{t} \leftarrow T^{\max}$;
do
  $\mathbf{x}'^{(t)} \leftarrow$ determine $(\mathbf{x}^{(t)}, T^{(t)})$;
  if (random([0, 1]) $< \exp((f(\mathbf{x}^{(t)}) - f(\mathbf{x}'^{(t)}))/T^{(t)})$)
    $\mathbf{x}^{(t+1)} \leftarrow \mathbf{x}'^{(t)}$;
  else
    $\mathbf{x}^{(t+1)} \leftarrow \mathbf{x}^{(t)}$;
  $T^{(t+1)} \leftarrow$ reduce $(T^{(t)})$;
  $t \leftarrow t + 1$;
while (stop condition).

The new solution is drawn from some neighbourhood of $\mathbf{x}^{(t)}$, and its value is proportional to the temperature $T^{(t)}$. Either a normal distribution with an expected value $\mathbf{x}^{(t)}$ and standard deviation $T^{(t)}$ or a uniform distribution is used.

The temperature is decreased in each iteration of the algorithm. There are a number of methods decreasing the temperature. However, the temperature is usually decreased linearly in the following way:

$$T^{(t+1)} = \alpha T^{(t)} \tag{3.22}$$

$$\alpha \in (0,1) \tag{3.23}$$

The heuristic algorithms described above have been used to solve engineering problems (Barabasz *et al.*, 2009; Paszyński *et al.*, 2009; Schaefer *et al.*, 2007; Sztangret *et al.*, 2009). The main advantage of these methods is their ability to search for optimal solutions for multimodal, discontinuous or non-linear objective functions. This is due to the fact that they usually work on a population of solutions. These methods do not need any additional information regarding the optimized function, except for the values of the objective function at selected points. On the other hand, this is a disadvantage, because these methods require numerous calls of the objective function, which results in a high computational cost for optimization problems with complex numerical models.

3.4 Application of metamodelling and optimization strategies in metal forming – case studies

Many different optimization problems are encountered in metal forming processes. Technologists are interested in minimization of the deformation energy, minimization of toolwear and maximization of the quality of the final product by achievement of the required mechanical properties, etc. Generally, optimization problems in the field of metal forming can be grouped into the following three categories (see Fig. 3.7):

- optimal design of the tool shape;
- optimization of the process technology;
- optimization of the quality of the final product.

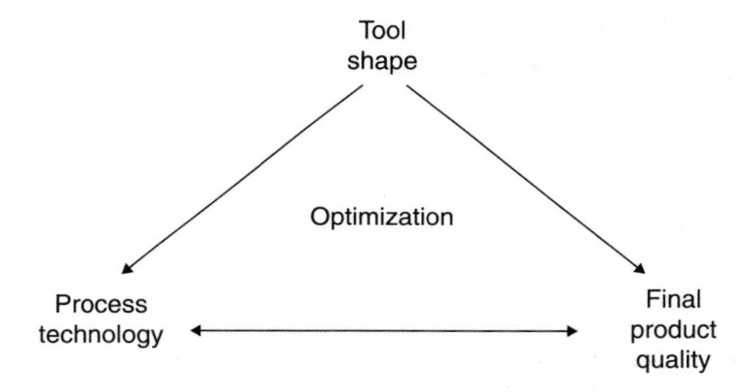

3.7 General classification of optimization problems related to metal forming processes.

Various examples of the solution of optimization problems in the field of metal forming can be found in the literature (e.g. Bonte *et al.*, 2005). Two selected examples of the optimization of forging processes will be presented as case studies in this chapter. The first is focused on the optimization of the microstructure of the final forged product, and the second on optimization of the parameters of a crankshaft forging process.

The application of optimization techniques to inverse analysis is a characteristic feature of various engineering problems. Since the identification of rheological models for metal forming seems to be the most frequent engineering application of inverse analysis (Szeliga *et al.*, 2006), this problem was selected as our third case study.

3.4.1 Optimization of the tool shape for metal forming

The optimization of the shape of tools for metal forming is still of considerable interest to engineers and technologists. To illustrate the application of the optimization strategies presented in this chapter, we have chosen a problem of optimization of the shape of a die for axisymmetric two-stage forging (Kusiak, 1996). The aim is to look for a shape for the die to be used in the first stage of forging that will allow optimization of the properties of the final product obtained after the final forging operation. To achieve this aim, it is necessary to find a correlation between the parameters of the deformation process and the microstructure development, especially with respect to the influence of the die shape on the microstructure of the forging. The forging technology considered is presented in Fig. 3.8. It is assumed that the shape of the preform upper die undergoes modification. In the analysis, we assumed that an austenitic steel sample containing 0.4% C and 0.15% Si was deformed, and the influence of the shape of the preform die on the final grain size and its distribution was evaluated. A uniform initial distribution of grains in the sample prior to deformation was assumed. A uniform grain size distribution throughout the whole volume of the final product was chosen as the objective for the purpose of validation of the optimization strategy. Thus, the aim of the optimization was to find a shape of the preform die for which the mean square deviation between the average and actual grain sizes reached a minimum. The objective function chosen for this optimization problem is expressed by the following equation:

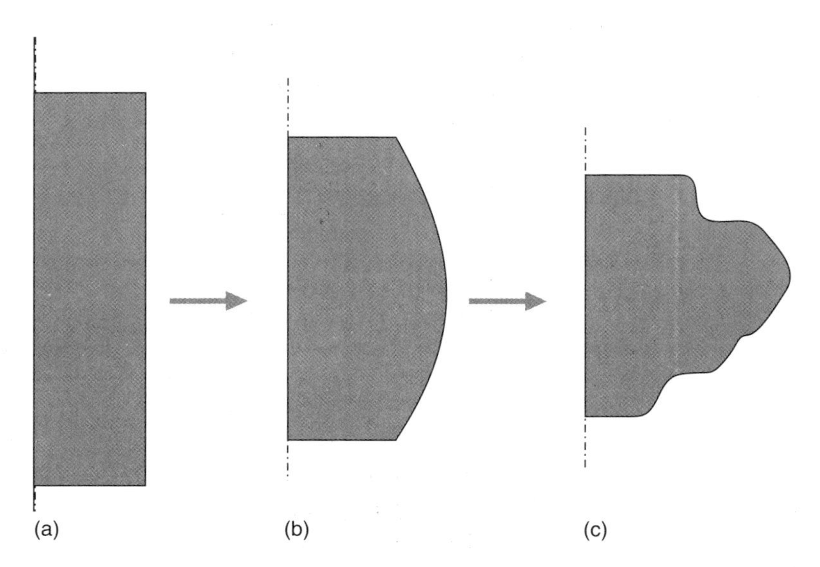

3.8 The two-stage axisymmetric forging process analysed in this chapter; (a) ingot, (b) preform, (c) final forging.

$$f(\mathbf{x}) = \min_{\mathbf{x}} \left[\frac{1}{n} \Sigma_{i=1}^{n} \left(D_i - D_{\text{ave}} \right)^2 \right] \qquad [3.24]$$

where $\mathbf{x}$ is the vector of the design (optimization) variables; D_i and D_{ave} are the local grain size in node i and the average grain size, respectively; and n is the number of grains.

The vector of optimization variables $\mathbf{x}$ contains the parameters of a function describing a segment of the profile of the upper preform die. The profile of this segment undergoes variation during the optimization procedure. The segment is described by a polynomial of second order containing two design variables, x_0 and x_1. An analysis of the metal flow during the forging process was performed using the FORGE2 code, which is a commercial finite element program designed for thermo-mechanical modelling of metal forming processes. The analysis of the metal flow was carried out assuming a visco-plastic model for the deformed material. The austenite grain size was calculated using a closed-form equation developed by Roberts *et al.* (1983), expressing the recrystallized grain size as a function of temperature and strain,

$$D_{\text{r}} = 6.2 + 55.7 D^{0.5} \varepsilon^{-0.65} e\ (-35000 / RT) \qquad [3.25]$$

where D_{r} is the recrystallized grain size, ε is the strain, T is the temperature and R is the gas constant.

This simplified approach assumes isothermal conditions during recrystallization. Its quantitative accuracy may be questioned, but it describes the microstructural phenomena qualitatively very well, as confirmed by experimental validation (Kusiak *et al.*, 1994). Consequently, this model can be regarded as satisfactory for the purpose of demonstrating an optimization method.

Searching for the optimal shape of the perform die was performed using the approximation-based optimization strategy described in Section 3.2.4. To begin the optimization procedure, a set of five vectors of design variables $\mathbf{x}$ was chosen arbitrarily. Each trial vector $\mathbf{x}^i$ generated a different profile of the upper preform die. These five die profiles were examined according to the ABO strategy, where the forging process was simulated using the FORGE2 program. The calculated values $f^{(\mathbf{x}^i)}$, $i = 1, 2, \ldots, 5$, of the objective function in Eq. 3.24 corresponding to these five profiles were approximated by a function g using a cubic spline method. The search for the minimum of the function g does not require FE recalculations of the forging process and can be performed by any optimization technique (Kowalik and Osborne, 1968; Kusiak *et al.*, 2009). A preform die shape was constructed corresponding to the minimal solution found for the approximation function g, and a simulation of the forging process was performed. The new value of the objective function calculated for this optimal solution was added to the set X of design vectors already calculated, and a new approximation function g was calculated. The whole procedure was continued until no improvement in the

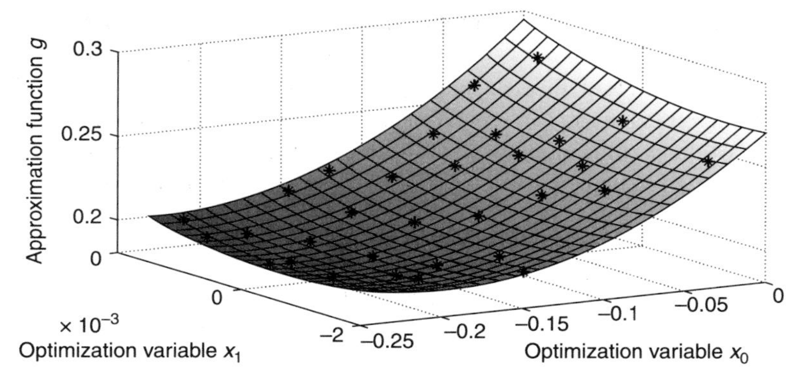

3.9 Final results for the approximation function g.

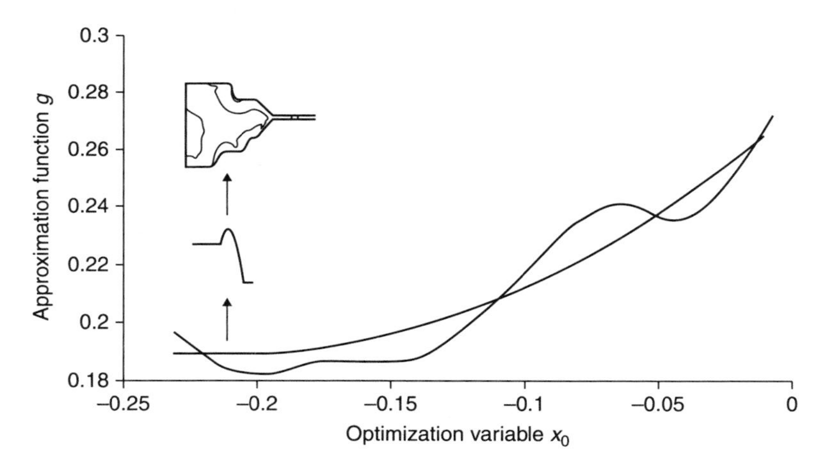

3.10 Cross section of the approximation surface g at the optimal values of the design variables.

optimum value was observed. The final results for the approximation to the objective function are shown in Fig. 3.9. The optimal shape of the preform die obtained and the corresponding grain size distribution in the final forging are shown in Fig. 3.10. The procedure described here can be applied to any objective function that is sensitive to the optimization variables of a forging process, as well as to other metal forming processes such as extrusion, rolling and drawing.

3.4.2 Optimization of a crankshaft forging process

The process considered was the forging of a crankshaft using the TR technology developed at the Metal Forming Institute in Poznan, Poland (the TR method takes its name from its inventors Tadeusz Rut (Rut and Walczyk, 2000)). This technology gives very good results when the shape of the shoulders of the crank is close to an ellipse. In the TR method, an additional forging operation is introduced. This is an unsymmetrical pre-upsetting of the stock material, which is aimed at obtaining the flow of material needed to obtain the required shape of the crank shoulders (see Fig. 3.11). The objective of the study was to design an optimal forging technology

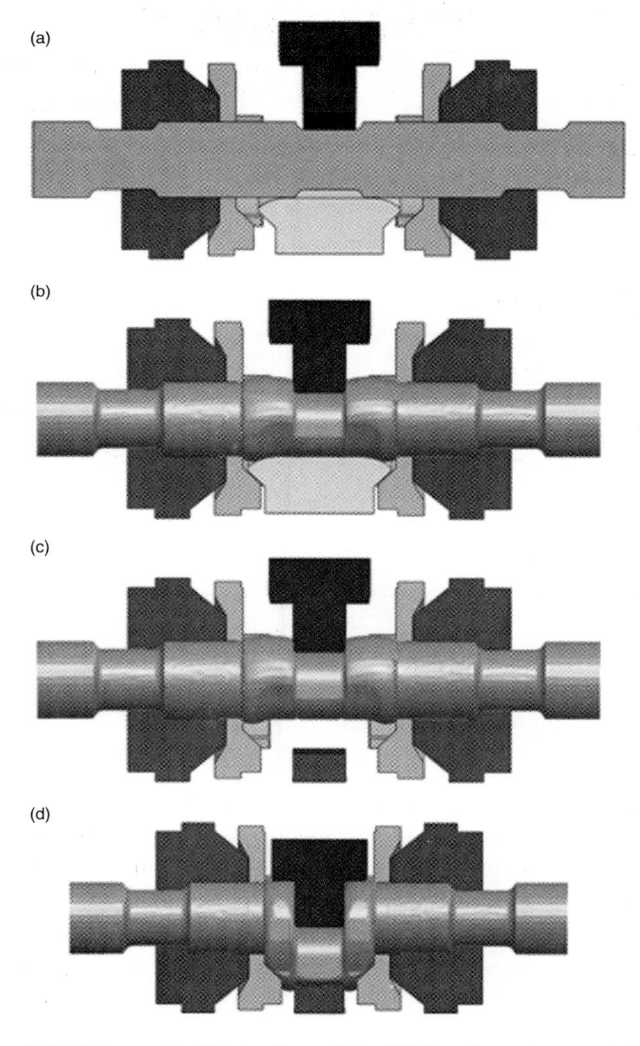

3.11 Schematic illustration of the TR forging process: (a) beginning of unsymmetrical pre-upsetting; (b) end of unsymmetrical pre-upsetting; (c) beginning of forging of the crank throw; (d) end of forging of the crank throw.

that gave a shape closest to the ideal shape of the crankshaft (Sztangret *et al.*, 2011). Two parameters were chosen as decision variables: the bending-tool displacement during forging of the crank throw, d_1, and the initial spacing of the face die inserts, d_2 (see Fig. 3.12).

Modification of the displacement of the bending tool during forging of the crank throw causes various displacements in the primary stage (the total shift must be constant). The velocity of the bending tool was assumed constant in the simulations. In the case of the spacing of the face die inserts, several different initial set-ups, causing various velocities of the tool, were considered. The percentage difference between the volume of the shape of the crankshaft obtained and the ideal shape was taken as the objective function. The values of the decision variables and the accuracy obtained for the forging scheme are presented in Table 3.2. The response surface methodology described in Section 3.2.2 was used to determine the optimal forging variant. A second-order response surface was constructed on the basis of simulations (see Fig. 3.13). The optimal values of the decision parameters found were d_1 = 283 mm for the bending-tool displacement

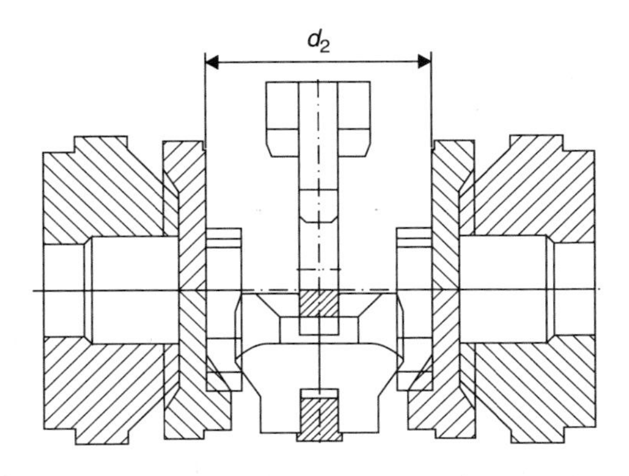

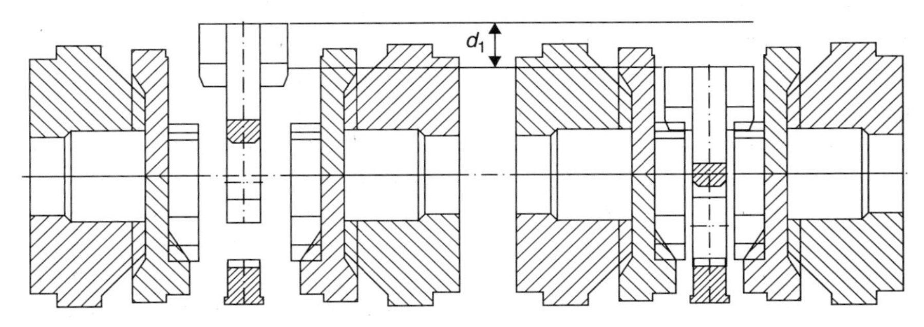

3.12 Bending-tool displacement during forging of the crank throw, d_1, and initial spacing of face die inserts, d_2.

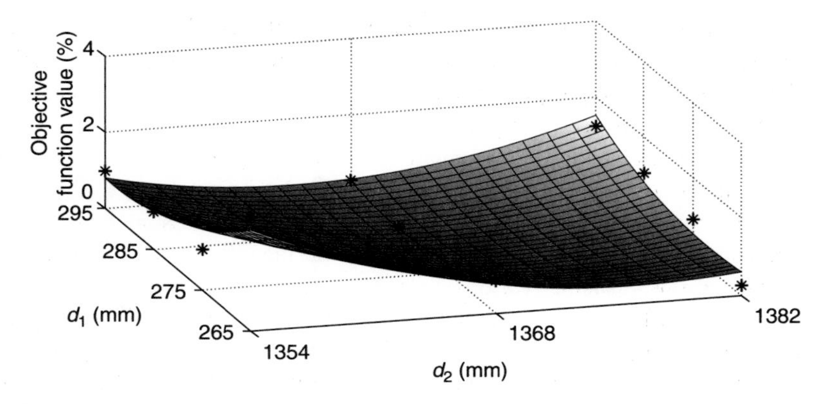

3.13 Second-order response surface constructed on the basis of simulations.

Table 3.2 Values of decision variables and accuracy obtained (%) for the forging scheme

d_2 (mm) \ d_1 (mm)	265	275	285	295
1354	3.02	1.07	0.99	0.98
1368	0.87	0.07	0.14	0.26
1382	0.25	0.93	1.05	1.22

during forging of the crank throw and $d_2 = 1370$ mm for the initial spacing of the face die inserts.

3.4.3 Application of metamodelling to inverse analysis

The accuracy of the predictions obtained from numerical simulations of metal processing depends on the mathematical model of the process, the numerical/analytical solution of the model, the definition of the boundary conditions, and the material data introduced into the model. The latter two factors are crucial to the accuracy of the predictions even if the model of the process is well constructed. The identification of the boundary conditions and the material data for the analysis of a metal forming problem is a type of problem called an inverse problem (Kirsch, 1996).

The applications of inverse analysis in engineering are numerous; see, for example, the identification of models for phase transformations in steels, which is described in Chapter 6 of this book. However, as has already been mentioned, the applications to metal forming were the earliest applications of inverse analysis among those in materials processing, and seem to be the most frequent (Gelin and Ghouati, 1994; Szeliga and Pietrzyk, 2002; Forestier *et al.*, 2002; Szeliga *et al.*, 2006).

An arbitrary metal forming process problem can be described using an operator equation of the form

$$K : X \to Y \qquad [3.26]$$

where K is the mapping between the normalized spaces X and Y.

The direct problem is defined as evaluating the value of $y \in Y$ for a given K and $x \in X$, and is equivalent to solving a boundary value problem for a differential equation or evaluating an integral. The inverse problem is defined as evaluating the value of $x \in X$ for a given K and $y \in Y$. Inverse problems for metal processing problems are ill-posed (Kirsch, 1996; Engl *et al.*, 1996). These problems require a regularization procedure and can be transformed to well-posed problems of the following type:

$$x \mapsto \| Kx - y \|^2 \qquad [3.27]$$

The form of Eq. 3.27 leads to minimization with respect to the parameters, which may be either boundary condition parameters or material parameters. In the terminology of optimization, the inverse problem is to find the minimum of the objective function

$$f(x) := \| Kx - y^{\delta} \|^2 \qquad [3.28]$$

where y^{δ} is the perturbed (measured) data such that $\|y^{\delta} - y\| \le \delta$, and $y \in K(X)$ is the exact solution of Eq. 3.26.

The objective function in Eq. 3.28 is usually defined as an average square root error (the error in terms of the Euclidean norm), but other norms can be used as well. A general flow chart of an inverse analysis algorithm is presented in Fig. 3.14. Regardless of the type of metal forming inverse problem and independently of the method of solution applied, the algorithm consists of three parts:

- A set of process outputs measured in experiment (real or virtual).
- A solver for the direct problem (in most cases, this involves high computation costs).
- An optimization procedure for the objective function. The objective function may be defined as a distance between the measured and calculated model outputs in a selected space norm, or it may be a Pareto set. Either gradient, non-gradient or nature-inspired optimization algorithms can be applied to determine the minimum of the objective function.

The optimization procedure in the last part of such an inverse algorithm requires multiple calls of the solver, which results in an unacceptably high computation time for the whole analysis. This problem can be solved by the replacement of calculations using a direct solver by a metamodel of the experiment considered (see Fig. 3.15). Such an approach allows significant reduction in computational cost.

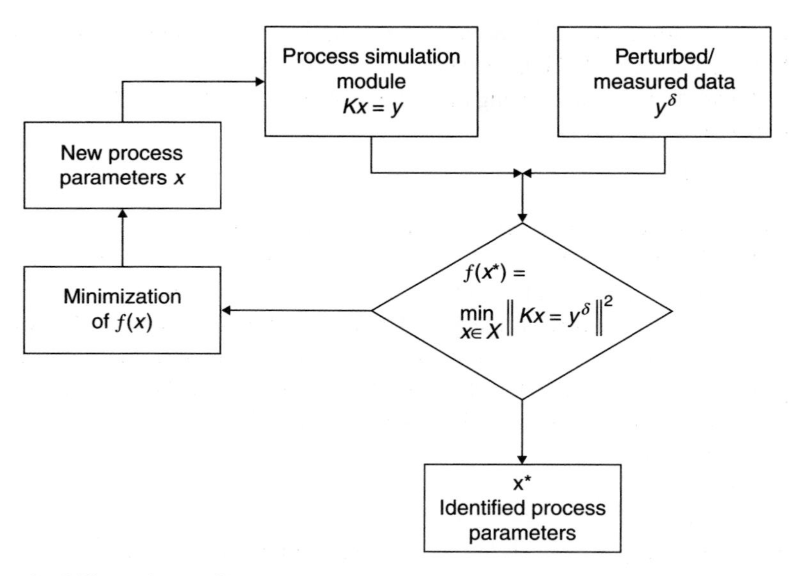

3.14 Flow chart of inverse analysis.

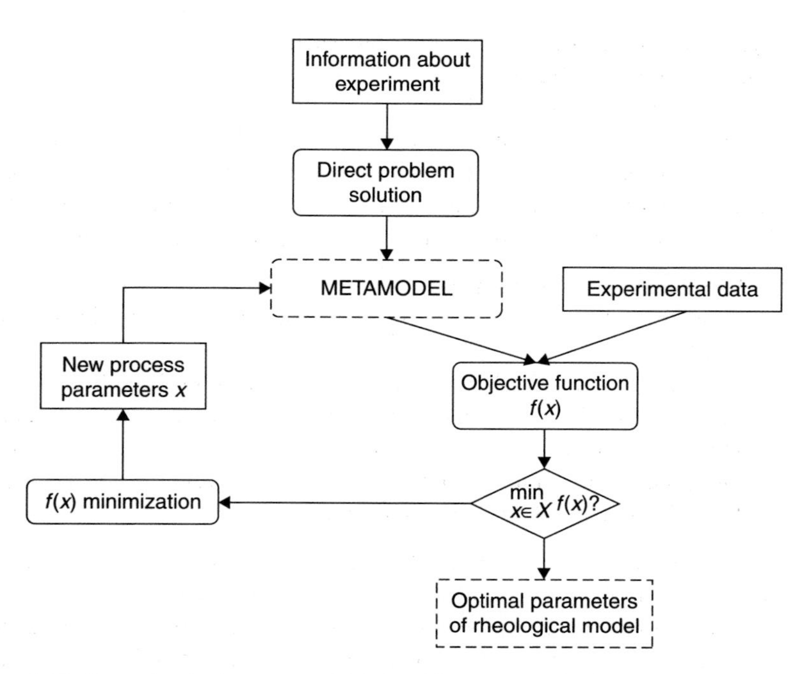

3.15 Flow chart of metamodel-based inverse analysis.

The identification of the flow properties of materials is still a challenge for researchers. The inhomogeneity of the strains, stresses and temperatures in the majority of the experimental plastometric tests that are performed to identify coefficients in material models is the main reason for the difficulties with the interpretation of the results of these tests. Inverse analysis is commonly used to overcome these difficulties. Therefore, the identification of a rheological model has been selected here to demonstrate the application of optimization strategies to inverse analysis.

Two approaches to the identification of model parameters formulated as an inverse problem are presented below. A classical inverse approach is applied in the first of these examples, and an inverse algorithm with a metamodel is applied in the second. This inverse algorithm is described in Szeliga and Pietrzyk (2002) and Szeliga *et al.* (2006). The results for identification of a rheological model for a dual-phase (DP) steel are shown as an example (Szeliga *et al.*, 2011). DP steels belong to a group of new generation steel grades used in the automotive industry. A two-phase microstructure with a specified relation between the volume fractions of ferrite and martensite is the basis of the special properties of these steels. Accurate identification of a rheological model for these steels is of particular importance in the design of processing technology (Pietrzyk *et al.*, 2009).

In the case considered here, a constitutive model of a deformed metal with the following form (Hansel and Spittel, 1979) was analysed:

$$\sigma_{\mathrm{p}} = A\varepsilon^{n} e^{-q\varepsilon} \dot{\varepsilon}^{m} e^{-BT} \qquad [3.29]$$

where ε is the strain; $\dot{\varepsilon}$ is the strain rate; T is the temperature (°C); R is the gas constant; and A, n, q, m and Q are parameters which have to be evaluated in the inverse analysis.

The coefficients in Eq. 3.29 for the flow stress were determined by searching for a minimum of the following objective function, defined as the Euclidean norm between the measured and predicted loads in compression tests:

$$f(\mathbf{x}) = \sqrt{\frac{1}{Npt}\sum_{i=1}^{Npt}\left[\frac{1}{Nps}\sum_{j=1}^{Nps}\left(\frac{F_{cji}(\mathbf{x},\mathbf{p}_i)-F_{mji}}{F_{mji}}\right)^2\right]} \qquad [3.30]$$

where F_{mji} and F_{cji} are the measured and calculated loads, respectively; Npt is the number of tests; Nps is the number of load measurements in one test; $\mathbf{p}$ is the vector of process parameters (strain rates and temperatures); and $\mathbf{x} = \{A, n, q, m, B\}$ is the vector of coefficients in the flow stress model.

The vector $\mathbf{x}$ of the parameters of Eq. 3.29 was evaluated using conventional inverse analysis. Details of this approach have been given by Szeliga *et al.* (2006) and will not be repeated here. The optimization that was performed, combined with FE simulations of plastometric tests, yielded the following values for the coefficients in Eq. 3.29: $A = 10\,375$ MPa, $n = 0.3768$, $q = 0.8345$, $m = 0.1297$ and

$B = 0.00371/°C$. The final (optimal) value of the objective function f, which can be treated as the accuracy of the solution, was 0.0965.

Next, assuming a standardization of the plastometric tests, a metamodel of a uniaxial compression test based on an artificial neural network was developed (Szeliga *et al.*, 2011). Training and testing data sets were generated by use of an FE code. The following input parameters were selected: (i) the test conditions, including the temperature, strain rate and friction coefficients; and (ii) the set of five coefficients of Eq. 3.30. The neural network developed replaced the conventional FE solver for the plastometric tests, as shown in Fig. 3.15. The optimization that was performed, combined with metamodel simulations of the plastometric tests, yielded the following values for the coefficients in Eq. 3.29: $A = 3255.3$ MPa, $n = 0.196$, $q = 0.2834$, $m = 0.119$ and $B = 0.0031/°C$. The final (optimal) value of the objective function f, which can be treated as the accuracy of the solution, was 0.0867.

A comparison of the results of the conventional and metamodel-based inverse analyses shows that different values of the coefficients were obtained, although the final values of the objective function were similar. This raises a question regarding the uniqueness of the solution and the existence of local minima, which is particularly important when the simple Eq. 3.29 is used. The problem of the uniqueness of the inverse solution will be investigated further by the authors.

The optimal solutions found were validated by comparison of measured and calculated forces. The calculations were performed using an FE simulation with the rheological model described by Eq. 3.29, in which the values of the parameters obtained using both conventional inverse analysis and inverse analysis with a metamodel were used. Selected results of this comparison are shown in Fig. 3.16. It can be seen that reasonably good agreement was obtained between the measured forces and the forces obtained from the two inverse approaches.

3.5 Conclusions and future trends

The computation time of simulations of real materials forming processes is still a crucial problem and appears often to be a barrier to the use of optimization procedures. Therefore, a lot of effort is being put into the improvement of computer models and the development of new optimization methods and strategies. The main goal of the present chapter was to demonstrate some ideas of strategies that allow one to decrease the computation time for optimization problems in the field of materials engineering. The results in the examples presented show that the optimization strategies described here can be useful for solving various optimization problems.

However, further research is needed that focuses on the implementation of sensitivity analysis in the ABO strategy and in inverse analysis. The implementation of sensitivity analysis should allow one to reduce the number of solver runs and avoid the problem of local minima (with unacceptable values of the objective

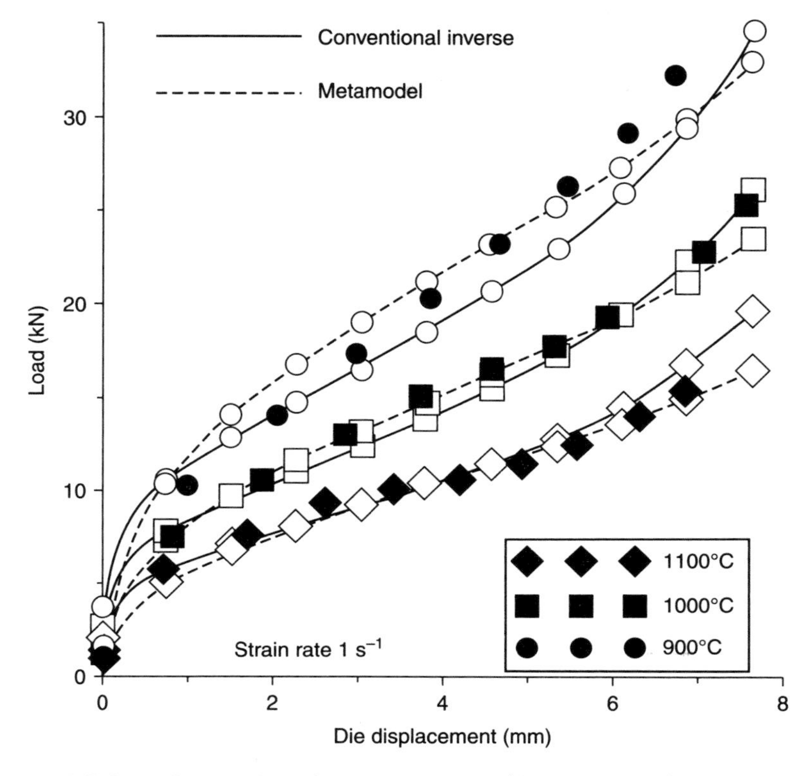

3.16 Selected examples of measured loads (filled symbols) and loads calculated using Eq. 3.32 (open symbols) (Szeliga *et al.*, 2011).

function), especially in the case of inverse algorithms. Sensitivity analysis studies the relationships between the information introduced into the model and the information obtained from the model. It studies how different sources of variation in the input data influence the output, either qualitatively or quantitatively. Sensitivity analysis methods (Kleiber *et al.*, 1997; Saltelli *et al.*, 2000) can be applied to the following:

- The pre-optimization step (Kopernik and Szeliga, 2007; Pietrzyk and Szeliga, 2004; Szeliga *et al.*, 2006; Trębacz *et al.*, 2006). In this case, sensitivity analysis can be used
  - to assess whether the material model resembles the process under study;
  - to verify the admissible domains of the parameters that have been defined.
- The optimization procedure in the case of weak convergence of the inverse algorithm. In this case, sensitivity analysis can be used
  - to verify the sensitivity of the objective function to the identified parameters in the domain explored;

- to investigate interactions with other model parameters (Szeliga and Madej, 2009);
- to eliminate dead zones in the optimization process.

Replacing a single objective function with a Pareto front, as a result of a multi-objective optimization (Steuer, 1986), is another approach to raising the efficiency of inverse calculations. This technique turns out to be useful if the objective function consists of more than one term and these terms are mutually exclusive or oppose each other.

3.6 Acknowledgements

The financial support of the Polish Ministry of Science and Higher Education project no. NR07 0006 10 is acknowledged.

3.7 References

Antony J (2003), *Design of Experiments for Engineers and Scientists*, Butterworth-Heinemann, Oxford.

Arabas J (2004), *Wykłady z algorytmów ewolucyjnych*, Wydawnictwa Naukowo-Techniczne, Warsaw (in Polish).

Barabasz B, Schaefer R and Paszyński M (2009), 'Handling ambiguous inverse problems by the adaptive genetic strategy hp-HGS', in: *Proceedings of the 9th International Conference on Computational Science*, Baton Rouge, 912–913.

Bishop C M (1995), *Neural Networks for Pattern Recognition*, Oxford University Press, Oxford.

Bonte M, van den Boogaard A and Huetink J (2005), 'Solving optimization problems in metal forming using finite element simulation and metamodelling techniques', in: *Proceedings of the 1st Invited COST526 Conference: APOMAT*, Morschach, Switzerland, 242–251.

Engl H W, Hanke M and Neubauer A (1996), *Regularization of Inverse Problems*, Kluwer Academic, Dordrecht.

Forestier R, Massoni E and Chastel Y (2002), 'Estimation of constitutive parameters using an inverse method coupled to a 3D finite element software', *Journal of Materials Processing Technology*, 125, 594–601.

Gelin J C and Ghouati O (1994), 'The inverse method for determining viscoplastic properties of aluminium alloys', *Journal of Materials Processing Technology*, 34, 435–440.

Hagan M T, Demuth H B and Beale M H (1996), *Neural Network Design*, PWS, Boston, MA.

Hansel A and Spittel T (1979), *Kraft- und Arbeitsbedarf Bildsamer Formgebungs-verfahren*, VEB Deutscher Verlag fur Grundstoffindustrie, Leipzig.

Haykin S (1999), *Neural Networks: A Comprehensive Foundation*, Macmillan, New York.

Karakasis M and Giannakoglou K (2005), 'Metamodel-assisted multi-objective evolutionary optimization', in: *Proceedings of Evolutionary and Deterministic Methods for Design, Optimization and Control with Applications to Industrial and Societal Problems*, EUROGEN, Munich, 1–11.

Kirsch A (1996), *An Introduction to the Mathematical Theory of Inverse Problems*, Springer, New York.

Kleiber M, Antunez H, Hien T D and Kowalczyk P (1997), *Parameter Sensitivity in Nonlinear Mechanics*, Wiley, New York.

Kopernik M and Szeliga D (2007), 'Modelling of nanomaterials – sensitivity analysis to determine the nanoindentation test parameters', *Computer Methods in Materials Science*, 7, 255–261.

Kowalik J and Osborne M (1968), *Methods for Unconstrained Optimization Problems*, Elsevier, New York.

Kusiak J (1996), 'Technique of the tool shape optimization in large scale problems of metal forming', *Journal of Materials Processing Technology*, 57, 79–84.

Kusiak J, Pietrzyk M and Chenot J-L (1994), 'Die shape design and evaluation of microstructure control in the closed-die axisymmetric forging by using FORGE2 program', *ISIJ International*, 34, 755–760.

Kusiak J, Danielewska-Tułecka A and Oprocha P (2009), *Optymalizacja. Wybrane metody z przykładami zastosowań*, Wydawnictwo Naukowe PWN, Warsaw (in Polish).

Mead R (1990), *The Design of Experiments*, Cambridge University Press, Cambridge.

Myers R H and Montgomery D C (1995), *Response Surface Methodology: Process and Product Optimization Using Designed Experiments*, Wiley, New York.

Paszyński M, Szeliga D, Barabasz B and Macioł P (2009), 'Inverse analysis with parallel 3D hp adaptive computations of the orthotropic heat transport and linear elasticity problems', in: *Proceedings of the 7th World Congress on Computational Mechanics*, Los Angeles.

Persson A, Grimm H and Ng A (2005), 'Simulation-based optimization using global search and neural network metamodels', in: *Proceedings of the 20th European Simulation and Modelling Conference*, Toulouse, 182–186.

Pietrzyk M and Szeliga D (2004), 'Analysis of sensitivity of loads and tendency to necking with respect to the tensile test parameters', *Metallurgy and Foundry Engineering*, 30, 117–127.

Pietrzyk M, Kusiak J, Kuziak R and Zalecki W (2009), 'Optimization of laminar cooling of hot rolled DP steels', in: *Proceedings of the XXVIII Verformungskundliches Kolloquium*, Planneralm, Austria 285–294.

Roberts W, Sandberg A, Siwecki T and Werlefors T (1983), 'Prediction of microstructure development during recrystallization hot rolling of Ti–V steels', in: *Proceedings of the Conference on Technology and Application of HSLA Steels*, Philadelphia, 67–84.

Rut T and Walczyk W (2000), 'Doskonalenie kucia wałów korbowych metodą TR', *Obróbka Plastyczna Metali*, 11, 5–8. (in Polish).

Saltelli A, Chan K and Scott E M (2000), *Sensitivity Analysis*, Wiley, New York.

Schaefer R, Barabasz B and Paszyński M (2007), 'Twin adaptive scheme for solving inverse problems', in: *Evolutionary Computation and Global Optimization*, Bedlęwo, 241–249.

Steuer R E (1986), *Multiple Criteria Optimization: Theory, Computations, and Application*, Wiley, New York.

Szeliga D and Madej Ł (2009), 'Sensitivity analysis of the cellular automata finite element model for the strain localization', *Computer Methods in Materials Science*, 9, 264–270.

Szeliga D and Pietrzyk M (2002), 'Identification of rheological and tribological parameters', in Lenard J G (ed.), *Metal Forming Science and Practice*, Elsevier, Amsterdam, 227–258.

Szeliga D, Gawąd J and Pietrzyk M (2006), 'Inverse analysis for identification of rheological and friction models in metal forming', *Computer Methods in Applied Mechanics and Engineering*, 195, 6778–6798.

Szeliga D, Sztangret Ł, Kusiak J and Pietrzyk M (2011), 'Two approaches to identification of the flow stress model – application of the metamodel', in: *Proceedings of the XXX. Verformungskundliches Kolloquium*, Planneralm, Austria.

Sztangret Ł, Stanisławczyk A and Kusiak J (2009), 'Bio-inspired optimization strategies in control of copper flash smelting process', *Computer Methods in Materials Science*, 9(3), 400–408.

Sztangret Ł, Milenin A, Sztangret M, Walczyk W, Pietrzyk M and Kusiak J (2011), 'Computer aided design of the best TR forging technology for crank shafts', *Computer Methods in Materials Science*, 11, 237–242.

Tadeusiewicz R (1993), *Siea neuronowe*, Akademicka oficyna wydawnicza, Warsaw (in Polish).

Trębacz L, Szeliga D and Pietrzyk M (2006), 'Sensitivity analysis of quantitative fracture criterion based on the results of the SICO test', *Journal of Materials Processing Technology*, 177, 296–299.

4
Recrystallisation and grain growth in hot working of steels

B. LÓPEZ and J. M. RODRIGUEZ-IBABE, CEIT
and Tecnun (University of Navarra), Spain

Abstract: This chapter analyses the hardening–softening mechanisms that operate during hot working of steels. Special attention is focused on such aspects as recrystallisation and strain-induced precipitation, which help to achieve refinement and conditioning of the austenite microstructure before transformation. An approach including both semi-empirical and physical models is described, followed by their application to selected industrial cases.

Key words: recrystallisation, grain growth, strain-induced precipitation, microalloying, austenite conditioning.

4.1 Introduction

During hot forming processes, several hardening–softening mechanisms can operate simultaneously. As deformation is applied, energy is stored in the form of dislocations, and this energy can be released by recovery, recrystallisation and grain growth (Doherty *et al.*, 1997).

These hardening and softening phenomena that occur during deformation cause dynamic structural changes which leave the steel in an unstable state. In addition, once the application of deformation has concluded, this unstable state, together with high-temperature conditions, favours the occurrence of softening processes. These recovery and recrystallisation mechanisms are called 'static' in order to distinguish them from those, called 'dynamic', which take place during deformation. Finally, once recrystallisation is completed, grain growth can occur before the next deformation step is applied.

The aforementioned softening mechanisms can undergo a significant change in their kinetics, and may indeed halt completely, if strain-induced precipitation occurs as a consequence of a microalloying addition. Similarly, the presence of very fine, numerous precipitates before deformation can modify the recrystallisation kinetics. In addition, the solute drag effect is another factor that has to be evaluated to model the overall kinetics of the softening process; this is important mainly as the temperature decreases during the hot working schedule.

In this chapter, the main aspects of softening mechanisms and their relevance to the evolution of austenite microstructures will be considered. Special attention is devoted to the interactions between softening and precipitation mechanisms and how they can determine the conditions required to achieve the proper austenite conditioning before transformation occurs during the subsequent cooling step. An approach including both semi-empirical and physical models will be described, followed by their application to some selected industrial cases.

4.2 Grain refinement due to recrystallisation

Recrystallisation can be considered as one of the most powerful tools for achieving significant grain size refinement during hot working. Recrystallisation is defined as a process in which the formation and migration of high-angle grain boundaries, driven by the stored deformation energy, take place. When this process occurs during deformation it is referred to as 'dynamic recrystallisation', whereas the term 'static' is applied when it happens after deformation.

During hot rolling and forging operations, each pass is characterised by the applied strain, the strain rate, the temperature and the interpass time. These process parameters, together with the material characteristics (chemical composition and initial grain size), have an effect on the recrystallisation kinetics and the resulting grain size. The relevance of each parameter has been extensively studied, and some of the most significant results related to dynamic, static and postdynamic recrystallisation kinetics will be described below.

4.2.1 Dynamic recrystallisation

When plastic deformation of austenite is carried out at high temperatures, competition between hardening (dislocation storage) and softening (dislocation elimination) mechanisms occurs (Jonas *et al.*, 1969). In the case of low-stacking-fault-energy materials, such as austenite, the recovery process is relatively slow and the dislocation density can attain a sufficiently high value to allow the nucleation of new recrystallised grains during deformation; this is a process called dynamic recrystallisation (DRX), with a typical stress–strain curve as indicated in Fig. 4.1. This figure shows some characteristic features associated with DRX, which can be summarised as follows (Sakai and Jonas, 1984):

- DRX starts once a critical strain ε_c is exceeded.
- Softening causes the flow stress to go through a maximum (denoted by the peak strain, ε_p) before falling to a steady state (ε_{ss}) characterised by a constant stress, σ_{ss}.

In dynamic recrystallisation, both nucleation and grain growth take place while the strain is being applied. The grains nucleate preferentially at existing grain boundaries and triple points. The applied strain causes the dislocation density to build up inside the new grains, gradually reducing the driving force for boundary migration and finally arresting it. Under these conditions, recrystallisation proceeds by means of the nucleation of new grains. These grains are, in turn, deformed until they once again reach the critical strain required to undergo recrystallisation,

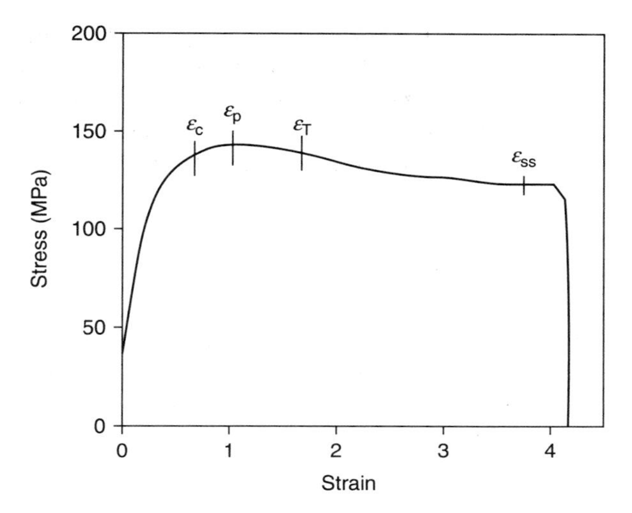

4.1 Stress–strain curve of hot-deformed austenite showing the important characteristic strain values.

resulting in a 'cascade' of nucleation and limited growth events (Sah *et al.*, 1974). This leads to a steady state, characterised by the maintenance of a structure of approximately equiaxed grains with a constant mean grain size D_{dyn}.

There are two practical aspects of DRX that need to be considered: the critical strain required for the onset of DRX, ε_c, and the resulting recrystallised grain size. The critical strain ε_c has been related to the peak strain ε_p, a parameter that is easier than ε_c to measure experimentally (see Fig. 4.1). The relationship between ε_c and ε_p has the form $\varepsilon_c = k\varepsilon_p$, where k is a constant, whose reported values range from 0.5 to 0.87, depending on the chemical composition of the steel (Sellars, 1980; Hodgson and Gibbs, 1992; Siciliano and Jonas, 2000; Xu *et al.*, 2010)).

The peak strain depends on the initial austenite grain size D_0 and the Zener–Hollomon parameter Z, as indicated in Eqs 4.1 and 4.2 (Sellars, 1980):

$$\varepsilon_p = A \cdot D_0^m \cdot Z^p \qquad [4.1]$$

$$Z = \dot{\varepsilon} \exp\left(\frac{Q_{def}}{RT}\right) \qquad [4.2]$$

where $\dot{\varepsilon}$ is the strain rate, Q_{def} is the apparent activation energy for deformation, R is the gas constant (8.31 J/K mol) and T is the absolute temperature. The activation energy Q_{def} and the coefficients of the equations (A, m and p) are dependent on the material.

Equation 4.1 shows that a coarse initial grain size delays the start of DRX. On the other hand, microalloying elements in solid solution may retard DRX. These elements, when in solution, are located in the vicinity of grain boundaries, reducing both the interfacial energy and the mobility of grain boundaries. The effect of 0.1% of each of several different microalloying elements, used singly, on the solute retardation parameter (SRP) has been quantified relative to what was observed for a plain C–Mn steel (Akben and Jonas, 1983; Jonas, 1984). This parameter is defined as

$$\mathrm{SRP} = \log\left(\frac{t_{\mathrm{x}}}{t_{\mathrm{ref}}}\right)\cdot\frac{0.1}{\mathrm{wt\%x}}\cdot 100\% \qquad [4.3]$$

where t_x is the time to peak strain for a steel containing the element x and t_{ref} is the equivalent time for the plain carbon steel. The SRP values assigned to the main microalloying elements are illustrated in Fig. 4.2. As observed, Nb is the most effective element in retarding recrystallisation.

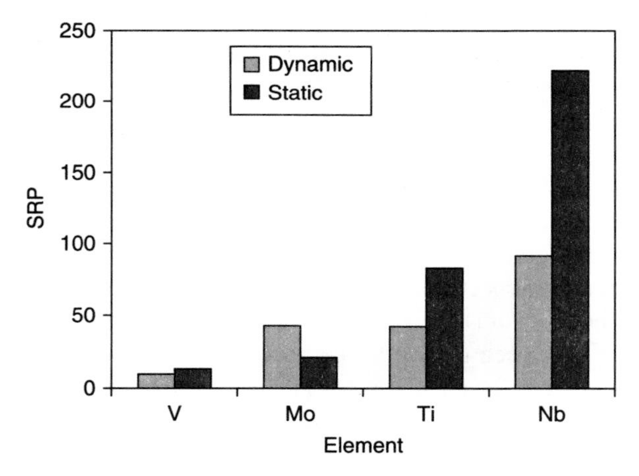

4.2 Effect of V, Mo, Ti and Nb on the solute retardation parameter (SRP) for dynamic and static recrystallisation.

Several equations that have been proposed to calculate ε_p are compiled in Table 4.1. These equations are drawn in Fig. 4.3 for the specific condition of $T = 1100°C$ and a strain rate of $5\ s^{-1}$. As observed, the addition of Nb has a major effect on the peak strain, larger than that assigned either to the carbon content or to the initial austenite grain size.

One of the main advantages of DRX is that the resulting grain size depends only on the parameter Z, being completely independent of the initial microstructure (Sah *et al.*, 1974; Sellars, 1980). This dependence follows a power relationship as follows:

$$D_{\mathrm{dyn}} = BZ^{-n} \qquad [4.4]$$

Table 4.1 Equations describing the peak strain (D_0 in μm)

Steel	Equation	Reference
	Low carbon steels	
C–Mn	$\varepsilon_p = 4.9\text{x}10^{-4} D_o^{0.5} Z^{0.15}$; $Q = 312\,\text{kJ/mol}$	Sellars (1980)
Nb	$\varepsilon_p = 2.8\text{x}10^{-4} \frac{\{1+20[\text{Nb}]\}}{1.78} D_o^{0.5} Z^{0.17}$; $Q = 375\,\text{kJ/mol}$	Minami *et al.* (1996)
Nb and Nb–Ti	$\varepsilon_p = 3.7\text{x}10^{-3} \frac{\{1+20([\text{Nb}]+0.02[\text{Ti}])\}}{1.78} D_o^{0.147} Z^{0.155}$ $Q = 325\text{kJ/mol}$	Fernández *et al.* (2003)
	Medium-high carbon steels	
Eutectoid	$\varepsilon_p = 4.5\text{x}10^{-4} D_o^{0.33} Z^{0.18}$; $Q = 315\,\text{kJ/mol}$	Kuziak *et al.* (1996)
0.23C–Mn–Cr	$\varepsilon_p = 5.24\text{x}10^{-5} D_o^{0.5} Z^{0.188}$; $Q = 378.6\,\text{kJ/mol}$	Wang *et al.* (2005)

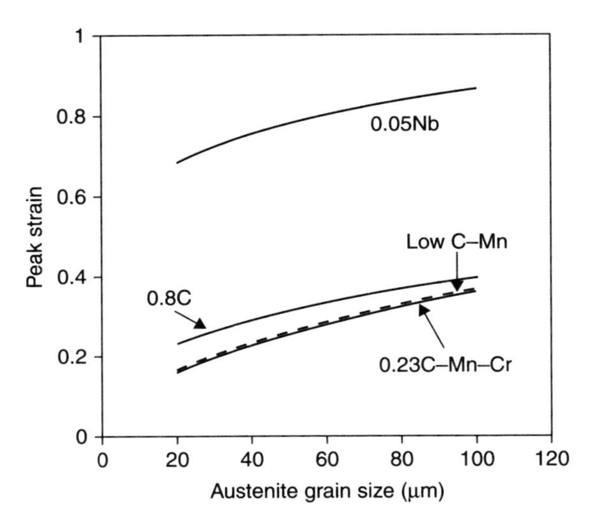

4.3 Influence of chemical composition and austenite grain size on peak strain at 1100°C and $5\,\text{s}^{-1}$ (from the equations in Table 4.1).

where B and n are two material constants. The value of the exponent n ranges from 0.11 to 0.35 (Nazabal *et al.*, 1987; Hodgson *et al.*, 1993). Several examples are shown in Fig. 4.4 (Pereda *et al.*, 2007a). As observed, very fine austenite grains can be achieved for high Z values.

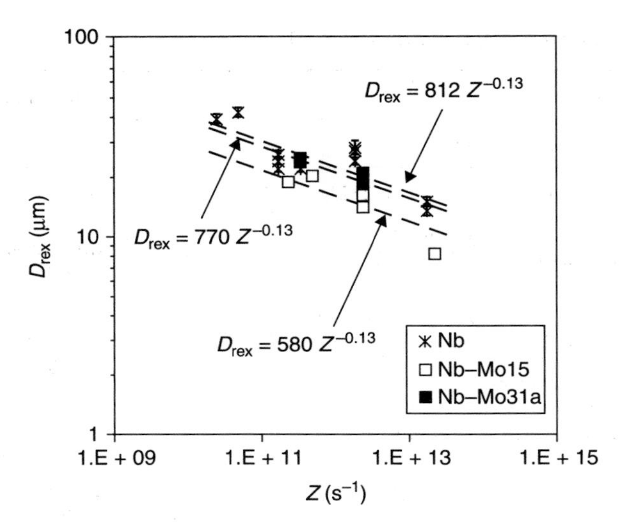

4.4 Dependence of dynamically recrystallised grain size on Z for various Nb and Nb–Mo microalloyed steels (Pereda *et al.*, 2007a).

4.2.2 Static and metadynamic recrystallisation

Under hot working conditions, softening of the austenite can occur after deformation during the interpass intervals. As recovery in austenite is very limited, recrystallisation is the main softening mechanism. Under conditions where DRX has not been activated ($\varepsilon < \varepsilon_c$), the recrystallisation that can occur in the interpass time is called 'static recrystallisation' (SRX). In contrast, once DRX has been activated, the postdynamic softening that occurs is known as 'metadynamic recrystallisation' (MDRX). The two softening mechanisms differ in several ways.

The evolution of the recrystallised fraction with time can be described by the Avrami equation,

$$X = 1 - \exp\left(-\ln\ 2 \cdot \left(\frac{t}{t_{0.5x}}\right)^n\right) \qquad [4.5]$$

where X is the recrystallised fraction after a time t, $t_{0.5x}$ is the time required to reach 50% recrystallisation and n is the Avrami exponent. In the case of static recrystallisation, the value of the exponent n ranges between 1 and 2 (Laasraoui and Jonas, 1991; Hodgson and Gibbs, 1992; Arribas *et al.*, 2005). However, other authors have suggested a dependence of this coefficient on temperature. For example, Medina and Quispe (2001) suggested a slight dependence of n on the temperature, with values between 0.62 and 1.4. They proposed the following expressions for low alloy steels (Eq. 4.6) and microalloyed steels (Eq. 4.7):

$$n = 2.93 \exp\left(-\frac{12500}{RT}\right) \qquad [4.6]$$

$$n = 28 \exp\left(-\frac{36000}{RT}\right) \qquad [4.7]$$

In the case of metadynamic recrystallisation, the Avrami exponent takes values close to unity (Roucoules *et al.*, 1994; Uranga *et al.*, 2003).

Static recrystallisation requires the nucleation and growth of new grains in the deformed austenite microstructure. In contrast, in metadynamic recrystallisation there is no nucleation, only growth of grains previously nucleated during DRX. This difference between the two softening mechanisms implies some peculiarities in the factors that have an effect on the recrystallisation kinetics. A general expression that takes into account the parameters that affect the value of $t_{0.5}$ is, for the case of static recrystallisation (Sellars, 1980),

$$t_{0.5\mathrm{srx}} = A\varepsilon^{-p}\,\dot{\varepsilon}^{-q}\,D_0^m \exp\left(\frac{Q_{\mathrm{srx}}}{RT}\right) \qquad [4.8]$$

where m, p and q are constants and Q_{srx} is the activation energy for SRX. For metadynamic recrystallisation, the corresponding expression is

$$t_{0.5\mathrm{mdrx}} = B\cdot Z^{-r}\cdot\exp\left(\frac{Q_{\mathrm{mdrx}}}{RT}\right) = B\dot{\varepsilon}^{-r}\exp\left(\frac{Q_{\mathrm{app}}}{RT}\right) \qquad [4.9]$$

where r is a constant and Q_{app} is an apparent activation energy, which is a function of both the activation energy for MDRX, Q_{mdrx}, and the activation energy of deformation, Q_{def}.

As in dynamic recrystallisation, microalloying elements in solution can retard SRX. In some cases, this effect can be quantified by introducing the amount of microalloying elements directly into the recrystallisation equation (Dutta and Sellars, 1987; Fernández *et al.*, 2000), whereas in other cases the effect is implicit in the value of the activation energy for recrystallisation (Medina and Quispe, 2001; Cho *et al.*, 2001). For example, Dutta and Sellars (1987) proposed an equation for the 50% recrystallisation time in Nb-bearing steels which took into account the effect of Nb as follows:

$$t_{0.5\mathrm{srx}} \propto \exp\left(\left\{\left[\frac{275000}{T}\right] - 185\right\}\cdot[C]\right) \qquad [4.10]$$

where $[C]$ is the Nb concentration in solution in wt%, and T is the deformation temperature. Equation 4.10 can also be applied to other microalloying elements if the solute concentration of each of these elements is modified by an appropriate multiplying factor. This can be done by considering the values of the solute retardation parameter for static recrystallisation shown in Fig. 4.2.

Several examples of empirical equations are shown in Table 4.2 for the case of static recrystallisation and in Table 4.3 for metadynamic recrystallisation.

Table 4.2 Equations to determine the time $t_{0.5srx}$ for static recrystallisation

Steel	Equation	Reference
	Low carbon steels	
C–Mn	$t_{0.5srx} = 2.5 \cdot 10^{-19} \varepsilon^{-4} D_o^2 \exp(300000/RT)$	Sellars (1980)
Ti–V	$t_{0.5srx} = 5 \cdot 10^{-18} (\varepsilon - 0.085)^{-3.5} D_o^2 \exp(280000/RT)$	Roberts *et al.* (1984)
Nb, Ti and Nb–Ti	$t_{0.5srx} = 9.92 \cdot 10^{-11} D_o \varepsilon^{-5.6 D_o^{-0.15}} \dot{\varepsilon}^{-0.53} \cdot$ $\cdot \exp(180000/RT) \exp[(275000/T - 185)([Nb] + 0.374[Ti])]$	Fernández *et al.* (2000)
Nb and Nb–V	$t_{0.5srx} = A D_o \varepsilon^{-4.3 D_o^{-0.169}} \dot{\varepsilon}^{-0.53} \exp\left(\frac{Q}{RT}\right)$ $A = 3.754 \cdot 10^{-4} \exp(-7869 \cdot 10^{-5} \cdot Q)$ Q = 148636.8 – 71981.3[C]+56537.6[Si]+21180[Mn]+121243.3[Mo]+64469.6[V]+109731.9[Nb]0.15	Medina and Quispe (2001)
	Medium-high carbon steels	
Eutectoid	$t_{0.5srx} = 2.4 \cdot 10^{-8} \varepsilon^{-1.006 \cdot D_o^{0.22}} \dot{\varepsilon}^{-0.29} D_o^{-0.2} \exp(160420/RT)$	Kuziak *et al.* (1996)

Table 4.3 Equations to determine the time $t_{0.5mdrx}$ for metadynamic recrystallisation

Steel	Equation	Reference
	Low carbon steels	
C–Mn	$t_{0.5mdrx} = 2.13 \cdot 10^{-6} \dot{\varepsilon}^{-0.67} \exp(133000/RT)$	Sun and Hawbolt (1997)
Nb	$t_{0.5mdrx} = 1.77 \cdot 10^{-6} \dot{\varepsilon}^{-0.62} \exp(153000/RT)$	Uranga *et al.* (2003)
Ti	$t_{0.5mdrx} = 8.9 \cdot 10^{-6} \dot{\varepsilon}^{-0.83} \exp(125000/RT)$	Roucoules *et al.* (1995)
	Medium-high carbon steels	
55Cr3	$t_{0.5mdrx} = 3.59 \cdot 10^{-2} \dot{\varepsilon}^{-0.785} \exp(24800/RT)$	Bianchi and Karjalainen (2005)

In static recrystallisation, the role of the applied strain is due to an increase in the dislocation density that provides a higher driving force for recrystallisation, which results in a decrease in the recrystallisation time. Concerning the influence of the initial austenite grain size, it must be remembered that the recrystallised grains nucleate mainly on austenite grain boundaries. When the grain size decreases, the specific grain boundary area increases, which provides more sites for the nucleation of new grains. In the equations in Table 4.2, the grain size exponent takes values between 1 and 2. According to Fernández *et al.* (2000), a quadratic exponent would agree better with the data in the range of grain sizes below about 200 μm (conventionally reheated austenite). However, in the range of larger grain sizes (thin-slab conditions and direct charging), a linear dependence represents the experimental behaviour better.

The kinetics of metadynamic recrystallisation is highly strain-rate dependent but depends only weakly on the amount of deformation and the temperature, whereas the opposite is the case for static recrystallisation. In the kinetics of SRX, a significant effect of the applied strain is observed. In contrast, once a steady state is achieved during DRX, the balance between the softening and hardening mechanisms causes the strain energy to remain constant with strain.

Finally, it should be noted that some recent publications disagree with the criterion of using the critical strain ε_c to distinguish between the dynamic–metadynamic and static recrystallisation regions (Uranga *et al.*, 2003; Cartmill *et al.*, 2005). In some cases, normally related to large differences between the initial austenite grain size D_0 and the dynamically recrystallised grain size D_{dyn}, the characteristics of an MDRX process are only achieved after a minimum strain, referred to as the transition strain, ε_T, is reached (see Fig. 4.1). Depending on the deformation and the microstructural conditions, this strain can be significantly larger than ε_c. In consequence, there will be a transition where metadynamic and, after that, static recrystallisation will take place. An example where this transition affects the dependence of the time $t_{0.5}$ on the strain is shown in Fig. 4.5.

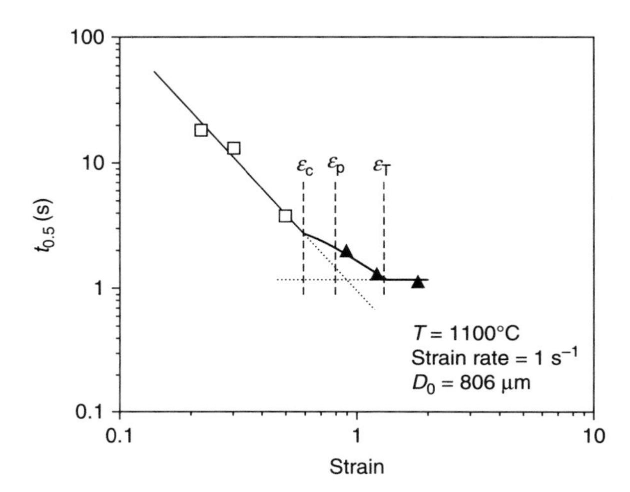

4.5 Dependence of $t_{0.5}$ on applied strain for the case of a 0.035% Nb microalloyed steel. In the interval between ε_c and ε_T, both metadynamic and static recrystallisation mechanisms operate (Uranga *et al.*, 2003).

4.2.3 Recrystallised grain size

One of the main characteristics of static recrystallisation is that, in the majority of cases, it allows refinement of the austenite grain size. The SRX grain size, D_{srx}, is related to the initial grain size and the deformation parameters. In contrast, the grain size produced by metadynamic recrystallisation, D_{mdrx}, is dependent only on the temperature and strain rate, and not on the strain. The following relationships have been proposed for each range:

$$D_{srx} = D \cdot D_0^k \cdot \varepsilon^m \qquad \text{for } \varepsilon < \varepsilon_c \tag{4.11}$$

$$D_{mdrx} = D' \cdot Z^s \qquad \text{for } \varepsilon < \varepsilon_c \tag{4.12}$$

A wide range of values has been given for the constants in these equations (see Table 4.4). In some cases, parameters other than those present in the above equations are also included.

Table 4.4 Equations describing the recrystallised austenite grain size (in μm)

Steel	Equation	Reference
	Static recrystallisation	
C–Mn	$D_{srx} = 0.743 D_o^{0.674} \varepsilon^{-1}$	Beynon and Sellars (1992)
Nb	$D_{srx} = 1.4 D_o^{0.56} \varepsilon^{-1}$	Abad *et al.* (2001a)
42CrMo4	$D_{srx} = 215.6 \cdot \varepsilon^{-0.48} \dot{\varepsilon}^{-0.144} D_o^{0.078} \exp\left(\frac{-28448}{RT}\right)$	Lin and Chen (2009)
	Metadynamic recrystallisation	
C–Mn	$D_{mdrx} = 2.6 \cdot 10^4 \cdot Z^{-0.23}$, Q= 300 kJ/mol	Hodgson and Gibbs (1992)
Nb	$D_{mdrx} = 1370 \cdot Z^{-0.13}$, Q = 375 kJ/mol	Roucoules *et al.* (1995)

4.3 Grain growth after recrystallisation

Once recrystallisation (static or metadynamic) is completed, the high temperature will favour austenite grain growth if the interpass time during hot working is long enough. The reason for this growth is the reduction of the internal energy obtained by decreasing the total austenite grain boundary area.

The evolution of the austenite grain size after recrystallisation, under isothermal conditions, is usually described as follows:

$$D^n = D_{rex}^n + B \cdot t_q \cdot \exp\left(-\frac{Q_{gg}}{RT}\right) \tag{4.13}$$

Table 4.5 Coefficients of Eq. 4.13 for various steel grades (grain size in μm and time in seconds)

Steel	n	B	Q_{gg} (kJ/mol)	Reference
C–Mn ($T > 1273$ K)	10	3.87×10^{32}	400	Sellars and Whiteman (1979)
C–Mn ($T < 1273$ K)	10	5.02×10^{53}	914	Sellars and Whiteman (1979)
C–Mn ($t < 1$ s)	2	4.0×10^{7}	113	Hodgson *et al.* (1995)
C–Mn ($t > 1$ s)	7	1.5×10^{27}	400	Hodgson *et al.* (1995)
C–Mn–Nb	4.5	4.1×10^{23}	435	Hodgson and Gibbs (1992)

where D_{rex} is the fully recrystallised grain size and t_q is the time after complete recrystallisation, normally taken as meaning a 95% recrystallised fraction ($t_q = t_{ip} - t_{0.95srx}$, t_{ip} being the interpass time). In this equation, Q_{gg} is the apparent activation energy for grain growth, and n and B are constants depending on the chemical composition. Some examples of values of the coefficients of Eq. 4.13 are listed in Table 4.5.

The time and the temperature are not the only parameters that affect grain growth. The recrystallised grain size is also an important variable, as large grains have a lower tendency to grow than smaller ones. Consequently, the kinetics of grain growth is strongly dependent on the recrystallised grain size.

4.4 Recrystallisation–precipitation interactions

Alloying and microalloying elements may produce a delay in the recrystallisation kinetics due to the solute drag effect. Moreover, when austenite containing microalloying elements in solution is thermomechanically worked while the temperature is falling, the degree of supersaturation may become high enough for dislocations generated during deformation to serve as sites for the precipitation of carbonitrides. These fine precipitates exert a pinning force on the moving boundaries during recrystallisation, and the softening process may be significantly retarded or even halted, the effect of these precipitates usually being stronger than that produced by solute drag.

4.4.1 Strain-induced precipitation kinetics of Nb

The solubility products of various nitrides and carbides are shown in Fig. 4.6. The evolution of the precipitated percentages of microalloying elements as a function of temperature, calculated with the help of the relationships shown in Fig. 4.6, is illustrated in Fig. 4.7 for several typical microalloying contents. This figure shows that the element most likely to precipitate during hot rolling is Nb. However, the solubility products do not provide information about precipitation kinetics. For example, it is well documented that in undeformed austenite, the precipitation of

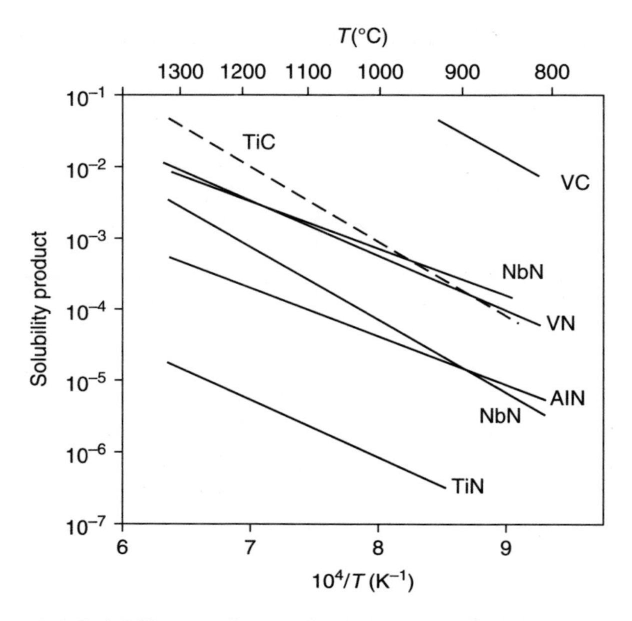

4.6 Solubility products of various nitrides and carbides in austenite.

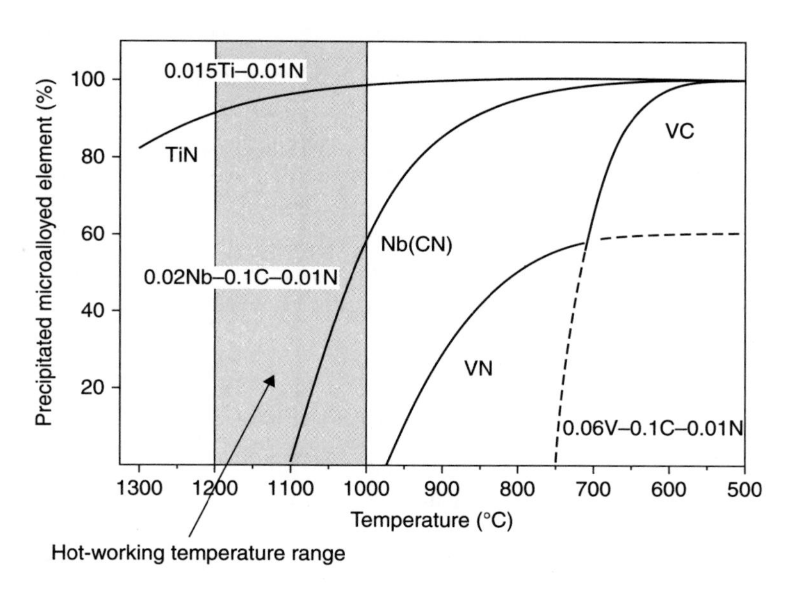

4.7 Amount of precipitation as a function of temperature, under equilibrium conditions, for three microalloyed steels containing Ti, Nb and V.

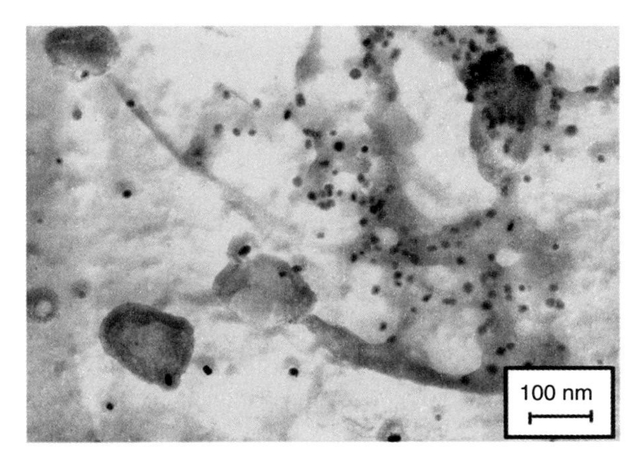

4.8 Strain-induced precipitates found in a 0.02% Nb microalloyed steel after deformation (Iparraguirre *et al.*, 2007).

Nb(C,N) is relatively slow. However, the application of strain during hot working causes an increase in the dislocation density, leading to faster precipitation, an effect known as strain-induced precipitation. An example is shown in Fig. 4.8 (Iparraguirre *et al.*, 2007). These precipitates are characterised by their small size, around 10–12 nm.

The effect of strain on the subsequent precipitation kinetics was quantified by Dutta and Sellars (1987), who determined the time required for 5% precipitation as follows:

$$t_{0.05\text{p}} = A \cdot \left[C\right]^{-1} \cdot \varepsilon^{-1} \cdot Z^{-0.5} \cdot \exp\left(\frac{Q_\text{d}}{RT}\right) \cdot \exp\left(\frac{B}{T^3 \cdot \left[\ln k_\text{s}\right]^2}\right) \qquad [4.14]$$

where $[C]$ represents the amount of Nb in solution available for precipitation, Z is the Zener–Hollomon parameter, Q_d is the activation energy for diffusion of the rate-controlling solute species in the matrix and k_s is the supersaturation ratio at the deformation temperature of the austenite.

In a recent publication, Pereda *et al.* (2008) have modified Eq. 4.14 by proposing a dependence on composition for the constants A and B. They observed that both C and N, together with the amount of Nb in solution, had an effect on the values of A and B.

4.4.2 Definition of non-recrystallisation temperature T_nr

The non-recrystallisation temperature T_nr is the temperature below which complete static recrystallisation does not occur between consecutive rolling passes. Above

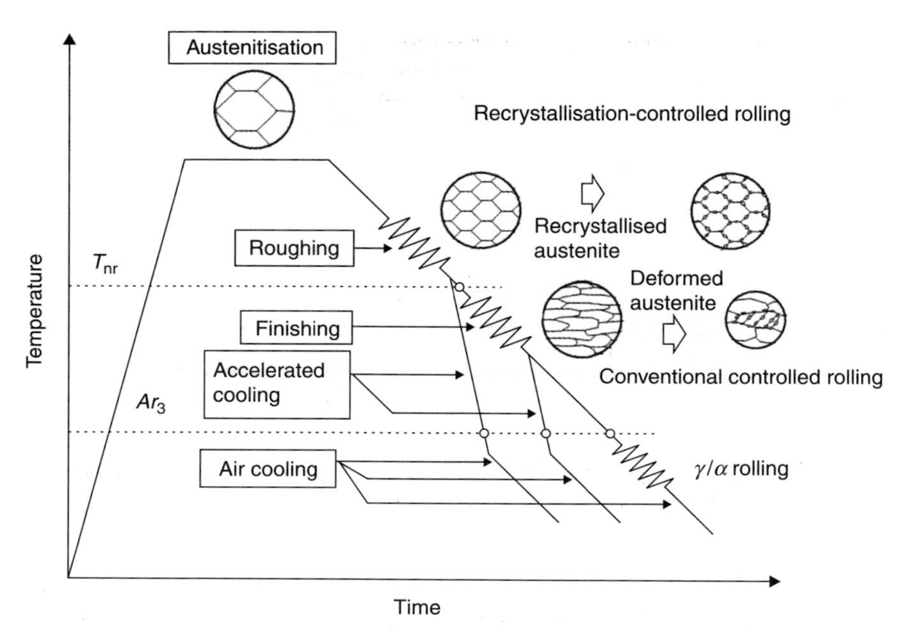

4.9 Schematic illustration of various thermomechanical processes.

this temperature, the recrystallisation is complete, which leads to the formation of relatively large polygonal grains. Below this temperature, the static recrystallisation is retarded and complete recrystallisation is no longer possible. The application of deformation passes below T_{nr} causes an accumulation of deformation, which results in the appearance of elongated grains (pancake structure) and the formation of deformation bands (see Fig. 4.9).

In clean steels, the most relevant nucleation sites for the austenite-to-ferrite transformation are the austenite grain boundaries. In the case of transformation in deformed austenite, dislocation bands within the grains can also act as nucleation sites. The ferrite grains are smaller when the transformation starts from a smaller austenite grain and especially from an elongated austenite grain, where the ratio S_v of the surface area to the volume of the grain is increased.

4.4.3 Physical metallurgical principles behind T_{nr}: interaction between deformation recovery, recrystallisation and precipitation

The temperature T_{nr} is influenced by the interaction of recovery, recrystallisation and precipitation. These three mechanisms influence each other and are all dependent on the parameters of the preceding deformation. A schematic overview of the various interacting parameters is shown in Fig. 4.10.

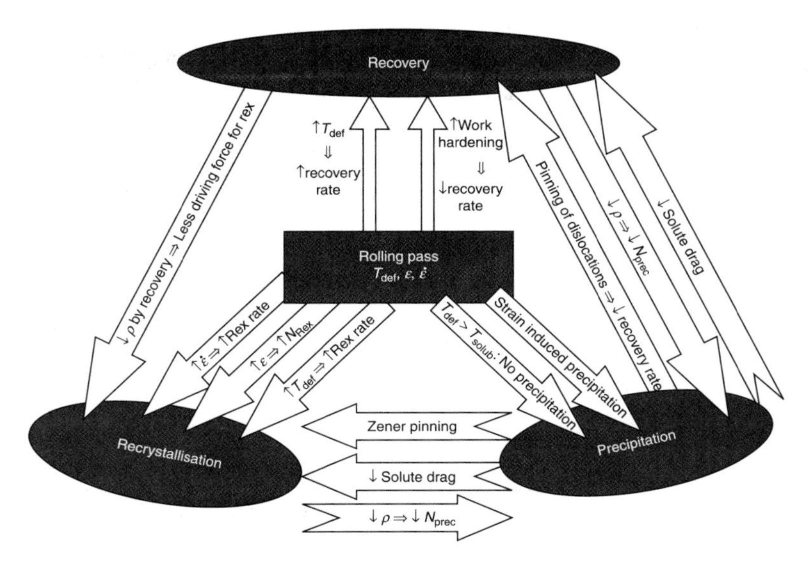

4.10 Schematic illustration of the interactions between deformation, recovery, recrystallisation and precipitation.

Interaction between recrystallisation and precipitation

In microalloyed steels, recrystallisation and precipitation may interact in at least three distinct ways (Zurob *et al.*, 2001; Zurob *et al.*, 2002):

- A decrease in the dislocation density ρ can occur as recrystallisation reduces the number of precipitate nucleation sites N_{prec} available and thus retards the onset of precipitation.
- A precipitate dispersion can provide a pinning force that can slow down or halt the progress of recrystallisation.
- The mobility of grain boundaries is known to be strongly affected by the solute content of the matrix. The progress of precipitation, which reduces the matrix solute content, also affects the mobility of grain boundaries and can enhance the progress of recrystallisation.

Kinetic equations (Eqs 4.8 and 4.14) for recrystallisation and precipitation can be used to predict the temperature at which, for a given strain, precipitation occurs sufficiently quickly to stop recrystallisation. This temperature is known as the recrystallisation stop temperature (RST). The criterion usually used to define this temperature is that 5% precipitation should be reached before 5% recrystallisation has occurred (Dutta and Sellars, 1987). This procedure is illustrated in Fig. 4.11 for the case of a 0.035% Nb microalloyed steel. It is possible to determine the temperature T_{nr} similarly. The criterion usually adopted is that 95% recrystallisation is reached before 5% precipitation occurs (see Fig. 4.11 again).

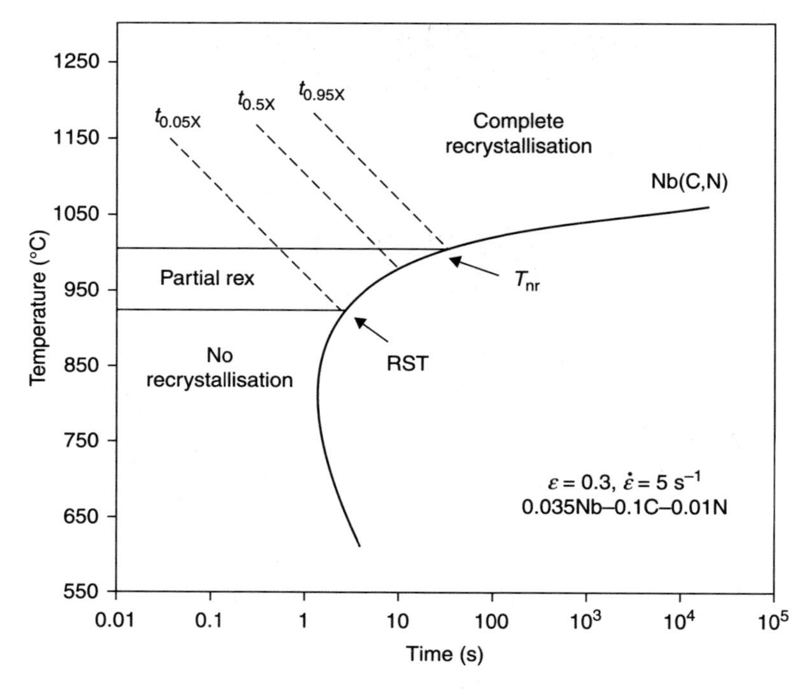

4.11 **Definitions of recrystallisation stop temperature (RST) and non-recrystallisation temperature T_{nr}.**

Taking Eqs 4.8 and 4.14 into account, it is possible to evaluate the variations that occur in T_{nr} when different Nb contents or processing conditions are considered. For example, Fig. 4.12 shows the decrease in T_{nr} as the strain increases from $\varepsilon = 0.2$ to 0.6.

When precipitation starts before recrystallisation is finished, there is a shift of the curve that indicates the end of recrystallisation (R_f) to longer times, as indicated schematically in Fig. 4.13. Thus the recrystallisation is retarded. If precipitation starts before the onset of recrystallisation, then the curves of the start (R_s) and end (R_f) of recrystallisation are shifted to longer times, and in some cases the recrystallisation is completely halted.

Interaction between recrystallisation and recovery

Although in the case of austenite, softening due to recovery is less important than that due to recrystallisation, the two phenomena occur at high temperatures and the interaction between them should also be considered. The driving force for both mechanisms is the internally stored deformation energy. The progress of

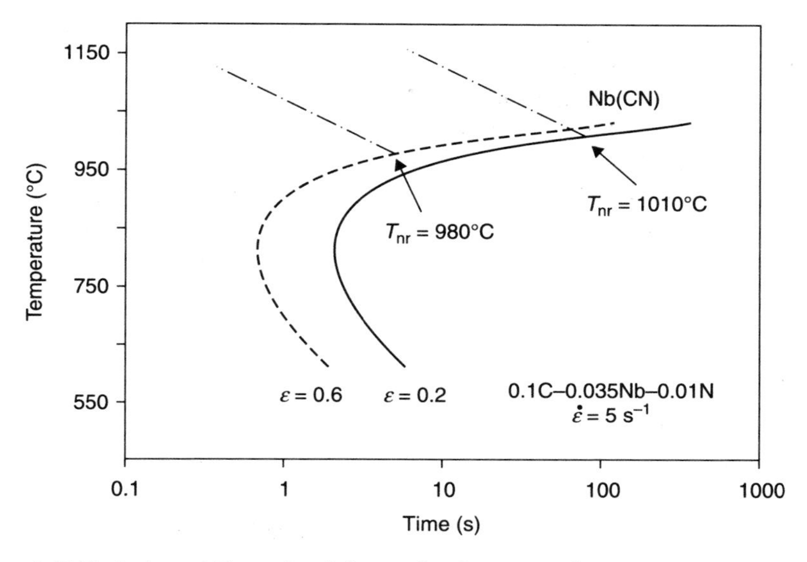

4.12 Variation of T_{nr} as the deformation increases from $\varepsilon = 0.2$ to 0.6.

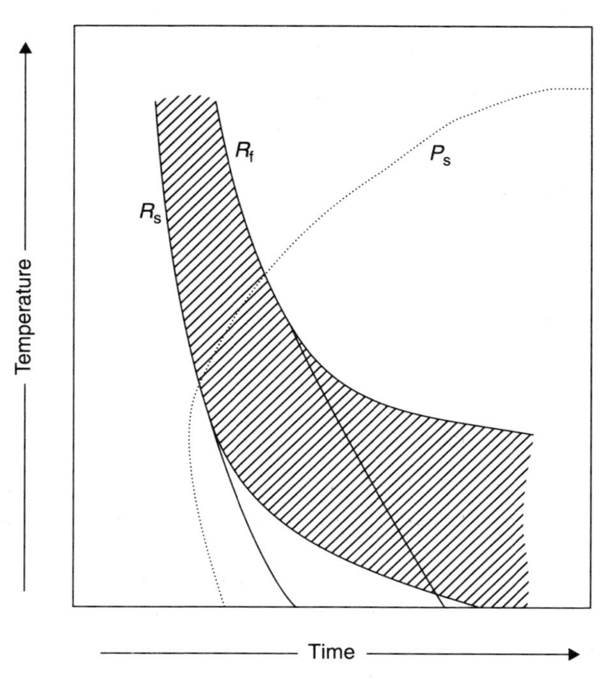

4.13 Interaction between recrystallisation and precipitation (P_s = precipitation start, R_s = recrystallisation start, R_f = recrystallisation finish).

recovery will reduce the driving force available for the migration of grain boundaries during recrystallisation and should slow down the recrystallisation process. Recrystallisation and recovery are competing processes.

Yoshie *et al.* (1996) stated that the amount of recovery is controlled mainly by the climb of dislocations in cases of larger deformation, higher temperature and lower Nb content, whereas it is controlled mainly by the annihilation of dislocations of opposite sign in cases of smaller deformation and higher Nb content. Further recovery is retarded by an increase in the austenite grain size, a decrease in the deformation temperature and an increase in the Nb content.

Interaction between recovery and precipitation

The potential interactions between recovery and precipitation are similar to those between recrystallisation and precipitation. At least three distinct interactions can be identified:

- Recovery may delay the progress of precipitation by lowering the number of available nucleation sites.
- The precipitation of fine particles can pin segments of the dislocation network and therefore retard recovery.
- Microalloying elements in solution are thought to retard recovery through the effect of solute drag on the dislocation mobility.

Most of the microalloying literature is centred on the Zener interaction between recrystallisation and precipitation, and the role of concurrent recovery is frequently overlooked. Nevertheless, depending on the deformation and the microstructural conditions, the contribution of recovery to softening sometimes cannot be ignored. Figure 4.14 shows an example of the evolution of the fractional softening and the metallographically measured recrystallised fraction with time for a Nb microalloyed steel deformed at 975°C, with a very large initial grain size (Iparraguirre *et al.*, 2007). The recrystallised fraction and the softening do not become equal until a very long time, about 10 000 s, is reached.

4.4.4 Parameters influencing T_{nr}

The temperature T_{nr} depends on the processing conditions, as well as on the composition of the steel. With respect to the interpass time, three regions can be distinguished (Abad *et al.*, 2001b), as illustrated in Fig. 4.15. For very short times, precipitation is unable to occur, and solute drag is the only mechanism that delays recrystallisation. At intermediate interpass times (second region), the occurrence of strain-induced precipitation retards recrystallisation and leads to an increase in T_{nr}, which then remains nearly constant. Finally, a further increase in the time means that precipitate coarsening is able to occur, weakening the retardation of recrystallisation, and T_{nr} decreases again (third region).

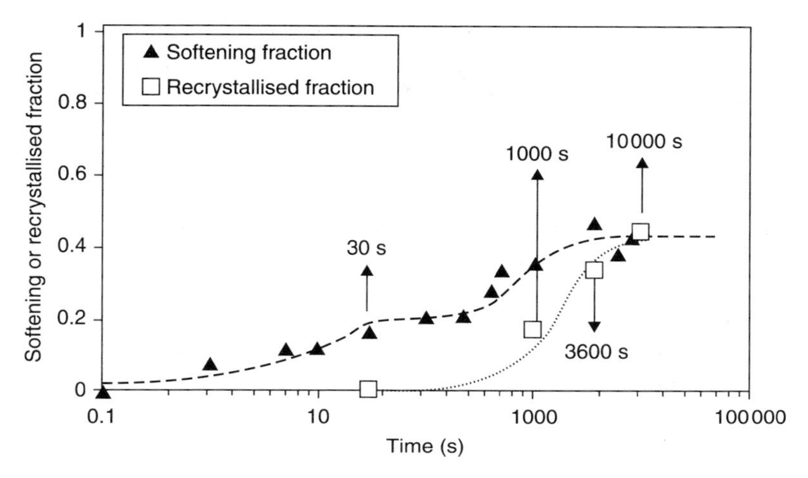

4.14 Evolution of the fractional softening and the recrystallised fraction with time determined for a 0.019 wt% Nb steel deformed at 975°C, with $\varepsilon = 0.3$ and $\dot{\varepsilon} = 1\,s^{-1}$, and with an initial grain size of 1000 µm (Iparraguirre *et al.*, 2007).

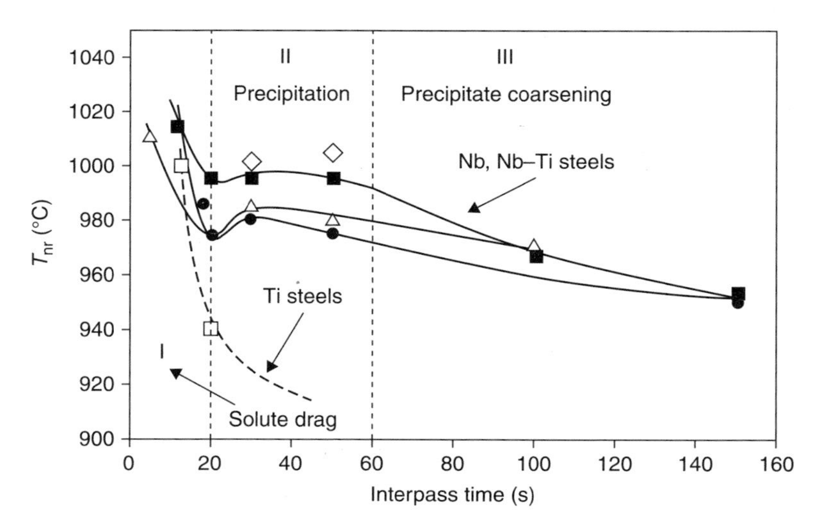

4.15 The three different ranges of the effect of interpass time on the non-recrystallisation temperature in Nb, Ti and Nb–Ti microalloyed steels (Abad *et al.*, 2001b).

Previous deformation significantly affects the temperature T_{nr}. As mentioned above, strain affects both recrystallisation and strain-induced precipitation (see Fig. 4.12). A relationship of the type $t_{0.05p} \approx \varepsilon^{-1}$ has been reported for the start time of strain-induced precipitation (see Eq. 4.14). On the other hand, a dependence of the recrystallisation time on strain of the type $t_{0.5x} \approx \varepsilon^{-p}$ is observed, with p varying between 2 and 4 (see Eq. 4.5). This means that the effect of strain on the recrystallisation kinetics is stronger than that on the precipitation kinetics. As a result, a decrease in T_{nr} with increasing pass strain is observed, as shown in Fig. 4.16 (Abad *et al.*, 1998). This behaviour is consistent with that previously reported for Nb microalloyed steels by Bai *et al.* (1993), who explained the decrease in T_{nr} with increasing pass strain by several contributing factors such as grain refinement, increased dislocation density and precipitate coarsening.

Finally, at higher strain rates, there is less restoration by dynamic recovery. Consequently, the less restored and more highly work-hardened austenite supplies a higher driving force for static recrystallisation, and therefore a decrease of T_{nr} with increasing strain rate is observed (Bai *et al.*, 1993).

The microalloying content significantly affects the value of T_{nr}. An increase in the content of microalloying elements, particularly Nb, Ti and V, can increase T_{nr}, with Nb being the most effective element (Cuddy, 1982).

The solubility product of TiN in austenite is very low (Fig. 4.6). As a consequence, Ti will remain nearly undissolved in the austenitic temperature range of the hot rolling process. This means that although Ti has a high solute drag potential (see Fig. 4.2), its real effect in retarding the recrystallisation kinetics will be small. On the other hand, as a consequence of the high solubility of VC in austenite, the role of VC in retarding recrystallisation and thus in increasing T_{nr} will also be of minor importance.

Nb, which forms carbides and carbonitrides that show intermediate solubility products, will be the most effective microalloying element for delaying austenite

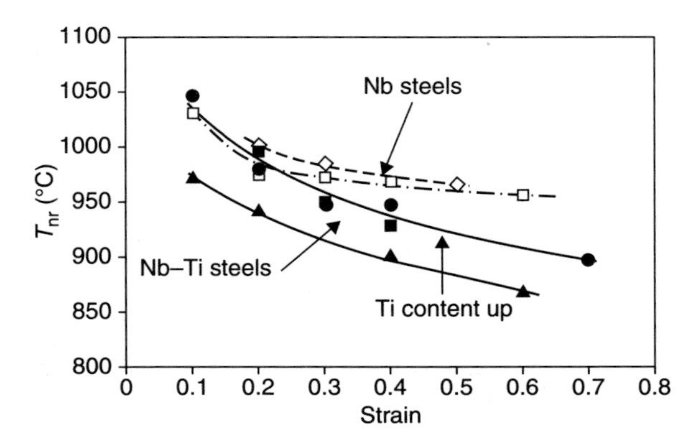

4.16 Influence of pass strain on T_{nr} for Nb and Nb–Ti steels (Abad *et al.*, 1998).

recrystallisation, owing to both a higher solute concentration and a higher driving force for precipitation. Moreover, Maruyama *et al.* (1998) gave another reason why Nb has a higher potential to retard softening kinetics. Nb atoms can trap or combine with interstitials and other defects faster than other elements and, additionally, Nb has a larger diffusivity, which is an important factor in its significant retarding effect on recovery and recrystallisation.

The effect of multiple microalloying elements is more complex. For example, for similar contents of Nb, T_{nr} for Nb–Ti steels is significantly lower than that observed for Nb steels (Fig. 4.16). This behaviour could be related to several factors contributing to a decrease in the supersaturation level when Ti is present in addition to Nb (Abad *et al.*, 1998).

The influence of other alloying elements, such as Mo, Ni and Cu, which are often used in microalloyed steels in order to increase the strength, should also be considered. The impact of these elements on the recrystallisation behaviour is still not completely understood and, to date, there is no systematic data on the influence of nickel and copper on recrystallisation behaviour.

The addition of Mo to Nb microalloyed steels may lead to significant changes in the microstructural evolution during hot working. For example, changes in the kinetics of austenite recrystallisation and strain-induced carbonitride precipitation have been reported in Nb–Mo steels (Akben *et al.*, 1983; Pereda *et al.*, 2007a). In this context, a quantification of the effect of Mo on the recrystallisation kinetics was done by Pereda *et al.* (2007b), indicating that the temperature T_{nr} increased by about 40°C when 0.3% Mo was added to a 0.03% Nb steel (Fig. 4.17).

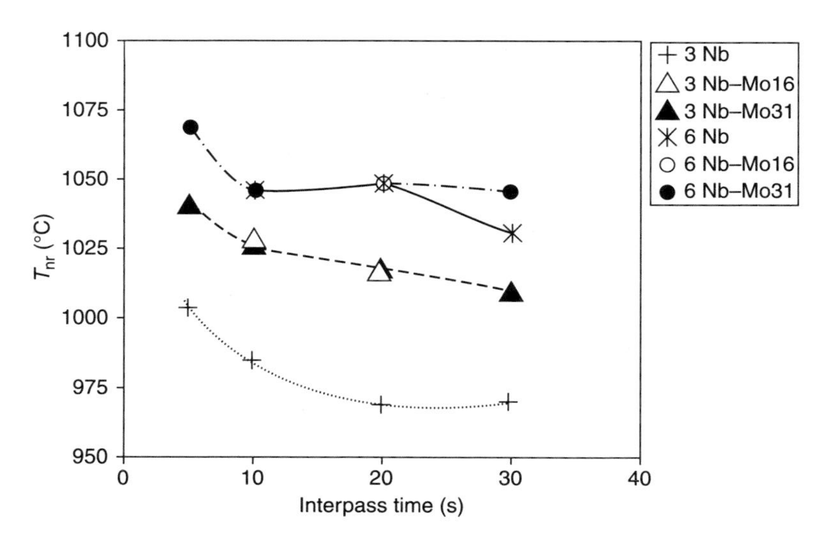

4.17 Changes in T_{nr} with the addition of Mo to Nb-bearing steels as a function of interpass time for a given pass strain of $\varepsilon = 0.4$ (Pereda *et al.*, 2007b).

However, increasing the Nb content to values as high as 0.06% makes the contribution of Mo less relevant.

4.5 Modelling methods

Over the last few years, much more attention has been paid to the development of mathematical models that predict the final microstructure of a steel product based on the processing conditions, in order to optimise the different variables and achieve the best combination of strength–toughness properties in as-rolled materials. Most of these models are based on semi-empirical equations that describe the softening and strengthening mechanisms governing the microstructural evolution of austenite (Sellars, 1990). Recently, there has been a trend towards developing more physically based models, as semi-empirical models are often limited to the conditions that they were developed for and extrapolation to other ranges may lead to significant errors (Dutta *et al.*, 2001; Zurob *et al.*, 2002). Nevertheless, one of the disadvantages of using physical models is that the values of the physical parameters required are not always well known.

This section will deal with the modelling of microstructural evolution during thermomechanical processing of steels, considering two approaches: one based on the semi-empirical equations described in the previous sections and on non-isothermal conditions, and the other, applied to isothermal conditions, which considers more fundamental studies based on physical concepts.

4.5.1 Semi-empirical models

In most cases, recrystallisation and strain-induced precipitation processes are modelled separately, and to cope with the interaction between them, the corresponding recrystallisation and precipitation times are compared. For modelling purposes, the following criteria are used to describe this interaction after a given deformation pass:

- If $t_{0.05p} < t_{0.05x}$, precipitation starts before recrystallisation and there is no recrystallisation at all.
- If $t_{0.05x} < t_{0.05p} < t_{0.95x}$, recrystallisation starts, but precipitation takes place before recrystallisation is finished, stopping its evolution, and thus partial recrystallisation occurs.
- If $t_{0.05p} > t_{0.95x}$, precipitation starts when recrystallisation is finished, and thus there is no interaction between the two processes and full recrystallisation occurs.

Modelling of static recrystallisation

In the case of $\varepsilon < \varepsilon_c$, the main softening mechanism that operates during the interpass interval between consecutive deformations will be static recrystallisation.

Therefore, the evolution of the austenite microstructure after each deformation pass will be determined by the kinetics of the corresponding static recrystallisation and grain growth processes.

The evolution of the recrystallised fraction with time is described by the Avrami equation (Eq. 4.5). In Table 4.2, a list of some of the equations reported in the literature for calculating $t_{0.5srx}$ is given. Similarly, the recrystallised grain size and its subsequent growth can be quantified with the help of Eqs 4.11–4.13. All of these equations were established for isothermal conditions. However, the temperature decreases continuously during hot working. In such cases, the principles of the additivity rule (Scheil, 1935) can be used to modify the isothermal recrystallisation model for use under continuous-cooling conditions.

In this approach, the cooling curve is divided into increments of time Δt, and the temperature drop for each step can be calculated as $\Delta T = \Delta t \cdot v$, v being the cooling rate. If the accumulated recrystallised fraction at the end of an interval of temperature T_j is X_j, the recrystallised fraction at the end of the subsequent interval of temperature $T_{j+1} < T_j$ is calculated by first determining the 'equivalent time' t_{eq} needed to reach the previous value of the recrystallised fraction X_j during isothermal annealing at T_{j+1}. This is schematically illustrated in Fig. 4.18. The real recrystallised fraction at T_{j+1} is then calculated as

$$X_{j+1}\left(\Delta t\right) = X_{j+1}\left(\Delta t + t_{eq}\right) - X_{j+1}\left(t_{eq}\right) \tag{4.15}$$

The recrystallised fraction is calculated and added from interval to interval. By adding the holding time for each increment, the time necessary to obtain a given recrystallised fraction can also be determined.

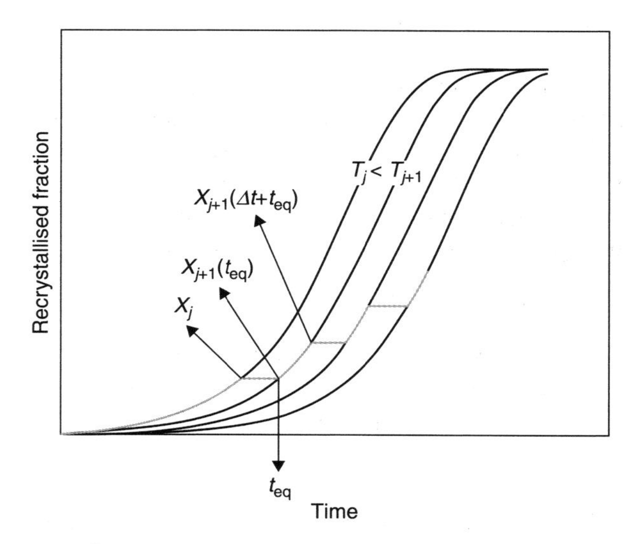

4.18 Schematic representation of the additivity principle.

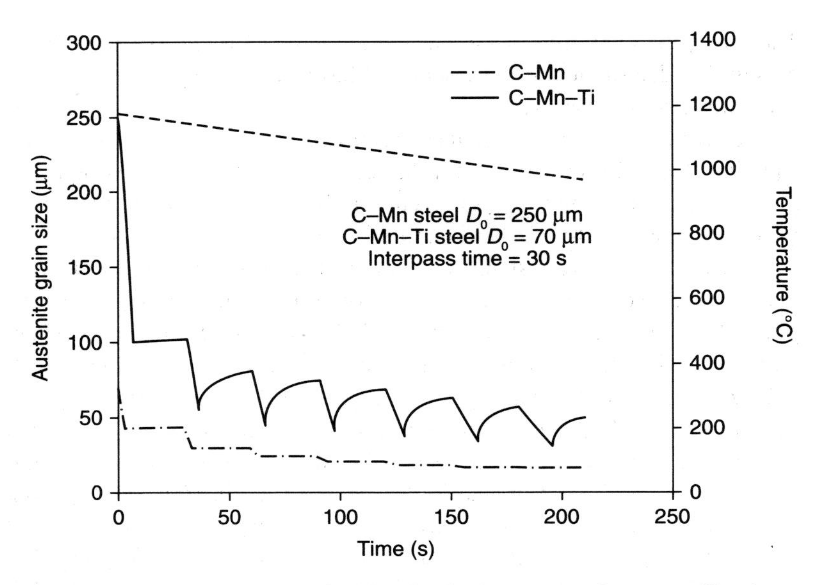

4.19 **Predicted evolution of grain size in the range of recrystallization for the case of C–Mn and C–Mn–Ti microalloyed steels.**

Figure 4.19 shows an example of the predicted evolution of the austenite grain size for a plain C–Mn steel and a Ti microalloyed steel. In this latter case, because grain growth between passes is completely avoided, a significant grain refinement is achieved at the exit of the last rolling pass compared with the C–Mn steel. This behaviour is related to the presence of very small TiN particles (Arribas *et al.*, 2008).

In multipass rolling, it is possible to have partial recrystallisation after a pass strain ε_i if the time between passes is not long enough for complete recrystallisation. This introduces a mixed microstructure before the next deformation pass, with a strain ε_{i+1}. To cope with this effect, the 'uniform softening method' (Choquet *et al.*, 1985) can be used. This method considers a single average microstructure, described by an average grain size, with an effective strain as follows:

$$\varepsilon_{\mathrm{eff}_i} = \varepsilon_i + \lambda\left(1 - X_{i-1}\right)\varepsilon_{i-1} \qquad [4.16]$$

where λ is a constant (0.5 for Nb steels (Perdrix, 1987)) and X_{i-1} is the recrystallised fraction between passes i and $i - 1$. This effective strain must be taken into consideration in calculations of the kinetics of recrystallisation and strain-induced precipitation after each pass.

The partially recrystallised microstructure is described by an average grain size, and several different expressions have been proposed to calculate it (Perdrix, 1987; Beynon *et al.*, 1988). The microstructure is composed of recrystallised and unrecrystallised grains, and the law of mixtures is usually applied. This method requires the determination of the instantaneous sizes of the recrystallised grains, d_r, and unrecrystallised grains, d_u. The evolution of the mean recrystallised grain size d_r with time can be described by the following equation (Orsetti Rossi and Sellars, 1997):

$$d_r(t) = X^{1/3}(t)\, d_{rex} \quad [4.17]$$

where d_{rex} represents the final recrystallised grain size. Figure 4.20(a) shows an example of the fitting of experimental measurements of d_r to the predictions of Eq. 4.17 (Fernández *et al.*, 2002).

The effective size of the unrecrystallised grains, d_u, can be modelled with the help of Eq. 4.18 below (Beynon and Sellars, 1992). A modified expression was proposed by Anelli (1992) (Eq. 4.19), taking into account both the flattening and the elongation of the original grains due to the deformation applied. Figure 4.20(b) shows a comparison between experimental measurements and the predictions of the two equations.

$$d_u(t) = \left(1 - X(t)\right) d_0 \quad [4.18]$$

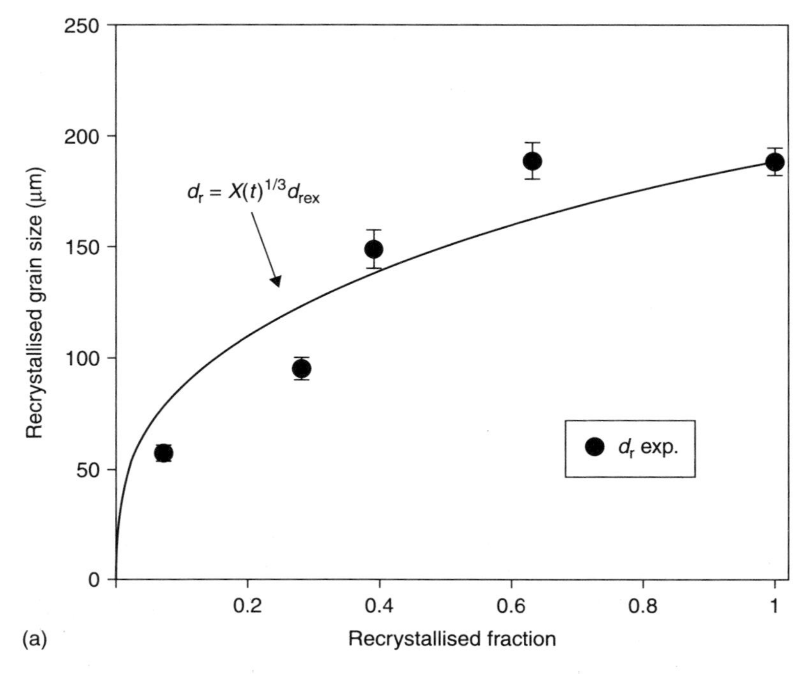

4.20 (a) Comparison between the experimental evolution of the mean recrystallised grain size d_r with the recrystallised fraction and the prediction of Eq. 4.17.

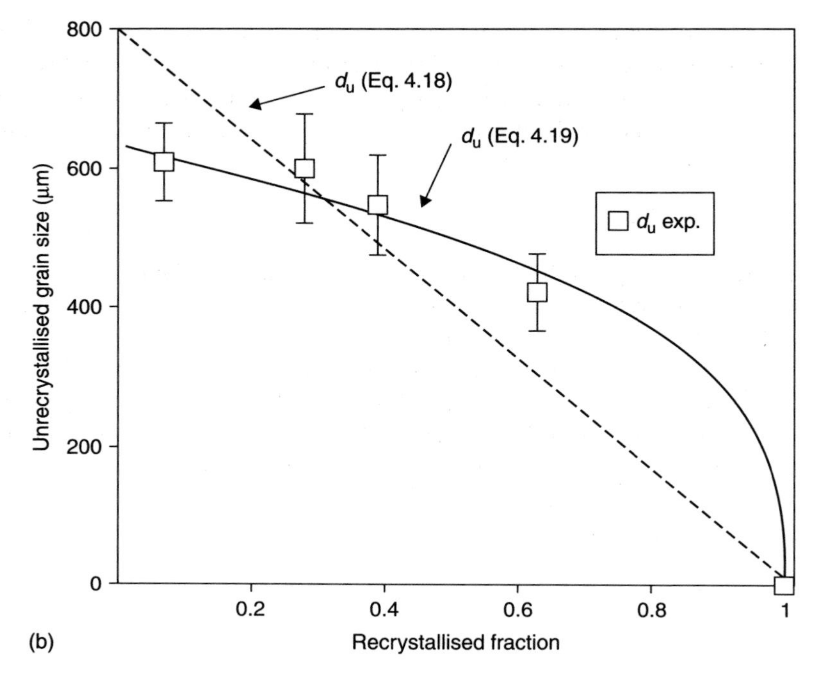

4.20 Continued. (b) Comparison between the experimental mean unrecrystallised grain size d_u and the predictions of Eqs 4.18 and 4.19 (Fernandez *et al.*, 2002).

$$d_u(t) = 1.06\,\exp(-\varepsilon)\,(1 - X(t))^{1/3}\,d_0 \tag{4.19}$$

The mean grain size d_m can be evaluated by directly applying the law of mixtures, using Eq. 4.17 for d_r and Eq. 4.18 or 4.19 for d_u, as follows:

$$d_m(t) = X^{4/3}(t)\,d_{rex} + (1 - X(t))^2\,d_0 \tag{4.20}$$

$$d_m(t) = X^{4/3}(t)\,d_{rex} + 1.06\,\exp(-\varepsilon)\,(1 - X(t))^{4/3}\,d_0 \tag{4.21}$$

An example of the application of these equations to the results shown in Fig. 4.20 is shown in Fig. 4.21. The predictions of the two equations are compared with the experimental mean total grain size d_T (the metallographically measured grain size, determined without distinguishing between recrystallised and unrecrystallised grains). The mean grain sizes d_m evaluated by applying the law of mixtures directly to the experimentally measured data for d_r and d_u are also plotted for comparison.

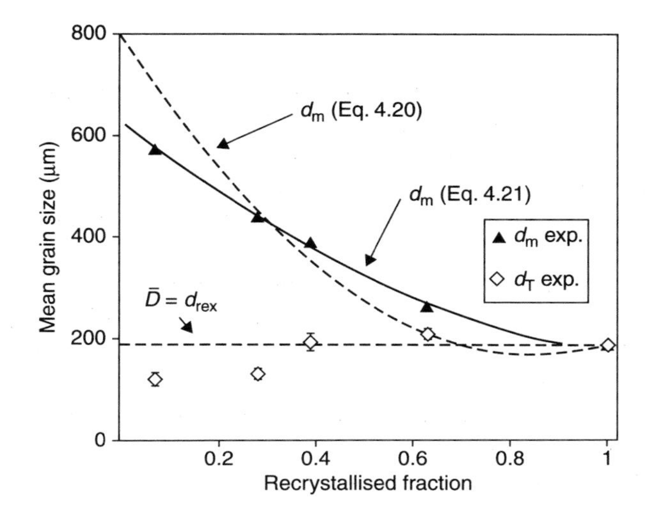

4.21 Comparison of the experimental mean total grain size d_T with values of d_m determined by applying the law of mixtures to experimental data for d_r and d_u, and with values calculated using Eqs 4.20 and 4.21. d_T is also compared with an average grain size taken as $\bar{D} = d_{rex}$ (Fernandez *et al.*, 2002).

Although some authors (Liu *et al.*, 1995) have considered the grain size determined by Eq. 4.20, others (Choquet *et al.*, 1985) have suggested that one should employ an approach that takes the average grain size as being equal to the fully recrystallised grain size, i.e. $\bar{D} = d_{rex}$. This provides a good correlation with experimental results once the recrystallised fraction has exceeded some minimum value.

Modelling of Nb(C,N) strain-induced precipitation

The kinetics of Nb(C,N) precipitation may be also modelled by an Avrami-type equation,

$$X_p = 1 - \exp\left(\ln(0.95) \cdot \left(\frac{t}{t_{0.05p}}\right)^{n_p}\right) \qquad [4.22]$$

where X_p is the precipitated fraction at a time t, n_p is the Avrami exponent for precipitation and $t_{0.05p}$ is given by Eq. 4.14. Several different values have been proposed for the exponent n_p (Herman *et al.*, 1992; Pereloma *et al.*, 2001).

For a constant temperature, the maximum precipitated volume fraction depends on the equilibrium conditions. The precipitated volume fraction is calculated as

$$f_v = f_{veq} \cdot X_p \qquad [4.23]$$

where f_{veq} is the equilibrium volume fraction at the temperature considered and X_p is the precipitated fraction calculated from Eq. 4.22.

All of these equations correspond to isothermal conditions. Their extension to conditions of continuously decreasing temperature can be done by applying the additivity rule, as indicated above for the case of static recrystallisation (Pereda *et al.*, 2008).

Equation 4.14 provides the conditions for the beginning of strain-induced precipitation after a given deformation pass. However, the evolution of the precipitated fraction (Eq. 4.22) must also be determined, since once precipitation starts, more and more Nb is taken out of solution, and therefore a smaller amount of Nb is available for further precipitation and the solute drag effect after the next pass. Similarly to the case of recrystallisation, the additivity rule can be applied to Eq. 4.22 to calculate the evolution of the precipitated fraction under continuous-cooling conditions.

Figure 4.22 shows a flow chart representing the structure of the overall model (Pereda *et al.*, 2008). The first step is the calculation of the actual values of the concentrations in solution [Nb] and [C+(12/14)N] at each reheating temperature according to the appropriate solubility product. For each deformation pass, the model predicts three possible final situations: partial recrystallisation without precipitation, grain growth after complete recrystallisation, and partial recrystallisation due to the onset of strain-induced precipitation. In the latter case, the precipitated volume fraction is calculated.

In the next pass, the new initial grain size, the concentrations in solution [Nb] and [C+(12/14)N], and the accumulated strain are considered in the calculations. The precipitation constants in Eq. 4.14 are taken to depend on the concentrations of Nb, C and N in solution, and are also recalculated for each pass. After the last pass, an output file containing the recrystallised fraction, the austenite grain size, the Nb(C,N) precipitated volume fraction, and the dissolved amounts [Nb] and [C+(12/14)N] calculated after each pass is generated.

Figure 4.23 shows an example of the application of a microstructural evolution model to a series of multipass torsion tests carried out with two different Nb microalloyed steels (Pereda *et al.*, 2008). The evolution of the interpass softening predicted by the model is compared with the experimental results. Once precipitation starts (the calculated temperature, P_s, is indicated in Fig. 4.23), the model predicts a faster fall in the fractional softening, in good agreement with the experimental observations. In the figure, the experimental values of the temperature T_{nr} are also indicated. It is interesting to note that in one case the temperatures T_{nr} and P_s are relatively close, whereas in the other the temperature T_{nr} appears significantly higher than P_s. Coincidence between the two temperatures indicates that strain-induced precipitation is the main mechanism, whereas when T_{nr} is higher than P_s, this means that solute drag itself is able to stop recrystallisation before precipitation occurs.

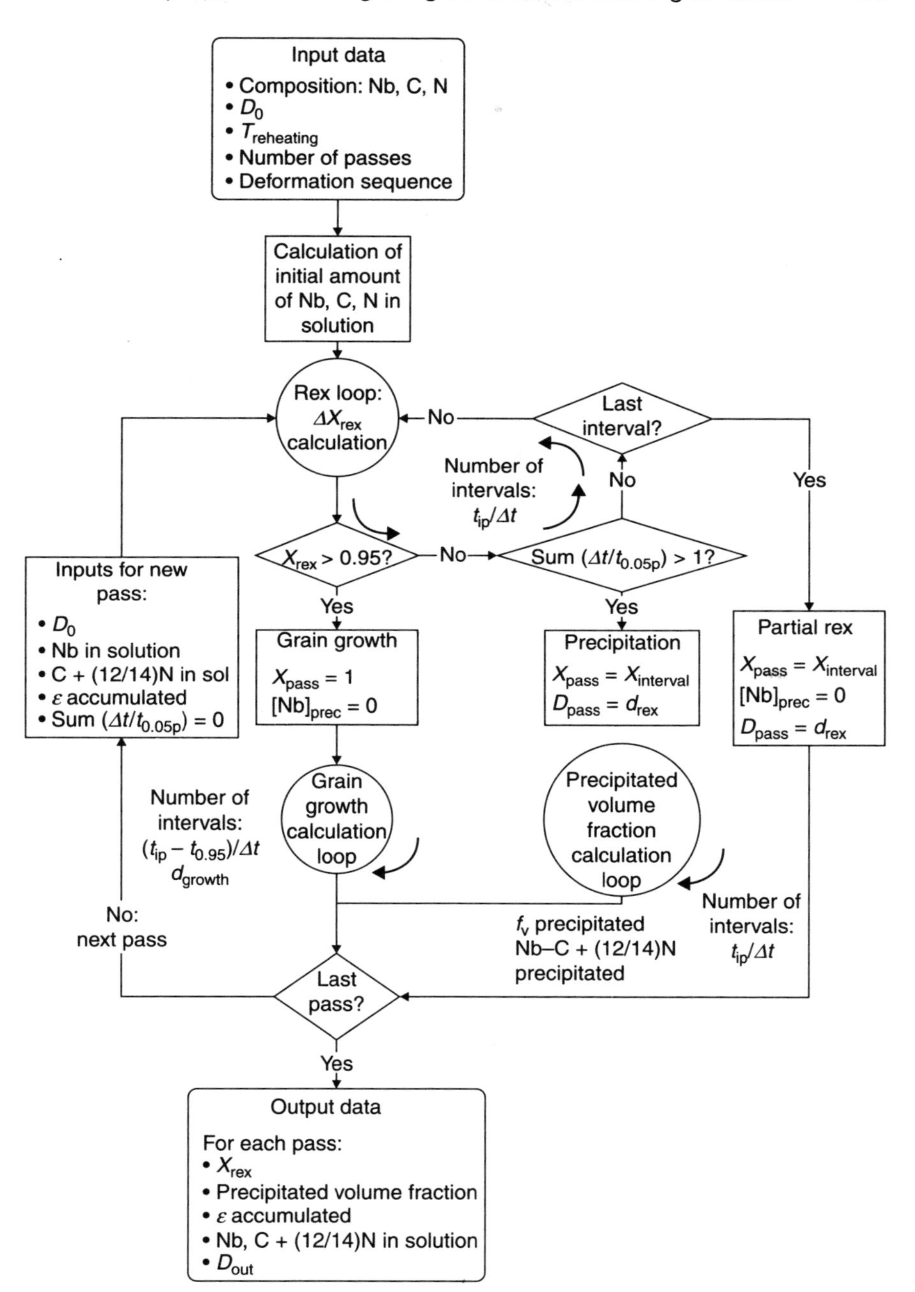

4.22 Flow chart representing the structure of the model described in Section 4.5.1 (Pereda *et al.*, 2008).

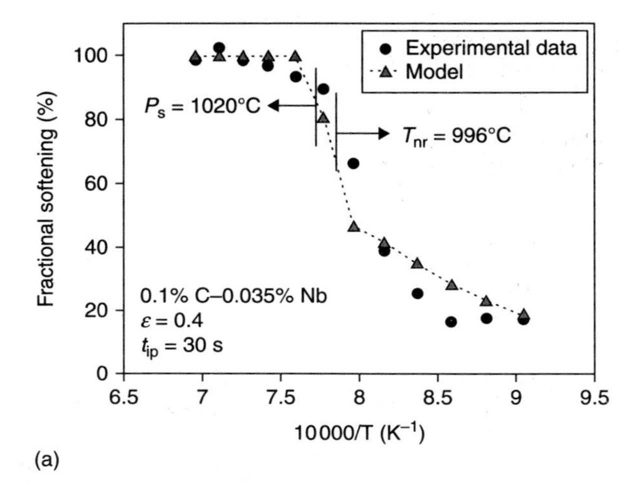

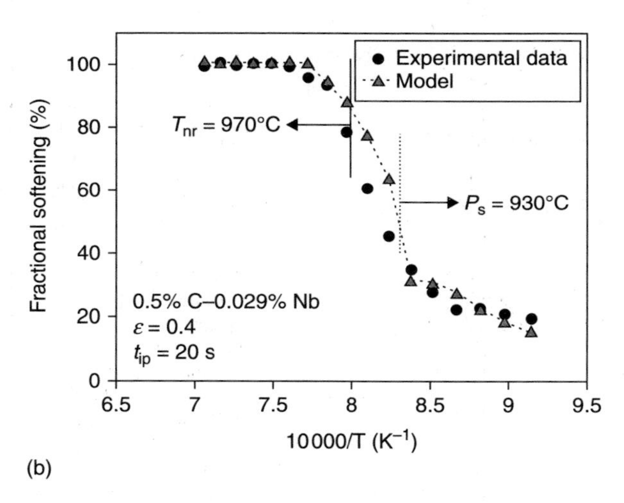

4.23 (a,b) Plots of interpass fractional softening vs mean interpass temperature. Comparison between experimental and predicted values (Pereda *et al.*, 2008).

4.5.2 Physical models

In recent years, the modelling of hot-deformation behaviour has increasingly been based on the physical phenomena occurring within the material. The model recently proposed by Zurob *et al.* (2002) is, to date, one of the most complete quantitative evaluations of the coupling between recovery, recrystallisation and strain-induced precipitation in microalloyed austenite. In this model, it is assumed that at low temperatures, recovery and strain-induced precipitation interact. At high temperatures, major softening occurs because of recrystallisation without any interaction with precipitation. In the intermediate range, all three mechanisms can operate simultaneously. For simplicity, it is assumed that recrystallisation does not influence the precipitation processes.

Individual processes are modelled separately, with subsequent consideration of the interaction between them. The model is divided into three elementary blocks or modules: recrystallisation, recovery and strain-induced precipitation. In what follows, the main equations used to describe each phenomenon are given.

Recrystallisation module

The progress of recrystallisation is based on the treatment by Zurob *et al.* (2001). Assuming site saturation (N_{rex} sites per unit volume), the recrystallised fraction is given by

$$X = 1 - \exp\left[-N_{rex}\left(\int_0^t M(t)G(t)\,dt\right)^3\right] \qquad [4.24]$$

where $G(t)$ is the net driving force for recrystallisation (Humphreys and Hatherly, 1996; Furu *et al.*, 1990) and $M(t)$ is the mobility of grain boundaries. Both are time-dependent.

The effect of concurrent precipitation on the driving force for recrystallisation may be expressed as a stored energy (given by the dislocation density) modified by a retarding Zener drag term representing the pinning force exerted by the precipitates:

$$G(t) = \frac{1}{2}\rho(t)\mu b^2 - \frac{3}{2}\frac{\gamma_{gb} f_v(t)}{R(t)} \qquad [4.25]$$

where $\rho(t)$ represents the instantaneous dislocation density, μ the shear modulus and b the Burgers vector. The Zener pressure exerted by the precipitates is described by the second term in Eq. 4.25, with γ_{gb}, $f_v(t)$ and $R(t)$ denoting the austenite grain boundary energy, the precipitate volume fraction and the mean precipitate radius, respectively.

The Zener pressure resulting from a precipitate distribution is found to vary from $3\gamma_{gb}f_v(t)/2R(t)$ to an upper limit equal to twice this value when the precipitate interphase energy exhibits a strong dependence on orientation (Jones and Ralph, 1975). It has been observed that in some cases, larger pinning forces are required to fit the experimental results. To cope with this, Iparraguirre *et al.* (2007) introduced a pinning factor to modulate the value of the pinning force calculated from the Zener pressure expression.

The effect of solute elements on the mobility $M(t)$ of grain boundaries has been treated by Cahn (1962). For low driving forces, Cahn's solution simplifies to

$$M(t) = \left(\frac{1}{M_{\text{pure}}} + \alpha C_{\text{Nb}} \right)^{-1} \qquad [4.26]$$

where M_{pure} is the intrinsic grain boundary mobility, C_{Nb} is the concentration of Nb in solution and α is a parameter depending on several factors. The value of M_{pure} is usually considered as one-half of the value estimated by Turnbull (1951).

Recovery module

It is assumed that the flow stress of the precipitate-free austenite, σ, is related to the total dislocation density, ρ, and the yield stress of the precipitate-free fully recrystallised austenite, σ_y, by a forest-type hardening relation,

$$\sigma = \sigma_y + M\alpha_T \mu b \sqrt{\rho} \qquad [4.27]$$

where M is the Taylor factor (≈3.1 for an FCC structure) and $\alpha_T \cong 0.15$. The recovery kinetics of the dislocation contribution to the hardening $\sigma - \sigma_y$ is described using the approach proposed by Verdier *et al.* (1999).

Figure 4.24 shows an example of the predicted stress relaxation in a Nb microalloyed steel (Iparraguirre *et al.*, 2007). At low temperatures, i.e. 850 and 875°C, recovery–precipitation interaction occurs. The stagnation of softening caused by the onset of precipitation is predicted reasonably well. At the highest temperature of 1060°C, recovery followed by recrystallisation is predicted to occur, without interaction with precipitation. The recrystallisation model is able to reproduce the general form of the stress relaxation curves observed in this range well. At intermediate temperatures, i.e. 990 and 1015°C, a more complex situation, where recovery, recrystallisation and precipitation interact, is predicted. In this case, larger deviations from the model are observed.

Precipitation module

The precipitation of Nb(C,N) in austenite is described using the approach of Deschamps and Brechet (1999). Precipitation is assumed to occur in two stages: nucleation and growth as concurrent processes, and, after that, growth and coarsening. Nucleation is assumed to occur exclusively on dislocations. The steady-state nucleation rate is given by

$$\frac{dN}{dt} = \left(1 - \frac{N}{N_{\text{total}}}\right) N_{\text{total}} Z\beta^* \exp\left(\frac{\Delta G_n}{\kappa_B T}\right) \qquad [4.28]$$

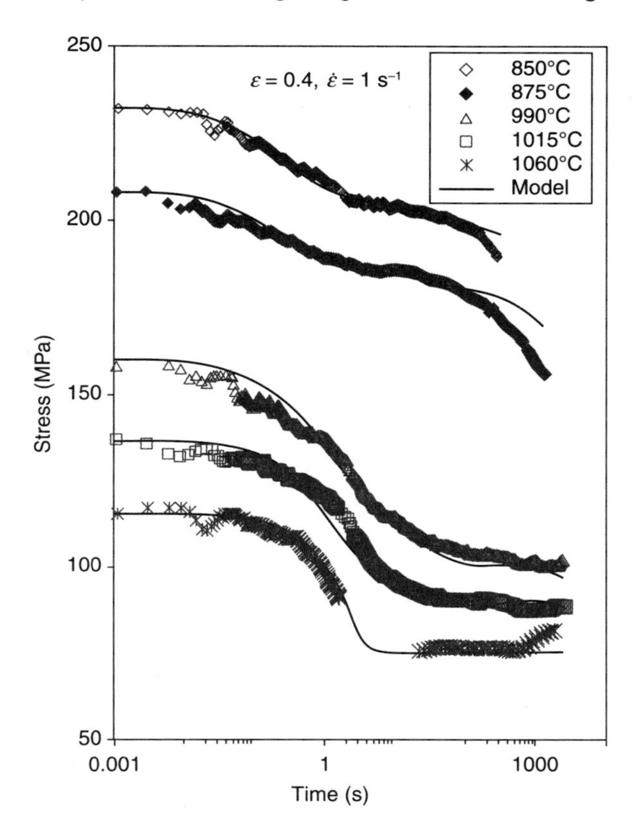

4.24 Stress relaxation curves obtained for a 0.051% Nb microalloyed steel deformed to $\varepsilon = 0.4$ at different temperatures, compared with the predictions of the model described in Section 4.5.2 (Iparraguirre *et al.*, 2007).

where Z is the Zeldovich factor, β^* is the rate at which atoms are added to the critical nucleus, N_{total} is the number of nucleation sites and ΔG_n is the activation barrier for the nucleation process. Zurob *et al.* (2002) approximated the density of nucleation sites as $F\rho/b$, where F was an adjustable factor smaller than unity. Similarly, Dutta *et al.* (2001) introduced a 'dislocation density factor' f_p to fit observed data and predictions. This factor is used to adjust the dislocation density.

Several different approaches can be used for the calculation of ΔG_n. Dutta *et al.* (2001) used the following energy balance:

$$\Delta G_n = \frac{16}{3}\pi \frac{\gamma^3}{\Delta G_V^2} - 0.8\mu b^2 \frac{\gamma}{|\Delta G_V|} \qquad [4.29]$$

whereas Zurob *et al.* (2002) estimated this energy balance as

$$\Delta G_{\text{n}} = V\Delta G_{\text{V}} + A\gamma - \frac{\mu b^2 R\ln(R/b)}{2\pi(1-\nu)} - \frac{\mu b^2 R}{5} \qquad [4.30]$$

where V and A are the volume and area of the nucleus, respectively; ΔG_{V} is the free energy change attendant upon nucleation; γ is the interphase energy; and ν is the Poisson's ratio.

A summary of the model is shown in Fig. 4.25, with those terms that provide the coupling between concurrent recovery, recrystallisation and precipitation (ρ, f_{v}, R and C_{Nb}) highlighted to emphasise the interdependency of these processes.

Figure 4.26 shows an example of the application of the model, where the evolution of the precipitate number density, volume fraction and particle size with the holding time were estimated in the case of a Nb microalloyed steel at two temperatures after deformation (Iparraguirre *et al.*, 2007). It can be observed that the experimental precipitation kinetics is reasonably well reproduced by the model.

Model parameters

In the above model, several simplifying assumptions have been made. It is assumed that the progress of recrystallisation does not influence the precipitation processes. In doing so, it is assumed that precipitation takes place mainly on dislocations in the unrecrystallised regions, and is not perturbed by the migration of the recrystallised front. On the other hand, the nucleation of recrystallisation is assumed to be site-saturated.

Moreover, the model makes use of several physical quantities and fitting parameters. Some of the physical quantities are known with a reasonable degree of accuracy, but others are not well known and there are discrepancies in the values proposed in the literature for the same quantity. For example, for the diffusion constants of Nb, a difference of nearly one order of magnitude is observed between values found in the literature (Zurob *et al.* 2002, Dutta *et al.*, 2001). Other important quantities for which there is also uncertainty are the interphase energy γ (Eqs 4.30 and 4.31), the activation energy U_{a} and the activation volume for recovery V_{a} (see Fig. 4.25). Estimates of the austenite–MC interphase energy vary from ~0.3 to more than 1.5 J/m^2 (Dutta *et al.*, 1992; Dutta *et al.*, 2001; Yang and Enomoto, 1999). However, the interphase energy is expected to exhibit both a temperature and a composition dependence (Zurob *et al.*, 2002; Iparraguirre *et al.*, 2007). It should be noted that small changes in this parameter significantly modify the calculations.

The activation energy for the recovery process is expected to lie between $0.6Q_{\text{diff}}$ and $1Q_{\text{diff}}$, where Q_{diff} is the activation energy for self-diffusion or solute diffusion, depending on the rate-controlling process (Nes, 1995). Finally, for the

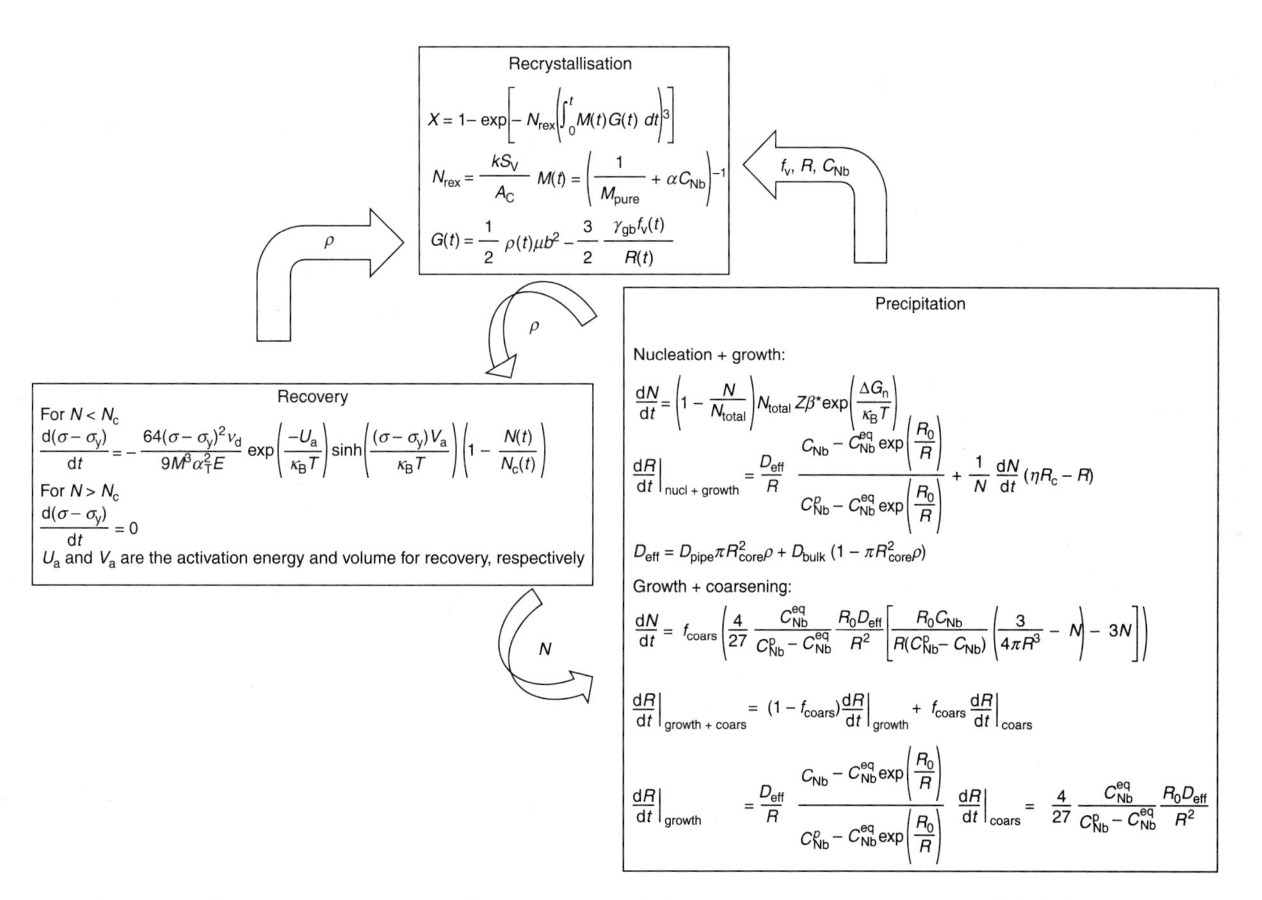

4.25 Summary of the recovery, recrystallisation and precipitation modules used in the model. Additional details can be found in Zurob *et al.* (2001), Zurob *et al.* (2002), Verdier *et al.* (1999) and Deschamps and Brechet (1999).

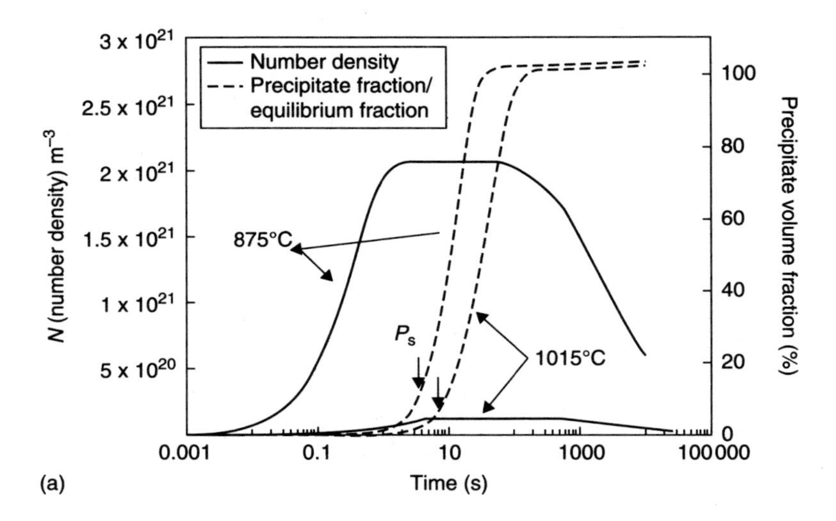

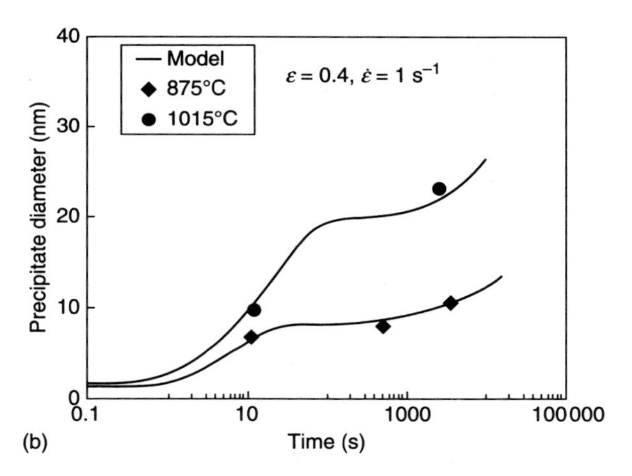

4.26 Evolution of (a) the particle number density and volume fraction and (b) the size, at two different temperatures in a Nb microalloyed steel after deformation (Iparraguirre *et al.*, 2007).

determination of the activation volume for recovery V_a, stress relaxation data have been used. The influence of this parameter on softening calculations can be observed in Fig. 4.27 (Iparraguirre and López, 2008), where the predictions of the model using different values of the recovery activation volume V_a are compared. It can be observed that this parameter may significantly affect the predicted evolution of the softening fraction.

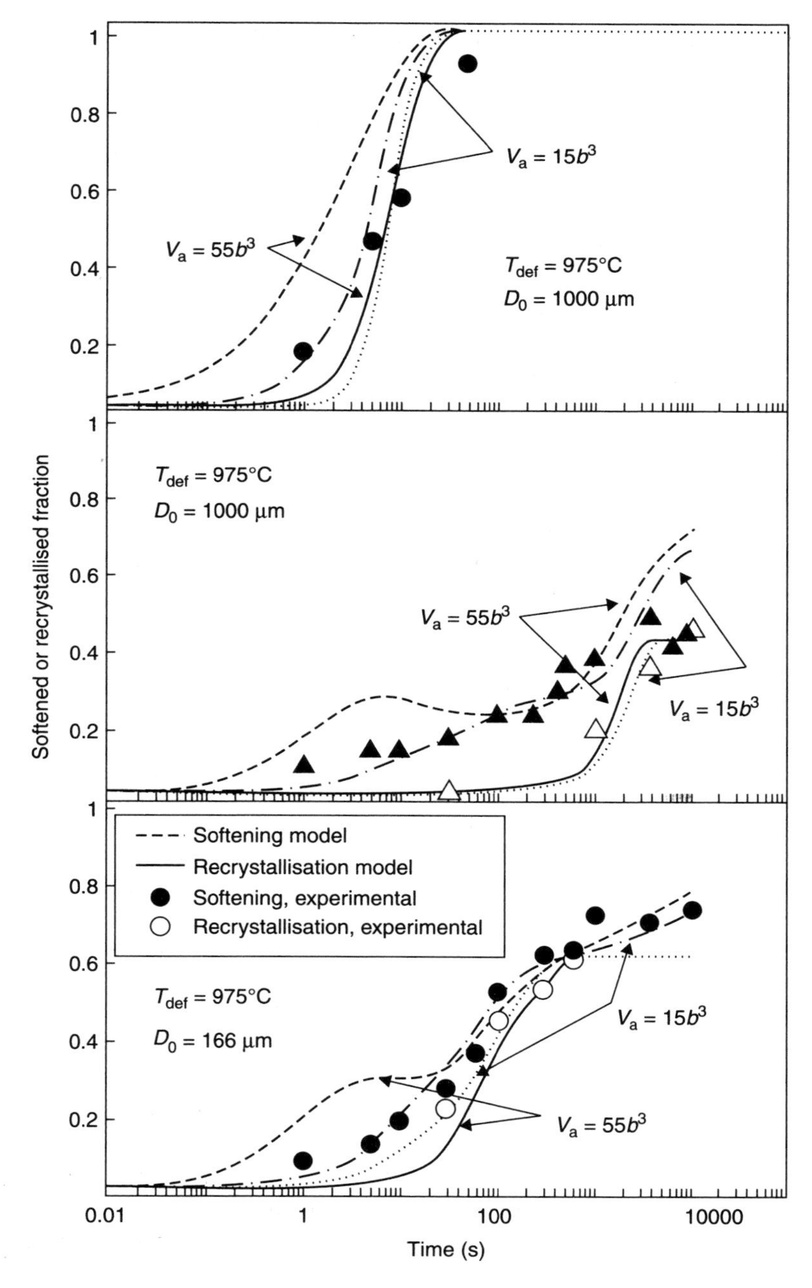

4.27 Comparison between the predictions of softening from the model for different values of the recovery activation volume $V_a = 55b^3$ and $15b^3$ (Iparraguirre and López, 2008).

4.6 Case studies in metal forming

In this section, two different cases will be considered as examples of grain refinement by recrystallisation. The first corresponds to the possible application of dynamic recrystallisation in long products as a procedure to achieve a very fine austenite grain size prior to transformation. In the second case, some peculiarities associated with thin-slab direct rolling will be analysed.

4.6.1 Thermomechanical processes based on dynamic recrystallisation

In rod rolling, the high strain rates in the finishing mill lead to very short interpass times (below 50 ms) compared with reversal or strip rolling. Under these conditions, the interpass time is not long enough for the microstructure to be statically recrystallised (Jonas, 1998). If this occurs, strain will accumulate from pass to pass, making it possible to reach the critical value ε_c for the onset of DRX, followed by metadynamic recrystallisation processes. As a result, there is the possibility of grain refinement by dynamic/metadynamic recrystallisation (as indicated in Eqs 4.4 and 4.12, D_{dyn} and D_{mdrx} depend only on the parameter Z). This approach has been extensively analysed by Jonas (1994).

Figure 4.28 corresponds to a simulation of laboratory torsion tests, comparing the flow stress for a single pass with that obtained after multiple passes at 950°C for a plain C–Mn steel. In the multipass test, the short time applied did not allow SRX to progress, and the stress–strain curve is very close to that for the monotonic case, confirming the occurrence of dynamic recrystallisation. The austenite grain sizes measured under the conditions were very similar: 7 μm for the single pass and 8.3 μm for the multipass condition. This indicates that the effect of accumulation of strain from pass to pass can be evaluated with equations determined from monotonic tests.

In Fig. 4.29, the dependence of the metadynamic grain size on the Z parameter is shown, considering the equation proposed by Hodgson and Gibbs (1992) (see Table 4.4). In the figure, the interval of the Zener–Hollomon parameter that may occur during the final rod-rolling passes is also indicated. Even assuming that some grain coarsening can take place before transformation begins (this will depend also on the possible inhibition of grain growth by second-phase particles or the solute drag effect), a very fine austenite microstructure results.

In summary, the combination of dynamic/metadynamic recrystallisation occurring in the finishing mill with low temperatures provides a very useful tool for achieving a fine austenite microstructure before transformation during cooling.

4.6.2 Thin-slab direct rolling

Thin-slab direct rolling (TSDR) is the most widely applied near-net-shape industrial process in the field of flat products. Compared with traditional routes,

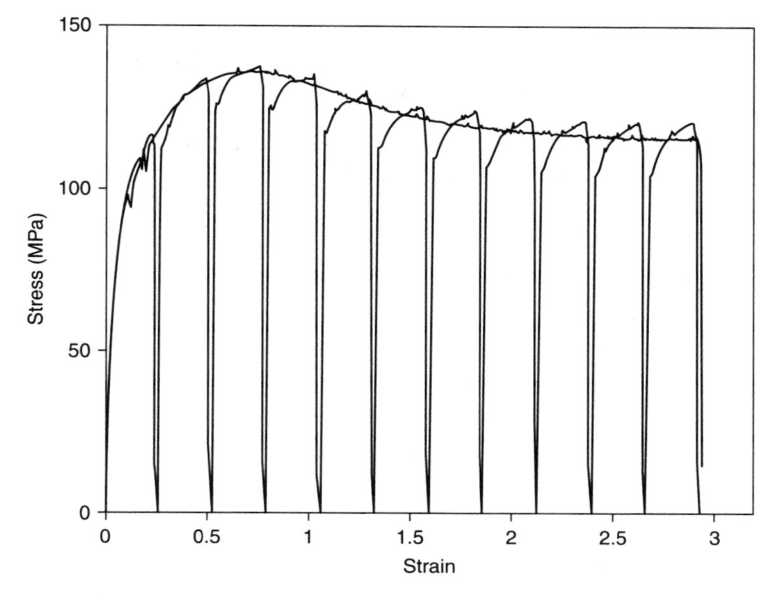

4.28 Simulation of torsion tests on a plain C–Mn steel, comparing stress–strain curves for the cases of monotonic and multipass schedules; $T = 950°C$, strain rate = $1\,s^{-1}$, $t_{ip} = 0.5\,s$.

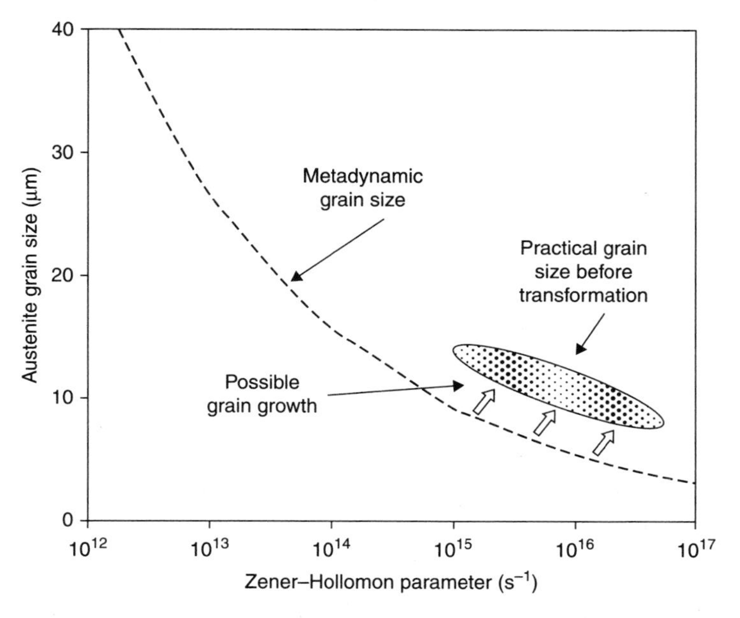

4.29 Evolution of metadynamic recrystallised grain size as a function of the Zener–Hollomon parameter. The range of Z values corresponding to rolling passes in industrial finishing is indicated.

TSDR introduces some peculiarities that may affect the final microstructural homogeneity, particularly in the case of Nb microalloyed steels. In this context, it is necessary to take account of the following facts:

- At entry to the rolling mill, the microstructure (as-cast) is characterised by a wide range of austenite grain sizes (a fraction of the grains is larger than 1 mm).
- The total reduction is significantly smaller than that applied in conventional rolling (assuming a slab thickness of 200–250 mm).

Taking these peculiarities into account, there are two main objectives that must be achieved during the rolling schedule: first, the coarse as-cast microstructure must be completely eliminated and refined and second, once this has been achieved, the remaining rolling passes should provide adequate austenite conditioning prior to transformation. If the above requirements are not completely fulfilled, the final rolled product can show microstructural heterogeneity. In the following, some aspects related to the first step, that is, the refinement of the as-cast austenite microstructure, will be considered, with the focus mainly on the case of Nb microalloyed steels.

As mentioned previously, Nb exerts an important solute drag effect on recrystallisation kinetics, together with the effects of strain-induced precipitation. Figure 4.30 shows the influence of solute drag on the recrystallisation time $t_{0.5}$ as a function of temperature for Nb and V microadditions. This influence has been quantified by considering the evolution of the multiplying factor affecting $t_{0.5}$, this factor being equal to unity for plain C–Mn steels. The figure confirms that Nb has a very important delaying effect on recrystallisation kinetics in the temperature range corresponding to the initial rolling passes of TSDR, whereas the behaviour with V microadditions is very close to that of plain C–Mn grades.

The combined effect of a coarse austenite grain size and Nb solute drag can significantly delay the recrystallisation kinetics and, in consequence, the required refinement of the as-cast microstructure. The interaction between the composition, the process parameters and the final gauge can be quite complex. In this context, the modelling of austenite grain size evolution during rolling becomes a very useful tool. Based on this, a semi-empirical model (as described in Section 4.5.1) adapted to TSDR conditions has been developed to evaluate the evolution of austenite microstructure (details of the model are given in Uranga *et al.*, 2004). Instead of the mean grain size, this model works with grain size distributions, providing the austenite grain size distribution after each interpass time, as well as the amount of accumulated strain corresponding to each stand.

With the help of the model, it is possible not only to predict process conditions that are prone to local heterogeneities but also to associate these heterogeneities with their corresponding mechanisms of origin, that is, delayed refinement due to solute drag or a complete stop of recrystallisation due to strain-induced precipitation. Examples of various recrystallisation/solute drag/precipitation

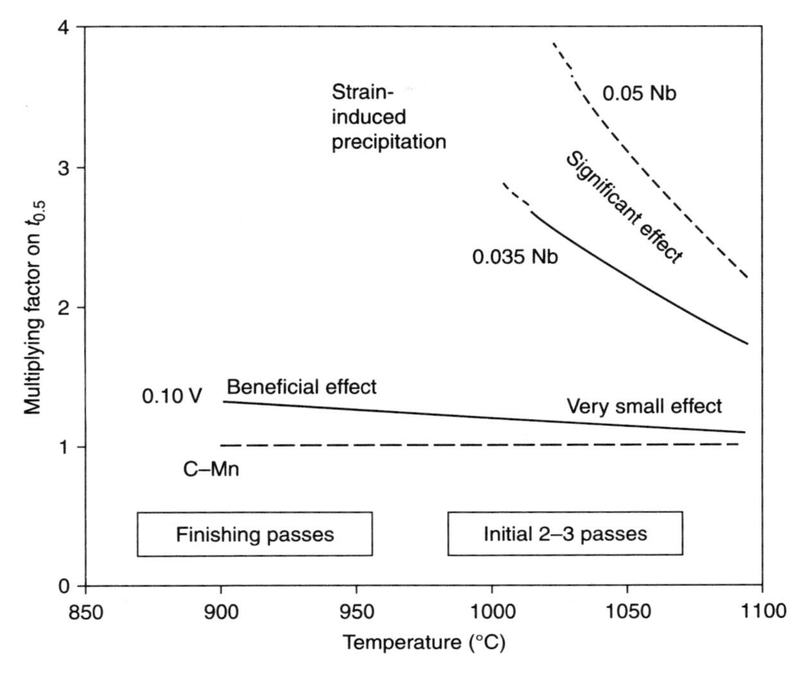

4.30 Influence of solute drag by V and Nb on static recrystallisation kinetics as a function of temperature.

interactions are shown in Figs 4.31 and 4.32. In these figures, the predicted austenite grain size distributions at the end of the second and third interpass times are shown for a 0.05% Nb steel during a deformation sequence of 0.5/0.5/0.45 (10 mm final thickness), for initial rolling temperatures T_i of 1060 and 1090°C.

The model predicts that when T_i = 1060°C, there is an important fraction of unrecrystallised grains at the end of the second interstand. Some of these grains are as-cast grains that have remained because of premature Nb(C,N) precipitation, and another fraction results from grains that recrystallised during the first interstand but then remained owing to a delay in the recrystallisation due to solute drag. After the third pass, the unrecrystallised fraction is mainly due to the Nb(C,N) precipitation pinning mechanism, and the as-cast fraction remains unchanged as a consequence of these precipitates. This fraction will result finally in a heterogeneous austenite grain size distribution before transformation.

If T_i = 1090°C is selected, the situation becomes different (Fig. 4.32). At the exit of the second interstand, the amount of as-cast grains is significantly smaller, and these grains disappear completely after the following rolling pass, as there is no Nb(C,N) precipitation. Similarly, the unrecrystallised fraction present at the end of the third interstand is due to Nb solute drag.

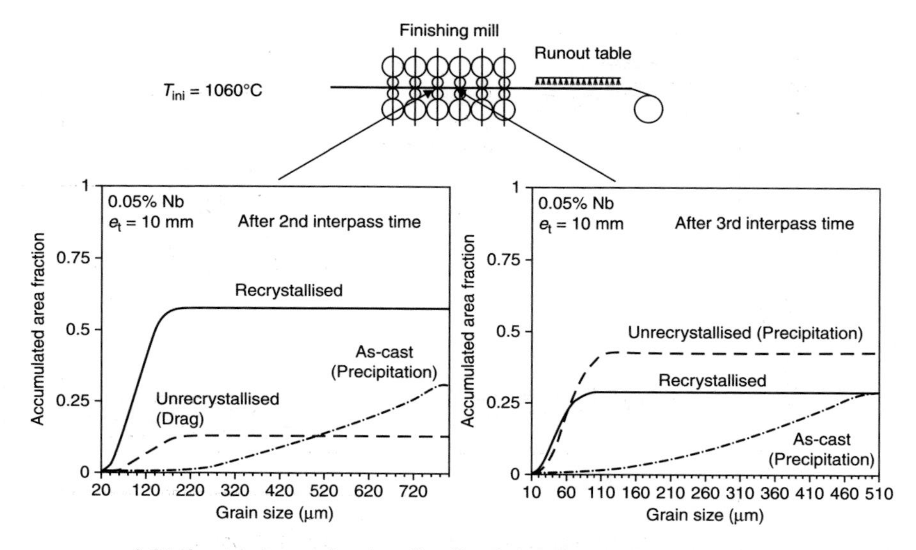

4.31 Austenite grain size distributions, subdivided as a function of their state (recrystallised and unrecrystallised), after the second and third interstands for T_i = 1060°C (Uranga *et al.*, 2009).

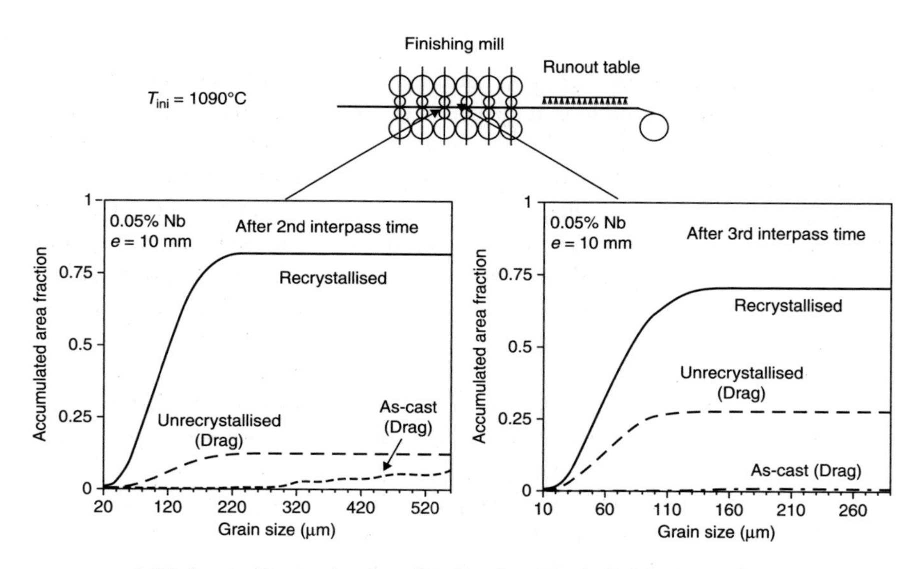

4.32 Austenite grain size distributions, subdivided as a function of their state (recrystallised and unrecrystallised), after the second and third interstands for T_i = 1090°C (Uranga et al., 2009).

In summary, with the help of austenite microstructure evolution models, it is possible to predict the potential presence of coarse austenite grains when the correct composition/processing parameters are not selected. In addition, the mechanisms associated with these heterogeneities (solute drag and premature Nb(C,N) precipitation) can be determined.

4.7 Sources of further information and advice

Baker T N (1994), *Titanium Technology in Microalloyed Steels*. London: Institute of Materials.

Gladman T (1997), *The Physical Metallurgy of Microalloyed Steels*. London: Institute of Materials.

Humphreys FJ and Hatherly M (1996), *Recrystallization and Related Annealing Phenomena*. Oxford: Pergamon Press.

4.8 References

Abad R, López B and Gutierrez I (1998), 'Combined effect of Nb and Ti on the recrystallisation behaviour of some HSLA steels', *Mater Sci Forum*, 284–286, 167–174.

Abad R, Fernández AI, López B and Rodriguez-Ibabe JM (2001a), 'Interaction between recrystallization and precipitation during multipass rolling in a low carbon Nb microalloyed steel', *ISIJ Int*, 41, 1375–1384.

Abad R, López B and Rodriguez-Ibabe JM (2001b), 'The effect of prior austenite microstructure and deformation variables on the recrystallization behaviour of some HSLA steels', *J Mater Process Technol*, 117/3, CD-ROM Section C4.

Akben MG and Jonas JJ (1983), 'Influence of multiple microalloy additions on the flow stress and recrystallization behavior of HSLA steels', in Korchynsky M (ed.), *HSLA Steels: Technology and Applications*. Metals Park, OH: ASM, 149–161.

Akben MG, Bacroix B and Jonas JJ (1983), 'Effect of vanadium and molybdenum addition on high temperature recovery, recrystallization and precipitation behavior of niobium-based microalloyed steels', *Acta Metall*, 31, 161–174.

Anelli E (1992), 'Application of mathematical modelling to hot rolling and controlled cooling of wire rods and bars', *ISIJ Int*, 32, 440–449.

Arribas M, López B and Rodriguez-Ibabe JM (2005), 'Influence of Ti on static recrystallization in near net shape steels', *Mater Sci Forum*, 500–501, 131–138.

Arribas M, López B and Rodriguez-Ibabe JM (2008), 'Additional grain refinement in recrystallization controlled rolling of Ti microalloyed steels processed by near-net-shape casting technology', *Mater Sci Eng A*, 485, 383–394.

Bai DQ, Yue S, Sun WP and Jonas JJ (1993), 'Effect of deformation parameters on the no-recrystallization temperature in Nb-bearing steels', *Metall Trans*, 24A, 2151–2159.

Beynon JH and Sellars CM (1992), 'Modelling microstructure and its effects during multipass hot rolling', *ISIJ Int*, 32, 359–367.

Beynon JH, Ponter ARS and Sellars CM (1988), 'Metallographic verification of computer modelling of hot rolling', in Chenot JL and Oñate E (eds), *Proc. Int. Conf. Modelling of Metal Forming Processes*. Dordrecht: Kluwer Academic, 321–328.

Bianchi CH and Karjalainen LP (2005), 'Modelling of dynamic and metadynamic recrystallisation during bar rolling of a medium carbon spring steel', *J Mater Process Technol*, 160, 267–277.

Cahn JW (1962), 'The impurity-drag effect in grain boundary motion', *Acta Metall*, 10, 789–798.

Cartmill MR, Barnett MR, Zahiri SH and Hodgson PD (2005), 'An analysis of the transition between strain dependent and independent softening in austenite', *ISIJ Int*, 45, 1903–1908.

Cho S-H, Kang K-B and Jonas JJ (2001), 'The dynamic, static and metadynamic recrystallization of a Nb microalloyed steel', *ISIJ Int*, 41, 63–69.

Choquet P, LeBon A and Perdrix C (1985), 'Mathematical model for prediction of austenite and ferrite microstructures in hot rolling processes', in McQueen HJ *et al.* (eds), *Proc. ICSMA 7*. Oxford: Pergamon Press, 2, 1025–1030.

Cuddy LJ (1982), 'The effect of microalloying concentration on the recrystallization of austenite during hot deformation', in DeArdo AJ, Ratz GA and Wray PJ (eds), *Thermomechanical Processing of Microalloyed Austenite*. Warrendale, PA: TMS-AIME, 129–140.

Deschamps A and Brechet Y (1999), 'Influence of predeformation and ageing of Al–Zn–Mg alloy – II. Modeling of precipitation kinetics and yield stress', *Acta Mater*, 47, 293–305.

Doherty RD, Hughes DA, Humphreys FJ, Jonas JJ, Juul Jensen D, *et al.* (1997), 'Current issues in recrystallization: a review', *Mater Sci Eng A*, 238, 219–274.

Dutta B and Sellars CM (1987), 'Effect of composition and process variables on Nb(C,N) precipitation in Nb microalloyed austenite', *Mater Sci Technol*, 3, 197–206.

Dutta B, Valdes E and Sellars CM (1992), 'Mechanisms and kinetics of strain induced precipitation of Nb(C,N) in austenite', *Acta Metall Mater*, 40, 653–662.

Dutta B, Palmiere EJ and Sellars CM (2001), 'Modelling the kinetics of strain induced precipitation in Nb microalloyed steels', *Acta Mater*, 49, 785–794.

Fernández AI, López B and Rodriguez-Ibabe JM (2002), 'Modeling of partially recrystallized microstructures for a coarse initial Nb microalloyed austenite', *Scr Mater*, 46, 823–828.

Fernández AI, Uranga P, López B and Rodriguez-Ibabe JM (2000), 'Static recrystallization behaviour of a wide range of austenite grain sizes in microalloyed steels', *ISIJ Int*, 40, 893–901.

Fernández AI, Uranga P, López B and Rodriguez-Ibabe JM (2003), 'Dynamic recrystallization behavior covering a wide austenite grain size range in Nb and Nb–Ti microalloyed steels', *Mater Sci Eng A*, 361, 367–376.

Furu T, Marthinsen K and Nes E (1990), 'Modelling recrystallization', *Mater Sci Technol*, 6, 1093–1102.

Herman JC, Donnay B and Leroy V (1992), 'Precipitation kinetics of microalloying additions during hot-rolling of HSLA steels', *ISIJ Int*, 32, 779–785.

Hodgson PD and Gibbs RK (1992), 'A mathematical model to predict the mechanical properties of hot rolled C–Mn and microalloyed steels', *ISIJ Int*, 32, 1329–1338.

Hodgson PD, Jonas JJ and Yue S (1993), 'Strain accumulation and post-dynamic recrystallization in C–Mn steels', *Mater Sci Forum*, 113–115, 473–478.

Hodgson PD, Hazeldon LO, Matthews DL and Gloss RE (1995), 'The development and application of mathematical models to design thermomechanical processes for long products', in Korchynsky M, De Ardo AJ, Repas P and Tither G (eds), *Microalloying 95*. Warrendale, PA: ISS, 341–353.

Humphreys FJ and Hatherly M (1996), *Recrystallization and Related Annealing Phenomena*. Oxford: Pergamon Press.

Iparraguirre C and López B (2008), 'Modelización basada en parámetros físicos de las cinéticas de ablandamiento y precipitación en aceros microaleados con niobio', in Hurtado I (ed.), *X Congreso Nacional de Materiales*, San Sebastian, 1, 435–438.

Iparraguirre C, Fernández AI and López B (2007), 'Effect of initial austenite microstructure on the softening–precipitation interaction in a low Nb microalloyed steel', *Mater Sci Forum*, 550, 429–434.

Jonas JJ (1984), 'Mechanical testing for the study of austenite recrystallization and carbonitride precipitation', in Dunne DP and Chandra T (eds), *High Strength Low Alloy Steels*. Wollongong: University of Wollongong, 80–91.

Jonas JJ (1994), 'Dynamic recrystallization – scientific curiosity or industrial tool?', *Mater Sci Eng A*, 184, 155–164.

Jonas JJ (1998), 'Effect of interpass time on the hot rolling behaviour of microalloyed steels', *Mater Sci Forum*, 284–286, 3–14.

Jonas JJ, Sellars CM and Tegart WJ McG (1969), 'Strength and structure under hot-working conditions', *Met Rev*, 14, 1–24.

Jones AR and Ralph B (1975), 'The influence of recrystallization on carbide particle distributions in a fully stabilized austenitic steel', *Acta Metall*, 23, 355–363.

Kuziak R, Glowacki M and Pietrzyk M (1996), 'Modelling of plastic flow, heat transfer and microstructural evolution during rolling of eutectoid steel rods', *J Mater Process Technol*, 60, 589–596.

Laasraoui A and Jonas JJ (1991), 'Prediction of steel flow stresses at high temperatures and strain rates', *Metall Trans*, 22A, 1545–1558.

Lin YC and Chen M-S (2009), 'Study of microstructural evolution during static recrystallization in a low alloy steel', *J Mater Sci*, 44, 835–842.

Liu X, Solberg JK, Gjengedal R and Kluken AO (1995), 'Modelling of the interaction between recrystallization and precipitation during multipass rolling of Nb microalloyed steels', *Mater Sci Technol*, 11, 469–473.

Maruyama N, Uemori R and Sugiyama M (1998), 'The role of Nb in the retardation of the early stage of austenite recovery in hot-deformed steels', *Mater Sci Eng*, 250A, 2–7.

Medina SF and Quispe A (2001), 'Improved model for static recrystallization kinetics of hot deformed austenite in low alloy and Nb/V microalloyed steels', *ISIJ Int*, 41, 774–781.

Minami K, Siciliano F, Maccagno TM and Jonas JJ (1996), 'Mathematical modeling of mean flow stress during the hot strip rolling of Nb steels', *ISIJ Int*, 36, 1507–1515.

Nazabal JL, Urcola JJ and Fuentes M (1987), 'The transition from multiple to single peak recrystallization during the hot working of austenite', *Mater Sci Eng*, 86, 93–103.

Nes E (1995), 'Recovery revisited', *Acta Metall Mater*, 43, 2189–2207.

Orsetti Rossi PL and Sellars CM (1997), 'Quantitative metallography of recrystallization', *Acta Mater*, 45, 137–148.

Perdrix C (1987), *Characteristic of Plastic Deformation of Metals during Hot Working*, ECSC Report, No. 7210 EA/31, IRSID.

Pereda B, Fernandez A, López B and Rodriguez-Ibabe JM (2007a), 'Effect of Mo on dynamic recrystallization of Nb–Mo microalloyed steels', *ISIJ Int*, 47, 859–867.

Pereda B, López B and Rodriguez-Ibabe JM (2007b), 'Increasing the non-recrystallization temperature of Nb microalloyed steels by Mo addition', in DeArdo AJ and Garcia IC (eds), *Proc. Int. Conf. on Microalloyed Steels*. Warrendale, PA: AIST, 151–159.

Pereda B, Rodriguez-Ibabe JM and López B (2008), 'Improved model of kinetics of strain induced precipitation and microstructure evolution of Nb microalloyed steels during multipass rolling', *ISIJ Int*, 48, 1457–1466.

Pereloma EV, Crawford BR and Hodgson PD (2001), 'Strain-induced precipitation behaviour in hot rolled strip steel', *Mater Sci Eng A*, 299, 27–37.

Roberts W, Sandberg A, Siwecki T and Werlefors T (1984), 'Prediction of microstructure development during recrystallization hot rolling of Ti–V steels', in Korchynsky M (ed.), *HSLA Steels: Technology and Applications*. Metals Park, OH: ASM, 67–84.

Roucoules C, Yue S and Jonas JJ (1995), 'Effect of alloying elements on metadynamic recrystallization in HSLA steels', *Metall Mater Trans*, 26A, 181–190.

Roucoules C, Hodgson PD, Yue S and Jonas JJ (1994), 'Softening and microstructural change following the dynamic recrystallization of austenite', *Metall Trans*, 25A, 389–400.

Sah JP, Richardson GJ and Sellars CM (1974), 'Grain-size effects during dynamic recrystallization of nickel', *Met Sci*, 8, 325–331.

Sakai T and Jonas JJ (1984), 'Dynamic recrystallization: mechanical and microstructural considerations', *Acta Metall*, 32, 189–209.

Scheil E (1935), 'Anlaufzeit der Austenitumwandlung', *Arch Eisenhüttenwes*, 8, 565–567.

Sellars CM (1980), 'The physical metallurgy of hot working', in Sellars CM and Davies GJ (eds), *Hot Working and Forming Processes*. London: Metals Society, 3–15.

Sellars, CM (1990), 'Modelling microstructural development during hot rolling', *Mater Sci Technol*, 6, 1072–1081.

Sellars CM and Whiteman JA (1979), 'Recrystallization and grain growth in hot rolling', *Met Sci*, 13, 187–194.

Siciliano F and Jonas JJ (2000), 'Mathematical modeling of the hot strip rolling of microalloyed Nb, multiply-alloyed Cr–Mo, and plain C–Mn steels', *Metall Mater Trans*, 31A, 511–530.

Sun WP and Hawbolt EB (1997), 'Comparison between static and metadynamic recrystallization an application to the hot rolling of steels', *ISIJ Int*, 37, 1000–1009.

Turnbull D (1951), 'Theory of grain boundary migration rates', *Trans AIME*, 191, 661–665.

Uranga P, López B and Rodriguez-Ibabe JM (2009), 'Microalloying and austenite evolution during hot working in near net shape processed steels', *Mater Sci Technol*, 25, 1147–1153.

Uranga P, Fernández AI, López B and Rodriguez-Ibabe JM (2003), 'Transition between static and metadynamic recrystallization kinetics in coarse Nb microalloyed austenite', *Mater Sci Eng A*, 345, 319–327.

Uranga P, Fernández AI, López B and Rodriguez-Ibabe JM (2004), 'Modeling of austenite grain size distribution in Nb microalloyed steels processed by thin slab casting and direct rolling (TSRD) route', *ISIJ Int*, 44, 1416–1425.

Verdier M, Brechet Y and Guyot P (1999), 'Recovery of AlMg alloys: flow stress and strain-hardening properties', *Acta Mater*, 47, 127–134.

Wang BX, Liu XL and Wang GD (2005), 'Dynamic recrystallization behavior and microstructural evolution in a Mn–Cr gear steel', *Mater Sci Eng A*, 393, 102–108.

Xu YB, Yu YM, Xiao BL, Liu ZY and Wan GD (2010), 'Modelling of microstructure evolution during hot rolling of a high-Nb HSLA steel', *J Mater Sci*, 45, 2580–2590.

Yang ZG and Enomoto M (1999), 'A discrete lattice plane analysis of coherent F.C.C./B1 interfacial energy', *Acta Mater*, 47, 4515–4524.

Yoshie A, Fujita T, Fujioka M, Okamoto K and Morikawa H (1996), 'Formulation of flow stress of Nb added steels by considering work-hardening and dynamic recovery', *ISIJ Int*, 36, 467–473.

Zurob HS, Brechet Y and Purdy G (2001), 'A model for the competition of precipitation and recrystallization in deformed austenite', *Acta Mater*, 49, 4183–4190.

Zurob HS, Hutchinson CR, Brechet Y and Purdy G (2002), 'Modeling recrystallization of microalloyed austenite: effect of coupling recovery, precipitation and recrystallization', *Acta Mater*, 50, 3075–3092.

5

Severe plastic deformation for grain refinement and enhancement of properties

A. ROSOCHOWSKI, University of Strathclyde, UK and
L. OLEJNIK, Warsaw University of Technology, Poland

Abstract: This chapter is a review of the technique of severe plastic deformation, which converts coarse-grained metals into ultrafine-grained metals possessing a range of enhanced properties useful in modern applications. The chapter first describes the effects of grain size and grain boundaries on the properties of metallic materials and discusses various approaches to grain refinement. It then concentrates on the mechanism and conditions of severe plastic deformation; this is followed by a discussion of various severe plastic deformation processes. Ultrafine-grained metals are attractive materials for a range of modern applications, which are also presented in this chapter.

Key words: ultrafine-grained metals, severe plastic deformation.

5.1 Introduction

5.1.1 Grain size/grain boundary effects on properties of metallic materials

One of the important structural features of metals is their grain size, which has an influence on the way certain micro-mechanisms operate. This is the case with plastic flow, where dislocation generation and mobility depend on the grain size. Based on the concept of dislocation pile-ups at grain boundaries, the Hall–Petch relationship was proposed (Hall, 1951; Petch, 1953), which predicts an increase in the yield stress σ_y with decreasing average grain size d (σ_0 is the friction stress opposing dislocation motion and k is the Hall–Petch slope):

$$\sigma_y = \sigma_0 + \frac{k}{\sqrt{d}} \qquad [5.1]$$

For grain sizes smaller than 25 nm, the relationship in Eq. 5.1 may lose its validity, leading to a yield plateau or even an inverse Hall–Petch effect, but this still remains to be proved (Meyers *et al.*, 2006). Nevertheless, for larger grains, it predicts properly an increase in the yield stress by a few times, as shown in Fig. 5.1 for low-carbon steel at room temperature (Shin and Park, 2004). This phenomenon is sometimes referred to as grain size hardening, by analogy to strain hardening or precipitation hardening. The related properties of ultimate tensile strength and hardness are also improved by reducing the grain size of the metal.

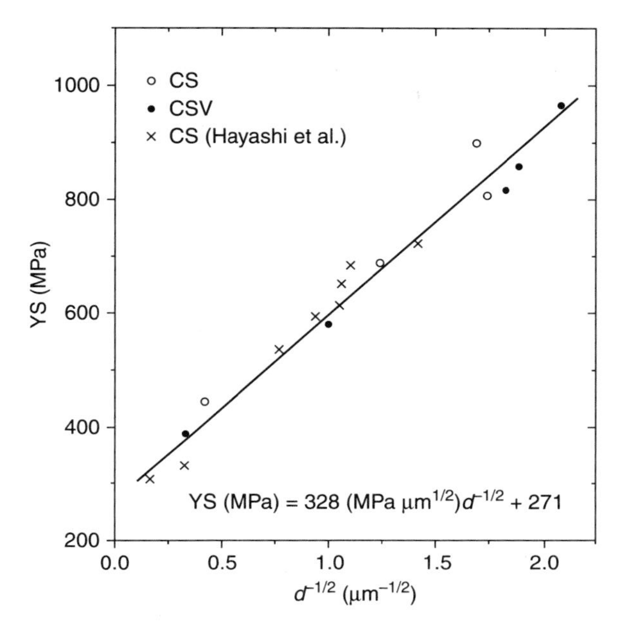

5.1 Yield stress (YS) vs ferrite grain size (*d*) in CS steel (Fe–0.15C–0.25Si–1.1Mn) and CSV steel (Fe–0.15C–0.25Si–1.1Mn–0.06V–0.008N) (Shin and Park, 2004).

There are also other aspects of metal behaviour which depend on the volume fraction and characteristics of grain boundaries. At high temperatures, the disordered nature of grain boundaries means that vacancies and atoms can diffuse more rapidly in the boundaries, leading to more rapid diffusion creep. With small grains, new mechanisms of plastic deformation such as grain boundary sliding may be activated. Since boundaries are regions of high energy, they make excellent sites for the nucleation of precipitates and other second phases, which in turn can slow down grain growth by pinning grain boundaries.

5.1.2 Traditional methods of refining grain structure

The attempts to refine the grain structure of metals often start at the casting stage. The common method used is based on adding nucleating agents, known as inoculants. For example, to refine the grain structure of cast aluminium, a so-called aluminium master alloy, with small additions of titanium and boron (and sometimes carbon), can be added to the melt (Cooper and Barber, 2003). Another method is related to the casting parameters; it is known that reduced superheat and an increased cooling rate both reduce the grain size (Ning *et al.*, 2007). In recent years, the application of intensive shear above or below the liquid's temperature has been explored as a method of helping traditional direct chill casting to achieve a substantially finer grain structure (Haghayeghi *et al.*, 2010). It should be emphasised that 'fine grain' in cast metals usually means a grain size in the range of tens of micrometres, so such grains are still relatively coarse.

Cast semi-products are usually hot formed by multistep processes of rolling, forging or extrusion to consolidate casting imperfections and refine the grain structure. This refinement is due to complex processes of recovery and recrystallisation, which can be followed by grain growth. These processes depend mainly on the temperature and strain rate. For lower process temperatures, characteristic of warm forming, strain hardening and recovery are the main competing phenomena. In cold forming, the dominating mechanism is strain hardening. Since cold and warm forming do not lead to the creation of a new grain structure, the existing grain size can be substantially reduced but only in one or two directions, as in the case of rolling of thin sheets or drawing of fine wires.

By combining classical forming processes with an appropriate thermal treatment, it is sometimes possible to reduce the grain size to less than 10 μm. For example, Verma (2004) continuously cast and immediately afterwards hot-rolled aluminium–magnesium alloys, which subsequently were annealed to promote precipitation of small dispersed particles. After cold rolling and another annealing step, he was able to obtain 5–10 μm recrystallised grains originating from intermetallic particles (an effect called particle-stimulated nucleation). Such small grains made the material highly superplastic: that is, when deformed at an elevated temperature and a low strain rate, it underwent a very large plastic extension without fracture.

5.1.3 New approaches to refining grain structure

It has been always interesting to consider the possible consequences of reducing the grain size below 1 μm. Bulk metals with such a structure are referred to as ultrafine-grained (UFG) metals. Finer structures, with grains smaller than 0.1 μm, are known as nanocrystalline (NC) metals. The micro- and nano-grain refinement technologies claiming bulk capability include electrodeposition and the crystallisation of initially amorphous materials. However, the two main competing technologies are compaction and sintering of NC/UFG powders and severe plastic deformation (SPD) of bulk coarse-grained (CG) metals. The former is an example of a bottom-up approach, while the latter represents a top-down approach.

A popular method of making NC powders is inert gas condensation (Siegel, 1993). In this method, vaporised atoms of precursor material agglomerate into small clusters by condensation on a liquid-nitrogen-cooled bar, in a low-pressure inert gas atmosphere. Since handling the powder involves ultrahigh vacuum, there is very little contamination. The smallest particle sizes produced by this method are 5–25 nm. However, the expensive equipment and the small quantities of powder produced limit the use of the method to laboratory applications. Another process capable of producing NC/UFG particles is mechanical alloying (Siegel,

1993). This involves high-energy ball milling with or without solid-state chemical reactions. In order to obtain fine, uniform particles, the milling time must be long. This time can be reduced for hard metals and low milling temperatures. The process is capable of producing a commercial quantity of powder in one batch. The powder is prone to contamination by the milling equipment and the environment. Generally, very fine powders are difficult to handle. In addition to increased reactivity, which may lead to oxidation of the powder, they are perceived as a health hazard. In order to obtain a bulk component, the powder has to be compacted and sintered. This may result in grain coarsening. Furthermore, there is always the issue of residual porosity, which adversely affects the mechanical properties of components. On the other hand, the powder metallurgy method offers the advantage of being able to combine different particles to create nanocomposites.

An alternative approach to obtaining bulk UFG metals is based on processing solid billets of CG metals (Azushima *et al.*, 2008). The method involves plastic deformation of the material to a very large strain, usually above 3 (true, logarithmic strain), without substantially changing the dimensions of the material. The term used in this context is 'severe plastic deformation'. Some pre- and/or post-processing heat treatment usually accompanies SPD, which justifies using the more general term of 'severe thermo-mechanical treatment'.

5.2 Principles of severe thermo-mechanical treatment

5.2.1 Mechanisms of refining grain structure by severe plastic deformation (SPD)

The exact mechanism responsible for the refinement of grain structure in severely deformed metals is still being debated. However, the general view is that this refinement results from a non-uniform distribution of dislocations, which tend to arrange themselves in low-energy configurations to form cell structures within the original coarse grains (Zhu and Langdon, 2004). Another opinion emphasises the role of shear bands (thin bands of localised shear deformation), where different bands cross each other to create a skewed chessboard pattern as shown in Fig. 5.2 (Richert *et al.*, 1998). The distance between shear bands is very small, which leads to the creation of dislocation cells having sub-micrometre dimensions. The dislocation cells, which do not initially have crystallographic orientation very different from their neighbours, are referred to as subgrains. Only when, owing to further deformation and rotation, their grain boundary misorientation angle exceeds 15° are they treated as distinct grains. These new small grains are often relatively free of dislocations.

Subgrains are created in the early stages of SPD, even for true strains smaller than three. Since the starting grain size of the annealed material can be in the range of 50–500 μm, the subgrain dimensions represent a reduction in size by two

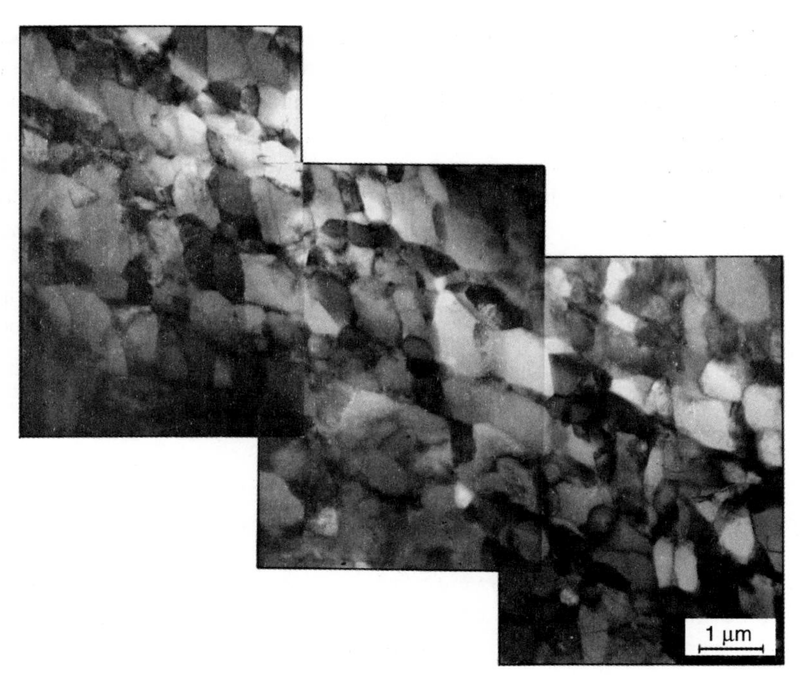

5.2 Skewed chessboard-like microstructure in Al99.992 after a true strain of 12.5 (Richert *et al.*, 1998).

or three orders of magnitude. Further deformation gradually converts subgrains into real grains, but the grain/subgrain size distribution does not change much as a result of this. Figure 5.3 illustrates these points by indicating the evolution of grain boundary misorientation angles (shown for a few grains in TEM images) and by comparing the grain/subgrain size distributions for AA1070 subjected to true strains of 4.6 and 9.2 (Rosochowski *et al.*, 2008).

5.2.2 Material-dependent results

The grain refinement mechanism described above is universal for all metals, but the results of the refinement can differ between metals. The average grain size of UFG metals varies from 0.1 to 1 μm. The grain/subgrain size in commercially pure aluminium is likely to be approximately 0.5 μm (Fig. 5.3), in pure nickel 0.35 μm (Zhilyaev *et al.*, 2002) and in pure copper, as shown in Fig. 5.4, 0.2 μm (Geißdörfer *et al.*, 2008). Pure (Armco) iron responds well to SPD, which enables the ferrite grain size to be reduced to approximately 0.2 μm (Suś-Ryszkowska *et al.*, 2004). Alloys are more responsive to SPD than pure metals, i.e. it results in finer grains. For example, Hasegawa *et al.* (1999) quoted average grain sizes of

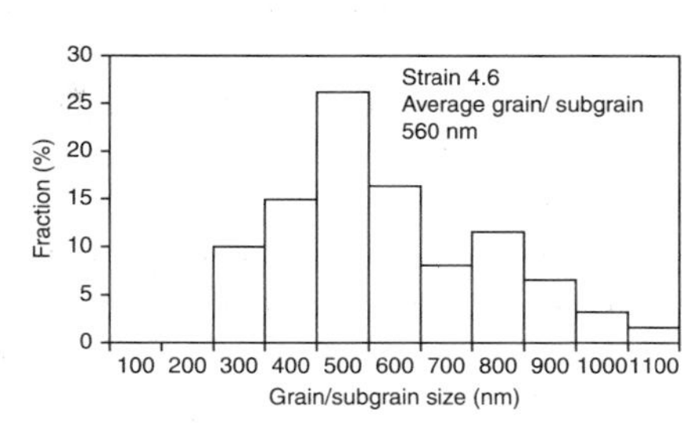

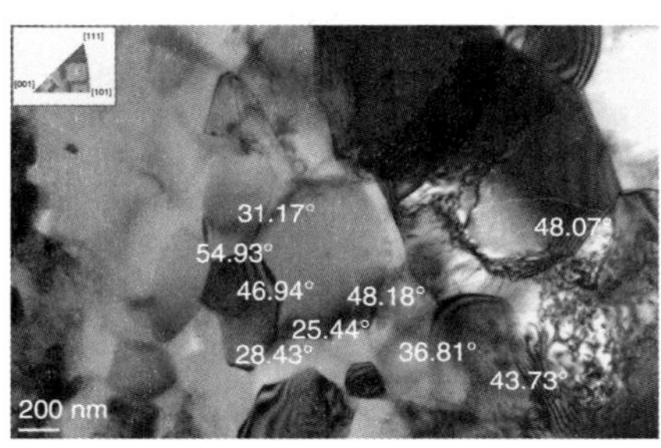

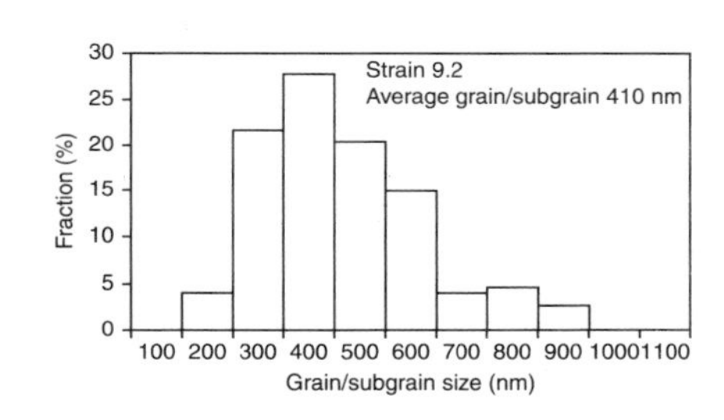

5.3 TEM images and grain/subgrain size distributions in AA1070 deformed to true strains of 4.6 (top) and 9.2 (bottom) (Rosochowski *et al.*, 2008).

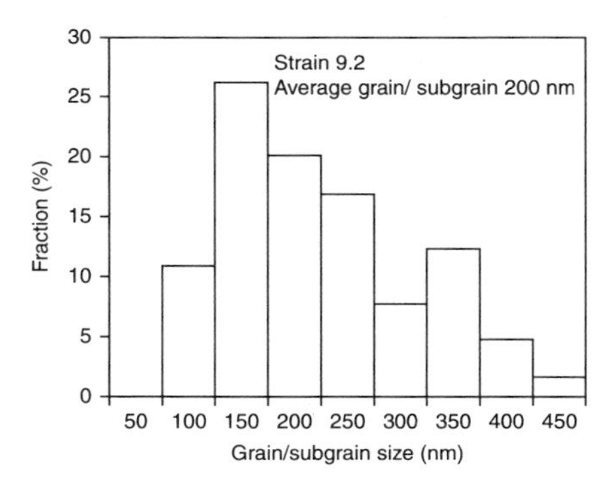

5.4 TEM image (Geißdörfer *et al.*, 2008) and grain/subgrain size distribution in Cu deformed to a true strain of 9.2.

1.3, 0.45 and 0.3 μm for pure Al, Al–1%Mg and Al–3%Mg, respectively. The rate of grain refinement can be increased by the presence of coarse second-phase particles (Apps *et al.*, 2004).

Iron alloys can have several different phase structures, which opens up new possibilities. One interesting point is that SPD of 0.15C–0.25Si–1.1Mn ferritic/pearlitic steel causes carbon from cementite to partially dissolve, which increases the carbon content in the ferrite above the equilibrium level. This, in turn, leads to nanoparticles of cementite precipitating at grain boundaries during annealing; such precipitates have a stabilising effect on grain growth. This effect is more visible for higher strains (Shin and Park, 2004). Park *et al.* (2004) subjected the same material to intercritical annealing (at 740°C for 10 min) after SPD. As a result of subsequent water quenching, a dual-phase (DP) ferrite/martensite structure was created with uniformly distributed, rounded martensite islands. The ferrite/martensite grains were slightly larger than 1 μm but the resulting UFG–DP steel had better mechanical properties, especially high strain hardening capability, than it had just after SPD or after intercritical annealing only. Similar results were obtained by Calgagnotto *et al.* (2008) for 0.17C–0.28Si–1.63Mn steel, by applying austenitisation, one-pass hot rolling, controlled cooling, warm rolling to 1.6 true strain, annealing, and intercritical annealing with quenching. Since the rolling operations changed the shape of billets and the cumulative strain was only 1.6, the authors did not refer to their process as SPD but instead referred to it as an advanced thermo-mechanical treatment.

Metals with a hexagonal closed-packed (HCP) structure, such as magnesium, have only a limited number of active dislocation slip systems at room temperature, which makes them brittle. In such materials, there is a good chance for twins to develop but, nevertheless, magnesium alloys cannot be subjected to SPD at room temperature. The solution is to increase the SPD temperature to increase the number of active dislocation slip systems. On the other hand, using a higher SPD temperature, for example 280°C for the AZ31 magnesium alloy, puts a limit of approximately 4 μm on the minimum average grain size that can be achieved (Kim and Sa, 2006). Grain size reduction below this limit requires lowering the SPD temperature, which is only possible by altering other process parameters. The lack of symmetry of the dislocation slip systems leads to a highly anisotropic plastic response of the material, so its post-SPD properties depend on the orientation of the material (Agnew *et al.*, 2004). Despite having the same HCP structure, titanium responds better to SPD than magnesium, allowing grain size reduction down to approximately 0.3 μm (Stolyarov *et al.*, 2001).

5.2.3 Effects of SPD conditions

SPD can be carried out at room temperature or, in the case of harder and more brittle materials, at elevated temperatures. In practice, this means that only pure metals such as aluminium, copper and nickel can be subjected to SPD at room temperature. In all other cases, the billet material and/or the SPD tools have to be heated to an elevated temperature. Process temperatures in excess of $0.3T_m$ should be avoided, since they would suppress grain refinement by recovery and recrystallisation (Valiev, 2004). The typical minimum SPD temperatures are 200°C for AA5083, 250°C for AZ31, 400°C for CP Ti and 600°C for Ti–6Al–4V.

Lower temperatures might be possible, provided other SPD parameters help reduce the tendency to strain localisation and fracture.

SPD processes are often repeated several times in order to accumulate a high strain in the material. In this context, it is helpful to use large strain increments in each SPD operation to reduce the number of operations required. It was found that using a smaller strain increment results in a smaller strengthening effect in AA1070 even for the same value of the accumulated strain (Rosochowski *et al.*, 2006a). On the other hand, smaller strain increments have been found to help CP Ti (grade 1) to be formed at room temperature without fracture (Zhao *et al.*, 2010). In some SPD processes, the strain path can be changed between consecutive processing steps. It has been found that certain alterations of the strain path have a positive effect on the ability to form small equiaxed grains with high misorientation angles (Furukawa *et al.*, 1998).

The strain rate has a negligible influence on grain refinement. However, it may affect the ability of materials such as magnesium and titanium to deform without fracture. Therefore, as shown for example by Semiatin *et al.* (1999) for CP Ti, it is recommended that the strain rate is reduced for such materials.

Applying a very high pressure (e.g. 6 GPa) during SPD enhances grain refinement; for example, the grain size obtained was reduced from the typical value of 0.35 μm to 0.17 μm for pure nickel (Zhilyaev *et al.*, 2002). Lower values of pressure (e.g. 200 MPa) do not have this effect but can suppress material fracture during SPD, as shown by Lapovok (2006) for aluminium alloys.

5.2.4 Modelling of SPD

The most popular way to model SPD processes is based on the solid mechanics/theory of plasticity approach, realised through finite element (FE) simulation. The earliest attempt to model an SPD process by the FE method was made by Prangnell *et al.* (1997); an up-to-date review of FE simulation results for various SPD techniques was given by Rosochowski and Olejnik (2007b). Despite its routine character, this approach is very useful in preliminary studies and optimisation of new SPD processes, especially for lower strains. Larger deformation leads to the creation of a UFG structure and accompanying substantial changes in the mechanical response of the material. One of these changes is initial material softening, observed during uniaxial compression tests of UFG AA1070 (Rosochowski *et al.*, 2006b). Taking this softening into account in an FE simulation of a post-SPD forward extrusion process led to the prediction of a localised shear developing in this process (Rosochowska *et al.*, 2010).

The modelling of UFG metals based on their microstructure is much more challenging because the underlying mechanisms are still not fully understood. What distinguishes UFG metals from their CG counterparts is an increasing presence of the grain boundary ‘phase’ and more difficult conditions for the classical dislocation mechanisms to operate under in the grain interiors. This led to a model of a UFG material based on a phase mixture and a critical grain size, in which the material is considered to be a composite of a crystalline phase and an intercrystalline grain boundary phase. Using a simple rule of mixtures, the grain size and the strain rate dependence of the stress in a fine-grained material were taken into account in a model proposed by Kim *et al.* (2001). This model was used to interpret various experimental phenomena, such as the breakdown of the Hall–Petch relation with decreasing grain size and the rate dependence of the deformation behaviour in copper. Khan *et al.* (2006) modified the so-called KHL (Khan, Huang and Liang) viscoplastic model, which already was an extension of the original Hall–Petch model, to include a bimodal stress response to the changing grain size in aluminium and iron. More complex materials such as HSLA steels, which feature fine precipitates, required further modification of this model by introducing an effect of the grain boundary misorientation angle (Muszka *et al.*, 2007).

Molecular-dynamics computer simulation (atomistic simulation) has revealed fundamental atomic-scale processes that have provided important information for mesoscopic models to describe plasticity in nanocrystalline materials. For example, it can analyse the nucleation and propagation of dislocations from grain boundaries (Van Swygenhoven *et al.*, 2006). This approach has great theoretical potential, but also inherent limitations such as atomic-scale spatial and temporal resolution.

5.3 Severe plastic deformation (SPD) processes

The amount of plastic strain produced by classical metal forming operations is often limited because of failure of the material or tool. In some sequential processes, such as rolling and drawing, large reductions in the material thickness can be achieved. However, the billet shapes produced by these processes are unsuitable for further conversion into bulk products. Thus new metal forming processes capable of generating very large, or severe, plastic deformation without damaging the material and without a major change in the billet geometry, have been developed. These processes are based on simple shear and/or repetitive reverse straining, and preserve the initial shape and dimensions of the billet. Despite the lack of shaping capability, SPD processes can be treated as just another branch of the discipline of metal forming. The number of SPD processes developed in laboratories worldwide exceeds 50. They can be divided into two groups, batch processes and continuous-billet processes.

5.3.1 Batch SPD processes

Batch processes use compact billets, which means that all billet dimensions are of the same order. The major batch SPD processes which have become prominent over the last 30 years are high-pressure torsion (HPT), equal-channel angular pressing (ECAP) and cyclic extrusion compression (CEC). In recent years, new processes, such as multiaxial forging (MF), repetitive corrugation and straightening (RCS), and twist extrusion (TE), have been developed. New batch SPD processes are still being developed in order to find the easiest and most efficient method of grain refinement. The SPD processes mentioned above are presented in Table 5.1.

High-pressure torsion was first investigated by Bridgman (1935). Bridgman used a pair of extremely thin discs (usually thinner than 0.1 mm), freely compressed and simultaneously twisted between outer cylinders and a central anvil. Bridgman's experiments did not shed much light on the microstructural changes taking place in severely deformed metals. Erbel (1979) seems to be the first who did this for a copper ring processed by HPT. In recent years, owing to the numerous investigations of Valiev and co-workers (e.g. Valiev *et al.*, 1991), as well as other researchers (e.g. Pippan *et al.*, 2008), the most popular scheme for such deformation has been the one shown schematically in Table 5.1. In this scheme, a very thin disc, with an aspect ratio of about 40:1, is compressed in a closed die by exerting a very high pressure of about 6 GPa. The torque is provided by the punch via the contact friction at the punch/disc interface. The high pressure applied is beneficial to grain refinement but can also cause tool problems. Another potential issue is the strain gradient in the radial direction. Fortunately, this becomes less important for very high strains, for which structural changes saturate.

The most popular SPD process is equal-channel angular pressing, also known as equal-channel angular extrusion; it was invented in 1972 in Russia and was described in the West by Segal *et al.* (1981). ECAP is based on simple shear taking place at the crossing plane of equal channels with a square or round cross section (Table 5.1). In order to achieve the required strain, the billet is subjected to consecutive passes of ECAP in the same die. The billet can be rotated about its axis between each pass. The three basic options for this rotation are called A, C and B_C (Fig. 5.5). Route A (without any rotation) seems to be especially attractive for obtaining a strong filamentary, textured structure (Segal, 1995). Langdon *et al.* (2000) established that the best route for obtaining a homogeneous microstructure of equiaxed grains separated by high-angle boundaries is B_C (90° rotations in the same sense). Route C (180° rotation) is likely to partially reverse the structural changes but it is more practical from the point of view of material handling and material utilisation. Other possible routes have been also proposed by Barber *et al.* (2004). In order to decrease the number of passes required, multiple turns of the ECAP channel can be used. For example, a two-turn S-shaped channel realises route C 'in-die' (Rosochowski and Olejnik, 2002). An in-die route B_C can also be achieved, as shown by Rosochowski *et al.* (2007b).

Richert invented cyclic extrusion compression in 1979 in Poland; the idea was first published in the West by Korbel *et al.* (1981). CEC involves a cyclic flow of metal between alternating extrusion and compression chambers as shown in Table 5.1. A recent implementation of the method is based on a stationary die and two horizontal punches with a controlled back pressure (Richert, 2010). The

Table 5.1 Batch SPD processes

Process name and origin	Schematic representation	Equivalent plastic strain
High-pressure torsion (HPT) (Valiev *et al.*, 1991)	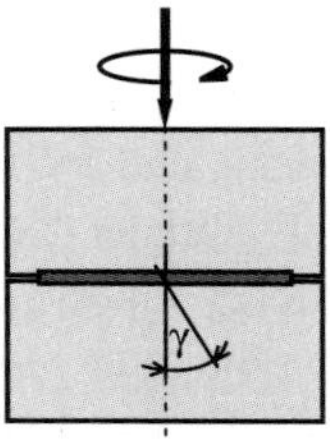	$\varepsilon = \frac{tg\gamma}{\sqrt{3}}$
Equal-channel angular pressing (ECAP) (Segal *et al.*, 1981)	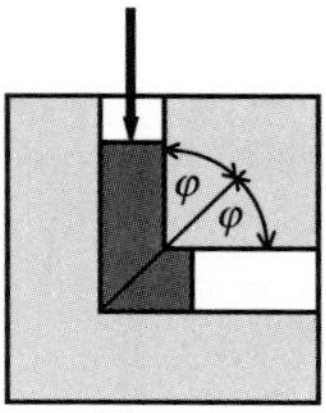	$\varepsilon = n\frac{2}{\sqrt{3}}\cot\varphi$
Cyclic extrusion compression (CEC) (Korbel *et al.*, 1981)	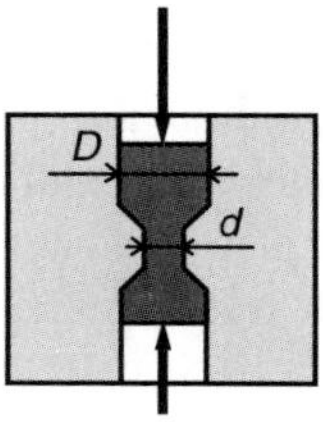	$\varepsilon = n4\ln\left(\frac{D}{d}\right)$
Multiaxial forging (MF) (Ghosh, 2000)	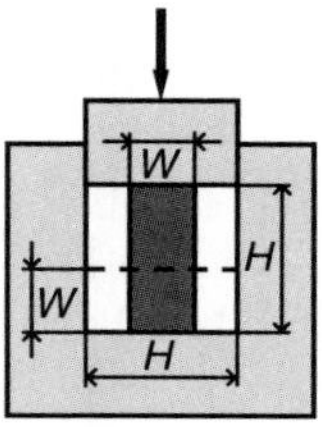	$\varepsilon = n\frac{2}{\sqrt{3}}\ln\left(\frac{H}{W}\right)$
Repetitive corrugation and straightening (RCS) (Huang *et al.*, 2001)	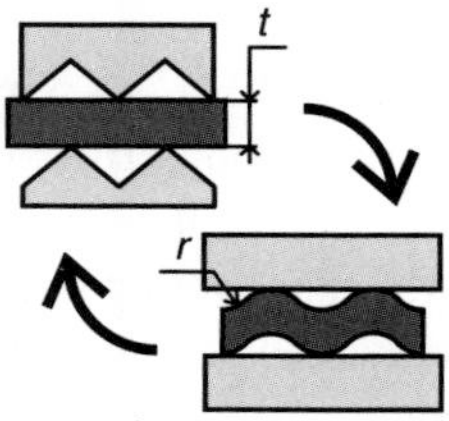	$\varepsilon = n\frac{4}{\sqrt{3}}\ln\left(\frac{r+t}{r+0.5t}\right)$
Twist extrusion (TE) (Beygelzimer *et al.*, 2002)	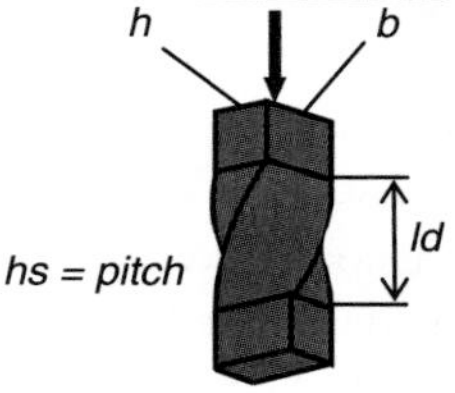	$\varepsilon = 3.46\left(\frac{hs}{0.5\sqrt{b^2+h^2}}\right)^{-0.47}\left(\frac{ld}{hs}\right)^{0.55}\left(\frac{h}{b}\right)^{-0.56}$

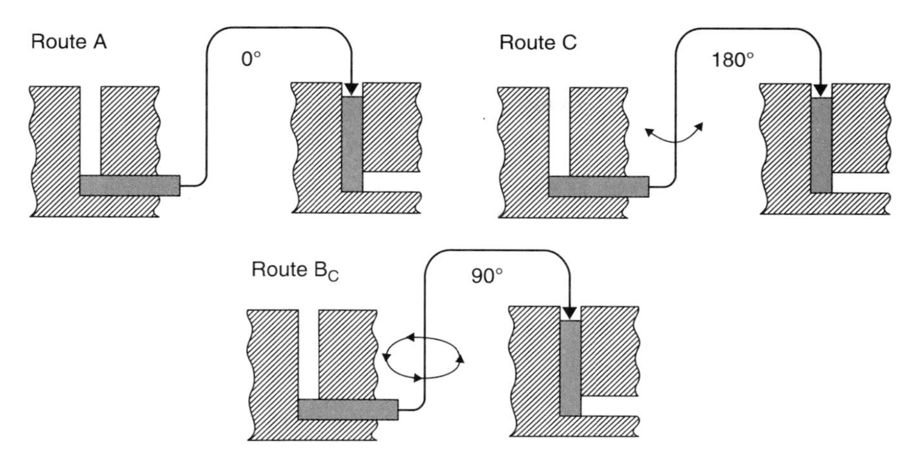

5.5 The three basic options for billet rotation between consecutive passes through an ECAP die.

mechanics of the process was analysed by Rosochowski (2005), who concluded that the stress path in CEC comprised primary yielding of the material due to extrusion, unloading into the elastic domain and secondary yielding by compression on the opposite side of the yield surface. Using a finite element simulation, issues such as the strain distribution, maximum forming force and tool pressure were also addressed by Rosochowski (2005). Except at the ends of the billet, the equivalent plastic strain was very uniform. The forming force and tool pressure depended very much on friction. Even a small value of the Coulomb friction coefficient of 0.06 caused a very high tool contact pressure of 4 GPa when a low carbon steel was processed at room temperature. This is why CEC is better suited for processing soft materials (e.g. aluminium alloys). Even for those materials, very good lubrication will be the key to success. On the other hand, the naturally high hydrostatic pressure in CEC may be an asset when processing brittle or particulate materials.

Multiaxial forging is known under various names and configurations, from free forging to closed-die forging. In 1988, Ghosh and Huang (2000) invented three-axis forging. As shown in Table 5.1, it was a process in which a billet was compressed along two or three axes as a result of one or two rotations between successive compressions. The strain distribution was non-uniform (the so-called forging cross distribution); however, after several billet rotations and strain accumulation, this did not affect the homogeneity of the microstructure. The process has been used by Vorhauer and Pippan (2004), Zherebtsov *et al.* (2004) and Kuziak *et al.* (2005).

Huang *et al.* (2001) invented an SPD method based on repetitive corrugation and straightening. This comprises bending of a straight billet with corrugated tools and then restoring the straight shape of the billet with flat tools (Table 5.1). Repetition of the process (usually with a rotation about the pressing axis between cycles) is required to obtain a large strain and the desired structural changes. It is worth noting that Ghosh and Huang (2000) had earlier suggested using grooved platens for the repetitive forging of plates. Since the process involved indentation rather than bending, it was called multipass coin-forging. Another implementation of the same idea was constrained groove pressing, proposed by Shin *et al.* (2002).

As shown in Table 5.1, twist extrusion is based on forcing a brick-shaped billet through a twisted die (Beygelzimer *et al.*, 2002). Since the billet leaving the die maintains its original shape, it can be extruded repeatedly in order to accumulate high strain. The process enables back pressure to be added. The strain distribution is not uniform in the transverse direction, with the minimum strain close to the billet centre. Finite element simulation has been used to calculate the average strain (Varyukhin *et al.*, 2011), as shown in Table 5.1.

5.3.2 SPD processes for continuous billets

Recently, interest is focusing on SPD processes which might be capable of refining the grain structure in long billets, for example long strips of sheet metal. Saito *et al.* (1998) suggested that a piece of sheet metal could be cut into two pieces, cleaned, stacked and hot rolled to bond two pieces together while reducing their thickness by 50%. This sequence can be repeated several times until a desired strain is achieved. The process, which is shown schematically in Table 5.2, has been called accumulative roll bonding (ARB). ARB is not a true continuous process, because it is limited by the manageable sheet length. Nevertheless, ARB was the first SPD process which dealt with the material in sheet form while retaining the sheet thickness. Research on ARB has also been undertaken by Heason and Prangnell (2004), Reis *et al.* (2004) and Höppel *et al.* (2004). The final success of the process depends critically on the quality of the bond. Good quality can be achieved by increasing the rolling temperature. However, the possibility of grain growth limits the applicability of this approach.

The batch process of repetitive corrugation and straightening also has a continuous version (Table 5.2). Continuous RCS (CRCS) uses a rotating corrugating tool and a straightening roll, which enables processing of long billets (Huang *et al.*, 2004). Since localised bending remains the main feature of the process, it will be difficult to achieve a uniform strain distribution (Głuchowski *et al.*, 2011). Owing to the limited range of rotations available to continuous billets, even multiple passes of CRCS will probably not help.

There are many developments of the ECAP process which are intended to make it suitable for processing continuous billets. Most of them, for example continuous SPD (CSPD), shown in Table 5.2 (Srinivasan *et al.*, 2006), rely on friction between the billet and a set of small and/or large rolls as the means to feed the billet into the ECAP die. A similar approach has been used in the continuous shear

Table 5.2 SPD processes for continuous billets

Process name and origin	Schematic representation	Equivalent plastic strain
Accumulative roll bonding (ARB) (Saito *et al.*, 1998)	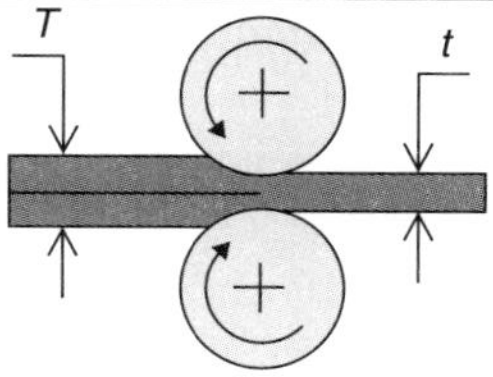	$\varepsilon = n\frac{2}{\sqrt{3}}\ln\left(\frac{T}{t}\right)$
Continuous repetitive corrugation and straightening (CRCS) (Huang *et al.*, 2004)	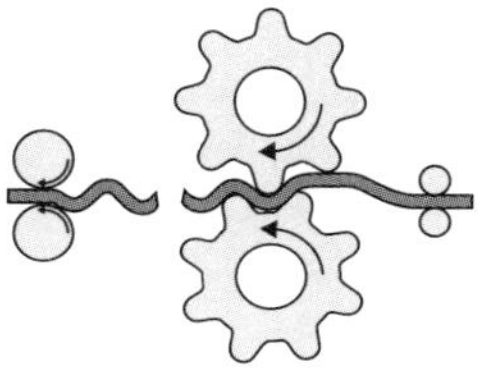	$\varepsilon = n\frac{4}{\sqrt{3}}\ln\left(\frac{r+t}{r+0.5t}\right)$
Continuous SPD (CSPD) (Srinivasan *et al.*, 2006)	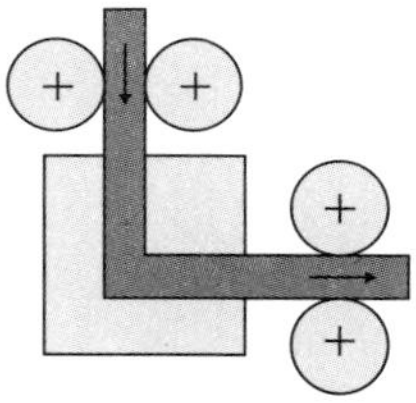	$\varepsilon = n\frac{2}{\sqrt{3}}\cot\varphi$
ECAP–Conform (ECAP–C) (Raab *et al.*, 2004)		$\varepsilon = n\frac{2}{\sqrt{3}}\cot\varphi$
Incremental ECAP (I-ECAP) (Rosochowski and Olejnik, 2007)	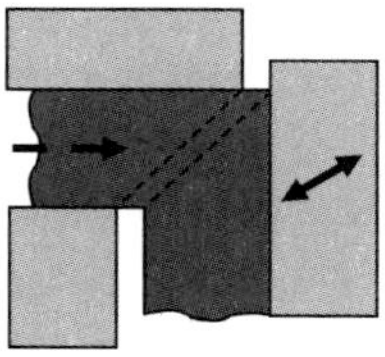	$\varepsilon = n\frac{2}{\sqrt{3}}\cot\varphi$

deformation (conshearing) process invented by Utsunomiya *et al.* (2001), the continuous confined strip shearing (C2S2) process introduced by Lee *et al.* (2001) and the ECAP–Conform process proposed by Raab *et al.* (2004). This last process (shown in Table 5.2) is a combination of ECAP and Conform, which is a continuous forming (extrusion) process invented in the 1970s in England. Despite the relatively low forming force required for ECAP, feeding the billet using only frictional force proves to be a problem because the frictional force is often not enough to do this job. Thus, to reduce the forming force, either the channel angle in the ECAP die has to be increased to more than 90°, which reduces the strain produced in one pass, or the forming temperature has to be increased, which increases the grain size.

One of the new SPD processes suitable for continuous billets is incremental ECAP (I-ECAP), developed by Rosochowski and Olejnik (2007a). It is explained in Fig. 5.6, where classical ECAP and I-ECAP are shown schematically. In classical ECAP, a square or cylindrical billet is pushed from one section of a constant-profile channel to the next section, orientated at an angle $\geq 90°$ to the previous one, in one go of a punch. The length of the leading channel limits the length of the billet processed to about six times the lateral dimension; it must not be too long, to avoid an excessive force caused by friction and the associated problems of punch design. The short length of the billet causes poor utilisation of the material due to end effects, which can be understood as relatively unstrained (and thus with CG structure) and geometrically distorted billet ends. In the I-ECAP process, the die is split into a fixed part, which leads the billet into the working zone, and a working part, which moves in a reciprocating manner at an appropriate angle to the billet. Feeding of the billet takes place when there is no contact between the billet and the working part of the die. When the billet stops moving and becomes fixed or clamped, the working part of the die deforms it plastically in a localised zone of simple shear. In the next cycle, the feeding and deformation stages are repeated and, provided the feeding stroke is not excessive, the consecutive shear zones overlap, giving a uniform strain distribution along the billet. The separation of the feeding and deformation stages reduces or eliminates friction during feeding; this enables processing of infinite billets. I-ECAP can be used for nanostructuring of long bars (Rosochowski *et al.*, 2008), plates (Olejnik *et al.*, 2008) and sheets (Rosochowski *et al.*, 2010), which makes it an attractive option for industrial implementation. The developments in I-ECAP have recently been summarised by Rosochowski and Olejnik (2010).

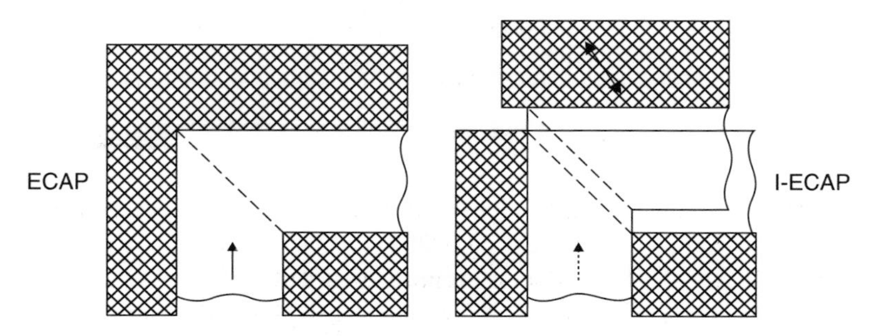

5.6 Schematic illustrations of classical ECAP and I-ECAP (Rosochowski and Olejnik, 2007a).

5.4 Properties of ultrafine-grained (UFG) metals produced by SPD

The main reason for refining the grain structure of metallic materials is the possibility of increasing their strength, which was explained in the introductory part of this chapter. However, there is a limit to the strength, which cannot be exceeded because of the observed saturation of grain size and strength regardless of the level of strain achieved in SPD. Interestingly, post-SPD metal forming processes such as rolling or extrusion, used to shape UFG metals into products, are capable of increasing the strength of UFG metals further, above the level achieved by SPD alone. For example, Olejnik *et al.* (2009) have shown that hydrostatic extrusion (HE) of UFG AA1070 produced by ECAP can increase the strength of the material by 25% over the level achieved by ECAP only (Fig. 5.7).

Figure 5.7 also illustrates the fact that the increased strength results in proportionally reduced ductility. To avoid problems related to low ductility when UFG metals are shaped into components, machining or metal forming processes that suppress material fracture can be used. Recently, the strength-versus-ductility issue had to be re-examined after the publication of experimental results where both increased strength and very high ductility were obtained. Valiev *et al.* (2003) subjected UFG CP Ti produced by HPT to annealing at 300°C for 10 min, which resulted in an increase in both the strength and the ductility. Interestingly, increasing the annealing temperature to 350°C did not have the same effect. Wang

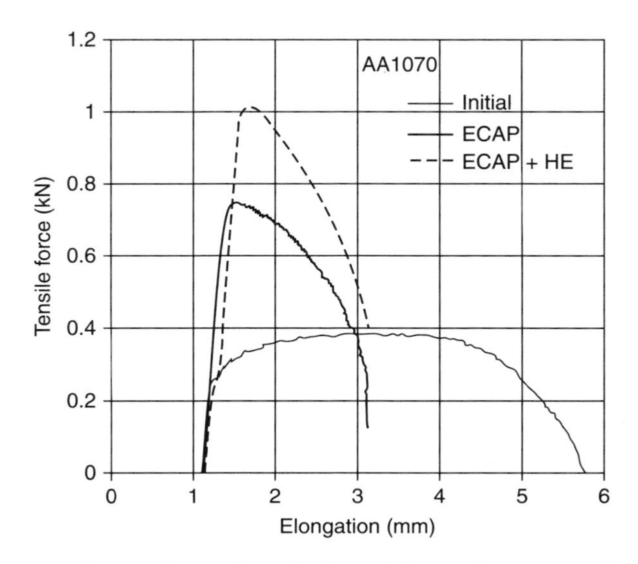

5.7 Comparison of tensile force/elongation curves obtained for CG AA1070, UFG AA1070 produced by ECAP, and the same material subjected to ECAP+HE (Olejnik *et al.*, 2009).

et al. (2002) used a thermo-mechanical process involving two stages: repetitive rolling of Cu to 93% reduction, carried out at about −125°C, and then annealing at 200°C for 3 min. The cryogenic forming suppressed dynamic recovery (allowing a high density of dislocations), while the annealing caused recrystallisation and abnormal grain growth, resulting in a bimodal structure containing 25% and 75% of large grains and ultrafine grains, respectively. This structure resulted in a unique combination of high yield stress and large elongation. There are also other examples where the ductility has been increased above the normal level. One of these is another case of metals deformed at cryogenic temperatures, where the UFG structure leads to the improvement of both the strength and the ductility (Wang *et al.*, 2004). Another example is provided by magnesium alloys, which are very brittle at room temperature but become substantially more ductile when converted into UFG materials (Mukai *et al.*, 2001).

Some mechanical properties depend on a combination of strength and ductility. If both parameters are high, the area under the stress/strain curve is large, which is indicative of high toughness. UFG metals exhibit increased fracture toughness at low temperatures. This was documented by measuring the impact energy in a Charpy test carried out on a low carbon steel subjected to various thermo-mechanical treatments (Hanamura *et al.*, 2004); the material with a UFG ferrite/cementite structure exhibited the lowest ductile–brittle transition temperature.

Fatigue strength is another property related to strength and ductility. The effect of grain refinement depends on whether one is concerned with high-cycle or low-cycle fatigue. When subjected to high-cycle fatigue, UFG metals produced by SPD exhibit properties 10% to 50% better than their CG counterparts (Semenova *et al.*, 2009). In contrast, the low-cycle fatigue properties of UFG metals are worse (Höppel *et al.*, 2006). The explanation for these effects is based on the underlying cause of material failure, which is stress-related in high-cycle fatigue and plastic-strain-related in low-cycle fatigue. Thus the majority of UFG metals with high strength and low ductility perform better when subjected to stress cycling and not so well when strain-cycled.

The thermal behaviour of UFG metals is complex. The common wisdom is that, owing to the high energy stored in the new, so-called non-equilibrium grain boundaries, UFG metals should be thermally unstable, which means that grain growth should occur at lower temperatures. However, there is experimental evidence, for example for CP Ti, that this is not necessarily the case (Valiev, 2004). It appears that perhaps another mechanism, for example greatly increased grain boundary diffusion, is responsible for the quick recovery of non-equilibrium grain boundaries. High thermal stability can also be achieved by adding precipitating elements, for example zirconium and/or scandium added to aluminium; such precipitates reduce the mobility of grain boundaries and suppress grain growth (Lee *et al.*, 2002). The enhanced diffusivity mentioned earlier is responsible for the increased creep of UFG metals (Sklenicka *et al.*, 2005), which is not good news for high-temperature applications. However, it is much appreciated in superplastic forming, where it promotes grain boundary sliding, which is thought to be responsible for the extraordinary tensile elongation achievable in superplastic forming (Figueiredo and Langdon, 2008).

UFG metals have a number of physical properties which are different from those of their coarse-grained counterparts. In addition to a higher diffusivity, UFG metals have a higher electrical resistance, an increased specific heat capacity, a lower thermal conductivity and improved magnetic properties.

5.5 Applications of UFG metals

Despite the range of improved properties of UFG metals, their uptake by industry has been rather slow. The main reason is the lack of industrial awareness of UFG metals. This is related to the scarcity of appropriately-sized UFG samples for industrial trials; those produced by laboratories are usually too small because they are intended for metallurgical observations or basic mechanical testing. Also, it is still not clear which of the numerous laboratory-based SPD methods will emerge as the most appropriate one for industrial implementation. As a result, potential producers of UFG metals are hesitating to commit themselves to any particular method. There is also a lack of knowledge regarding post-SPD processing or shaping of UFG metals. In addition to the technical aspects, there are concerns about the commercial viability of UFG metals, which depends on the demand from potential markets and production costs.

5.5.1 Structural elements

The obvious applications seem to be structural elements, which, owing to the increased strength of UFG metals, can be made lighter. However, in reality, the increased cost of producing UFG metals makes them less attractive for mass market components. This is why the automotive industry is still waiting for its first application of UFG metals. A different situation exists in the sports equipment industry, where performance is sometimes more important than cost. Therefore, initial attempts have been made to use UFG metals in high-performance bicycles (Reshetnikova *et al.*, 2008) and in the faceplate of a golf club head (Safiullin *et al.*, 2009). The next few examples refer to advantages of UFG metals other than high strength; high strength may still be desirable, but it is not the main reason for using UFG metals in these applications.

5.5.2 Sputtering targets

The first commercial application of bulk UFG metals was in sputtering targets for physical vapour deposition (Fig. 5.8). Honeywell Electronic Materials, a division of Honeywell International Inc., offers UFG Al and Cu sputtering targets up to 300 mm in diameter, which are produced from plates by ECAP (Ferrase *et al.*,

5.8 UFG sputtering target (Ferrase *et al.*, 2003).

2003). These targets are used for metallisation of silicon wafers in the production of semiconductor devices. The main advantage of UFG sputtering targets, compared with their CG counterparts, is a more uniform coating, which results from reduced arcing during physical vapour deposition. Also, UFG sputtering targets are strong enough to be designed as monolithic parts, which increases their life.

5.5.3 Medical implants

Medical devices are another example where higher performance can justify an increased cost. The product family which can benefit from the use of UFG metals is medical implants. These include dental, hip and knee implants, as well as various screws, plates and meshes used in orthopaedic applications. Implants are usually made of cobalt–chrome alloys, stainless steel and titanium alloys. Titanium alloys are used for implants because of their strength, low modulus of elasticity (which provides a better match with that of bones), corrosion resistance and good biocompatibility. CP titanium has even better compatibility than titanium alloys, but it is not used for load-bearing implants because it is not strong enough. However, when nanostructured by SPD and subjected to further thermo-mechanical treatment, CP titanium can be strengthened to achieve a yield stress of 1100 MPa, which is comparable with the yield strength of titanium alloys (Salimgareeva *et al.*, 2005). Traditional titanium implants do not perform well with respect to wear resistance and fatigue life. Therefore, improvements in these properties, which have been reported for UFG titanium, will also be appreciated.

Valiev (2006) reported that UFG CP titanium implants were being already tried (Fig. 5.9). However, the first commercial application was reported by a Czech company, Timplant, which makes UFG titanium dental screws (Valiev *et al.*, 2008). Good mechanical properties of the material enable reduction of the screw diameter from 3.5 mm to 2.4 mm, making it more appropriate for front teeth and dental implants for children. Valiev *et al.* (2008) also reported dramatically

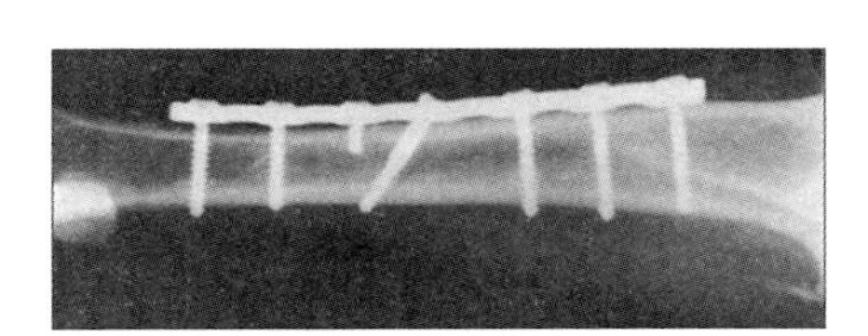

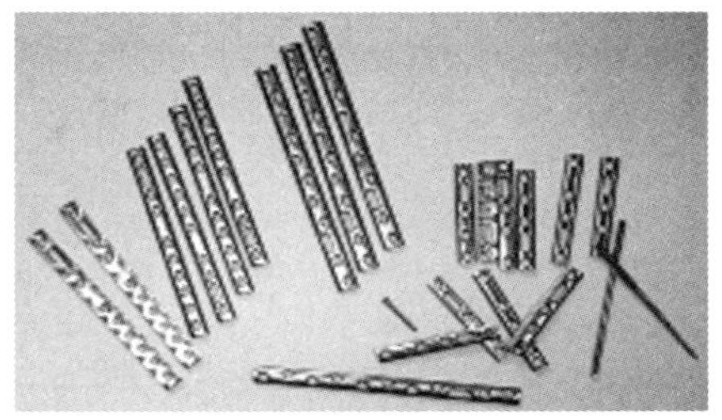

5.9 Plate implants made of nanostructured titanium (Valiev, 2006).

increased fibroblast colonisation of UFG CP Ti compared with standard CG CP Ti. This should result in shorter post-surgery healing times and more reliable integration of implants with the patient's body.

5.5.4 Superplastic forming/diffusion bonding of aerospace components

Boeing has started using nearly UFG Ti–6Al–4V sheet (grain size approximately 1 μm) to improve their superplastic forming (SPF) operations. SPF is a popular technology in the aerospace industry, because it enables complex aircraft panels to be manufactured in just one operation. Unfortunately, owing to the average grain size of 6–10 μm for a standard SPF grade of Ti–6Al–4V, the temperature required for SPF is 900°C and the forming time is 1–2 h. This leads to a very short life of the dies, press platens and heating elements, adversely affects the surface quality of the SPF components, and produces a brittle alpha case, which has to be etched. The process is characterised by low productivity and high energy consumption, and therefore is expensive. By using a UFG version of Ti–6Al–4V, Boeing managed to reduce the SPF temperature to 775°C and increase the speed of the process, which resulted in a dramatic improvement of the process conditions (Comley, 2004). Similarly, a diffusion bonding (DB) process, which often accompanies SPF of panels, could be improved by reducing its temperature from 925°C to 775°C. Interestingly, this temperature is also appropriate for diffusion bonding of UFG Ti–6Al–4V with CG Ti–6Al–4V (Hefti, 2007). Another example of a component made by DB and SPF is a hollow, wide-chord fan blade used in aerospace engines (Fig. 5.10). Valiakhmetov *et al.* (2010) have said that by employing a UFG Ti–6Al–4V core membrane, it should be possible to reduce the DB/SPF temperature for this component by 200°C.

5.5.5 Micro-manufacturing

There is increased interest in metal components in the sub-millimetre range, for which the volume of the material becomes so small that the components consist

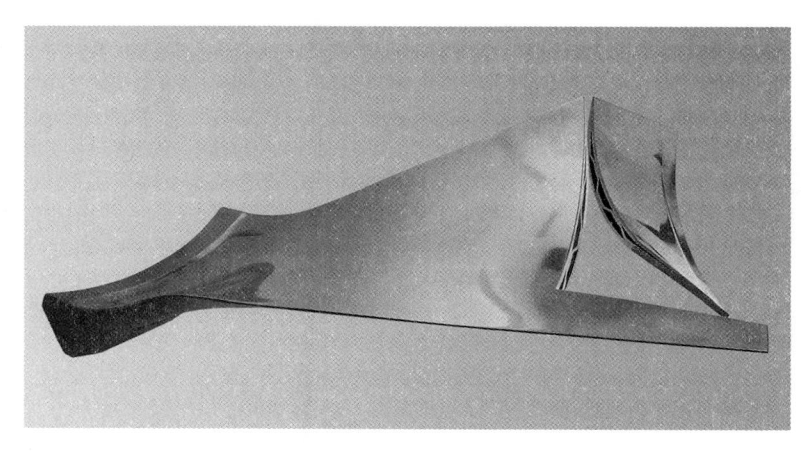

5.10 Experimental diffusion-bonded, superplastically formed wide-chord fan blade (Valiakhmetov *et al.*, 2010).

of only a few coarse grains. The departure from the polycrystalline nature of the material causes problems with process variability and product quality. Using UFG metals in micro-manufacturing of such components allows micro-billets to behave as polycrystalline bodies. This refers to both the inner structure and the surface of the billet. The latter is illustrated in Fig. 5.11 by a substantial reduction of the orange peel effect that occurs in the micro-bulging of AA1070 sheet (Presz and Rosochowski, 2006). A 30–50% reduction in surface roughness was also observed in micro-milling (Popov *et al.*, 2006) and in diamond turning of optical surfaces (Osmer *et al.*, 2007). The benefits that have been realised with backward micro-extrusion include increased and more uniform mechanical properties such as hardness (Rosochowski *et al.*, 2007a), as well as better dimensional accuracy of micro-parts (Geißdörfer *et al.*, 2008).

5.6 Sources of further information and advice

The discipline of UFG metals obtained by SPD has grown substantially in the last ten years. It is dominated by research related to the microstructural aspects of UFG metals, but properties and methods are also represented. The number of researchers involved worldwide has reached a few hundred; they have published more than 3000 publications on UFG metals and SPD. There have been a few monographs published on this subject as well (Altan, 2006; Zehetbauer and Zhu, 2009; Segal *et al.*, 2010). There are well-established conferences dedicated to UFG metals produced by SPD. The series of international NanoSPD conferences was initiated in 1999 in Moscow, followed by conferences in Vienna in 2002, Fukuoka in 2005, Goslar in 2008 and Nanjing in 2011. In parallel, there have been

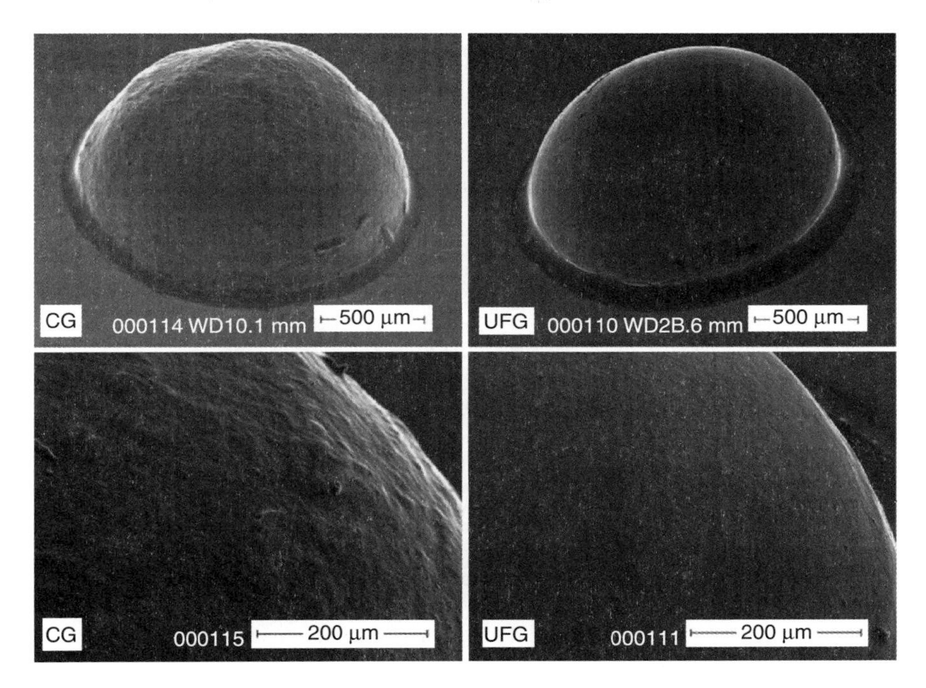

5.11 SEM images of a dome component made by micro-bulging using CG and UFG AA1070 sheet (Presz and Rosochowski, 2006).

TMS meetings on ultrafine-grained materials organised in the USA every two years since 2000. A similar series of biennial events devoted to bulk nano-metals was initiated in 2007 in Ufa, Russia. There have also been sporadic sessions devoted to nanomaterials at the European EMRS conferences, and more regular but small minisymposia on nanostructured materials and micro-forming at the European ESAFORM conferences. In addition, local conferences, for example UFG-2006 in Germany, have been organised from time to time.

A good source of information on the resources available is the website http://www.nanospd.org, established by the International NanoSPD Steering Committee.

5.7 References

Agnew S R, Horton J A, Lillo T M and Brown D W (2004), 'Enhanced ductility in strongly textured magnesium produced by equal channel angular processing', *Scripta Materialia*, 50, 377–81.

Altan B S (ed.) (2006), *Severe Plastic Deformation: Toward Bulk Production of Nanostructured Materials*. Nova Science Publishers, New York.

Apps P J, Bowen J R and Prangnell PB (2004), 'The effect of second-phase particles on the severe deformation of aluminium alloys during equal channel angular extrusion', in: Zehetbauer M and Valiev R Z (eds), *Proceedings of the Conference on Nanomaterials by*

Severe Plastic Deformation: NANOSPD2, Vienna, 9–13 December 2002. Wiley-VCH, Weinheim, 138–44.

Azushima A, Kopp R, Korhonen A, Yang D Y, Micari F, *et al.* (2008), 'Severe plastic deformation (SPD) processes for metals', *CIRP Annals: Manufacturing Technology*, 57, 716–35.

Barber R E, Dudo T, Yasskin P B and Hartwig K T (2004), 'Product yield of ECAE processed material', in: Zhu Y T, Langdon T G, Valiev R Z, Semiatin S L, Shin D H and Lowe T C (eds), *Ultrafine Grained Materials III, 2004 TMS Annual Meeting*, Charlotte, NC, March 14–18, 2004, 667–72.

Beygelzimer Y, Orlov D and Varyukhin V (2002), 'A new severe plastic deformation method: twist extrusion', in: Zhu Y T, Langdon T G, Mishra R S, Semiatin S L, Saran, M J and Lowe T C (eds), *Ultrafine Grained Materials II*, Proceedings of a symposium held during the 2002 TMS Annual Meeting in Seattle, WA, February 17–21, 2002. The Minerals, Metals, and Materials Society, Warrendale, PA, 297–304.

Bridgman P W (1935), 'Effects of high shearing stress combined with high hydrostatic pressure', *Physical Review*, 48(10), 825–47.

Calgagnotto M, Ponge D and Raabe D (2008), 'Ultrafine grained ferrite/martensite dual phase steel fabricated by large strain warm deformation and subsequent intercritical annealing', *ISIJ International*, 48(8), 1096–101.

Comley P N (2004), 'Manufacturing advantages of superplastically formed fine-grain Ti–6Al–4V alloy', *Journal of Materials Engineering and Performance*, 13(6), 660–64.

Cooper P and Barber A (2003), 'Review of the latest developments and best use of grain refiners', in: *2nd International Melt Quality Workshop*, Prague, 16–17 October 2003.

Erbel S (1979), 'Mechanical properties and structure of extremely strainhardened copper', *Metals Technology*, 12, 482–86.

Ferrase S, Alford F, Grabmeier S, Düvel A, Zedlitz R, *et al.* (2003), Technology White Paper, http://www.honeywell.com/sites/docs/doc128e30a–f9d1a68f6a–e0df9bfada07602278603c6cb43673fb.pdf.

Figueiredo R B and Langdon T G (2008), 'Record superplasticity in a magnesium alloy processed by equal channel angular pressing', *Advanced Engineering Materials*, 10, 37–40.

Furukawa M, Iwahashi Y, Horita Z, Nemoto M and Langdon T G (1998), 'The shearing characteristics associated with equal-channel angular pressing', *Materials Science and Engineering A*, 257, 328–332.

Geißdörfer S, Rosochowski A, Olejnik L, Engel U and Richert M (2008), 'Micro-extrusion of ultrafine grained copper', *International Journal of Material Forming*, 1 (Suppl. 1), 455–58.

Ghosh A K and Huang W (2000), 'Severe deformation based process for grain subdivision and resulting microstructures', in: Lowe T C and Valiev R Z (eds), *Investigations and Applications of Severe Plastic Deformation*. Kluwer, Dordrecht, 29–36.

Głuchowski W, Stobrawa J, Rdzawski Z and Malec W (2011), 'Ultrafine grained copper alloys processes by continuous repetitive corrugation and straightening method', *Materials Science Forum*, 674, 177–88.

Haghayeghi R, Zoqui E J, Green N R and Bahai H (2010), 'An investigation on DC casting of a wrought aluminium alloy at below liquidus temperature by using melt conditioner', *Journal of Alloys and Compounds*, 502, 382–86.

Hall E O (1951), 'The deformation and aging of mild steel: III. Discussion of results', *Proceedings of the Physical Society of London, Section B*, 64(9), 747–53.

Hanamura T, Fuxing Y and Nagai K (2004), 'Ductile–brittle transition temperature of ultrafine ferrite/cementite microstructure in a low carbon steel controlled by effective grain size', *ISIJ International*, 44 (3), 610–17.

Hasegawa H, Komura S, Utsunomiya A, Horita Z, Furukawa M, *et al.* (1999), 'Thermal stability of ultrafine-grained aluminum in the presence of Mg and Zr additions', *Materials Science and Engineering A*, 265, 188–96.

Heason C P and Prangnell P B (2004), 'Comparative study and texture modeling of accumulative roll bonding (ARB)', in: Zehetbauer M and Valiev R Z (eds), *Nanomaterials by Severe Plastic Deformation*. Wiley-VCH, Weinheim, 498–504.

Hefti L D (2007), 'Advances in the superplastically formed and diffusion bonded process', *Materials Science Forum*, 551–552, 87–93.

Höppel H W, May J and Göken M (2004), 'Enhanced strength and ductility in ultrafine-grained aluminium produced by accumulative roll bonding', *Advanced Engineering Materials*, 6(4), 219–22.

Höppel H W, Kautz M, Xu C, Murashkin M, Langdon T G, *et al.* (2006), 'An overview: fatigue behavior of ultrafine-grained metals and alloys', *International Journal of Fatigue*, 28(9), 1001–10.

Huang J Y, Zhu Y T, Jiang H and Lowe T C (2001), 'Microstructures and dislocation configurations in nanostructured Cu processed by repetitive corrugation and straightening', *Acta Materialia*, 49, 1497–505.

Huang J, Zhu Y T, Alexander D J, Liao X, Lowe T C and Asaro R J (2004), 'Development of repetitive corrugation and straightening', *Materials Science and Engineering A*, 371, 35–9.

Khan A S, Suh Y S, Chen X, Takacs L and Zhang H (2006), 'Nanocrystalline aluminum and iron: mechanical behavior at quasi-static and high strain rates, and constitutive modeling', *International Journal of Plasticity*, 22(2), 195–209.

Kim H S, Estrin Y and Bush M B (2001), 'Constitutive modelling of strength and plasticity of nanocrystalline metallic materials', *Materials Science and Engineering A*, 316, 195–99.

Kim W J and Sa Y K (2006), 'Micro-extrusion of ECAP processed magnesium alloy for production of high strength magnesium micro-gears', *Scripta Materialia*, 54, 1391–95.

Korbel A, Richert M and Richert J (1981), 'The effects of very high cumulative deformation on structure and mechanical properties of aluminium', in: Hansen N, Horsewell A, Leffers T and Lilholt H (eds), *Proceedings of the 2nd Risø International Symposium on Metallurgy and Material Science*, Roskilde, Denmark, 14–18 September, 1981. Risø National Laboratory, 445–50.

Kuziak R, Zalecki W, Weglarczyk S and Pietrzyk M (2005), 'New possibilities of achieving ultra-fine grained microstructure in metals and alloys employing MaxStrain technology', *Solid State Phenomena*, 101–102, 43–8.

Langdon T G, Furukawa M, Nemoto M and Horita Z (2000), 'Using equal-channel angular pressing for refining grain size', *JOM*, 52(4), 30–3.

Lapovok R (2006), 'The positive role of back-pressure in equal channel angular extrusion', *Materials Science Forum*, 503–504, 37–44.

Lee J C, Seok H K, Han J H and Chung Y H (2001), 'Controlling the textures of the metal strips via the continuous confined strip shearing (C2S2) process', *Materials Research Bulletin*, 36(5–6), 997–1004.

Lee S, Utsunomiya A, Akamatsu H, Neishi K, Furukawa M, *et al.* (2002), 'Influence of scandium and zirconium on grain stability and superplastic ductilities in ultrafine-grained Al–Mg alloys', *Acta Materialia*, 50(3), 553–64.

Meyers M A, Mishra A and Benson D J (2006), 'Mechanical properties of nanocrystalline materials', *Progress in Materials Science*, 51, 427–556.

Mukai T, Yamanoi M, Watanabe H and Higashi K (2001), 'Ductility enhancement in AZ31 magnesium alloy by controlling its grain structure', *Scripta Materialia*, 45(1), 89–94.

Muszka K, Majta J and Hodgson P D (2007), 'Modeling of the mechanical behavior of nanostructured HSLA steels', *ISIJ International*, 47(8), 1221–7.

Ning Z, Cao P, Wang H, Sun J and Liu D (2007), 'Effect of cooling conditions on grain size of AZ91 alloy', *Journal of Materials Science and Technology*, 23(5), 645–9.

Olejnik L, Rosochowski A and Richert M (2008), 'Incremental ECAP of plates', *Materials Science Forum*, 584–586, 108–13.

Olejnik L, Kulczyk M, Pachla W and Rosochowski A (2009), 'Hydrostatic extrusion of UFG aluminium', *International Journal of Material Forming*, 2 (Suppl. 1), 621–4.

Osmer J, Riemer O, Brinksmeier E, Rosochowski A, Olejnik L and Richert M (2007), 'Diamond turning of ultrafine grained aluminium alloys', *Proceedings of the 7th Euspen International Conference*, Bremen, Germany, May 20–24, 2007, 2, 316–9.

Park K-T, Han S Y, Ahn B D, Shin D H, Lee Y K and Um K K (2004), 'Ultrafine grained dual phase steel fabricated by equal channel angular pressing and subsequent intercritical annealing', *Scripta Materialia*, 51, 909–13.

Petch N J (1953), 'The cleavage strength of polycrystals', *Journal of the Iron and Steel Institute*, 74(1), 25–8.

Pippan R, Scheriau S, Hohenwarter A and Hafok M (2008), 'Advantages and limitations of HPT: a review', *Material Science Forum*, 584–586, 16–21.

Popov K B, Dimov S S, Pham D T, Minev R M, Rosochowski A and Olejnik L (2006), 'Micromilling: material microstructure effects', *Proceedings of IMechE Part B: Journal of Engineering Manufacture*, 220(11), 1807–13.

Prangnell P B, Harris C and Roberts S M (1997), 'Finite element modelling of equal channel angular extrusion', *Scripta Materialia*, 37, 983–89.

Presz W and Rosochowski A (2006), 'The influence of grain size on surface quality of microformed components', in: Juster N and Rosochowski A (eds), *Proceedings of the 9th International Conference on Material Forming: ESAFORM 2006*, Glasgow, UK, April 26–28, 2006. Publishing House Akapit, Krakow, 587–90.

Raab G J, Valiev R Z, Lowe T C and Zhu Y T (2004), 'Continuous processing of ultrafine grained Al by ECAP-Conform', *Materials Science and Engineering A*, 382, 30–4.

Reis A C C, Tolleneer I, Barbé L, Kestens L and Houbaert Y (2004), 'Ultra grain refinement of Fe-based alloys by accumulater roll bonding', in: Zehetbauer M and Valiev R Z (eds), *Nanomaterials by Severe Plastic Deformation*. Wiley-VCH, Weinheim, 530–6.

Reshetnikova N A, Salakhova M R, Safargalina Z A and Scherbakov A V (2008), 'R&D of nanoSPD materials in Ufa via international cooperation', *Materials Science Forum*, 584–586, 9–15.

Richert J (2010), 'Strain–stress conditions of shear band formation during CEC processing on a new machine with control back pressure', *Archives of Metallurgy and Materials*, 55(2), 391–408.

Richert M, McQueen H J and Richert J (1998), 'Microband formation in cyclic extrusion compression of aluminium', *Canadian Metallurgical Quarterly*, 37(5), 449–57.

Rosochowska M, Rosochowski A and Olejnik L (2010), 'FE simulation of micro-extrusion of a conical pin', *International Journal of Material Forming*, 3 (Suppl. 1), 423–6.

Rosochowski A (2005), 'Processing metals by severe plastic deformation', *Solid State Phenomena*, 101–102, 13–22.

Rosochowski A and Olejnik L (2002), 'Numerical and physical modelling of plastic deformation in 2-turn equal channel angular extrusion', *Journal of Materials Processing Technology*, 125–126, 309–16.
Rosochowski A and Olejnik L (2007a), 'FEM simulation of incremental shear', in: Cueto E and Chinesta F (eds), *Proceedings of the 10th International Conference on Material Forming: Esaform 2007*, Zaragoza, Spain, April 18–20, 2007, American Institute of Physics, New York, 907, 653–8.
Rosochowski A and Olejnik L (2007b), 'Finite element simulation of severe plastic deformation processes', *Proceedings of the Institution of Mechanical Engineers, Part L, Journal of Materials: Design and Applications*, 221(4), 187–96.
Rosochowski A and Olejnik L (2010), 'Incremental equal channel angular pressing for grain refinement', *Materials Science Forum*, 674, 19–28.
Rosochowski A, Olejnik L and Richert M (2006a), 'Channel configuration effects in 3D-ECAP', *Materials Science Forum*, 503–504, 179–84.
Rosochowski A, Olejnik L, Gagne J, Ladeveze N and Rosochowska M (2006b), 'Compression behaviour of UFG aluminium', in: Juster N and Rosochowski A (eds), *Proceedings of the 9th International Conference on Material Forming: ESAFORM 2006*, Glasgow, UK, April 26–28, 2006. Scientific Publishing House Akapit, Krakow, 543–6.
Rosochowski A, Presz W, Olejnik L and Richert M (2007a), 'Micro-extrusion of ultra-fine grained aluminium', *International Journal of Advanced Manufacturing Technology*, 33(1–2), 137–46.
Rosochowski A, Olejnik L and Richert M (2007b), '3-D ECAP of square aluminium billets', in: Banabic, D. (ed.), *Advanced Methods in Material Forming*. Springer, Berlin, 215–32.
Rosochowski A, Olejnik L and Richert M (2008), 'Double-billet incremental ECAP', *Materials Science Forum*, 584–586, 139–44.
Rosochowski A, Rosochowska M, Olejnik L and Verlinden B (2010), 'Incremental equal channel angular pressing of sheets', *Steel Research International*, 81(9), 470–73.
Safiullin A R, Safiullin R V and Kruglov A A (2009), 'Application of nanostructural Ti alloy for producing a face plate for a golf club', in: *Congress of Nanotechnologies, Second International Symposium on Bulk Nanostructured Materials: From Fundamentals to Innovations, BNM2009*, Ufa, Russia, 22–26 September 2009, 30–1.
Saito Y, Tsuji N, Utsunomiya H, Sakai T and Hong R G (1998), 'Ultra-fine grained bulk aluminium produced by accumulative roll-bonding (ARB) process', *Scripta Materialia*, 39(9), 1221–7.
Salimgareeva G H, Semenova I P, Latysh V V and Valiev R Z (2005), 'Nanostructuring of Ti in long-sized Ti rods by severe plastic deformation', in: Banabic, D. (ed.), *Proceedings of the 8th International ESAFORM Conference on Material Forming*, Cluj-Napoca, Romania. Publishing House of the Romanian Academy, Bucharest, 661–4.
Segal V M (1995), 'Materials processing by simple shear', *Materials Science and Engineering A*, 197, 157–64.
Segal V M, Reznikov V I, Drobyshevskiy A E and Kopylov V I (1981), 'Plastic working of metals by simple shear', *Russian Metallurgy (Metally)*, 1, 99–105.
Segal V M, Beyerlein I J, Tome C N, Chuvil'deev V N and Kopylov V I (2010), *Fundamentals and Engineering of Severe Plastic Deformation*. Nova Science Publishers, New York.
Semenova I P, Salimgareeva G Kh, Latysh V V, Lowe T and Valiev R Z (2009), 'Enhanced fatigue strength of commercially pure Ti processed by severe plastic deformation', *Materials Science and Engineering A*, 503, 92–5.

Semiatin S L, Segal V M, Goforth R E, Frey N D and DeLo D P (1999), 'Workability of commercial-purity titanium and 4340 steel during equal channel angular extrusion at cold-working temperatures', *Metallurgical and Materials Transactions*, 30A, 1425–35.

Shin D H and Park K-T (2004), 'Microstructural stability and tensile properties of nanostructured low carbon steels processed by ECAP', in: Zehetbauer M and Valiev R Z (eds), *Proceedings of the Conference on Nanomaterials by Severe Plastic Deformation: NANOSPD2*, 9–13 December 2002, Vienna. Wiley-VCH, Weinheim, 616–22.

Shin D H, Park J-J, Kim Y-S and Park K-T (2002), 'Constrained groove pressing and its application to grain refinement of aluminum', *Materials Science and Engineering A*, 328, 98–103.

Siegel R W (1993), 'Synthesis and properties of nanophase materials', *Materials Science and Engineering A*, 168, 189–97.

Sklenicka V, Dvorak J, Kral P, Stonawska Z and Svoboda M (2005), 'Creep process in pure aluminium processed by equal-channel angular pressing', *Materials Science and Engineering A*, 410–411, 408–12.

Srinivasan R, Chaudhury P K, Cherukuri B, Han Q, Swenson D and Gros P (2006), 'Continuous severe plastic deformation processing of aluminum alloys', Final Technical Report, Wright State University, http://www.osti.gov/bridge/purl.cover.jsp?purl=/885079–37CRhi.

Stolyarov V V, Zhu Y T, Alexandrov I V, Lowe T C and Valiev R Z (2001), 'Influence of ECAP routes on the microstructure and properties of pure Ti', *Materials Science and Engineering A*, 299, 59–67.

Suś-Ryszkowska M, Wejrzanowski T, Pakieła Z and Kurzydlowski K J (2004), 'Microstructure of ECAP severely deformed iron and its mechanical properties', *Materials Science and Engineering A*, 369, 151–6.

Utsunomiya H, Saito Y, Suzuki H and Sakai T (2001), 'Development of the continuous shear deformation process', *Proceedings of the Institution of Mechanical Engineers, Part B: Journal of Engineering Manufacture*, 215(7), 947–57.

Valiakhmetov O R, Galeyev R M, Ivan'ko V A, Imayev R M, Inozemtsev A A, *et al.* (2010), 'The use of nanostructured materials and nanotechnologies for the elaboration of hollow structures', *Nanotechnologies in Russia*, 5(1–2), 108–22.

Valiev R Z (2004), 'Paradoxes of severe plastic deformation', in: Zehetbauer M J and Valiev R Z (eds), *Proceedings of the Conference 'Nanomaterials by Severe Plastic Deformation: NanoSPD2'*, Vienna, December 9–13, 2002. Wiley-VCH, Weinheim, 109–17.

Valiev R Z (2006), 'The new trends in SPD processing to fabricate bulk nanostructured materials', in: Juster N and Rosochowski A (eds), *Proceedings of the 9th International Conference on Material Forming: ESAFORM 2006*, Glasgow, UK, April 26–28, 2006. Scientific Publishing House Akapit, Krakow, Poland, 1–9.

Valiev R Z, Krasilnikov N A and Tsenev N K (1991), 'Plastic deformation of alloys with submicro-grained structure', *Materials Science and Engineering A*, 137, 35–40.

Valiev R Z, Sergueeva A V and Mukherjee A K (2003), 'The effect of annealing on tensile deformation behavior of nanostructured SPD titanium', *Scripta Materialia*, 49(7), 669–74.

Valiev R Z, Semenova I P, Jakushina E, Latysh V V, Rack H, *et al.* (2008), 'Nanostructured SPD processed titanium for medical implants', *Materials Science Forum*, 584–586, 49–54.

Van Swygenhoven H, Derlet P M and Frøseth A G (2006), 'Nucleation and propagation of dislocations in nanocrystalline fcc metals', *Acta Materialia*, 54(7), 1975–83.

Varyukhin V, Beygelzimer Y, Kulagin R, Prokofeva O and Reshetov A (2011), 'Twist extrusion: fundamentals and applications', *Materials Science Forum*, 667–669, 31–7.

Verma R (2004), *Method for Processing of Continuously Cast Aluminum Sheet*, General Motors Corporation, US Patent No. 6 811 625, November 2, 2004.

Vorhauer A and Pippan R (2004), 'The influence of type and path of deformation on the microstructural evolution during severe plastic deformation', in: Zehetbauer M and Valiev R Z (eds), *Nanomaterials by Severe Plastic Deformation*. Wiley-VCH, Weinheim, 684–90.

Wang Y, Chen M, Zhou F and Ma E (2002), 'High tensile ductility in a nanostructured metal', *Nature*, 419, 912–5.

Wang Y, Ma E, Valiev R Z and Zhu Y (2004), 'Tough nanostructured metals at cryogenic temperatures', *Advanced Materials*, 16(4), 328–31.

Zehetbauer M J and Zhu Y T (eds) (2009), *Bulk Nanostructured Materials*. Wiley-VCH, Weinheim.

Zhao X, Yang X, Liu X, Wang X and Langdon T G (2010), 'The processing of pure titanium through multiple passes of ECAP at room temperature', *Materials Science and Engineering A*, 527, 6335–9.

Zherebtsov S V, Salishchev G A, Galeyev R M, Valiakhmetov O R and Semiatin S L (2004), 'Formation of submicrocrystalline structure in large-scale Ti–6Al–4V billets during warm severe plastic deformation', in: Zehetbauer M and Valiev R Z (eds), *Nanomaterials by Severe Plastic Deformation*. Wiley-VCH, Weinheim, 835–40.

Zhilyaev A P, Kim B-K, Nurislamova G V, Baró M D, Szpunar J A and Langdon T G (2002), 'Orientation imaging microscopy of ultrafine-grained nickel', *Scripta Materialia*, 46, 575–80.

Zhu Y T and Langdon T G (2004), 'The fundamentals of nanostructured materials processed by severe plastic deformation', *JOM*, 56(10), 58–63.

Part II

Microstructure evolution in the processing of steel

6 Modelling phase transformations in steel

M. PIETRZYK, AGH – University of Science and Technology, Poland and R. KUZIAK, Institute for Ferrous Metallurgy, Poland

Abstract: This chapter discusses aspects of the physical and numerical modelling of phase transformations in steels. The basic features of these phase transformations are described. Dilatometric tests, which are performed to identify the parameters of phase transformation models, are explained. Four models of phase transformations of various complexity and various predictive capability are described. The chapter includes a case study of the simulation and optimization of two industrial processes for dual-phase steel strips: laminar cooling after hot rolling, and cooling after continuous annealing.

Key words: phase transformations, steels, physical modelling, numerical modelling, identification of models, sensitivity analysis.

6.1 Introduction

The multitude of structures that can be produced during thermomechanical processing or heat treatment has made steels very attractive engineering materials. The various structures in steels are developed primarily because of phase transformations in austenite, which, under controlled conditions, produce substantial variation in properties. Steels have been developed largely on the basis of experiments conducted by metallurgists. However, a thorough understanding of the atomistic mechanisms of phase transformations, combined with advances in mathematical methods, has in recent times led to new tools for predicting and controlling the properties of steel. These advances have contributed to a substantial increase in the competitiveness of steels in comparison with other engineering materials. In this chapter, some models that describe the most important transformations in steels are presented for use in combination with thermal and mechanical finite element (FE) codes to predict the microstructure and properties of steel products. The capability of these models is demonstrated for dual-phase (DP) steels, which have recently been growing in demand from the automotive industry.

6.2 Phase transformation in steels

The structure of steels depends on the nucleation and growth rates of ferrite, pearlite and bainite in austenite. The aim of phase transformation modelling is to develop closed-form equations that account for the incubation and growth stages of phase transformations without distinguishing between different types of phase constituents. Some ferritic structures, for example Widmanstätten ferrite, are dominated by the strain energy generated by the deformation of the parent lattice, and others are closer to equilibrium because the transformation is achieved with the aid of diffusion. The latter transformations include the formation of allotriomorphic and idiomorphic ferrite, as well as pearlite formation, during which all elements must diffuse to produce the structural change.

The pearlitic transformation is accomplished through an eutectoid decomposition of austenite, which leads to the formation of parallel plates of ferrite and cementite. In this case, the ferrite and carbide phases grow cooperatively at a common transformation front with the austenite. Some other transformations are produced by a displacive mechanism. To this category belongs the formation of Widmanstätten ferrite, bainite, acicular ferrite and martensite.

The transformation of austenite into these structures involves invariant-plane-strain deformation caused by the transformation, in which the large shear is the main cause of the plate-like structure of the transformation product. Substitutional solutes do not partition between the parent and product phases, and, as a result, the transformation proceeds under non-equilibrium conditions. However, the lengthening of the plates is controlled by the diffusion of carbon in austenite. Moreover, carbon must partition into the austenite during the nucleation of Widmanstätten ferrite and bainite, but there is probably no carbon diffusion during transformation. Nevertheless, excess carbon is rejected into the remaining austenite soon after transformation. The partitioned carbon may then precipitate as carbides at lath boundaries, giving the classical upper bainitic microstructure. At low temperatures, where the partitioning of carbon is slower, the excess carbon precipitates inside the bainitic ferrite, leading to the lower bainitic microstructure. Acicular ferrite is frequently referred to as an alternative to bainite with a more chaotic morphology. The plates of acicular ferrite are nucleated intragranularly on non-metallic inclusions. As opposed to the ferritic and bainitic transformations, the martensitic transformation is entirely diffusionless, which means that diffusion of neither alloying elements nor carbon takes place during the transformation.

6.3 Experimental techniques

6.3.1 Selection of material for validation of models

After identification of the coefficients on the basis of dilatometric tests, a model that has been developed can be applied to steel of any grade. In the present work, the procedure for the identification and validation of models is demonstrated using a DP steel as an example. The structure of DP steels is composed of ductile ferrite and hard martensite. The volume fraction of martensite F_m does not exceed 30% in the majority of products. Such a phase composition and second-phase morphology give high strength and ductility in products. The soft matrix (ferrite) contributes to the ductility of the steel. Deformation of this structure leads to strain concentration in the ferrite, which gives better hardening characteristics in comparison with high-strength low-alloy steels. The main features of DP steels are:

- the lack of the field point;
- a low $R_{0.2}/R_m$ ratio, in the range of 0.50–0.75, which prevents steep gradients of stresses in cold-formed products;
- a high hardening coefficient, particularly at the beginning of deformation, which lowers the residual stresses in cold-formed products;
- a large, uniform total elongation.

Hot-rolled DP steels are classified as follows:

- Si–Mn steels, which are subjected to further processing after hot rolling;
- steels with high Si, Cr and Mo contents, processed using conventional hot rolling technologies;
- Si–Mn steels with simple compositions, which are subject to thermomechanical rolling followed by controlled cooling and coiling at low temperatures.

The two-phase structure of DP steel strips is obtained by control of the ferritic transformation. Accelerated cooling is applied at the temperature of the maximum rate of the ferritic transformation. The material is maintained at this temperature until the required volume fraction of ferrite is obtained. Accelerated cooling is then applied, and the remaining austenite is transformed into martensite. This controlled cooling is applied either during laminar cooling after hot rolling (for thicker strips) or during cooling after continuous annealing (for thinner strips). The practical realization of this scheme of cooling is difficult. Therefore, the application of optimization techniques to design a cooling schedule for DP steels is presented in this work.

6.3.2 Dilatometric tests

Tests for identification of the phase transformation models were performed with a DIL 805 dilatometer, shown in Fig. 6.1, which is capable of deforming the sample prior to cooling. In this case the sample is cylindrical, having dimensions of $\phi 4 \times 7$ mm. Tubular samples, of dimensions $\phi 4/2 \times 10$ mm, were used for experiments not involving deformation.

The dilatometer was used to determine continuous-cooling transformation (CCT), deformation continuous-cooling transformation (DCCT) and time–temperature–transformation (TTT) diagrams, all of which are referred to in this chapter as phase transformation diagrams. To determine a CCT diagram, a series of samples were austenitized under defined conditions (of temperature and time) and cooled at different rates to ambient temperature. If the cooling stage was preceded by deformation, the phase transformation diagram was called a DCCT diagram. In contrast, to determine a TTT diagram, the sample was cooled at a high cooling rate to an isothermal holding temperature and then held at this temperature until completion of the phase transformation. During a phase transformation, the

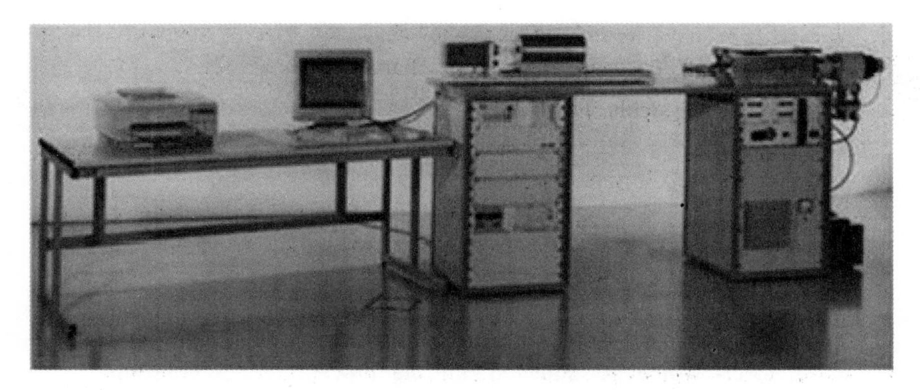

6.1 The DIL 805 dilatometer used in the present work.

crystallographic structure of the parent phase changes, which results in a change in specific volume. The dilatation curves were then used for identification of the phase transformation. An example of a dilatation signal obtained during a dilatometric test is shown in Fig. 6.2. The deviation of the curve from linearity is connected to the onset of a phase transformation. The so-called lever rule $X = (L_3 - L_1)/(L_2 - L_1)$ can be used to assess the parent phase fraction X as a function of temperature.

The lever rule gives only an approximate value of the volume transformed. This is due to the fact that the method neglects the differences in the specific volumes of different phases and the partition of atoms between phases. Another source of uncertainty is connected with the superposition of dilation signals originating from several phases.

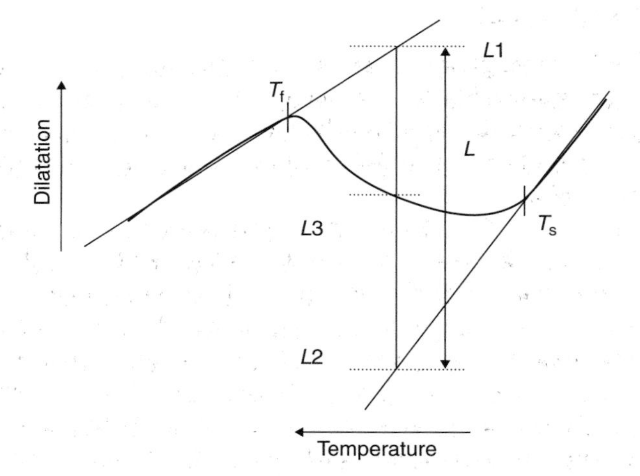

6.2 Dilatation curve obtained in a dilatometric test: T_s, transformation start temperature; T_f, transformation finish temperature.

The dilatometric tests supplied information for identification of a phase transformation model based on the Avrami equation, as described in Section 6.4.4.

6.3.3 Continuous annealing

The validation of the phase transformation model for DP steel was performed using a Gleeble 3800 simulator to simulate continuous annealing. The experimental set-up is shown in Fig. 6.3. A sheet sample 55 mm in width and 250 mm in length was resistance heated and then cooled, either with an inert gas or with a water/air mist. The maximum heating rate was around 2000°C/s and the cooling rate was around 100°C/s. The 1 mm thick sheet of DP60 steel was subject to a continuous-annealing simulation using the temperature profile shown in Fig. 6.4(a). The thermal profile consists essentially of two parts, namely, heating to 800°C, which lies in the $\alpha + \gamma$ intercritical region, and cooling. The calculated changes in the volume fractions of austenite and ferrite during heating are shown in Fig. 6.4(b). One can read from the graph that the amount of austenite in the sample at 800°C is approximately 70%. The cooling part of the thermal profile is composed of two slow cooling stages followed by a fast cooling stage. The aim of the slow cooling is to produce a suitable amount of ferrite in the structure, typically 70–80%. During the ferrite transformation, carbon diffuses into austenite, which increases the hardenability of this phase. The aim of the fast cooling stage is to transform the remaining austenite into martensite. The structure of a sample subjected to a continuous-annealing simulation is shown in Fig. 6.5(a). LePera etchant and colour metallography were used to distinguish the phase constituents.

6.3 View of an experimental set-up installed in a Gleeble 3800 simulator for the physical modelling of continuous annealing.

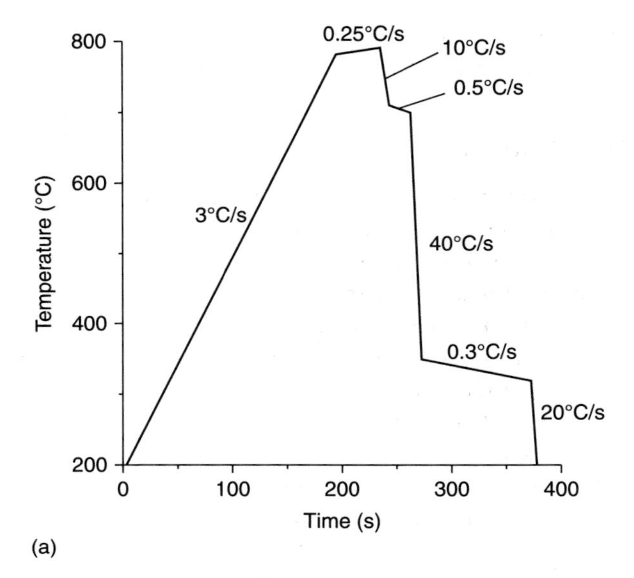

(a)

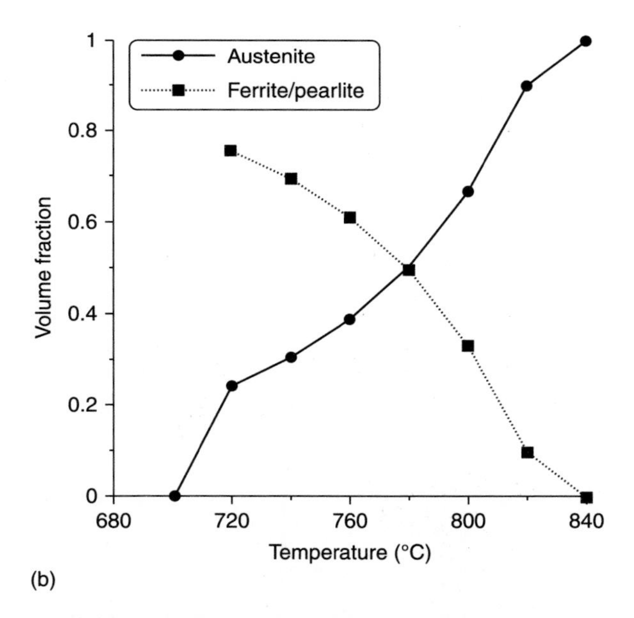

(b)

6.4 (a) Typical thermal cycle for the continuous annealing of DP steels, and (b) volume fractions of phases during heating.

It can be seen that the microstructure of the DP600 strip is composed of ferrite, martensite, bainite and a small content of retained austenite, as revealed by light optical microscopy and field-emission gun scanning electron microscopy (FEG SEM), as shown in Fig. 6.5. The fractions of ferrite, bainite and martensite can be determined from the micrographs.

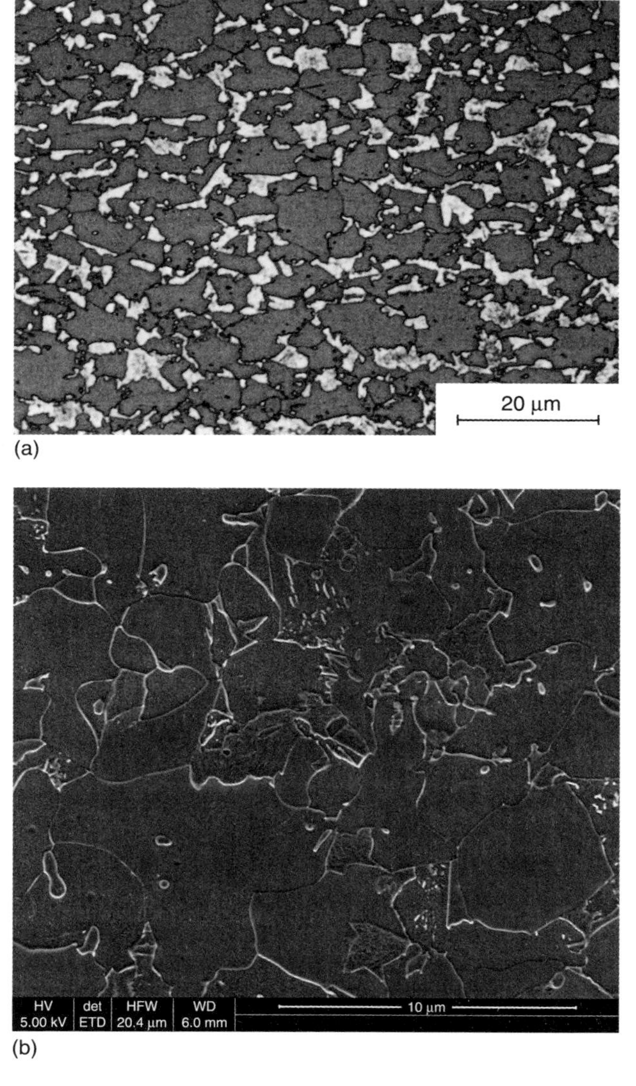

6.5 (a) Dual-phase structure of a sheet sample after simulation of continuous annealing in the Gleeble simulator: white phase, martensite; grey phase, ferrite (light optical microscopy); (b) characterization of microstructure of DP600 steel with FEG SEM;

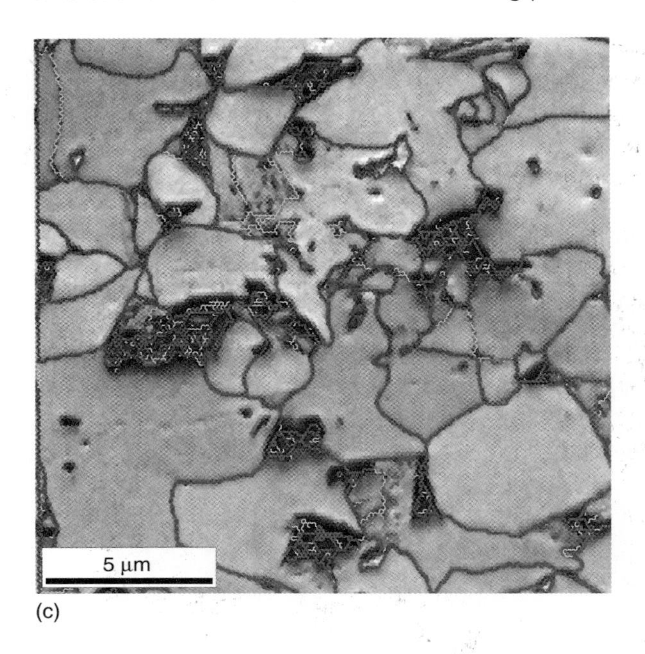

(c)

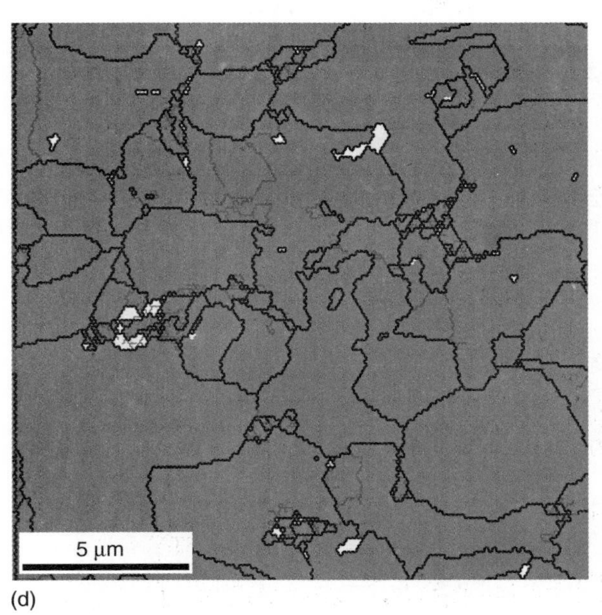

(d)

6.5 Continued. (c) image quality map; (d) distribution of retained austenite.

6.4 Modelling methods

6.4.1 Managing multi-component systems using predictions of thermodynamics

In order to fulfil the requirements on mechanical properties, modern alloys have to be based on very complex multi-component systems. However, experimentally determined phase diagrams are usually available only for binary systems and, to a limited extent, for ternary systems. The thermodynamic properties controlling the driving forces for phase transformations, the boundary conditions and the kinetic parameters are complex functions of temperature and chemical composition. A computational approach has emerged and become established over last thirty years as the most efficient way to design and study complex systems.

The most advanced method of calculating phase transformations in complex systems is CALPHAD (CALculation of PHAse Diagrams). The CALPHAD method is based on the fact that a phase diagram is a representation of the thermodynamic properties of a system. Thus, if the thermodynamic properties are known, it should be possible to calculate multi-component phase diagrams (Hack, 1996). Computational thermodynamics enables a number of materials-dependent engineering problems to be modelled and numerically solved. Some of these problems may be very complex, for example the analysis of all possible interactions in a system composed of ten to twenty alloying elements at a given temperature. Although this problem can be solved experimentally, the time required to do so can be significantly longer than the time required to solve it using computer simulations. The application of tools such as Thermo-Calc, DICTRA, JMatPro, FactSage, Pandat and MTDATA reduces the need for costly experiments. Phase transformation models for conditions of varying temperature use equilibrium data predicted by the thermodynamic software described above.

6.4.2 Phase transformation models

General information

The modelling of phase transformation is a wide topic, which cannot be discussed even briefly in a complete form in this chapter. Depending on the objective of the modelling, various models and techniques are used, ranging from models based on first principles and thermodynamic relations to experimental closed-form equations describing the kinetics of transformations. The present chapter is limited to a description of phenomenological models for engineering applications. The emphasis is put on the identification of parameters in these models and on implementation of the models into FE software.

The modelling of phase transformations for engineering applications has for years been based on the JMAK approach (named after Johnson, Mehl, Avrami and

Kolmogorov). The general Avrami equation for an arbitrary transformation is (Avrami, 1939, 1940, 1941):

$$X = 1 - \exp\left[a\left(\frac{t}{t_X}\right)^n\right] \quad [6.1]$$

where X is the volume fraction of the new phase, and n is the Avrami exponent. The value of the coefficient a depends on the time t_X, which is a 'basic time'. It can be shown that $a = \ln(1 - X_b)$, where X_b is the volume fraction of the new phase after the basic time.

In the modelling of recrystallization, the time for 50% recrystallization $t_{0.50}$ is usually used as the basic time (Sellars, 1979); in this case, $X_b = 0.5$ and $a = \ln(0.5) = 0.693$. However, this basic time is not used in the modelling of phase transformations in steels, and one Avrami coefficient k is used instead:

$$X = 1 - \exp(-kt^n) \quad [6.2]$$

where k is the Avrami coefficient and n is the Avrami exponent. Equation 6.2, with constant coefficients k and n, was used effectively for modelling phase transformations in the second half of the last century. The Scheil additivity rule (Scheil, 1935) was used to take account of changes in temperature. More advanced models, with a varying coefficient k, were introduced in the last decade of the twentieth century (Ronda and Oliver, 2000).

Additivity rule

As mentioned above, the classical modelling of phase transformations is based on the Avrami theory (Avrami, 1939, 1940, 1941), represented by Eq. 6.2. This equation was derived on the assumption of a uniform distribution of nucleation sites. This means that a constant ratio between the growth rate and nucleation rate is also assumed and that the progress of the transformation does not depend on the history of temperature changes. These conditions were called isokinetic conditions by Avrami. Assuming that isokinetic conditions are fulfilled, Eq. 6.2 for kinetics under isothermal conditions can be also applied to simulations of a transformation which takes place during continuous changes of temperature.

The assumption that nucleation is homogeneous in the volume of the considered material constrains the possibility of the analysis of phase transformations to certain cases that have no application to the majority of phase transformations in alloys. In general, the nucleation sites are saturated in the early stages of the transformation and the further progress of the transformation is due to growth. This observation led to the following condition, which defines when the additivity rule can be applied:

$$\frac{G(T)t_{0.50}}{D_\gamma} < 0.5 \quad [6.3]$$

where T is the temperature, D_γ is the austenite grain size and $t_{0.50}$ is the time for 50% transformation.

In the majority of industrial processes, transformations occur under non-isothermal conditions. Thus, the model of the transformation kinetics has to take account of changes in temperature. The additivity rule proposed by Scheil (1935) was applied in the study described in this chapter to predict the incubation time and the kinetics of the pearlitic and bainitic transformations. According to this rule, the transformation begins when a sum of certain fractions reaches unity:

$$\int_0^\tau \frac{dt}{\tau_n(T)} = \int_0^\tau \frac{1}{\tau_n(T)} \frac{dt}{dT} dT = \int_0^\tau \frac{1}{v\tau_n(T)} dT = 1 \qquad [6.4]$$

where $\tau_n(T)$ is the incubation time at constant temperature T, and v is the cooling rate.

The additivity rule can be applied under certain conditions, which are discussed in Scheil (1935) and Cahn (1956), and which are fulfilled in the industrial processes of cooling of products after hot rolling or annealing. Briefly, if the progress of the phenomenon at any instant depends only on the temperature, the process is additive. Thus, transformations during laminar cooling after hot rolling or during cooling after continuous annealing can also be modelled using the additivity rule. Since only k in Eq. 6.2 is temperature-dependent in the models discussed below, Eq. 6.2 can be written as

$$X = 1 - \exp\left[-\int_{T_0}^{T} \frac{k(T)^{1/n}}{v} dT\right] \qquad [6.5]$$

where T_0 is the temperature at the end of the incubation time.

Equation 6.5 is valid for the continuous-cooling stage of the process considered here. The derivation of this equation, as well as its extension to the holding and heating stages of the process, is described in Cahn (1956). The dependence of the coefficient k on the temperature is discussed in the next section.

Dependence of the Avrami coefficient k on temperature

In the remainder of this chapter, the model described in this section will be referred to as 'model A'. Theoretical considerations show that, according to the type of transformation (a nucleation and growth process or a site saturation process), a constant value of the coefficient n can be used in Eq. 6.2. In contrast, the value of the coefficient k must vary with the temperature in a way linked to the form of the TTT diagram. The form of the function $k = f(T)$ must be chosen carefully to describe properly the temperature dependence of the transformation kinetics. The present author has investigated the following two functions $k = f(T)$:

$$k = \frac{k_3}{D_\gamma^{k_4}} \exp(k_2 - k_1 T) \qquad [6.6]$$

$$k = k_3 \exp(k_2 - k_1 T) \tag{6.7}$$

Investigation has shown that Eq. 6.6, with a dependence on the grain size, describes the pearlitic transformation well. The coefficients in this equation were taken as optimization variables as follows: $k_1 = a_{12}$, $k_2 = a_{13}$, $k_3 = a_{14}$ and $k_4 = a_{16}$. Equation 6.7 was used for the bainitic transformation. The optimization variables were $k_1 = a_{21}$, $k_2 = a_{22}$ and $k_3 = a_{23}$,

A modified Gaussian function, which was proposed by Donnay *et al.* (1996), was selected for the ferritic transformation:

$$k = k_{max} \exp\left[-\left(\frac{T - T_{nose}}{a_7}\right)^{a_8}\right] \tag{6.8}$$

The four coefficients k_{max}, T_{nose}, a_7 and a_8 allow description of all shapes of TTT curves in a quite intuitive way. Thus, k_{max} is the maximum value of k, T_{nose} is the temperature position of the 'nose' of the Gaussian function, a_7 is proportional to the width of the nose at mid-height and a_8 is related to the sharpness of the curve. The equations which were used to calculate the coefficients k_{max} and T_{nose} were:

$$k_{max} = \frac{a_5}{D_\gamma} \quad T_{nose} = Ae_3 + \frac{400}{D_\gamma} - a_6 \tag{6.9}$$

Incubation times τ_p and τ_b were introduced for the pearlitic and bainitic transformations. The following equations were used:

$$\tau_p = \frac{a_9}{(Ae_1 - T)^{a_{11}}} \exp\left[\frac{a_{10} \times 10^3}{R(T + 273)}\right] \tag{6.10}$$

$$\tau_b = \frac{a_{17} k_b}{(T_b - T)^{a_{19}}} \exp\left[\frac{a_{18} \times 10^3}{R(T + 273)}\right] \tag{6.11}$$

The incubation time is negligible in the case of the ferritic transformation for a lean chemical composition of the steel. Equation 6.2 with the modified Gaussian function in Eq. 6.8 describes properly the nucleation and growth process and the site saturation process. In calculations of the CCT diagram, an assumption was made that 5% of ferrite corresponded to the beginning of the transformation.

The start temperatures for the bainitic and martensitic transformations were functions of chemical composition:

$$T_b[°C] = a_{20} - 425[C] - 42.5[Mn] - 31.5[Ni]$$

$$T_m[°C] = a_{26} - a_{27} C_\gamma$$

The fraction of austenite which transformed into martensite, calculated according to the model of Koistinen and Marburger (described in Suehiro *et al.*, 1992; Senuma *et al.*, 1992), was:

$$X_m = 1 - \exp[-0.011(T_M - T)] \quad [6.12]$$

Equation 6.12 represents the volume fraction of martensite with respect to the whole volume of austenite, which remains at the temperature T_M. The volume fraction of martensite with respect to the whole volume of the material is

$$F_m = (1 - F_f - F_p - F_b)\{1 - \exp[-0.011(T_M - T)]\} \quad [6.13]$$

where F_f, F_p and F_b are the fractions of ferrite, pearlite and bainite, respectively, with respect to the whole volume of material.

Differential equation model

A more physically formulated model was proposed by some Japanese scientists (Suehiro *et al.*, 1992; Senuma *et al.*, 1992). This model is based on a differential equation, and therefore it can be applied easily to varying temperature conditions. In addition, several physical parameters are introduced into this model. In the remainder of this chapter, the model based on differential equations will be referred to as 'model B'. The basic principles of this model are explained using the ferritic transformation as an example.

Two differential equations are used to describe austenite–ferrite phase transformation kinetics:

- for nucleation and growth,

$$\frac{dX_f}{dt} = a_5(SIG^3)^{0.25}\left[\ln\left(\frac{1}{1-X_f}\right)\right]^{0.75}(1-X_f) \quad [6.14]$$

- for site saturation,

$$\frac{dX_f}{dt} = a_6 \times 10^{-12}\exp\left(\frac{a_8}{RT}\right)\frac{6}{D_\gamma}G(1-X_f) \quad [6.15]$$

where X_f is the transformed volume fraction, I is the rate of nucleation, G is the rate of transformation, S is the specific area of grain boundaries and D_γ is the austenite grain size.

The equations used to calculate the parameters in Eq. 6.14 are given in Table 6.1 (Suehiro *et al.*, 1992; Senuma *et al.*, 1992; Pietrzyk and Kuziak, 1999), with the following notation: R, gas constant; T, temperature in °C; $\hat{T}$, absolute temperature in K; $\Delta T = Ae_3 - T$, temperature drop below Ae_3; ΔG, Gibbs free energy calculated using ThermoCalc; r, radius of curvature of the advancing phase; D, diffusion coefficient of carbon in austenite; C_γ, average carbon content in austenite; C_α, carbon content in ferrite; C_0, initial carbon content in the steel; $C_{\gamma\alpha}$, carbon content in austenite at the γ–α phase boundary; $C_{\gamma\beta}$, carbon content in austenite at the γ–cementite phase boundary; G_p, velocity of the motion of the austenite/pearlite boundary; Q, heat generated during transformation; and ρ,

Table 6.1 Equations describing the parameters in the differential equation model

$I=\frac{D}{\sqrt{T}}\exp\left(\frac{-\alpha_7\times 10^9}{RT\,\Delta G^2}\right)$	$G=\frac{1}{2r}D\frac{C_{\gamma\alpha}-C_\gamma}{C_\gamma-C_\alpha}$	$S=\frac{6}{D_\gamma^4}$
$C_\gamma=\frac{(C_0-X_fC_\alpha)}{1-X_f}$	$r=\frac{1.14D^{0.5}(C_\gamma-C_0)}{\sqrt{(C_\gamma-C_\alpha)(C_0-C_\alpha)}}t^{0.5}$	$Q=\rho\,\Delta H\frac{\Delta X}{\Delta t}$
$X_{f0}=\frac{C_{\gamma\alpha}-C_0}{C_{\gamma\alpha}-C_\alpha}$		

density. All of the thermodynamic parameters, such as the equilibrium temperatures Ae_1 and Ae_3 and the relations between the compositions $C_{\gamma\alpha}$, $C_{\gamma\beta}$, C_α and the temperature, were determined using ThermoCalc for the chemical composition of the steel.

Numerical solution

The simulation of the ferritic transformation started when the temperature dropped below Ae_3. In the case of model A, the calculations were performed with Eq. 6.2. In the case of model B, the calculations started with Eq. 6.14 and, when the value of the derivative calculated from Eq. 6.15 became larger than that determined from Eq. 6.14, the simulation continued with Eq. 6.15. In both models, the transformed volume fraction X_f was calculated with respect to the maximum volume fraction of ferrite X_{f0} at the current temperature. Thus, this volume fraction of ferrite with respect to the whole volume of the body is $F_f = X_{f0}X_f$. During numerical simulation at a varying temperature, the current value of X_f calculated from Eq. 6.2, 6.14 or 6.15 had to be corrected to account for the change in the equilibrium (maximum) volume fraction of ferrite X_{f0}, which, according to the equations in Table 6.1, is a function of temperature. The simulation continued until the transformed volume reached 1. However, when the carbon content in austenite reaches the limiting value $C_{\gamma\beta}$ (see Table 6.1), the austenite–pearlite transformation begins in the remaining volume of austenite.

There are several coefficients in the models. These coefficients are gathered together in a vector $\mathbf{a} = \{a_1, \ldots, a_{27}\}$. The values of the coefficients were determined using inverse analysis of dilatometric tests.

Model based on the solution of the diffusion equation

The controlling role of diffusion in the austenite–ferrite transformation is well known and it is accounted for in all models, usually indirectly. Even if the transport of carbon is modelled, however, an analytical solution of Fick's second law is used, with many assumptions and simplifications. On the other hand, models of the diffusion of carbon in austenite based on the solution of a partial differential equation with a moving boundary (Segal *et al.*, 1998; Savović and Caldwell, 2003) have been well investigated and tested, and advanced numerical techniques to solve such equations are available. Contemporary numerical methods allow accurate solution of partial differential equations, accounting for nonlinearity, complex initial and boundary conditions, and complex shapes of the processed area. The application of this approach to the simulation of the decomposition of austenite into ferrite and to testing the predictive capability of such a model is described below.

The model that was used is based on the assumption that the diffusivity of carbon is the main controlling parameter in the phase transformation. The finite difference and finite element methods were used to solve Fick's second law in the case of a moving boundary (the Stefan problem).

The mathematical model is based on the solution of Fick's second law:

$$\frac{\partial c}{\partial t} = \nabla \cdot D \nabla c \qquad [6.16]$$

where D is the diffusion coefficient, c is the carbon concentration and t is the time. Solutions of Eq. 6.1 were presented by Pernach and Pietrzyk (2008) for the 1D case, for a circle in a circle and for a regular hexagon in a regular hexagon in 2D, and for a sphere in a sphere in 3D. The solution for a hexagonal austenite grain and a round ferrite grain only is presented below. The initial and boundary conditions at the interface are:

$$\begin{aligned} & c(x,y,0) = c_0 \\ & c(x,y,t) = c_{\gamma\alpha}, (x,y) \in \Gamma_0 \\ & \frac{\partial c}{\partial \mathbf{n}}(x,y,t) = 0, (x,y) \in \Gamma_1, \Gamma_2, \Gamma_3 \end{aligned} \qquad [6.17]$$

where c_0 is the carbon content in the steel; x, y are coordinates; Γ_0 is the phase boundary; Γ_1, Γ_2, Γ_3 are the edges of the hexagon which represent the austenite grain; and $\mathbf{n}$ is the unit vector normal to the surface.

The volume fraction of ferrite F_f is calculated as the ratio of the area covered by ferrite to the area of the primary austenite grain. The volume fraction of ferrite X_f with respect to the equilibrium volume fraction of ferrite at the temperature considered is calculated from the equations in Table 6.1. A finite difference solution of Eq. 6.16 with a moving boundary has been described (Pernach and Pietrzyk, 2008). The finite element solution is presented briefly below. A typical FE discretization $c(x, y) = \mathbf{n}^T\mathbf{c}$ is introduced, where $\mathbf{n} = \{n_1, n_2, \ldots, n_n\}^T$ is the vector of shape functions, $\mathbf{c} = \{c_1, c_2, \ldots, c_n\}^T$ is the vector of carbon concentrations in the nodes of the FE mesh, and n is the number of nodes in one element. The solution follows the one presented by Lenard *et al.* (1999) for the thermal problem. It is based on a variational principle, which states that the function of the concentrations c which solves Eq. 6.16 gives a minimum of a general functional, for which Eq. 6.16 is the Euler equation. The following functional meets this condition:

$$J = \int_{\Omega} \frac{D}{2}\left[\left(\frac{\partial c}{\partial x}\right)^2 + \left(\frac{\partial c}{\partial y}\right)^2\right] \mathrm{d}V - \int_{\Gamma} qc \, \mathrm{d}S \qquad [6.18]$$

where Ω is the domain of the solution, which is the remaining austenite grain, and $\Gamma = \Gamma_1 + \Gamma_2 + \Gamma_3$ is a part of the boundary of the domain Ω on which a Neumann boundary condition is applied.

Equation 6.18 describes a stationary problem when $\partial c/\partial t = 0$. Substitution of the relationship $c(x, y) = \mathbf{n}^{\mathrm{T}}\mathbf{c}$ into Eq. 6.18 and minimization of the discretized steady-state functional results in the following set of equations:

$$\mathbf{Hc} = \mathbf{b} \qquad [6.19]$$

where

$$H_{ij} = \int_{\Omega} (\nabla n_i)^{\mathrm{T}} D(\nabla n_j) \, \mathrm{d}V \qquad [6.20]$$

The vector $\mathbf{b}$ contains terms representing a Dirichlet boundary condition on Γ_0. The assumption that the concentration is a function of time leads to:

$$\mathbf{Hc} + \mathbf{C}\frac{\partial}{\partial t}\mathbf{c} = \mathbf{b} \qquad [6.21]$$

where

$$C_{ij} = \int_{V} n_i n_j \, \mathrm{d}V \qquad [6.22]$$

The assumption of a quasi-stationary state during a time step Δt and application of the Galerkin time integration scheme yields the final set of equations:

$$\widehat{\mathbf{H}}\mathbf{c} = \widehat{\mathbf{b}} \qquad [6.23]$$

where

$$\widehat{\mathbf{H}} = \left[2\mathbf{H} + \frac{3}{\Delta t}\mathbf{C}\right] \qquad \widehat{\mathbf{b}} = \left[-\mathbf{H} + \frac{3}{\Delta t}\mathbf{C}\right]\mathbf{c}_i - 3\mathbf{b} \qquad [6.24]$$

and where $\mathbf{c}_i$ is the vector of concentrations in the nodes at the beginning of the time interval.

The set of equations in Eq. 6.23 is solved for each time step, and the interface boundary is moved according to the mass balance. This motion is described by the 'Stefan problem', which is generally used for heat and mass transfer problems with phase changes, such as recrystallization, melting and solidification. This term also refers to the mathematical model describing the position of the interface, which varies with time (Pernach and Pietrzyk, 2008).

The front-tracking method was used to solve the Stefan problem numerically. In this approach, the position of the interface is continuously tracked and the diffusion equation is solved on a stationary mesh. A schematic illustration of the solution domain is presented in Fig. 6.6. Owing to symmetry, only 1/12 of the hexagon is considered.

In this model, the growth of the regular hexagon or of round ferrite in the corner of the regular hexagon of austenite is considered. Lines of constant carbon concentration in front of the phase boundary calculated using a model based on the solution of the diffusion equation for round ferrite grains are presented in Fig. 6.7 for several different cooling rates. Similar results for a regular hexagon of ferrite were published by Pernach and Pietrzyk (2008). The character of the carbon segregation can be observed in Fig. 6.7. Zero mass flux is assumed through the boundaries Γ_1, Γ_2 and Γ_3 (Eq. 6.5), which represents the influence of the adjacent growing ferrite grain (Γ_1 and Γ_2) or the adjacent austenite grain (Γ_3). The boundaries Γ_1 and Γ_2 are axes of symmetry. A Dirichlet boundary condition with the equilibrium carbon concentration is applied at the austenite–ferrite boundary (Γ_0).

The model predicts changes in the concentration as a function of time in the remaining austenite. Selected results for points located along the boundary Γ_1 at distances of 10, 20, 30 and 40 μm from the centre of the nucleus of the ferrite grain are shown in Fig. 6.8. To make the results clearer, a reasonably large austenite grain of size 100 μm was assumed. The simulations were continued until the beginning of the pearlitic (Fig. 6.8(a)) or the bainitic (Figs 6.8(b) and (c)) transformation.

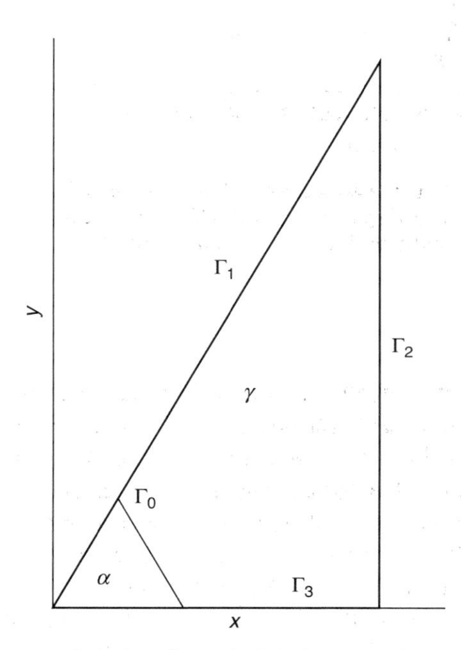

6.6 Solution domain for the two-dimensional diffusion problem.

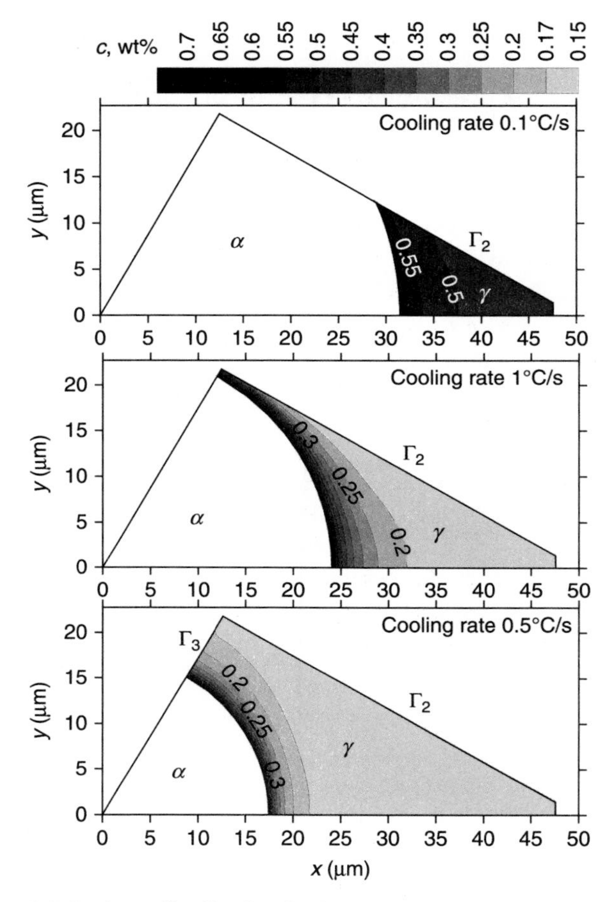

6.7 Carbon distribution in the austenite in front of a growing ferrite grain after continuous cooling at rates of 0.1°C/s, 1°C/s and 5°C/s, calculated using a model based on the diffusion equation.

Cellular automata model

The technique of cellular automata (CA) was originally developed in the early 1960s by John Von Neumann to simulate the behaviour of discrete and complex systems. The main idea of the cellular automata technique is to divide a part of a material into a one-, two- or three-dimensional lattice of finite cells. Each cell in this space is called a cellular automaton, and the lattice of cells is called a cellular automata space. Each cell in this CA space is surrounded by neighbours, which affect one another. Neighbourhoods can be specified in one, two- and three-dimensional spaces. The interactions of the cells within the CA space are based on knowledge defined by studying some particular phenomenon. At every time step,

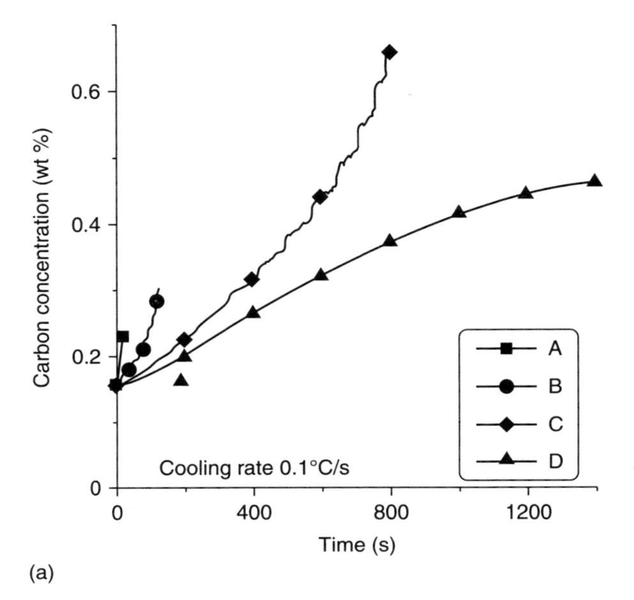

(a)

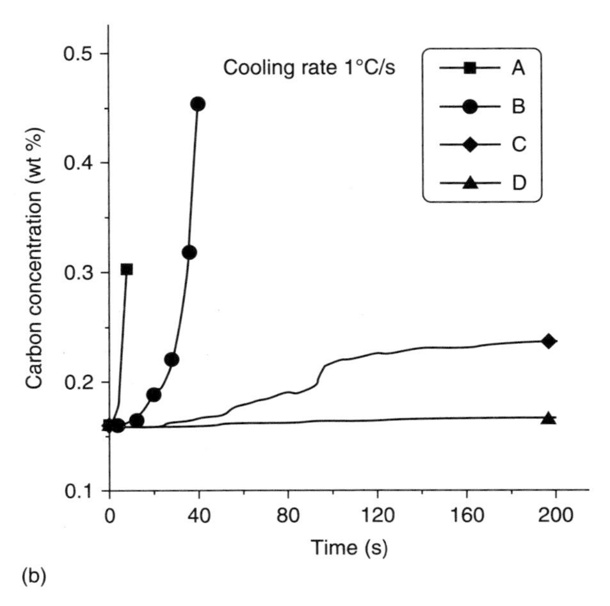

(b)

6.8 Calculated changes in the carbon distribution in the austenite at points located at distances of 10 μm (A), 20 μm (B), 30 μm (C) and 40 μm (D) from the centre of the nucleus of the ferrite grain, during continuous cooling at rates of (a) 0.1°C/s, (b) 1°C/s and (c) 5°C/s.

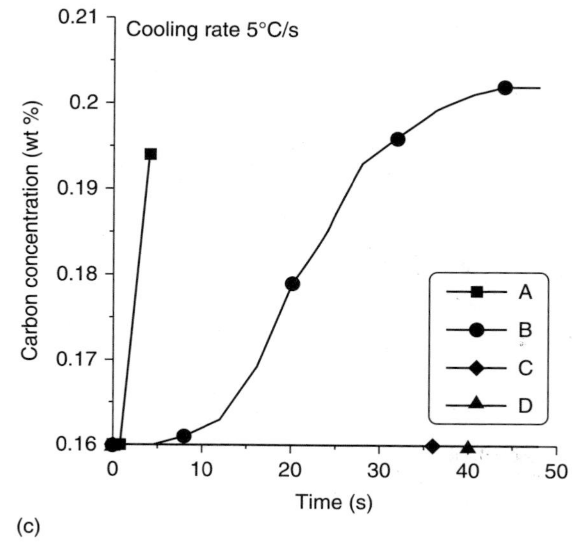

(c)

6.8 Continued.

the state of each cell in the lattice is determined by the previous states of its neighbours and of the cell itself, by a set of precisely defined transition rules:

$$Y_i^{t+1} = \varphi(Y_j^t) \qquad \text{where} \qquad j \in N(i) \tag{6.25}$$

and where $N(i)$ is the surroundings of the ith cell, Y_i is the state of the ith cell, and φ is a logical function which defines a new state of the cell i on the basis of the state of the neighbouring cells and on the basis of the values of the internal and external variables (see below).

Applications of the CA model to simulations of phase transformations can be found in the literature (Li *et al.*, 2007; Lan *et al.*, 2004); therefore, this model is presented only very briefly below. Since the transition rules control the behaviour of the cells during the calculations (i.e. during the cooling process), the definition of these rules in the process of designing a CA model critically affects the accuracy of this approach. Thus, a CA phase transformation model for the austenite–ferrite transformation has to be based on knowledge regarding two main mechanisms: the nucleation and the subsequent growth of ferrite grains. Each CA cell is described by several states and internal variables in order to properly describe the state of the material. The cell can be in three different states: ferrite (α), austenite (γ) and ferrite–austenite (α/γ), as shown in Fig. 6.9. The last of these states is used to describe CA cells located at an interface between austenite and ferrite grains. Additionally, a series of internal variables is defined to describe other necessary microstructural information. The cells contain information about, for example, how much ferrite phase is present in a particular cell, what the carbon concentration in the cell is, the length l of growth of a ferrite cell into a ferrite–austenite cell and the growth velocity v of an interface cell. These internal variables are used in the transition rules to replicate the mechanisms leading to phase transformation.

Two major transition rules are defined to describe nucleation and growth of the ferrite phase. The nucleation mechanism is a process that is stochastic in nature. To replicate this character, at the beginning of each time step a number of nuclei N_{nuc} is calculated in a probabilistic manner. Additionally, the locations of the grain nuclei are generated randomly on the grain boundaries. When a cell is selected as a nucleus, the state of that cell changes from austenite (γ) to ferrite (α). At the same time, all the neighbouring cells of the ferrite cell change their state to ferrite–austenite (α/γ), as seen in Fig. 6.9.

After a nucleus appears in the CA space, the growth of the ferrite phase is calculated in the following steps. However, the nucleation process continues, and it occurs during the entire CA simulation until the end of the transformation.

The transition rules describing the growth of ferrite grains are designed to replicate experimental observations of the mechanisms responsible for this process (Christian, 1975). The velocity of the γ/α interface is assumed to be the product of a mobility M and a driving force for interface migration F, i.e. $v = MF$. The mobility of the γ/α interface is described by:

$$M = M_0 D(T) \qquad [6.26]$$

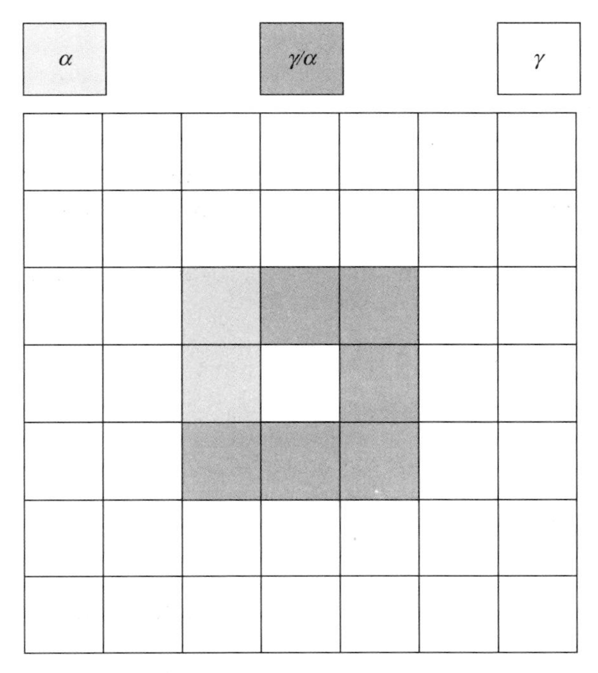

6.9 Illustration of a nucleus of the a phase and the surrounding cells in the ferrite–austenite (α/γ) state.

where M_0 is the mobility coefficient, T is the temperature and D is the diffusion coefficient. The driving force for the phase transformation is defined as:

$$F = F_{\text{chem}} + F_{\text{mech}} \tag{6.27}$$

where F_{chem} and F_{mech} are the chemical and mechanical driving forces, respectively. The influence of F_{mech} is often neglected in CA models. The chemical force is due to differences in the chemical potential of an iron atom in the austenite and ferrite phases at the interface:

$$F_{\text{chem}} = \mu^{\gamma}_{\text{Fe}} - \mu^{\alpha}_{\text{Fe}} \tag{6.28}$$

where μ^{γ}_{Fe}, μ^{α}_{Fe} are the chemical potentials of an iron atom in the austenite and ferrite phases, respectively.

The transition rules state that when the ferrite phase is present in the material, the ferrite cells grow into the austenite phase. This is realized by the following algorithm. In the current time step t, the growth length $l^{i}_{i,j}$ of a ferrite cell with indices (i, j) towards a neighbouring ferrite–austenite cell with indices (k, l) is described by (Li *et al.*, 2007):

$$l^{t}_{\text{i,j}} = \int_{t_0}^{t} v_{\text{i,j}}\, \mathrm{d}t \tag{6.29}$$

where t_0 is the time when the CA cell (i, j) changed into the ferrite state, and $v_{i,j}$ is the growth velocity of the CA cell (i, j). The ferrite volume fraction in the CA cell (k, l) calculated as a result of the ferrite growth is:

$$F^{t}_{k,l} = \sum_{1}^{N_{\text{neigh}}} f^{t}_{k,l} = \sum_{1}^{N_{\text{neigh}}} \frac{l^{t}_{i,j}}{L_{\text{CA}}} \tag{6.30}$$

where $F_{k,l}$ is the total ferrite volume fraction in the CA cell (k, l), considered as a contribution from all the neighbouring ferrite CA cells, and L_{CA} is the dimension of a CA cell in the space.

Based on these calculations, the transition rules are defined as follows (Pietrzyk *et al.*, 2010a,b):

$$Y^{t+1}_{k,l} = \begin{cases} \alpha \Leftrightarrow Y^{t}_{k,l} = \alpha/\gamma \wedge F^{t}_{k,l} > F_{\text{cr}} \\ Y^{t}_{k,l} \end{cases} \tag{6.31}$$

where $Y^{t}_{k,l}$ is the state of the cell (k, l), and

$$Y^{t+1}_{k,l} = \begin{cases} \alpha/\gamma \Leftrightarrow Y^{t}_{k,l} = \gamma \wedge Y^{t}_{i,j} = \alpha \\ Y^{t}_{k,l} \end{cases} \tag{6.32}$$

where $Y^{t}_{k,l}$ is the state of the cell (k, l) and $Y^{t}_{i,j}$ is the state of the neighbouring cell (i, j).

The CA cell changes its state from a γ/α to an α cell when the ferrite volume fraction F in the cell exceeds a critical value F_{cr}. Otherwise, the cell remains in the

γ/α state. When the cell changes its state to α, all the neighbouring cells in the γ state change their states to the γ/α state. When a change in the cell state occurs, the corresponding carbon concentration changes according to the Fe-C diagram. In the present model, the carbon concentration in the γ CA cells increases uniformly, because the diffusion problem is not considered directly. The effect of diffusion is accounted for by the relation between the phase boundary mobility and the diffusion coefficient (see Eq. 6.26).

6.4.3 Identification of the coefficients in the phase transformation models

Among the models presented here, identification was performed only for model A, based on the Avrami equation (Eq. 6.2) and described in Section 6.4.2. This model contains several coefficients, which are grouped together in the vector **a**. The values of these coefficients are different for different steels. In addition, they depend on the microstructure at the beginning of transformation and on the deformation of the austenite. The difficulties connected with the determination of these coefficients are the main factor which limits the wide application of the model to the simulation and control of phase transformations in industrial processes. A method which allows fast and easy determination of the components of the vector **a** is described below.

Inverse analysis

Identification can be performed using inverse analysis. The basic principles of this method are described in a number of publications; see, for example, Szeliga *et al.* (2006). The main aspects of inverse analysis and the results of recent research are discussed in Chapter 3 of this book. The most frequent applications of this method are connected with the determination of coefficients in rheological models of materials subjected to plastic deformation (Szeliga *et al.*, 2006; Forestier *et al.*, 2002). Torsion or compression plastometric tests are used as experiments in this analysis. Details of the application of the inverse approach to phase transformation models have been given by Kondek *et al.* (2003) and Krzyżanowski *et al.* (2006), and the general idea of the algorithm is described below. A mathematical model of an arbitrary phase transformation can be described by a set of equations:

$$\mathbf{d} = F(\mathbf{a},\mathbf{p}) \tag{6.33}$$

where $\mathbf{d} = \{d_1, \ldots, d_r\}$ is the vector of the start and finish temperatures of the transformations and the volume fractions of the structural components at room temperature, which are measured in dilatometric tests performed at a constant cooling rate; $\mathbf{a} = \{a_1, \ldots, a_l\}$ is the vector of the coefficients of the model; and $\mathbf{p} = \{p_1, \ldots, p_k\}$ is the vector of the process parameters such as the cooling rates, the austenite grain size and the deformation of the austenite.

When the vectors **p** and **a** are known, the solution of the problem in Eq. 6.33 is called a direct solution. The inverse solution of the problem in Eq. 6.33 is defined as the determination of the components of the vector **x** for known vectors **d** and **p**. If the problem is linear, the inverse function can be found, and the problem can be often solved analytically. For phase transformations, however, the relations are strongly nonlinear, and optimization techniques must be used to solve the inverse problem.

The objective of the inverse analysis is the determination of the optimum components of the vector **a**. This is achieved by searching for the minimum, with respect to the vector **a**, of the objective function defined as the square root error between the measured and calculated components of the vector **d**:

$$\Phi(\mathbf{a},\mathbf{p}) = \sum_{i=1}^{n} \beta_i [\mathbf{d}_i^{\mathrm{c}}(\mathbf{a},\mathbf{p}_i) - \mathbf{d}_i^{\mathrm{m}}]^2 \qquad [6.34]$$

where $\mathbf{d}^{\mathrm{m}}_i$ is a vector containing the measured values of the output parameters, $\mathbf{d}^{\mathrm{c}}_i$ is a vector containing the calculated values of the output parameters, the β_i are the weights of the points ($i = 1, \ldots, n$) and n is the number of measurements. The measurements $\mathbf{d}^{\mathrm{m}}_i$ are obtained from dilatometric tests carried out at constant cooling rates. The components $\mathbf{d}^{\mathrm{c}}_i$ are calculated using one of the models of the direct problem described above.

The identification of the parameters of a phase transformation model is composed of two parts. The first is solution of the direct problem, based on the model. The second part is solution of the inverse problem, in which optimization techniques are used.

The results of dilatometric tests (see Section 6.3.2), including measurements of the start and end temperatures of the transformations and the volume fractions of the phases after cooling to room temperature, are used as an input to the inverse analysis. Thus, in the particular case of a phase transformation model, the objective function in Eq. 6.34 is defined as:

$$\Phi(\mathbf{a},\mathbf{p}) = \sqrt{\frac{1}{n}\sum_{i=1}^{n}\left(\frac{T_{im}-T_{ic}}{T_{im}}\right)^2 + \frac{1}{k}\sum_{i=1}^{k}\left(\frac{X_{im}-X_{ic}}{X_{im}}\right)^2} \qquad [6.35]$$

where T_{im} and T_{ic} are the measured and calculated start and end temperatures of the phase transformations, n is the number of temperature measurements, X_{im} and X_{ic} are the measured and calculated volume fractions of the phases at room temperature, and k is the number of measurements of the volume fractions of phases.

Sensitivity analysis

Any phase transformation model can be identified using the technique described in the previous section; see, for example, Kondek *et al.* (2003) and Kuziak and Pietrzyk (2000). The identification of the model A described in Section 6.4.2 is presented below. Owing to the large number of coefficients in this model, problems of the effectiveness and efficiency of optimization techniques need to be considered. The uniqueness of the solution needs to be discussed as well. Some light can be shed on the solution of these problems by performing a sensitivity analysis of the output of the model with respect to the coefficients of the model. This method deals with the question of which factors in the physical model or computer simulation are really important. Various aspects of the application of sensitivity analysis to materials science are discussed in Chapter 3 of this book. The application of this method to the evaluation of phase transformation models is described briefly below.

The finite difference method was used in the present work. The elementary effect of the ith factor at a given point $\mathbf{a}$ in the domain Ω is defined as:

$$\varsigma_i(\mathbf{a}) = \frac{a_i}{d(\mathbf{a})} \frac{d(a_1,\ldots,a_{i-1},a_i+\Delta,a_{i+1},\ldots,a_k) - d(\mathbf{a})}{\Delta} \qquad [6.36]$$

where d is one of the outputs of the model. The vector $\mathbf{a}$ is any point from the region Ω such that the perturbed point $\mathbf{a} + \Delta$ is still in Ω.

The finite difference algorithm was run for all parameters of the phase transformation model. The model output was either the start and end temperatures of the transformations or the volume fractions of phases after cooling. The sensitivity of the output with respect to the coefficients $\mathbf{a}$ in the model was determined for various cooling rates in the range 0.02–500°C/s. These results give local values of sensitivities; owing to lack of space, they will not be presented here. In order to obtain a better impression of the effect of the coefficients on the model output, global values of the sensitivity were calculated as:

$$\tilde{\varsigma} = \int_{C_r} \varsigma \, \mathrm{d}C_r \qquad [6.37]$$

where $\tilde{\varsigma}$ is the global sensitivity factor, ζ is the sensitivity factor calculated from Eq. 6.36 and C_r is the cooling rate.

The global sensitivity factors account for both the values of the sensitivity and the range of cooling rates in which the model output is sensitive to the coefficient considered. The values of the sensitivity factors ς are scaled and represent the sensitivity quantitatively, but the global sensitivity factors $\tilde{\varsigma}$ can be used for comparisons only. Figure 6.10 shows the absolute values of the sensitivity factors for the start and end temperatures of the transformations with respect to all coefficients in the phase transformation model. Similar results for the volume fractions of the structural components are shown in Fig. 6.11. Since the sensitivity factors for some coefficients are much greater than others, the scale has been cut off at a certain level and the values of the sensitivities exceeding this level are given as numbers above the bars.

It can be seen that, as far as the start and end temperatures of the transformations are considered, the coefficients a_4–a_8 control the ferritic and pearlitic transformations, the coefficients a_9–a_{16} control the pearlitic transformation (the

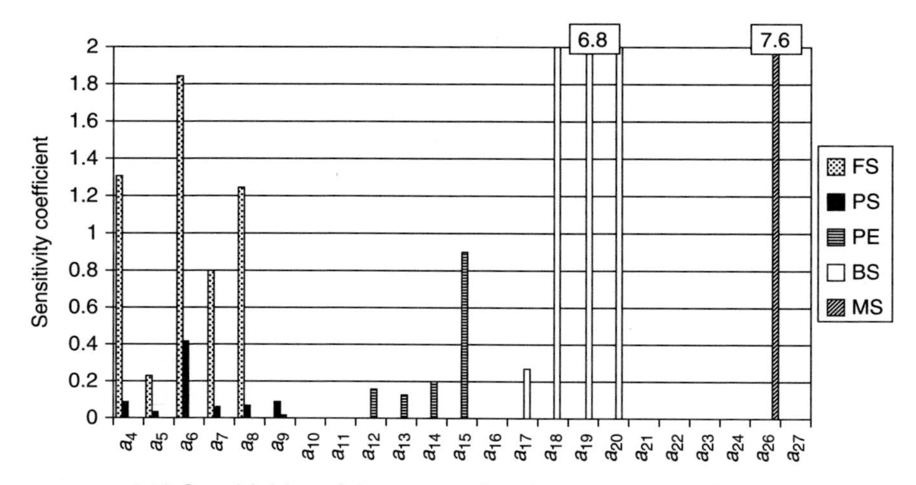

6.10 Sensitivities of the start and end temperatures of the transformation with respect to the coefficients in the model. (FS, ferrite start; PS, pearlite start; PE, pearlite end; BS, bainite start; MS, martensite start).

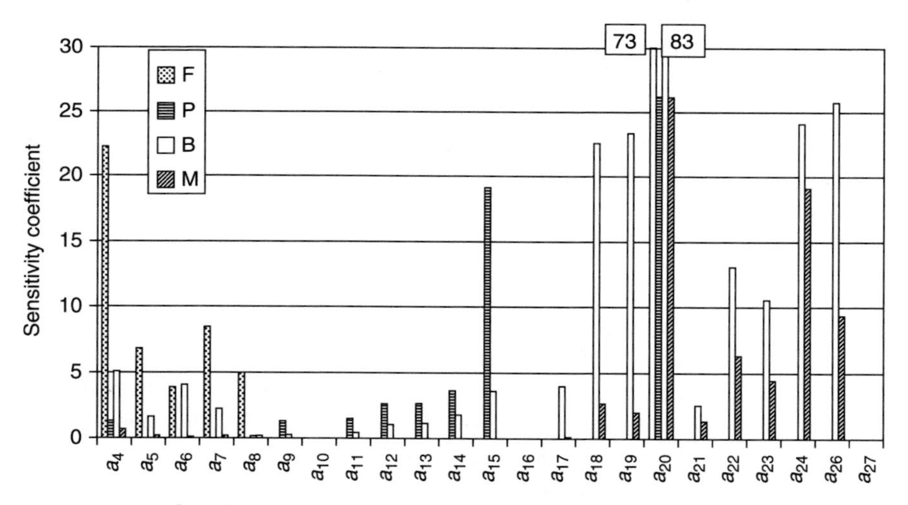

6.11 Sensitivities of the volume fractions of structural components with respect to the coefficients in the model (F, ferrite; P, pearlite; B, bainite; M, martensite).

influence of the coefficients a_{10}, a_{11} and a_{16} is negligible), the coefficients a_{17}–a_{24} control the bainitic transformation and the coefficients a_{26} and a_{27} control the start temperature of the martensitic transformation. There are some cross-relations, but the sensitivities of one transformation with respect to the coefficients of the model of another transformation are small.

Figure 6.11 shows the sensitivity factors for the volume fractions of the structural components with respect to the coefficients of the phase transformation model. The relations here are more complex than those observed for the temperatures. In addition, the sensitivity factors are much higher. The coefficient a_{20} has an influence on the volume fractions of all structural components. Since the sum of the volume fractions has to be equal to 1, each coefficient influences at least two structural components.

The coefficient a_{20} has a particularly strong influence on the volume fractions of ferrite and bainite. The sensitivity factor in this case is about three times higher than the average value for the remaining coefficients. The coefficients a_{10}, a_{16} and a_{27} show no influence on the volume fractions of the structural components.

6.4.4 Identification, testing and validation of models

Dilatometric tests were performed on the DP steel, and identification of the phase transformation model was performed using inverse analysis. The values of the coefficients **a** obtained from this analysis for the objective function in Eq. 6.35 are given in Table 6.2. The general model for all transformations contains 27 coefficients altogether. However, only 23 of these coefficients are active in model A for the DP steel considered. A comparison of predicted and calculated CCT diagrams is shown in Fig. 6.12, where A = austenite, F = ferrite, P = pearlite, B = bainite and M = martensite. A similar comparison for the volume fractions of the structural components is shown in Fig. 6.13.

6.5 Application in rolling and annealing of dual-phase steels

Simulations of laminar cooling and continuous annealing of a DP steel are presented in this section.

6.5.1 Laminar cooling

Simulations of the whole process of continuous hot rolling of DP steel strips were described by Pietrzyk *et al.* (2009), and this description will not be repeated in the

Table 6.2 Coefficients of the phase transformation model determined using inverse analysis

a_4	a_5	a_6	a_7	a_8	a_9	a_{10}	a_{11}	a_{12}	a_{13}	a_{14}	a_{15}
1.69	0.858	188	39.06	1.78	64.76	1.106	0.618	0.153	0.0085	0.001	1.28
a_{16}	a_{17}	a_{18}	a_{19}	a_{20}	a_{21}	a_{22}	a_{23}	a_{24}	a_{26}	a_{27}	
0.007	1600	64.64	3.495	669	0.118	0.074	0.344	1.037	421.7	1.83	

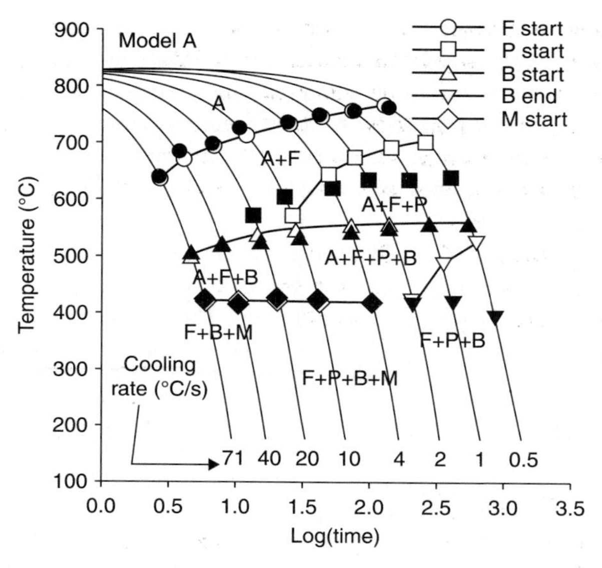

6.12 Comparison of CCT diagrams obtained from measurements (filled symbols) and calculated from model A with the coefficients in Table 6.2 (open symbols).

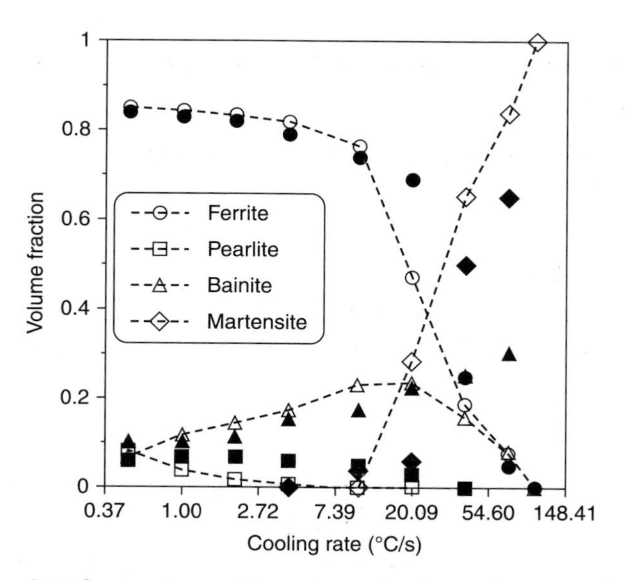

6.13 Comparison of the volume fractions of structural components obtained from measurements (filled symbols) and calculated from model A with the coefficients in Table 6.2 (open symbols).

present work. The finite element code described in Lenard *et al.* (1999) was used to simulate the hot-strip rolling process in a six-stand finishing train in an industrial steel plant. A microstructure evolution model based on the fundamental work of Sellars (1979) was implemented in an FE code and used to predict the grain size at the beginning of the transformation. The data obtained from these FE calculations were used as input parameters for simulations of laminar cooling.

An arbitrary laminar cooling system, composed of $n_1 = 40$ boxes in the first section and $n_2 = 40$ boxes in the second, was considered in this analysis. The length of each box was 1 m. The heat transfer coefficient for laminar water cooling was calculated on the basis of an equation proposed by Hodgson *et al.* (1991). The objective function was formulated as the square root error between the predicted and expected volume fractions of martensite in the steel:

$$\Phi(\mathbf{a}) = \sqrt{\left(\frac{F_{mr} - F_{mc}(\mathbf{a})}{F_{mr}(\mathbf{a})}\right)^2} \tag{6.38}$$

where F_{mr} and F_{mc} are the required and calculated volume fractions of martensite.

The vector **a** contains the optimization variables. In an initial approach, two variables were considered: a_1, the number of active boxes in the first section of the laminar cooling system, and a_2, the number of the first active box in the second section. The latter variable determines the time interval between the two sections when the strip is cooled slowly in air.

The primary optimization when the exit velocity from the last stand, v_6, was equal to 8 m/s was described by Pietrzyk *et al.* (2009) and Madej *et al.* (2010). A higher velocity, $v_6 = 10$ m/s, was considered in the present work. An increase in this velocity requires an increase in the distance between the active sections where accelerated cooling occurs.

Since only two optimization variables are considered, it is possible to plot the objective function given in Eq. 6.38. This plot, obtained for a required volume fraction of martensite $F_m = 0.28$, is shown in Fig. 6.14. It can be seen that the objective function is very sensitive to the number of active boxes in the first

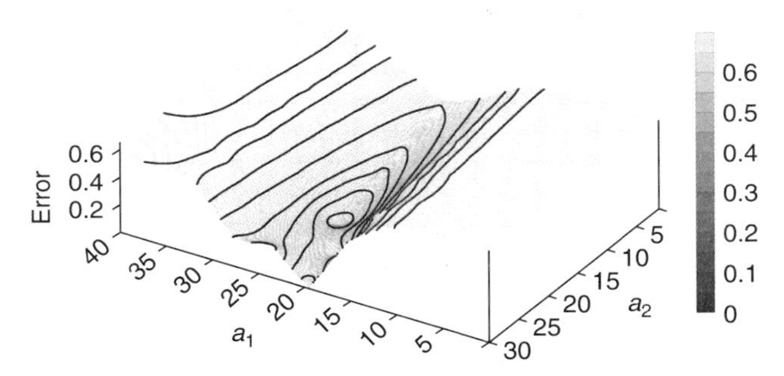

6.14 Plot of the cost function for a required volume fraction of martensite $F_{mr} = 0.28$.

section, a_1. A very pronounced minimum is observed for about 20–22 active boxes in this section. The minimum is less sensitive to the number of the first active box in the second section, a_2. The required volume fraction of martensite, in the range $F_m = 0.26–0.3$, was obtained for $a_2 = 15–25$. However, application of the simplex method to search for the minimum of the objective function gave the optimum value $\Phi = 0.025$ for the values $a_1 = 24$ and $a_2 = 21$.

6.5.2 Continuous annealing

Controlled cooling after continuous annealing after cold rolling is the second alternative for obtaining a dual-phase microstructure. This technology is used for thinner strips. A typical thermal cycle for this process is shown in Fig. 6.4(a). The modelling of continuous annealing involves recrystallization of ferrite, a ferrite–austenite transformation during heating and an austenite–ferrite transformation during cooling. The Avrami equation is used to describe the kinetics of recrystallization of ferrite:

$$X = \left[1 - \exp\left(-\frac{t}{\tau}\right)^n\right] \quad \text{where} \quad \frac{1}{\tau} = A\exp\left(-\frac{Q}{RT}\right) \qquad [6.39]$$

X is the recrystallized volume fraction, t is the time, R is the gas constant and T is the absolute temperature. The coefficients in Eq. 6.39 for the DP steel investigated were determined on the basis of experimental data, and the following values were obtained: $n = 1.98$, $A = 3.7 \times 10^{11}$ and $Q = 226\,000$ J/mol. These coefficients gave very good agreement with measurements in laboratory tests.

The results of a simulation of the whole continuous annealing cycle are presented in Fig. 6.15. It can be seen in this figure that the volume fraction of

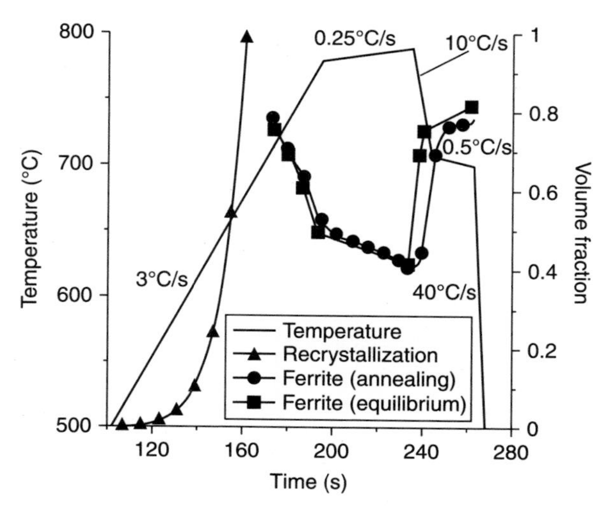

6.15 Temperature, ferrite recrystallization kinetics and changes in the volume fraction of ferrite during continuous annealing of DP steel.

ferrite is close to the equilibrium value during the heating stage of the process, which is characterized by relatively slow changes in the temperature. A non-equilibrium state is observed during fast heating or cooling. In the case presented here, the final volume fraction of martensite was 22%. The model presented here can be applied to the optimization of the heating schedule, using the objective function formulated in Eq. 6.38, similarly to what was done above for laminar cooling.

6.6 Discussion and future trends

Several models for the characterization of phase transformation kinetics, of various complexity and level of predictive capability, have been presented in this chapter. The model based on the Avrami equation (Eq. 6.2), where the coefficient k is a function of temperature, is the simplest one and is reasonably easy to use for identification. This model gives good results in simulations of industrial processes involving phase transformations. The accuracy and the fitting of experimental data can be improved by proper selection of the function $k(T)$. The sensitivity analysis described in Section 6.4.3 is helpful in selection of the best function.

The model based on the differential equations (Eqs 6.14 and 6.15) is more flexible and has better predictive capability. The physical meaning of some of the coefficients is another advantage of this model. Numerous tests and applications of this model to various industrial processes (Pietrzyk and Kuziak, 1999; Kuziak and Pietrzyk, 2000; Kondek *et al.*, 2003) have shown, however, that the slight improvement in the accuracy and flexibility of the solution does not compensate for the greater complexity of the computer program and problems with the identification of the model.

The model based on the finite element solution of the diffusion equation (Eq. 6.16) seems to be the most promising. It simulates, with reasonable accuracy, the kinetics of the ferritic transformation. The accuracy may be improved in the future by a more realistic description of the mobility of the phase boundary, depending on mechanisms controlling phase transformations. Beyond the kinetics of the transformation, this model supplies information about the distribution of the carbon concentration in the remaining austenite. This information can then be used for local prediction of such parameters as the retained austenite content, the start temperatures of the bainitic and martensitic transformations, and local properties (e.g. hardness) of the bainite and martensite. It creates interesting capabilities regarding prediction of the morphology and hardness of the martensite islands in DP steels, which are parameters that influence the behaviour of these steels in use. This model will be developed extensively in the future by the present authors.

The model based on cellular automata, and a method based on connection of the cellular automata model with the finite element method (CAFE), can deliver a new quality of information regarding phase transformations. The capability to account for the state of the microstructure, including the size, shape and orientation of grains, and inclusions, strain localization, microshear bands and shear bands, with local inhomogeneity of these features, is the main advantage of this model. This approach will be improved further and a CA-based model of phase transformations will become part of a digital material representation, the idea of which was developed at AGH and was described by Madej (2010) and Rauch and Madej (2010). This idea is under extensive development by researchers in other laboratories as well (see, for example, Bernacki *et al.*, 2007). The 'DMR' approach provides the possibility to take the features of a complex microstructure into account (see Fig. 6.16). A detailed numerical analysis of the inhomogeneities in a microstructure can be performed without the need to perform costly laboratory analysis, such as optical, scanning electron or transmission electron microscopy.

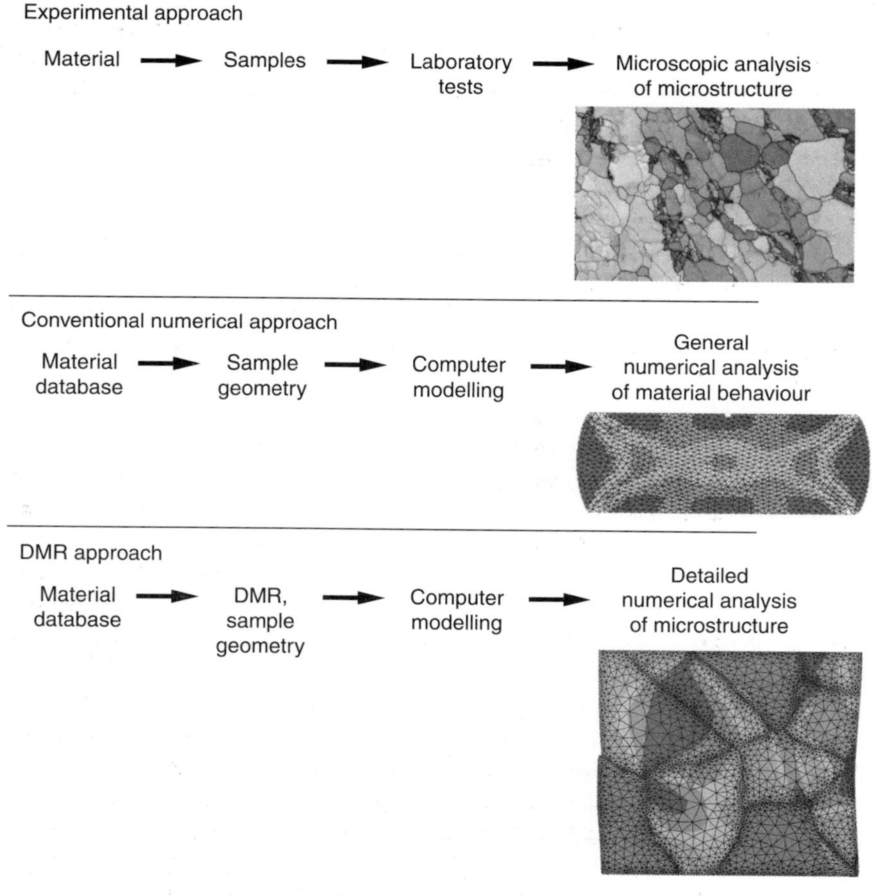

6.16 The basic concept of the DMR approach in comparison with the conventional approach and experimental investigation (Madej, 2010).

6.7 Sources of further information and advice

As has already been stated, the modelling of phase transformation is a wide topic, which is based on thermodynamics and includes a variety of methods. Recapitulating the author's research in this field, it can be concluded that the approaches presented in this chapter are limited to four approaches of different complexity and predictive capability. The models based on the Avrami equation (Eq. 6.2) and on differential equations (Eqs 6.14 and 6.15) are reasonably well researched, and the possibilities for their further development are limited. Research should instead focus on applications of these methods to new materials. In contrast, large prospects for further development of the models based on the diffusion equation (Eq. 6.16) and on cellular automata still exist. The former method should be improved by a better description of the influence of the coherence of interfaces on the progress of phase transformations and the morphology of the resulting phases. Even more possibilities for improvement exist for the latter models. Accounting for new features of the microstructure and making multiscale connections to the nanoscale are the most important suggestions for further research.

A new trend in the steel industry is based upon controlling the carbon distribution during phase transformations occurring under non-equilibrium conditions. This refers specifically to the third generation of AHSS for the automobile industry. In consequence, the development of models of phase transformations under approximations of local equilibrium, paraequilibrium or negligible partitioning (Van der Vent and Delaey, 1996) is specifically required. This concept is very promising, since it may lead to new technologies resulting in the development of steels characterized by the combination of high strength and extremely good ductility, which has not been achieved so far.

6.8 References

Avrami M (1939), Kinetics of phase change. I. General theory, *Journal of Chemical Physics*, 7, 1103–1112.

Avrami M (1940), Kinetics of phase change. II. Transformation–time relations for random distribution of nuclei, *Journal of Chemical Physics*, 8, 212–224.

Avrami M (1941), Kinetics of phase change. III. Granulation, phase change, and microstructure, *Journal of Chemical Physics*, 9, 177–184.

Bernacki M, Chastel Y, Digonnet H, Resk H, Coupez T, Logé RE (2007), Development of numerical tools for the multiscale modelling of recrystallization in metals, based on a digital material framework, *Computer Methods in Materials Science*, 7, 141–149.

Cahn JW (1956), Transformation kinetics during continuous cooling, *Acta Metallurgica*, 4, 572–575.

Christian JW (1975), *The Theory of Transformations in Metals and Alloys*, Pergamon Press, Oxford, 2nd edn.

Donnay B, Herman JC, Leroy V, Lotter U, Grossterlinden R and Pircher H (1996), Microstructure evolution of C–Mn steels in the hot deformation process: the STRIPCAM model, in Beynon JH, Ingham P, Teichert H and Waterson K, *Proceedings of the Conference on Modelling of Metal Rolling Processes*, London, 23–35.

Forestier R, Massoni E and Chastel Y (2002), Estimation of constitutive parameters using an inverse method coupled to a 3D finite element software, *Journal of Materials Processing Technology*, 125, 594–601.

Hack K (1996), *The SGTE Casebook: Thermodynamics at Work*, Materials Modelling Series, Institute of Materials, London.

Hodgson PD, Browne KM, Collinson DC, Pham TT and Gibbs RK (1991), A mathematical model to simulate the thermomechanical processing of steel, in: *Quenching and Carburizing*, Melbourne, 139–159.

Kondek T, Kuziak R and Pietrzyk M (2003), Identification of parameters of phase transformation models for steels, *Steel GRIPS*, 1, 59–66.

Krzyżanowski M, Beynon JH, Kuziak R and Pietrzyk M (2006), Development of technique for identification of phase transformation model parameters on the basis of measurement of dilatometric effect – direct problem, *ISIJ International*, 46, 147–154.

Kuziak R and Pietrzyk M (2000), Physical and mathematical simulation of phase transformation during accelerated cooling of eutectoid steel rods, in *Proceedings of the 42nd MWSP Conference*, Toronto, 101–110.

Lan YJ, Li DZ and Li YY (2004), Modeling austenite decomposition into ferrite at different cooling rate in low-carbon steel with cellular automaton method, *Acta Materialia*, 52, 1721–1729.

Lenard JG, Pietrzyk M and Cser L (1999), *Mathematical and Physical Simulation of the Properties of Hot Rolled Products*, Elsevier, Amsterdam.

Li D, Xiao N, Lan Y, Zheng C and Li Y (2007), Growth modes of individual ferrite grains in the austenite to ferrite transformation of low carbon steels, *Acta Materialia*, 55, 6234–6249.

Madej Ł (2010), Digital material representation – new perspectives in numerical simulations of inhomogenous deformation, *Computer Methods in Materials Science*, 10, 143–155.

Madej Ł, Rauch Ł and Pietrzyk M (2010), Hybrid knowledge system for optimization of rolling process for DP steels, *Transactions of NAMRI/SME*, 38, 475–482.

Pernach M and Pietrzyk M (2008), Numerical solution of the diffusion equation with moving boundary applied to modeling of the austenite–ferrite phase transformation, *Computational Materials Science*, 44, 783–791.

Pietrzyk M and Kuziak R (1999), Coupling the thermal–mechanical finite-element approach with phase transformation model for low carbon steels, in Covas J (ed.), *Proceedings of the 2nd ESAFORM Conference on Material Forming*, Guimaraes, Portugal 525–528.

Pietrzyk M, Kusiak J, Kuziak R and Zalecki W (2009), Optimization of laminar cooling of hot rolled DP steels, in *XXVIII Verformungskundliches Kolloquium*, Planneralm, 285–294.

Pietrzyk M, Madej Ł, Rauch Ł, Spytkowski P and Kusiak J (2010a), Conventional and multiscale modelling of austenite decomposition during laminar cooling of hot rolled DP steels, in *XXIX Verformungskundliches Kolloquium*, Planneralm, 41–46.

Pietrzyk M, Madej Ł, Rauch Ł and GoŁąab R (2010b), Multiscale modelling of microstructure evolution during laminar cooling of hot rolled DP steel, *Archives of Civil and Mechanical Engineering*, 10, 57–67.

Rauch Ł and Madej Ł (2010), Application of the automatic image processing in modeling of the deformation mechanisms based on the digital representation of microstructure, *International Journal for Multiscale Computational Engineering*, 8, 343–356.

Ronda J, Oliver GJ (2000), Consistent thermo-mechanical-metallurgical model of welded steel with unified approach to derivation of phase evolution laws and transformation induced plasticity, *Computer Methods in Applied Mechanics and Engineering*, 189, 361–417.

Savović S and Caldwell J (2003), Finite difference solution of one-dimensional Stefan problem with periodic boundary conditions, *International Journal of Heat and Mass Transfer*, 46, 2911–2916.

Scheil E (1935), Anlaufzeit der Austenitumwandlung, *Archiv Eisenhüttenwesen*, 12, 565–567.

Segal G, Vuik K, Vermolen F (1998), A conserving discretization for the free boundary in a two-dimensional Stefan problem, *Journal of Computational Physics*, 141, 1–21.

Sellars CM (1979), Physical metallurgy of hot working, in Sellars CM and Davies GJ (eds), *Hot Working and Forming Processes*, Metals Society, London, 3–15.

Senuma T, Suehiro M and Yada H (1992), Mathematical models for predicting microstructural evolution and mechanical properties of hot strips, *ISIJ International*, 32, 423–432.

Suehiro M, Senuma T, Yada H and Sato K (1992), Application of mathematical model for predicting microstructural evolution to high carbon steels, *ISIJ International*, 32, 433–439.

Szeliga D, Gawąd J and Pietrzyk M (2006), Inverse analysis for identification of rheological and friction models in metal forming, *Computer Methods in Applied Mechanics and Engineering*, 195, 6778–6798.

Van der Vent A and Delaey L (1996), Models for precipitate growth during the $\gamma \rightarrow \alpha + \gamma$ transformation in Fe–C and Fe–C–M alloys, *Materials Science*, 40, 181–264.

7

Determining unified constitutive equations for modelling hot forming of steel

J. LIN, Imperial College London, UK,
J. CAO, RTC Innovation Ltd, UK
and D. BALINT, Imperial College London, UK

Abstract: The chapter begins by introducing unified viscoplastic constitutive equations for metals deforming at high temperature. Numerical methods for solving the equations are presented. Advanced optimisation methods and procedures, including the formulation of objective functions, for the determination of constitutive equations from experimental data are described. The chapter includes case studies of the determination of constitutive equations from experimental data for a number of materials deforming under different conditions. A summary is given at the end of the chapter.

Key words: viscoplastic constitutive equations, optimisation, objective function, hot forming, warm forming.

7.1 Introduction

During industrial hot metal processing operations, such as multipass hot rolling, hot forging and hot extrusion, deformation takes place in a series of passes separated by intervals of time (Lin and Dean, 2005). The dynamic microstructural changes which occur during deformation are dependent on the strain rate and temperature and the initial microstructure of the material. These determine the flow stress and hence the working forces and also control the stored energy present at the end of deformation. This energy drives the static microstructural changes of recovery and recrystallisation, thereby influencing the kinetics of these thermally-activated changes. Grain growth may follow recrystallisation, and the static microstructure changes between passes determine the initial microstructure for entry to the next pass. During a hot deformation process, micro-voids and micro-defects, or damage, may be created, which is related to the stress state, deformation rate and microstructure of the deforming material (Lin, 2003; Lin *et al.*, 2005). Therefore, unified viscoplastic constitutive equations are required to model the interactions between the deformation rate, microstructure and damage evolution.

The constitutive equations used to describe the viscoplastic behaviour of materials are a system of ordinary differential equations (ODEs) (Cao *et al.*, 2008), which become more complex as additional state variables are introduced to accurately describe the physical and mechanical phenomena occurring in materials. Normally, there are no analytical solutions, and a numerical integration method is required to solve the resulting initial value problems (Cao *et al.*, 2008).

The material constants arising in the equations need to be determined from experimental data using optimisation techniques, such as the evolutionary programming (EP) method (Li *et al.*, 2002). This is difficult and problematic, and the aim of this chapter is to introduce a technique for determining unified constitutive equations from experimental data. This chapter first introduces techniques for formulating mechanism-based viscoplastic constitutive equations for hot forming applications. Numerical integration methods for solving ODE-type constitutive equations are then presented. The unitless universal objective function for optimisation and a combined efficient searching method to find a global minimum are presented; these can be used to determine the constitutive equations from experimental data efficiently. At the end of this chapter, a few sets of constitutive equations that have been determined are provided for a number of applications.

7.2 The form of unified constitutive equations for hot metal forming

7.2.1 Deformation mechanisms in hot metal forming

When metals deform at a temperature above about one third of the absolute melting temperature, the flow stress of the material is sensitive to the strain rate. This phenomenon is known as viscoplasticity. The theory of viscoplasticity describes the flow of matter by creep, which, in contrast to plasticity, depends on time. For metals and alloys, creep corresponds to mechanisms linked to the movement of dislocations in grains – climb, deviation and polygonisation – with superposed effects of intercrystalline gliding, i.e. grain rotation and grain boundary sliding. These deformation mechanisms begin to arise as soon as the temperature is greater than approximately one third of the absolute melting temperature. Viscoplastic behaviour of metals can be observed in warm and hot forming processes (Lin, 2003).

7.2.2 Modelling of viscoplastic flow, recrystallisation and hardening

In the case of viscoplasticity, the introduction of a flow rule needs the concept of the 'elastic domain' (Chaboche, 1977), defined by

$$G = J_2\left(\sigma_{ij} - X_{ij}\right) - R - k \le 0 \quad i, j = 1,2,3 \qquad [7.1]$$

where the components X_{ij} are related to the state of 'kinematic hardening', which displaces the centre of the yield surface (Armstrong and Frederick, 1966), as shown in Fig. 7.1. The scalar variable R is the amount of 'isotropic hardening', which expands the yield surface (Chaboche and Rousselier, 1983; Lin *et al.*, 1996), and k is the 'initial threshold stress', which describes the initial yield surface (Karim and Backofen, 1972). In plasticity, the associated flow rule requires the orthogonality of the plastic strain rate $\dot{\varepsilon}^{p}_{ij}$ to the yield surface at the point σ_{ij} that belongs to that surface (Skrzypek, 2000). In the case of viscoplasticity, the point σ_{ij} lies outside the actual yield surface. In other words, $G > 0$ (Skrzypek, 2000). To accommodate the existence of a viscoplastic potential as the extension of G beyond the yield surface, Rice (1970) suggested that the viscoplastic potential $\psi = \psi(\sigma_{ij})$ should be expressed as a power function of G:

$$\psi\left(\sigma_{ij}\right)=\left(\frac{A_1}{A_2}\right)\cosh\left\{A_2\left[J_2\left(\sigma_{ij}-X_{ij}\right)-R-k\right]_+\right\} \quad [7.2]$$

where A_1 and A_2 are material parameters and $\langle..\rangle_+$ denotes the positive part of $\langle..\rangle$; $\langle..\rangle_+$ equals zero when $\langle..\rangle < 0$. In view of the normality rule, the 'Mises-type viscoplastic flow rule for isotropic materials' is given as (Chaboche and Rousselier, 1983; Lin *et al.* 1993; Lin, 2003):

$$\frac{d\varepsilon^{p}_{ij}}{dt = \dot{\lambda}}\left(\frac{\partial\psi}{\partial\sigma_{ij}}\right)=\frac{3}{2}\frac{\left(\sigma'_{ij}-X'_{ij}\right)}{J_2\left(\sigma_{ij}-X_{ij}\right)}\dot{\varepsilon}^{p}_{e} \quad [7.3]$$

where

$$\dot{\varepsilon}^{p}_{e} = A_1\sinh\left\{A_2\left[J_2\left(\sigma_{ij}-X_{ij}\right)-R-k\right]_+\right\} \quad [7.4]$$

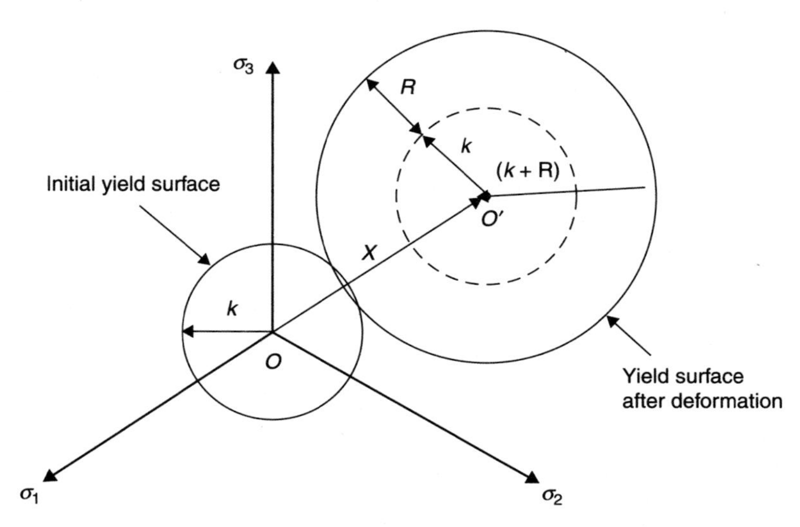

7.1 Schematic diagram of cyclic hardening: plastic yield locus, with isotropic and kinematic hardening.

is the effective viscoplastic strain rate, and σ'_{ij} and X'_{ij} are the deviatoric parts of the components σ_{ij} and X'_{ij}, respectively. $\sigma'_{ij} - X'_{ij} = (\sigma_{ij} - X_{ij}) - (\sigma_{kk} - X_{kk})/3$ and $J_2(\sigma_{ij} - X_{ij}) = (\sigma_{ij} - X_{ij}):(\sigma_{ij} - X_{ij})$ are the deviatoric and effective parts of the stress, respectively.

Kinematic hardening is essential for modelling unloading and strain reversals, especially for the purpose of predicting cyclic plasticity (Chaboche and Rousselier, 1983; Lin *et al.*, 1996). In hot and warm metal forming processes, however, owing to the occurrence of grain rotation, which tends to eliminate the development of anisotropy, it is commonly recognised that it is not necessary to model kinematic hardening (Cheong *et al.*, 2000). Therefore, X_{ij} and X'_{ij} X'_{ij} in Eq. 7.4 can be disregarded:

$$\dot{\varepsilon}_e^p = A_1 \sinh A_2 \langle \sigma_e - R - k \rangle_+ \qquad [7.5]$$

Another important attribute of hot metal working is that the material behaviour is sensitive to grain growth (Arieli and Rosen, 1977; Ghosh and Hamilton, 1979), since grain rotation and boundary sliding is the dominant deformation mechanism. Small grains make grain rotation easier, and thus the flow stress is reduced. Thus the effects of the grain size d need to be incorporated into the formulation of the corresponding flow rule. This can be achieved by the following definition (Lin and Dunne, 2001):

$$\dot{\varepsilon}_e^p = \frac{1}{d^{\gamma_1}} A_1 \sinh A_2 \langle \sigma_e - R - k \rangle_+ \qquad [7.6]$$

where γ_1 is the grain-size exponent. This is consistent with the experimental observations that $\dot{\varepsilon}_e^p \propto 1/d^{\gamma_1}$ (Edington *et al.*, 1976; Ghosh and Hamilton, 1979).

7.2.3 Unified uniaxial viscoplastic constitutive equations

When typical physical phenomena are considered, a set of unified superplastic constitutive equations can be written as follows in uniaxial form (Lin *et al.*, 2005):

$$\dot{\varepsilon}^p = \frac{A_1}{d^{\gamma_1}} \sinh A_2 \langle \sigma - R - k \rangle_+ \qquad [7.7]$$

$$\dot{S} = H[x\bar{\rho} - \bar{\rho}_c(1-S)](1-S)^{\gamma_s} \qquad [7.8]$$

$$\dot{x} = \mathrm{A}_3(1-x)\bar{\rho} \qquad [7.9]$$

$$\dot{\bar{\rho}} = (d/d_0)^{\gamma_d}(1-\bar{\rho})\dot{\varepsilon}^p - c_1\bar{\rho}^{\gamma_2} - [c_2\bar{\rho}/(1-S)]\dot{S} \qquad [7.10]$$

$$\dot{R} = B\dot{\bar{\rho}}^{n_0} \qquad [7.11]$$

$$\dot{d} = \alpha_1 d^{-\gamma_3} - \alpha_2 \dot{S}^{\gamma_6} d^{\gamma_5} \qquad [7.12]$$

$$\dot{\sigma} = E(\dot{\varepsilon}^T - \dot{\varepsilon}^p) \qquad [7.13]$$

where $\dot{\varepsilon}^{\mathrm{T}}$ is the rate of total deformation. S in Eq. 7.8 represents the recrystallised volume fraction. Its evolution is directly related to the normalised dislocation density $\bar{\rho}$ (Eq. 7.10). Once the dislocation density has accumulated to a critical value $\bar{\rho}_c$, recrystallisation will start if sufficient time is allowed for incubation. The incubation time for recrystallisation is controlled by the variable x in Eq. 7.9. The normalised dislocation density in Eq. 7.10 is defined by $\bar{\rho} = 1 - \rho_i/\rho$, where ρ_i is the initial dislocation density and ρ is the dislocation density in the deformed material. The normalised dislocation density varies from 0 (the initial state) to 1 (the saturated state of a dislocation network). The first term in Eq. 7.10 models the dislocation accumulation due to the plastic deformation $\dot{\varepsilon}^{\mathrm{P}}$ and dynamic recovery. A large average grain size d results in dislocations being accumulated more quickly, since less grain boundary sliding takes place. The second term in Eq. 7.10 represents the static recovery of dislocations at high temperature. Recrystallisation creates dislocation-free grains, which reduces the average dislocation density. This is modelled by the third term in Eq. 7.10. The isotropic hardening R of the material under plastic deformation is directly related to the dislocation density, which is described by Eq. 7.11. Equation 7.12 characterises the evolution of the average grain size. The first term represents static grain growth, which is especially useful for modelling grain growth during the interpass times in multi-stage hot deformation processes. The second term describes grain refinement due to recrystallisation. Dynamic grain growth is not modelled here, as dynamic recrystallisation normally takes place at a low strain under conditions of hot deformation. A detailed explanation of the above equations was given by Lin *et al.* (2005).

The quantities γ_1, k, H, γ_s, c_1, γ_2, c_2, d_0, γ_d, A_3, B, α_1, γ_3, α_2, γ_6 and γ_5 are material constants, which need to be determined from experimental data. This is the key problem considered in this chapter. Equations 7.7–7.13 are a set of ODEs and need to be solved using a numerical integration method. This is also a multi-objective optimisation problem, and thus the selection of suitable objective functions is also a difficult problem. This will be discussed in Section 7.4.

The state variables S, x, $\bar{\rho}$, R and d are used in the set of unified constitutive equations to represent individual physical phenomena in the material. For convenience of discussion in later sections, the above set of equations can be written in a general form as below:

$$\frac{\mathrm{d}y_1}{\mathrm{d}t} = f_1\left(y_1, \quad \ldots, y_i, T, \frac{\mathrm{d}y_2}{\mathrm{d}t}, \ldots, \frac{\mathrm{d}y_i}{\mathrm{d}t} \right) \qquad [7.14]$$

$$\ldots$$

$$\frac{\mathrm{d}y_i}{\mathrm{d}t} = f_i\left(y_1, \quad \ldots, y_i, T, \frac{\mathrm{d}\varepsilon}{\mathrm{d}t}, \frac{\mathrm{d}y_1}{\mathrm{d}t}, \ldots, \frac{\mathrm{d}y_{i-1}}{\mathrm{d}t} \right) \qquad [7.15]$$

$$\ldots$$

where y_i represents the plastic strain rate and the state variables, describing such things as dislocations, hardening and grain size.

7.3 Methods for integrating constitutive equations

As mentioned in the previous section, the constitutive equations applied to describe the superplastic behaviour of materials are a system of ordinary differential equations, which, normally, cannot be solved analytically. A numerical integration method is required to solve the resulting initial value problems.

7.3.1 Difficulties of numerically integrating unified constitutive equations

The multiple equations with different unit scales that arise, such as Eqs 7.7–7.13, need to be integrated simultaneously. The existence of different unit scales compounds the difficulty of obtaining a solution. Different tolerances are required to control the integration accuracy for individual equations, which are difficult to specify. Using the method of evolutionary programming (EP) (Li *et al.*, 2002) to determine the constitutive equations, which will be introduced in Section 7.5, thousands of generations are needed to get a solution through the whole optimisation process, and in each generation, hundreds of populations are normally generated. For each population within a generation, the unified constitutive equations are required to be accurately integrated, and the fitness (error) is assessed by examining the difference between the integrated and the experimental data (Li *et al.*, 2002; Lin et al., 2002). Such numbers of integrations take a huge amount of computational time, and the error estimations must be done accurately. To solve the problem efficiently, one must use the minimum number of time increments required to integrate the equations with a controlled accuracy. Thus, the integration needs to be performed accurately and efficiently. Because they have more stability when solving stiff problems (Cormeau, 1975; Cao *et al.*, 2008), lower-order implicit integration methods are considered in this work.

7.3.2 Forward integration

The vector y_i in Eqs 7.14 and 7.15 represents the integrated variables of the equations, and the vector $\dot{y}_i$ in the equations indicates differentiation with respect to time, for example $[\dot{y}_i] = dy_i/dt$. We now study the numerical solution of the first-order initial-value problem, that is, when $t = 0$, $y_i = y_{0,}i$. For example, for Eqs 7.7–7.13, when $t =$ 0, we have $\varepsilon^{\mathrm{p}} = 0$, $S = 0$, $x = 0$, $r = 0$, $\bar{\rho} = 0$, $d = d_0$ (the grain size before deformation) and $\sigma = 0$, and the loading condition can be, for example, $\dot{\varepsilon}^{\mathrm{T}} = 0.1\ \mathrm{s}^{-1}$. Using the explicit Euler method, the ith equation can be integrated as follows:

$$y_{1,i} = y_{0,i} + \dot{y}_{0,i}\ \Delta t \quad \text{and} \quad y_{2,i} = y_{1,i} + \dot{y}_{1,i}\ \Delta t. \tag{7.16a}$$

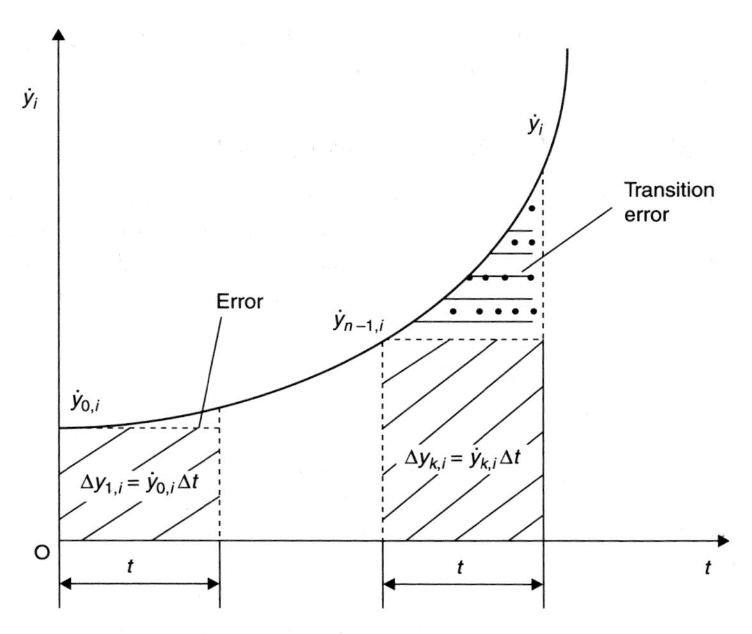

7.2 The explicit forward Euler method.

Expressed as a numerical algorithm, shown in Fig. 7.2, the explicit (or forward) Euler method is therefore as follows:

$$y_{k,i} = y_{k-1,i} + \dot{y}_{k-1,i}\,\Delta t \quad (k = 1,2,3,\ldots) \tag{7.16}$$

The Euler method is also known as the tangent line method, since the first straight-line segment of the approximate solution is tangent to the exact solution y_i at $y_{t=0,i}$ (= $y_{0,i}$), and each subsequent segment emanating from ($y_{k-1,i}$, t) is tangent to the solution curve through that point. The local truncation error is calculated from

$$e_{k,i} = \ddot{y}_{k-1,i}\,\Delta t^2 / 2 \text{ or } e_{k,i} = O(\Delta t_k^2) \tag{7.17}$$

The local truncation error can be used to control the integration accuracy and step sizes of the numerical integration procedure.

Higher accuracy of the numerical integration can be achieved using higher-order methods, such as the midpoint method and the fourth-order Runge–Kutta method. Details can be found in any book on numerical integration, such as Simos (2000).

7.3.3 Implicit numerical integration

With more stability when solving stiff problems (Cormeau, 1975), lower-order implicit integration methods are considered in this kind of work. Although these require the Jacobian matrices to be calculated, the overall efficiency can be improved significantly, since the numbers of iterations required for integration are reduced and convergence is guaranteed. The implicit Euler method is commonly used to integrate a set of stiff ODEs. The general form of the first order of the implicit method is given by Willima *et al.* (2002) for the ith equation and kth increment as follows:

$$[y_{k,i}^{(1)}]=[y_{k-1,i}]+\Delta t_k[\dot{y}_{k,i}] \qquad [7.18]$$

The superscript $^{(1)}$ represents the first-order integration. By linearising the first-order implicit method as in Newton's method (Willima *et al.*, 2002), we obtain

$$[\dot{y}_{k,i}]=[\dot{y}_{k-1,i}]+\left[\frac{\partial \dot{y}_j}{\partial y_i}\right]\left([y_{k,i}^{(1)}]-[y_{k-1,i}]\right) \qquad [7.19]$$

Thus, the general form of the implicit integration (Eq. 7.18) is rearranged into the form

$$[y_{k,i}^{(1)}]=[y_{k-1,i}]+\Delta t_k\left[[I]-\Delta t_k\left[\frac{\partial \dot{y}_j}{\partial y_i}\right]\right]^{-1}[\dot{y}_{k-1,i}] \qquad [7.20]$$

where $[y_{k,i}^{(1)}]$ represents the variables integrated from the ith ordinary differential equations $\dot{y}_{k,i}$ using the first-order implicit Euler method. k is the current iteration of the integration, and i and j vary from 1 to NE, where NE is the total number of differential equations in an equation set. Δt_k is the current step size. $[\partial \dot{y}_j/\partial y_i]$ is a matrix of order $NE \times NE$ and is known as the Jacobian matrix at the current iteration in the implicit numerical integration.

In attempting to integrate the equation sets implicitly (see Eqs. 7.14 and 7.15), one of the key problems is to develop a method to calculate the Jacobian matrix $[\partial \dot{y}_j/\partial y_i]$ in Eq. 7.20 accurately and efficiently. The partial derivatives $[\partial \dot{y}_j/\partial y_i]$ are defined as derivatives of a function of multiple variables when all variables but the one of interest are held at fixed values during the differentiation (Abramowitz and Stegun, 1972). An analytical Jacobian matrix is difficult to obtain for complex unified constitutive equations, which constrains the use of implicit numerical integration methods.

The numerical Jacobian matrix extends the idea of analytical partial derivatives and can be used to automatically generate information for a number NE of equations at the current (kth) iteration, for example. The partial derivatives in the Jacobian matrix can be calculated using (Cao *et al.*, 2008)

$$\left[\frac{\partial \dot{y}_j}{\partial y_i}\right]=\lim_{h_i\to 0}\frac{[\dot{y}_j]\big|_{y_i+h_i}-[\dot{y}_j]\big|_{y_i}}{[h_i]} \qquad [7.21]$$

where h_i is a small increment of the variable $[y_i]$, which is a fraction of the current increment $[\Delta y_i]$ for the ith equation and is defined as

$$[h_i] = \alpha \cdot [\Delta y_i] \qquad [7.22]$$

where α ($\approx$ 0–1) is a factor. Theoretically, $[\partial \dot{y}_j / \partial y_i]$ can be calculated easily from Eq. 7.21 as $h_i \to 0$ but, in practice, it is difficult to calculate for multiple complex equations, in which $[\dot{y}_j|_{yi+hi}]$ cannot be calculated simply from $[\dot{y}_j](y_1, \ldots, y_i + h_i, \ldots, y_{NE})$, since a small increment in the variable y_i affects the values of the other variables in a complex equation set. A flow chart for numerically calculating the Jacobian matrix is shown in Fig. 7.3. This describes the overall structure of the process of defining the Jacobian matrix over one increment of integration. During the calculation process, particular attention needs to be paid to the value of h_i. When $[\Delta y_i] = 0$, $[\dot{y}]|_{y_i+hi} = [\dot{y}]|_{yi}$, so the partial derivatives obey the equation $[\partial \dot{y}_j / \partial y_i] = 0$. The key points in the calculation of $[\partial \dot{y}_j / \partial y_i]$ using Eq. 7.21 are (i) the determination of the value of α and (ii) the calculation of each partial derivative in the equation set for every iteration. The value of α (see Eq. 7.22) has a significant effect on the accuracy of the partial derivatives $\partial \dot{y}_j / \partial y_i$ calculated numerically using Eq. 7.21, and thus affects the computational results and efficiency. Sensitivity studies have been carried out for a wide range of α values varying from 10^{-14} to 10 by Cao *et al.* (2008). It was found that the number of iterations of the numerical integration remained acceptable and almost unchanged as the value of α varied from 10^{-12} to 1.0. Such a large range of α values indicates that it is likely that a suitable value will be chosen and the value of the factor has a good chance of being selected safely. The choice $\alpha = 0.01$, which is suitable for most applications, was recommended.

Based on the implicit Euler scheme, stability can be obtained, but only first-order polynomials can be integrated exactly using a first-order method. Higher accuracy of the integration can be achieved by averaging the explicit and implicit Euler methods according to the implicit trapezoid rule (Willima *et al.*, 2002), which is given by

$$[y_{k,i}^{(2)}] = [y_{k-1,i}] + \frac{1}{2}\Delta t_k \left([\dot{y}_{k-1,i}] + [\dot{y}_{k,i}]\right) \qquad [7.23]$$

This method uses the average of the derivatives at $[\dot{y}_{k,i}]$ and $[\dot{y}_{k-1,i}]$ to compute $[y_{k,i}]$. The derivatives $[\dot{y}_{k,i}]$ are unknown, but can be obtained by linearisation using Newton's method as in Eq. 7.19. Then the second-order implicit integration, using the trapezoid rule, can be represented by

$$[y_{k,i}^{(2)}] = [y_{k-1,i}] + \Delta t_k \left[[I] - \frac{1}{2}\Delta t_k \left[\frac{\partial \dot{y}_j}{\partial y_i} \right] \right]^{-1} [\dot{y}_{k-1,i}] \qquad [7.24]$$

This method has a second-order truncation error from the Taylor series expansions, and thus it is more accurate than Eq. 7.19. Both implicit integration methods have good convergence, but the disadvantage of these methods is that in each step, the Jacobian matrix $[\partial \dot{y}_j / \partial y_i]$ must first be evaluated and then inverted to find $[y_{k,i}]$.

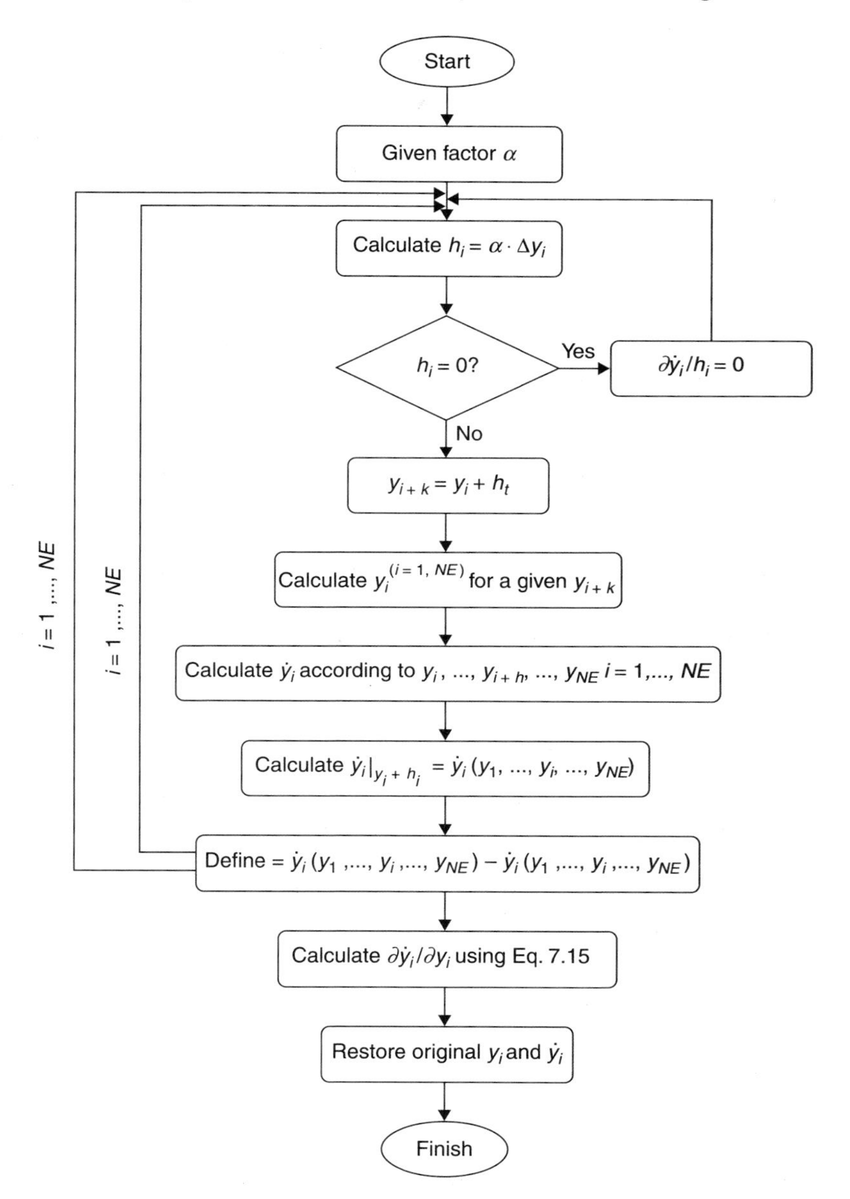

7.3 Flow chart showing the algorithm for numerically calculating the Jacobian matrix for an increment of the integration.

7.3.4 Error control methods for unified constitutive equations

The allowable truncation error, known as the tolerance, is normally used to control the integration accuracy. It can be measured by the difference between the first- and second-order solutions to the ODEs. The following general form of the truncation error is commonly used (Dmitri *et al.*, 2002; Dmitri *et al.*, 2004):

$$O_i(\Delta t_k^2) = \left|[y_{k,i}^{(2)}] - [y_{k,i}^{(1)}]\right| \quad [7.25]$$

where $O_i(\Delta t_k^2)$ is the local truncation error for the ith equation at the kth iteration. This defines how far the numerical approximation is from the solution and can be used with a defined tolerance to dictate the maximum allowable error. In the present case, $[y_k^{(2)}]$ is the solution obtained using a second-order implicit scheme (Eq. 7.24), as this improves the precision, while $[y_k^{(1)}]$ is the solution obtained using the first-order implicit Euler method (Eq. 7.20).

To integrate a set of unified constitutive equations using the minimum number of iterations, adaptive step size control is required. The step size is normally determined according to the truncation errors for the individual equations. The truncation error of the first-order approximations to the ODEs obtained from Eqs 7.20 and 7.24 can be defined as follows:

$$O_i(\Delta t_k^2) = \left|\frac{1}{2}\Delta t_k\left([\dot{y}_{k-1,i}^{(2)}] + [\dot{y}_{k,i}^{(2)}]\right) - \Delta t_k[\dot{y}_{k,i}^{(1)}]\right| \quad [7.26]$$

The truncation error $O_i(\Delta t_k^2)$ for the ith equation is the first-order implicit Euler approximation to the ODEs through Taylor series expansions (Dmitri *et al.*, 2002). For the $(k+1)$th (next) increment, the step size is normally estimated according to the truncation error in the kth (current) iteration. Replacing $O(\Delta t_k^2)$ by Tol_i in Eq. 7.26 results in

$$\Delta t_{k+1,i} = \frac{[Tol]_i}{\left|\frac{1}{2}\left([\dot{y}_{k-1,i}^{(2)}] + [\dot{y}_{k,i}^{(2)}]\right) - [\dot{y}_{k,i}^{(1)}]\right|} \quad [7.27]$$

$$\Delta t_{k+1} = \mathrm{Min}\{\Delta t_{k+1,i}\} i = 1,\ldots,NE \quad [7.28]$$

where Tol_i is the tolerance specified for the ith equation in the equation set. Δt_{k+1} is the estimated step size for the next iteration of the integration. After the truncation error of the integration at the $(k+1)$th iteration has been estimated, a decision can be made as to whether to accept or reject the step size.

In this error control method, the principal local truncation error remains less than a prescribed tolerance, so that the allowable error is under control. However, it is difficult to specify a tolerance for each equation in a complex equation set. It is unlikely to be possible to compare the error values, because they may have different unit scales. Therefore, the inclusion of a tolerance as a restraint on the size of the truncation error is ineffective, as each ODE would require a different value, and it is difficult to define the minimum step size uniquely owing to the problem of different units. The unit-related values associated with the integrated variables need to be transformed to dimensionless numbers. However, this is beyond the scope of this chapter. More detailed numerical techniques to deal with step size control problems have been given by Cao *et al.* (2008).

7.4 Objective functions for optimisation

7.4.1 Least squares method

Readings recorded from tests or experiments normally include errors of various kinds, and therefore the points plotted from these data are scattered about the positions they should occupy. Unless very few readings are taken, it can be assumed that the inherent errors will be of a random nature, resulting in some of the values being slightly too high and some slightly too low, such as in the case of the stress–strain experimental data (symbols) shown in Fig. 7.4. The experimental stress–strain data for an elastic–plastic problem can be often modelled using a simple power law equation

$$\sigma = k\varepsilon^n \qquad [7.29]$$

where σ is the true stress and ε the true strain, and k and n are material constants, which need to be determined from the experimental data using an optimisation method. The conventional least squares method has been widely used to define an objective function in which the sum of the squares of the errors between the experimental and computed data is minimised. The error for the ith data points

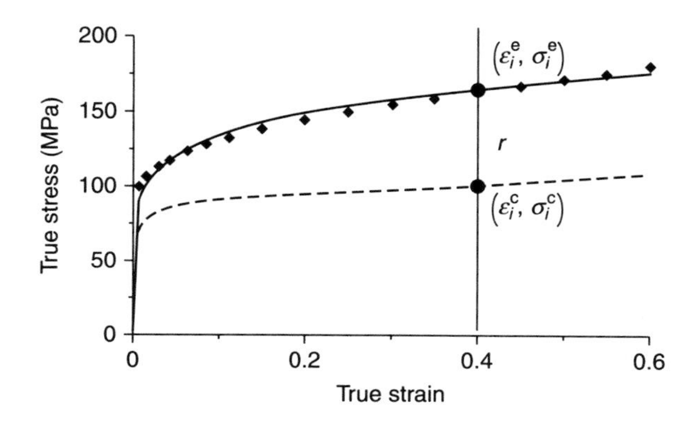

7.4 Definition of errors and the determination of k and n in the equation $\sigma = k\varepsilon^n$ for a pearlitic steel. The stress–strain curves were fitted to experimental data (symbols) (Zener and Hollomon, 1944).

$(\varepsilon_i^e, \sigma_i^e)$ and $(\varepsilon_i^c, \sigma_i^c)$, where the superscript 'c' represents the computed data and 'e' the experimental data, as shown in Fig. 7.4 for $\varepsilon_i^c = \sigma_i^e$, can be defined as

$$r_i^2 = w_i(\sigma_i^c - \sigma_i^e)^2 \qquad [7.30]$$

where w_i is a weighting factor for the ith data point. The sum of the errors is

$$f(x) = \sum_{i-1}^{N} r_i^2 \qquad [7.31]$$

where x represents the material constants k and n, which will be optimised so that the accumulated error $f(x)$ can be minimised for the purpose of fitting. The fitting process may start with initial values of k and n, which give the dashed curve shown in Fig. 7.4. During the searching process, the values of k and n are optimised so that the accumulated error $f(x)$ is minimised. The results of the fitting are shown in Fig. 7.4 by the solid curve, and the best fit is given by $k = 190.45$ MPa, $n = 0.15$. It can be seen from Fig. 7.4 that the experimental data can be modelled very well using a simple equation with values of the constants determined in this way. However, for complicated multi-objective fitting processes, it is difficult to form an effective objective function to guide the optimisation process.

Many optimisation methods, such as gradient-based methods (Zhou and Dunne, 1996; Kowalewski *et al.*, 1994; Lin and Hayhurst, 1993) and methods based on evolutionary algorithms (EAs) (Lin and Yang, 1999; Li *et al.*, 2002), have been developed for determining material constants in constitutive equations from experimental data. Various objective functions have been formulated to assess the errors between the experimental and computed data (Li *et al.*, 2002; Lin *et al.*, 2002). The objective function should be able to 'guide' the optimisation process efficiently to find the best fit to the experimental data. The commonly used objective function is in least squares form. An ideal objective function should obey the following criteria (Cao and Lin, 2008):

Criterion 1: for a single curve. All the experimental data points on the curve should be involved in the optimisation and have an equal opportunity to be optimised, assuming that the errors in the experimental data have first been eliminated.

Criterion 2: for multiple curves. All experimental curves should have an equal opportunity to be optimised. The performance of the fitting process should not depend on the number of data points on each experimental curve.

Criterion 3: for multiple sub-objectives. The objective function should be able to deal with multi sub-objective problems, in which the units of the sub-objectives may be different but all the sub-objectives should have an equal opportunity to be optimised. The presence of different units and the number of curves in each sub-objective should not affect the overall performance of the fitting process.

Criterion 4: weighting factors. The above criteria should be achieved automatically without choosing weighting factors manually, since these are difficult to choose in practice.

The objective functions are normally formulated based on minimising the sum of the squares of the differences between the computed and experimental data at the same strain level. For example, an experimental stress–strain curve (symbols) and four curves derived using different sets of material constants for a set of superplastic constitutive equations are shown in Fig. 7.5(a) (Cao and Lin, 2008). For a superplastic material model, strains to failure are implicitly characterised by some of the material parameters mentioned by Lin *et al.* (2002). The result of an assessment of the residuals between the computed (solid lines) and experimental (symbols) stress–strain curves for a viscoplastic-damage constitutive model is shown in Fig. 7.5(a). Four computed stress–strain curves are given in the figure, in which it can be seen that the computed curve 1 has a bad fit, with a low strain to failure. Curve 2 fits the experimental data best. Curve 3 has quite a close fit to the experimental data, but beyond the data, the curve has a much higher strain to failure. Curve 4 is far away from the experimental data, with the highest strain to failure.

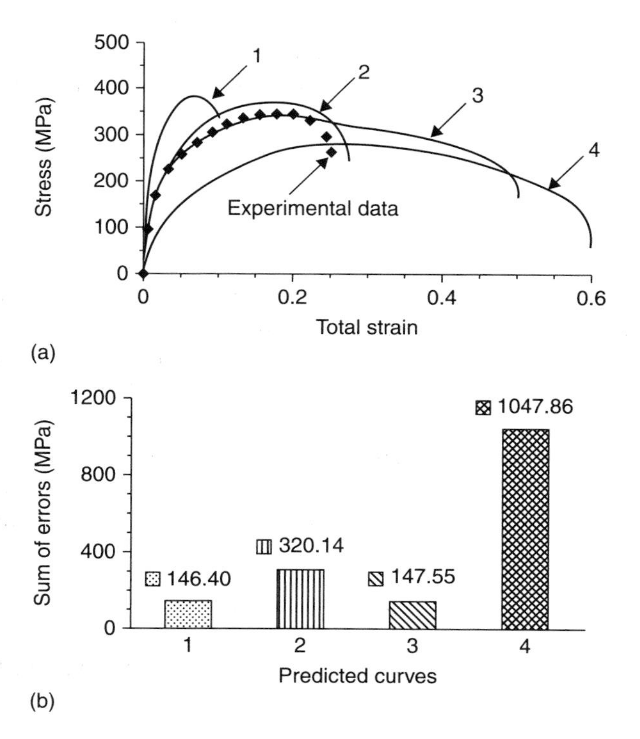

7.5 (a) Comparison of experimental (symbols) and computed (curves 1, 2, 3 and 4) stress–strain curves. (b) Sums of the squares of the errors in the stress (Cao and Lin, 2008).

Figure 7.5(b) shows the corresponding sums of errors (differences in the stress at various strains) for the computed curves in Fig. 7.5(a) when the difference between the computed and experimental stresses is assessed for the same strain level. From Fig. 7.5(b), it can be seen that the computed curve 4 has the largest error and curve 1 the smallest. However, it is obvious from Fig. 7.5(a) that curve 2 fits the experimental data best 'by eye'. The wrong assessment of the errors indicates that if the predicted strain to failure is less than the value indicated by the experimental data points, such as in curve 1 in Fig. 7.5(a), some of the experimental data should not be included in the error estimation. If the computed strain to failure is higher than the experimental value, such as in curves 3 and 4 in Fig. 7.5(a), the result of the fitting process can be misleading as well. It can be seen that curve 2 gives the best fit among the four curves according to observation 'by eye', and an appropriate objective function needs to be formulated to represent this fact. Thus, the definition of an objective function for optimisation should reflect both observation 'by eye' and human judgement.

If the calculation is done from both directions (strain and stress, as shown in Fig. 7.5(a)), another situation needs to be taken into account, in which the unit scales of strain and stress are significantly different. Thus, the most difficult task encountered in developing a universal objective function is to deal with the problems highlighted above. To overcome these problems, a number of objective functions have been formulated. The features of some recently developed objective functions are introduced below.

7.4.2 Objective functions

Shortest-distance method (OF_I)

This method was developed by Li *et al.* (2002) and used for the determination of a set of creep damage constitutive equations. In this method, the errors are defined by the shortest distance between the experimental and the corresponding computed creep curves, and are given by $r^2 = \Delta\varepsilon^2 + \Delta t^2$ as shown in Fig. 7.6. Here, $\Delta\varepsilon$ is the difference between the experimental and computed values of the creep strain (expressed in %), and Δt is the difference between the creep times (in hours). In order to compensate for the different scales of strain and time, two weighting parameters α and β were introduced (Li *et al.*, 2002). The error for the ith data point of the jth curve is defined by

$$r_{ij}^2 = \alpha\, \Delta\varepsilon_{ij}^2 + \beta\, \Delta t_{ij}^2 = \alpha(\varepsilon_{ij}^{c} - \varepsilon_{ij}^{e})^2 + \beta(t_{ij}^{c} - t_{ij}^{e})^2 \qquad [7.32a]$$

By summing r^2_{ij} for all the data points of the curves, the objective function is expressed as

$$f(x) = \sum_{j=1}^{M}\sum_{i=1}^{N_j} w_{ij} r_{ij}^2 + \sum_{j=1}^{M} W_j (t_{N_j j}^{c} - t_{N_j j}^{e})^2 \qquad [7.32]$$

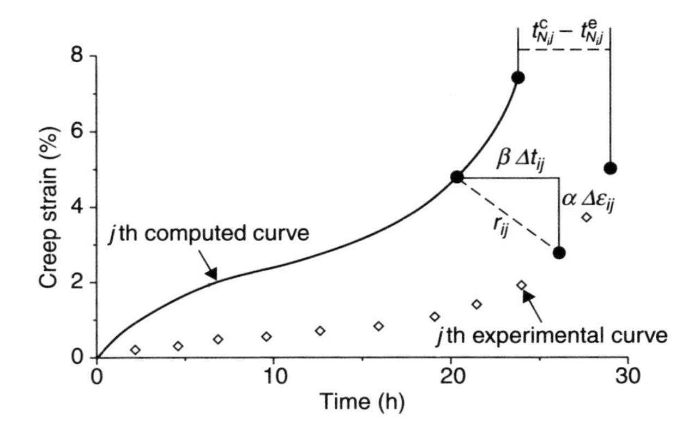

7.6 **Definition of errors using the shortest-distance method for OF_I.**

where w_{ij} is the relative weight for the ith data point of the jth experimental creep curve, and W_j is the relative weight for curve j. The superscripts 'c' and 'e' represent the computed and experimental values, respectively. The second term in Eq. 7.32 was introduced by Kowalewski *et al.* (1994) for determining creep damage constitutive equations, in order to increase the sensitivity of the creep lifetime in the error estimation. In the calculation of the values of r_{ij}, two scaling factors are introduced to keep the unit scales compatible and, at the same time, to increase the sensitivity of the objective function. By choosing the weighting parameters properly, a reasonably good fit can be obtained. Results have been presented by Li *et al.* (2002) for a set of creep damage constitutive equations. However, obtaining such a good result is dependent on experience with the particular problem. Weighting factors are difficult to choose for new problems. Thus, the objective function has a generic nature, but the weighting factors are problem-dependent and difficult to choose.

Universal multi-objective function (OF_II)

To overcome the difficulties associated with differences between the predicted and experimental strains to failure and the variation in unit scales, a dimensionless objective function was introduced by Lin *et al.* (2002) to determine a set of superplastic damage constitutive equations in which multiple objectives were involved. The error definition for the multi-objective problem is shown in Fig. 7.7, and the objective function takes the form

$$f(x) = f_\sigma + f_d \qquad [7.33]$$

$$f_\sigma = +\sum_{j=1}^{M}\left\{\left(\frac{1}{N_j}\right)\sum_{i=1}^{N_j}\left\{\frac{\sigma^e_{ij} - \sigma^c(\varepsilon^c_{N_j j}\ \varepsilon^e_{ij}\ \varepsilon^e_{N_j j})}{s_{\sigma,ij}}\right\}^2 + \left[\frac{\varepsilon^e_{N_j j} - \varepsilon^c_{N_j j}}{s_j}\right]^2\right\} \qquad [7.34]$$

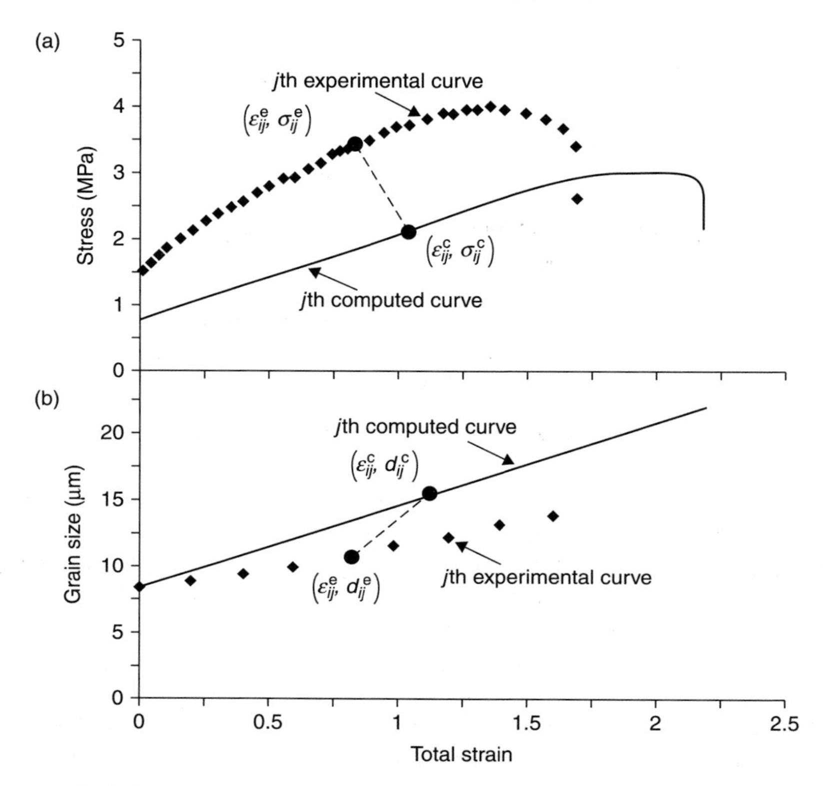

7.7 Definition of errors for the universal multi-objective function (OF_II): errors in (a) stress (MPa) and (b) grain size (μm).

$$f_d = +\sum_{j=1}^{M}\left\{\left(\frac{1}{N_j}\right)\sum_{i=1}^{N_j}\left\{\frac{d^{e}_{ij} - d^{c}(\varepsilon^{c}_{N_j j}\,\varepsilon^{e}_{N_{ij}} / \varepsilon^{e}_{N_j j})}{s_{d,ij}}\right\}^2 + \left[\frac{\varepsilon^{e}_{N_j j} - \varepsilon^{c}_{N_j j}}{s_j}\right]^2\right\} \quad [7.35]$$

where f_σ is the residual for the stress and f_d is the residual for the grain size; $s_{\sigma,ij} = 0.1\sigma^{e}_{ij}$ and $s_{d,ij} = 0.1d^{e}_{ij}$ for the ith data point of the jth curve, and $s_j = 0.1\varepsilon^{e}_{N_j j}$ for the jth curve. N_j is the number of data points for the jth experimental curve, and M is the number of experimental curves involved in the optimisation. In the first terms of Eqs 7.34 and 7.35, the experimental and computed data are normalised by the experimental data and the errors are transformed to dimensionless values. This approach enables multiple objectives to be dealt with, in such a way that the residuals can be combined directly to form a multi-objective function without introducing weighting factors. The second terms in Eqs 7.34 and 7.35, which have a similar function to those in Eq. 7.32 (OF_I), are normalised by the largest experimental strain values. This again results in a dimensionless error value, and ensures that the predicted strains to failure are close to the corresponding experimental values. The effect is all-pervasive, because pushing ($\varepsilon^{c}_{N_jj}$) towards ($\varepsilon^{e}_{N_jj}$) ultimately means pushing all the predicted points towards their corresponding counterparts in the model. This ensures that the two terms play the same important role in normalising the sum of errors with respect to the number of data points.

This objective function was successfully used to determine the material parameters arising in a set of superplastic-damage constitutive equations (Lin *et al.*, 2002). It is important to obtain the computed data accurately for all individual experimental data points. With this function, for each pair of experimental and computed curves, the same fraction is used to choose the corresponding data value σ^{c} for the *i*th experimental data point of the *j*th curve. This ensures that all the experimental data are included in the optimisation. Moreover, the predicted data will be pushed towards the corresponding experimental data.

General objective function (OF_III)

(i) *Definition of 'true' error (logarithmic error).* The assessment of the residual between the computed *i*th data point of the *j*th curve and the corresponding experimental point can be done on the basis of $\sigma^{c}_{ij}/\sigma^{e}_{ij}$. This quantity has the advantage of being dimensionless, which naturally avoids the weighting parameters used in OF_I and the normalisation in OF_II. The ratio of the computed data value to the experimental data value provides a measure of the relative error. To consider the data on a logarithmic scale, the square of the logarithm of $\sigma^{c}_{ij}/\sigma^{e}_{ij}$ was studied. When σ^{c}_{ij} is far from σ^{e}_{ij}, the value of $(\ln(\sigma^{c}_{ij}/\sigma^{e}_{ij}))^2$ is very large. This significantly increases the sensitivity when the value of $\sigma^{c}_{ij}/\sigma^{e}_{ij}$ becomes large, similarly to the definition of true strain. The definition of the 'true' error for the *i*th data point of the *j*th experimental stress–strain curve can be expressed as $E = (\ln(\sigma^{c}_{ij}/\sigma^{e}_{ij}))^2$, in which the squaring is used to increase the sensitivity and ensures that all errors are positive. This definition provides an assessment of the true error between the computed and experimental data.

(ii) *Weighting factors.* An automatic weighting factor for the *i*th experimental data point of the *j*th curve is introduced in accordance with the above definition of the error, which may be expressed as follows:

$$\omega_{ij} = \frac{\phi \cdot \varepsilon^{e}_{ij}}{\sum_{j=1}^{M}\sum_{i=1}^{N_j}\varepsilon^{e}_{ij}} \qquad (\omega_{11} + \omega_{12} + \cdots + \omega_{N_j M} = \phi) \tag{7.36}$$

where $\phi = \sum_{j=1}^{M} N_j$ is a scaling factor, which is related to the total number of data points to increase the sensitivity of the objective function.

(iii) *Objective function.* The same method for finding the corresponding computed and experimental data points as that defined for OF_II can be used. Taking account of the definition of the true error and the weighting factors defined in Eq. 7.36, the residuals in the fitting process are defined using a 'weighted distance' in the true-error coordinate system:

$$r_{ij}^2 = \omega 1_{ij}\left(\ln\frac{\varepsilon^{c}(\varepsilon^{c}_{N_j j}\,\varepsilon^{e}_{ij}/\varepsilon^{e}_{N_j j})}{\varepsilon^{e}_{ij}}\right)^2 + \omega 2_{ij}\left(\ln\frac{\sigma^{c}(\varepsilon^{c}_{N_j j}\,\varepsilon^{e}_{ij}/\varepsilon^{e}_{N_j j})}{\sigma^{e}_{ij}}\right)^2 \qquad [7.37]$$

where $\omega 1_{ij}$ and $\omega 2_{ij}$ are relative weighting factors for ε^{e}_{ij} and σ^{e}_{ij}, respectively. Based on Eq. 7.36, the factor for the *i*th experimental data point of the *j*th curve relative to the whole sum of data points is calculated automatically. This provides an equal opportunity for each data point to be optimised. The loss of information due to the use of a logarithmic scale can be compensated, and all points play equal roles. Taking the sum of the residuals r_{ij}^2 calculated in this novel way and the automatic weighting factors together, a general objective function is formulated as follows:

$$f(x) = \frac{1}{M}\cdot\frac{1}{N_j}\cdot\sum_{j=1}^{M}\sum_{i=1}^{N_j} r_{ij}^2 \qquad [7.38]$$

The sum of the residuals is normalised with respect to the numbers of experimental data points and curves. This ensures that the assessment of the residuals is based on an average of dimensionless data points. Equation 7.38 essentially provides a natural dimensionless average error and can easily be used for multi-objective problems. It makes it possible to deal with multiple sub-objectives, in which different numbers of curves and different units may be involved in the optimisation. Together with the automatic weighting factors, it enforces compatibility with each data point, curve and objective when one is dealing with multiple objectives. All objectives play equally important roles. An objective that has more experimental curves does not lead to a representation biased towards a criterion that has been assessed on the basis of more data points.

7.5 Optimisation methods for determining the material constants in constitutive equations

7.5.1 Introduction and background

One of the most difficult tasks encountered in materials modelling is to accurately determine the material constants in equations from experimental data using current optimisation techniques. Lin and Hayhurst (1993) developed an optimisation technique and successfully determined the material constants for some constitutive equations developed for leather, where the stress can be expressed explicitly as a function of strain. Kowalewski *et al.* (1994) developed a three-step method to determine the starting values for optimisation of the constants in a set of creep damage equations. Zhou and Dunne (1996) proposed a four-step method to determine the material constants for a particular set of superplastic constitutive equations. That study was carried out using a gradient-based optimisation method, and the difficulties associated with choosing proper starting values for the constants were highlighted by the investigators. In addition, extra difficulties were met when the number of constants to be determined was large, for example more than five (Zhou and Dunne, 1996).

The introduction of an optimisation method based on a genetic algorithm (GA) by Lin and Yang (1999) to these problems overcame the difficulty of choosing the starting values, and this method successfully solved a set of superplastic equations proposed by Zhou and Dunne (1996), which contain 13 constants. The method is based on the classic GA and is inefficient in solving continuous problems (Fogel and Atmar, 1990). To improve the efficiency, Li *et al.* (2002) used an optimisation technique based on evolutionary programming (EP) introduced by Yao *et al.* (1999). This algorithm has been applied with success to many constitutive equations and has been found particularly useful for determining unified creep damage constitutive equations and general viscoplastic constitutive equations (Li *et al.*, 2002). Details of some modern optimisation methods have been presented by Lin and Yang (1999) and Li *et al.* (2002). Many methods for searching for optima have been developed; it is not possible to describe them all in detail here. However, two commonly used optimisation methods are described below.

7.5.2 Gradient-based method (GBM) in multiple dimensions

The conjugate gradient method can be applied to the problem of minimising a non-linear multidimensional function $f(x)$ by considering its gradient $\nabla f(x)$ in a certain number of iterations and has the following form (Willima *et al.*, 2002):

$$x_{i+1}(k) = x_i(k) + \theta_i d_i(k) \qquad [7.39]$$

$$d_{i+1}(k) = g_{i+1}(k) + \lambda_i d_i(k) \qquad [7.40]$$

where i denotes the ith iteration, which is initialised by selecting a starting value of the kth material constant $x_0(k)$ and choosing the initial search direction $d_0(k) = g_0(k) = -\nabla f(x_0(k))$, where $k \in \{1, \ldots, \mu\}$, μ is the number of material constants, $\nabla f(x_0(k))$ is the gradient of the kth material constant for the first iteration and θ_i is a step size obtained by a golden section search in one dimension (Polak, 1971). In the calculation of $d_{i+1}(k)$ for the next iteration in Eq. 7.40, the search direction and a scalar (proposed by Polak, 1971) can accomplish the transition to further iterations more gracefully (Dennis and Schnabel, 1983), and are given by

$$g_{i+1}(k) = -\nabla f(x_{i+1}(k)) \text{ and } \lambda_i = \frac{(g_{i+1}(k) - g_i(k)) \cdot g_{i+1}(k)}{g_i(k) \cdot g_i(k)}. \qquad [7.41]$$

This optimisation method is based on conjugate search directions and is in the spirit of the steepest descent method. The convergence is faster than that of the steepest descent method because the previous directions are used as a part of a new direction. It is an effective search method for minimising objective functions if the number of parameters to be optimised is less than five (Zhou and Dunne, 1996). It is difficult to choose the starting values of material constants (Li *et al.*, 2002), and the method is not suitable for solving advanced material models that contain a large number of material constants.

7.5.3 Fast evolutionary programming (FEP)

Evolutionary programming was first proposed as an approach to artificial intelligence (Fogel *et al.*, 1966). It has been recently applied with success to determine material constants in unified constitutive equations from experimental data (Li *et al.*, 2002). The process of optimisation by EP can be summarised in two major steps:

- Mutate the solutions in the current population.
- Select the next generation from the mutated and current solutions.

FEP is an advanced optimisation technique and can be described by the following equations (Bäck and Schwefel, 1993; Yao *et al.*, 1999; Li *et al.*, 2002):

$$x_k'(j) = x_k(j) + \eta_k(j)\delta_j \quad [7.42]$$

$$\eta_k'(j) = \eta_k(j)\exp(\tau' N(0,1) + \tau N_j(0,1)) \quad [7.43]$$

where x_k is the kth material constant, and η_k is the standard deviation for Gaussian mutation (also known as a strategy parameter in the context of self-adaptive evolutionary algorithms) of that material constant. Each parent (x_k, η_k), where $k = 1, \ldots, \mu$ denotes an individual, creates a single offspring (x_k', η_k') in the jth population, $j = 1, \ldots, n$, where n is the specified population size. $N(0, 1)$ represents a normally distributed one-dimensional random number with mean zero and standard deviation one. $N_j(0, 1)$ indicates that the random number is generated anew for each value of j. The factors τ and τ' are parameters, which Bäck and Schwefel (1993) suggested should be set equal to $\left(\sqrt{2\sqrt{n}}\right)^{-1}$ and $\left(\sqrt{2n}\right)^{-1}$, respectively. The global factor $\tau' \cdot N(0, 1)$ allows an overall change in the mutability, and $\tau \cdot N_j(0, 1)$ allows individual changes in the 'mean step sizes' η_k. The quantity δ_j is a Cauchy random variable and is generated anew for each value of j. The one-dimensional Cauchy density function centred at the origin is defined by

$$f_c(x) = \frac{1}{\pi}\frac{1}{1+x^2}, \text{ where } -\infty < x < \infty \quad [7.44]$$

The variance of the Cauchy distribution is infinite, and therefore Cauchy mutation starting from Eq. 7.42 is more likely to generate an offspring further away from its parent, owing to the long, flat tails of this distribution. This method is therefore expected to have a higher probability of escaping from a local optimum or moving away from a plateau. This search method has been successfully used for the determination of creep damage constitutive equations in which the number of material constants is less than 10 (Li *et al.*, 2002).

EP-based optimisation techniques have been developed and programmed for determining the material constants in the set of constitutive equations given in Eqs 7.7–7.13 for a steel deforming under hot forming conditions. The experimental data reported by Medina and Hernandez (1996) and Hernandez *et al.* (1996), which were used for the optimisation here, was for a micro-alloyed steel with an average initial grain size of 189 μm. The material constants determined are listed in Table 7.1 and the results are shown in Fig. 7.8.

Experimental data (Medina and Hernandez, 1996; Hernandez *et al.*, 1996) (symbols in Fig. 7.8) for the evolution of the equivalent stress, recrystallisation and grain size at two strain rates $\dot{\varepsilon}$ = 0.544 and 5.224 s^{-1} were used for the optimisation. The solid curves in the figure, which were plotted using the above constitutive equations with the optimised constants listed in Table 7.1, approximate the experimental data. Close agreement was obtained in all cases. In addition, two other strain rates, $\dot{\varepsilon}$ = 1.451 and 3.628 s^{-1}, were used to predict the evolution of the flow stress, recrystallisation and grain size; the curves for these strain rates basically lie between the curves for the highest and lowest strain rates in the figure. Experimental data values for the peak stress were available for these two strain rates. These indicate that the set of constitutive equations enables the mechanical and physical behaviour to be well modelled.

7.6 Case studies

7.6.1 Determination of viscoplastic-damage constitutive equations for a steel and an Al alloy

Viscoplastic-damage constitutive equations have been developed by many researchers for many engineering materials, and have been used to model a wide range of time-dependent phenomena, such as strain rate effects, creep, recrystallisation and recovery. In these equations, the hardening of a material

Table 7.1 Constants determined for the unified viscoplastic constitutive equations (Eqs 7.7–7.13)

A_1 (s^{-1}) 1.81×10^{-6}	A_2 (MPa^{-1}) 3.14×10^{-1}	γ_1 1.00	H 30.00	$\bar{\rho}_c$ 1.84×10^{-1}	γ_s 1.02
c_1 16.00	γ_2 1.44	c_2 8.00×10^{-2}	d_0 (μm) 36.38	γ_d 1.02	A_3 40.96
B (MPa) 75.59	$\alpha 1$ (μm) 1.44	γ_3 3.07	α_2 (μm) 78.68	γ_6 1.20×10^{-1}	γ_5 1.06

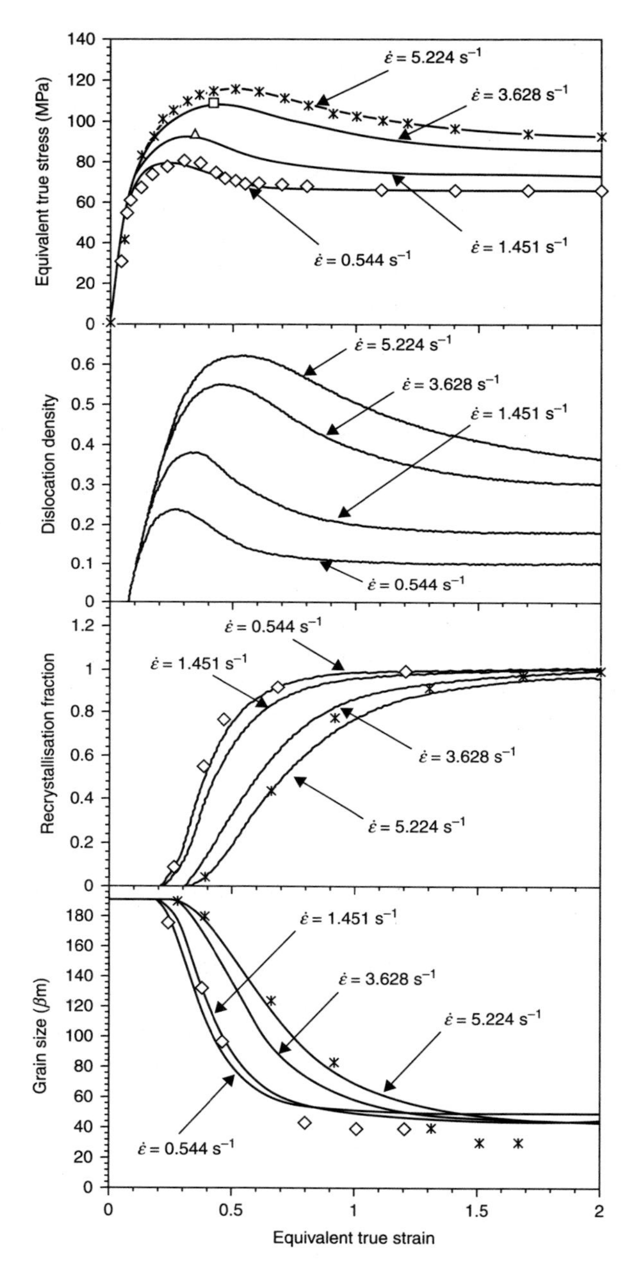

7.8 Comparison of experimental data (symbols) with computed results (curves) at different strain rates for the variation of stress, dislocation density, recrystallised volume fraction and grain size with equivalent true strain.

during viscoplastic deformation is modelled according to the accumulation of plastic strain only. Recently, dislocation-based hardening constitutive equations have been developed by Lin *et al.* (2005), where the recovery of dislocations due to annealing and recrystallisation under hot forming conditions is included. Based on the work described above, a new set of unified viscoplastic-damage constitutive equations was formulated for boron steel and aluminium alloy, and is given below:

$$\text{Plastic strain } (\mathrm{s}^{-1}): \dot{\varepsilon}^{\mathrm{p}} = \dot{\varepsilon}^{0} \cdot \left(\frac{\left| \frac{\sigma}{1-f_d} \right| - R - k}{K} \right)_{+}^{n_1} \cdot \frac{1}{\left(1-f_d\right)^{\gamma_1}}; \qquad [7.45]$$

$$\text{if} \left| \frac{\sigma}{1-f_d} \right| - R - k \leq 0, \dot{\varepsilon}^{\mathrm{p}} = 0$$

$$\dot{\varepsilon}^{0} = \begin{cases} 1, & \sigma > 0 \\ -1, & \sigma > 0 \end{cases}$$

$$\text{Dislocation hardening } (\mathrm{MPa/s}): \dot{R} = n_0 \cdot B \cdot \bar{\rho}^{(n_0-1)} \cdot \dot{\bar{\rho}} \qquad [7.46]$$

$$\text{Dislocation } (\mathrm{s}^{-1}): \dot{\bar{\rho}} = A \cdot (1-\bar{\rho}) \cdot \left| \dot{\varepsilon}^{\mathrm{p}} \right| - C \cdot \bar{\rho}^{n_2} \qquad [7.47]$$

$$\text{Damage } (\mathrm{s}^{-1}): \begin{cases} \text{Boron steel: } \dot{f}_d = D_1 \cdot \dfrac{\sigma \cdot \left| \dot{\varepsilon}^{\mathrm{p}} \right|}{(1-f_d)^{\gamma_2}} \\ \text{Al alloys: } \dot{f}_d = D_1 \cdot f_d^{\gamma_2} \cdot \left| \dot{\varepsilon}^{\mathrm{p}} \right|^{d_1} + D_2 \cdot \left| \dot{\varepsilon}^{\mathrm{p}} \right|^{d_2} \cdot \cosh\left(D_3 \left| \varepsilon^{\mathrm{p}} \right| \right) \end{cases} \qquad [7.48]$$

$$\text{Flow stress } (\mathrm{MPa}): \sigma = E \cdot (1-f_d) \cdot (\varepsilon^{\mathrm{t}} - \varepsilon^{\mathrm{p}}) \qquad [7.49]$$

where ε^{p} in Eq. 7.45 is the traditional power-law formulation for viscoplastic flow. The material hardening R in Eq. 7.46 due to plastic deformation is calculated according to the accumulation of dislocation density (see Eq. 7.47). In hot metal forming processes, in the late stages of deformation, softening due to damage decreases the flow stress, which can be modelled based on mechanisms of void nucleation and growth. The effective evolution of the damage is defined in Eq. 7.48, where the damage is 0 in the initial state. The constants k, K, n_1, B, C, E, D_1 and D_2 are temperature-dependent parameters, and A, n_2, γ_1, γ_2, d_1, d_2 and D_3 are material constants. E is the Young's modulus. The temperature-dependent parameters can be represented by the Arrhenius equations below, where R is the universal gas constant and Q is the activation energy:

$$k = k_0 \exp\left(\frac{Q_k}{RT}\right), \quad C = C_0 \exp\left(-\frac{Q_C}{RT}\right), \quad K = k_0 \exp\left(\frac{Q_K}{RT}\right),$$

$$E = E_0 \exp\left(\frac{Q_E}{RT}\right),$$

$$n_1 = n_{10} \exp\left(\frac{Q_n}{RT}\right), \quad D_1 = D_{10} \exp\left(\frac{Q_1}{RT}\right), \quad B = B_0 \exp\left(\frac{Q_B}{RT}\right),$$

$$D_2 = D_{20} \exp\left(\frac{Q_2}{RT}\right)$$

The constants in the viscoplastic-damage model developed were determined by fitting to experimental stress–strain curves for different strain rates and forming temperatures. Tensile tests were carried out using a Gleeble 3800 materials simulator on USIBOR 1500P boron steel and AA6082 Al alloy. The constants in the equations were determined by fitting the corresponding experimental data using the optimisation methods discussed in this chapter. Figures 7.9 and 7.10

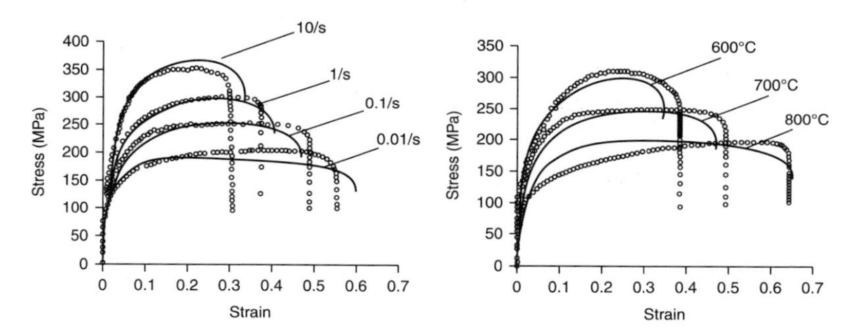

7.9 Comparison of computed (solid curves) and experimental (symbols) stress–strain relationships for boron steel deformed at different strain rates and temperatures.

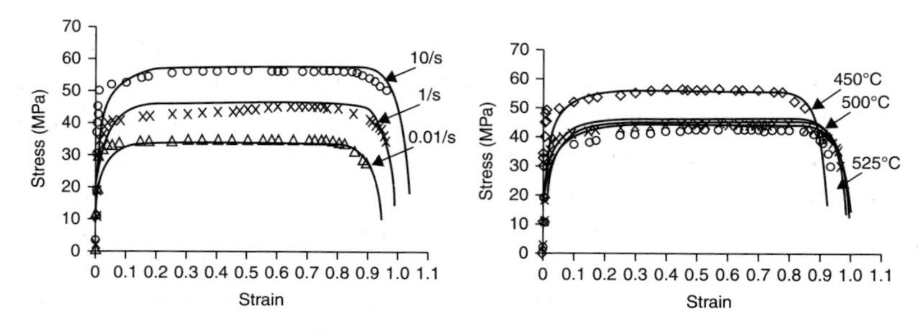

7.10 Comparison of computed (solid curves) and experimental (symbols) stress–strain relationships for AA6082 alloy deformed at different strain rates and temperatures.

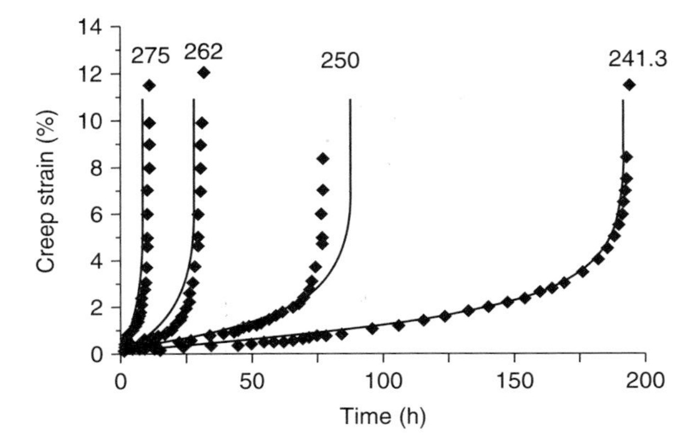

7.11 Comparison of experimental (symbols) and computed (solid lines) creep curves for Eqs 7.50–7.53. The numbers above each curve give the stress levels in MPa.

show a comparison between the computed (solid curves) and experimental (symbols) stress–strain relationships for the boron steel and aluminium alloy, respectively, and close agreement was obtained in both cases. The values of the constants in the equations determined are listed in Table 7.2.

7.6.2 Determination of creep damage constitutive equations

The use of continuum damage mechanics with two damage variables was justified by Kowalewski *et al.* (1994) and used to model tertiary creep softening in an aluminium alloy, caused by nucleation and growth of grain boundary cavities, and the ageing of some particular microstructures. The form of the constitutive equations (with creep curves shown in Fig. 7.11) was given by Kowalewski *et al.* (1994) as

Creep strain rate (h^{-1})
$$\dot{\varepsilon} = \frac{A}{(1-\omega_2)^n}\sinh\left(\frac{B\sigma(1-H)}{1-\phi}\right) \quad [7.50]$$

Primary hardening rate (h^{-1})
$$\dot{H} = \frac{h_0}{\sigma}\left(1-\frac{H}{H^*}\right)\dot{\varepsilon} \quad [7.51]$$

Ageing softening rate (h^{-1})
$$\dot{\phi} = \frac{K_c}{3}(1-\phi)^4 \quad [7.52]$$

Grain boundary damage rate (h^{-1})
$$\dot{\omega}_2 = \frac{DA}{(1-\omega_2)^n}\sinh\left(\frac{B\sigma(1-H)}{1-\phi}\right) \quad [7.53]$$

Table 7.2 Material constants in the viscoplastic-damage constitutive equations for boron steel and AA6082

Constants	A	n_2	γ_1	γ_2	d_1	d_2	D_3	k_0 (MPa)	n_0
Boron steel	5.222	1.54	3.1	17.5	–	–	–	12.4	0.4
AA6082	13	1.8	0	1.2	1.0101	0.5	26.8	0.89	0.5
Constants	K_0 (MPa)	n_{10}	B_0 (MPa)	C_0	E_0 (MPa)	D_{10}	D_{20}	Q_k (J/mol)	Q_K (J/mol)
Boron steel	30	0.0068	80	55500	1100	1.39×10^{-4}	–	8400	8400
AA6082	0.219	5	4.91	0.26	322.8191	10.32	5.49×10^{-19}	6679	27687.1
Constants	Q_n (J/mol)	Q_B (J/mol)	Q_C (J/mol)	Q_E (J/mol)	Q_1 (J/mol)	Q_2 (J/mol)	R (J/mol)		
Boron steel	50000	8400	99900	17500	10650	–	8.3		
AA6082	0	11625.8	3393.4	12986.7	6408.4	119804.6	8.3		

Table 7.3 Material constants determined for an Al alloy at 150°C

A (h^{-1})	B (MPa^{-1})	h_0 (MPa)	H^*	K_c (h^{-1})	D
4.04×10^{-15}	0.11	2.95×10^4	0.11	1.82×10^{-4}	2.75

where A, B, h_0, H^*, K_c and D are material constants, and n is given by

$$n = \frac{B\sigma(1-H)}{1-\phi}\coth\left(\frac{B\sigma(1-H)}{1-\phi}\right)$$

Equation 7.50 describes the evolution of creep for the aluminium alloy concerned. Equation 7.51 describes primary creep. The equation set contains two damage variables to model tertiary softening mechanisms. The first damage variable is described by Eq. 7.52, and the second damage variable is defined by Eq. 7.53. Detailed descriptions of the creep damage constitutive equations were given by Kowalewski *et al.* (1994). The constants in the equation set were determined for an aluminium alloy at 150°C using an optimisation method. The material constants in the equations are listed in Table 7.3.

7.7 Conclusion

Process modelling techniques can be used to predict forming processes, microstructure evolution and the mechanical properties of formed parts, provided that the constitutive equations for the materials used contain the necessary features. This can be achieved by the introduction of state variables into the equations. Each state variable represents a particular physical phenomenon that occurs when the material deforms in a particular hot-forming temperature range.

The unified viscoplastic constitutive equations cannot be solved analytically, and need to be integrated numerically. The difficulties associated with integrating the equations accurately and efficiently include control of the step size, since different equations in an equation set may contain different units. This makes error assessment difficult. The constants in the unified viscoplastic constitutive equations need to be determined from experimental data using an optimisation method. In addition to new search algorithms, the selection of a suitable objective function for optimisation is also very important.

7.8 References

Abramowitz M and Stegun I A (1972), *Handbook of Mathematical Functions with Formulas, Graphs, and Mathematical Tables*, New York: Dover.

Arieli A and Rosen A (1977), 'Superplastic deformation of Ti-6Al-4V alloy', *Metallurgical Transactions A*, 8A (10), 1591–1596.

Armstrong P J and Frederick C O (1966), 'A mathematical representation of the multiaxial Bauschiger effect', C. E. G. B. Report RD/B/N, 731.

Bäck T and Schwefel H-P (1993), 'An overview of evolutionary optimisation for parameter optimisation', *Evolutionary Computation*, 1(1), 1–23.

Cao J and Lin J (2008), 'A study on formulation of objective functions for determining material models', *International Journal of Mechanical Sciences*, 50, 193–204.

Cao J, Lin J and Dean T A (2008), 'An implicit unitless error and step-size control method in integrating unified viscoplastic/creep ODE-type constitutive equations', *International Journal of Numerical Methods in Engineering*, 73, 1094–1112.

Cormeau I (1975), 'Numerical stability in quasi-static elasto/visco-plasticity', *International Journal of Numerical Methods in Engineering*, 9, 109–127.

Chaboche J L (1977), 'Viscoplastic constitutive equations for the description of cyclic and anisotropic behaviour of metals', *Bulletin of the Polish Academy of Sciences, Series on Science and Technology*, 25(1), 33–42.

Chaboche J L and Rousselier G (1983), 'On the plastic and viscoplastic constitutive equations: part 1 – rules developed with internal variable concept', *Journal of Pressure Vessel Technology*, 105(5), 153–158.

Cheong B H, Lin J and Ball A A (2000), 'Modelling of the hardening characteristics for superplastic materials', *Journal of Strain Analysis*, 35(3), 149–157.

Dennis J E and Schnabel R B (1983), *Numerical Methods for Unconstrained Optimization and Nonlinear Equations*, Englewood Cliffs, NJ: Prentice-Hall.

Dmitri K, Philip B and Scott W S (2002), 'Adaptive backward Euler time stepping with truncation error control for numerical modelling of unsaturated fluid flow', *International Journal of Numerical Methods in Engineering*, 53, 1301–1322.

Dmitri K, Philip B and Scott W S (2004), 'Truncation error and stability analysis of iterative and non-iterative Thomas–Gladwell methods for first-order non-linear differential equations', *International Journal of Numerical Methods in Engineering*, 60, 2031–2043.

Edington J W, Melton K N and Cutler C P (1976), 'Superplasticity', *Progress in Materials Science*, 21, 63–169.

Fogel D B and Atmar J W (1990), 'Comparing genetic operators with Gaussian mutations in simulated evolutionary process using linear systems', *Biological Cybernetics*, 63(2), 111–114.

Fogel L J, Owens A J and Walsh M J (1966), *Artificial Intelligence through Simulated Evolution*, New York: Wiley.

Ghosh A K and Hamilton C H (1979), 'Mechanical behaviour and hardening characteristics of a superplastic Ti-6Al-4V alloy', *Metallurgical Transactions A*, 10A(6), 699–706.

Hernandez *et al.* (1996)

Karim A U and Backofen W A (1972), 'Some observations of diffusional flow in a superplastic alloy', *Metallurgical Transactions*, 3(3), 702–712.

Kowalewski Z L, Hayhurst D R and Dyson B F (1994), 'Mechanisms-based creep constitutive equations for an aluminium alloy', *Journal of Strain Analysis*, 29, 309–316.

Li B, Lin J and Yao X (2002), 'A novel evolutionary algorithm for determining unified creep damage constitutive equations', *International Journal of Mechanical Sciences*, 44(5), 987–1002.

Lin J (2003), 'Selection of material models for predicting necking in superplastic forming', *International Journal of Plasticity*, 19(4), 469–481.

Lin J and Dean T A (2005), 'Modelling of microstructure evolution in hot forming using unified constitutive equations', *Journal of Materials Processing Technology*, 167, 354–362.

Lin J and Dunne F P E (2001), 'Modelling grain growth evolution and necking in superplastic blow-forming', *International Journal of Mechanical Sciences*, 43(3), 595–609.

Lin J and Hayhurst D R (1993), 'Constitutive equations for multi-axial straining of leather under uni-axial stress', *European Journal of Mechanics A: Solids*, 12 (4), 471–492.

Lin J and Yang J (1999), 'GA based multiple objective optimization for determining viscoplastic constitutive equations for superplastic alloys', *International Journal of Plasticity*, 15, 1181–1196.

Lin J, Cheong B H and Yao X (2002), 'Universal multi-objective function for optimising superplastic-damage constitutive equations', *Journal of Materials Processing Technology*, 125–126, 199–205.

Lin J, Dunne F P E and Hayhurst D R (1996), 'Physically-based temperature dependence of elastic viscoplastic constitutive equations for copper between 20 and 500 °C', *Philosophical Magazine A*, 74(2), 655–676.

Lin J, Hayhurst D R and Dyson B F (1993), 'A new design of uniaxial testpiece with slit extensometer ridges for improved accuracy of strain measurement', *International Journal of Mechanical Sciences*, 35(1), 63–78.

Lin J, Liu Y and Dean T A (2005a), 'A review on damage mechanisms, models and calibration techniques', *International Journal of Damage Mechanics*, 14, 299–319.

Lin J, Liu Y, Farrugia D C J and Zhou M (2005b), 'Development of dislocation based-unified material model for simulating microstructure evolution in multipass hot rolling', *Philosophical Magazine A*, 85(18), 1967–1987.

Medina S F and Hernandez C A (1996), 'General expression of the Zener–Hollomon parameter as a function of the chemical composition of low alloy and micro-alloyed steels', *Acta Metallurgica*, 44(1), 137–148.

Polak E (1971), *Computational Methods in Optimization*, New York: Academic Press.

Rice J R (1970), 'On the structure of stress–strain relations for time-dependent plastic deformation in metals', *Transactions of the A. S. M. E., Journal of Applied Mechanics*, 37, 728–737.

Simos T E (2000), 'Exponentially fitted Runge–Kutta methods for the numerical solution of the Schrödinger equation and related problems', *Computational Materials Science*, 18, 315–332.

Willima H P, Brian P F, Saula A and Willima T V (2002), *Numerical Recipes in C: The Art of Scientific Computing*, London: Cambridge University Press.

Yao X, Liu Y and Lin G (1999), 'Evolutionary programming made faster', *IEEE Transactions on Evolutionary Computation*, 3 (2), 82–102.

Zhou M and Dunne F P (1996), 'Mechanism-based constitutive equations for the superplastic behaviour of a titanium alloy', *Journal of Strain Analysis*, 31(3), 187–196.

8 Modelling phase transformations in hot stamping and cold die quenching of steels

J. CAI and J. LIN, Imperial College London, UK
and J. WILSIUS, ArcelorMittal, France

Abstract: This chapter introduces the phase transformations occurring in the hot stamping and cold die quenching of steels. Both experimental and modelling techniques are described. Material models of the phase transformations, e.g. austenitization during heating and soaking and the bainite transformation during cooling, are presented in the form of unified constitutive equations. Future trends in the forming process and the relevant modelling techniques are discussed.

Key words: phase transformation, hot stamping, cold die quenching, material modelling, unified constitutive equations.

8.1 Introduction

8.1.1 Introduction to the process

Ideally, components are formed from materials that initially have high ductility and low strength, but are hardened during processing. It is common knowledge that, in general, metals are softened at elevated temperatures. But in hot/warm forming processes, the original microstructure of the material may be destroyed, and subsequent restorative heat treatment for formed sheet-metal components can cause significant distortion. In order to solve these problems, a forming process termed 'hot stamping and cold die quenching' is being used to meet the requirement of a high strength/weight ratio in the production of high-strength auto panels.

Figure 8.1 schematically illustrates the process of hot stamping and cold die quenching, with the corresponding temperature profile. As shown in the figure, the sheet steel blank is first austenitized in a furnace. The furnace temperature is above 900°C. When the steel is fully austenitized, the hot blank is quickly transferred to a cold die for simultaneous forming and quenching. A cooling system is often used to guarantee a sufficient cooling rate between the parts of the closed die. The formed component is held firmly in the closed die until the temperature of the blank drops below 200°C, when the phase transformation ceases. The microstructure of the formed part comprises largely martensite.

Hot stamping of sheet metals enables metals to be formed at low strength and high ductility. When a metal, for example a steel with high hardenability, is formed and held in a cold die set, a good cooling rate is achieved, which enables a transformation from austenite to pearlite, bainite or even martensite, with very high strength. In addition, springback and distortion due to rapid cooling and phase transformation

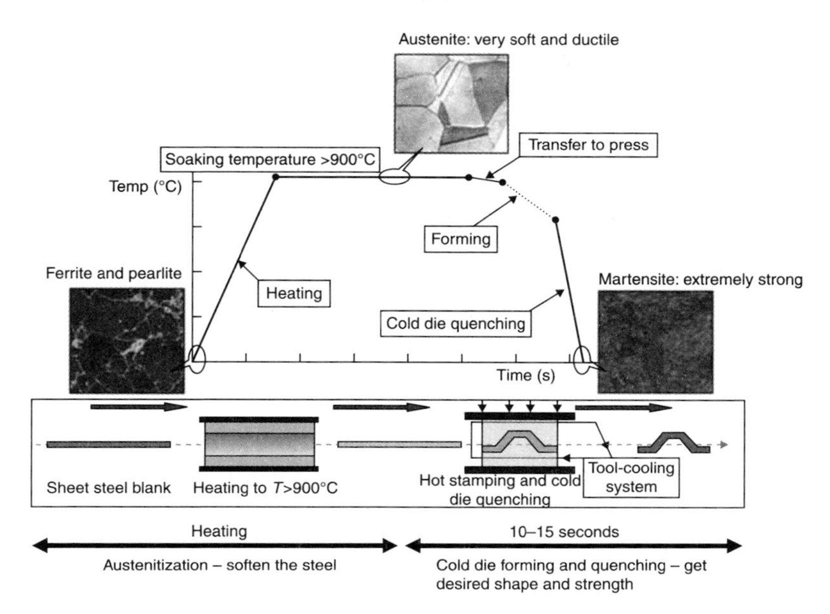

8.1 Illustration of the hot stamp and cold die quenching process and the temperature profile. Microstructures at different stages are indicated.

can be eliminated. The geometrical accuracy of the product can be significantly improved. In order to maximize the potential of the process, full transformation to martensite is required, so that the strongest components can be formed.

In this chapter, the boron steel USIBOR 1500 P® is used as an example material. Figure 8.2 shows the continuous-cooling temperature (CCT) diagram of USIBOR 1500 P, developed by ArcelorMittal particularly for the hot stamping and cold die quenching process. The alloy composition is given in Table 8.1. As shown in Fig. 8.2, the critical martensitic cooling rate is 27°C/s for this boron steel. However, a cooling rate of 300°C/s is required to ensure full transformation to martensite for some normal low carbon steels, which is unachievable by cold die quenching.

The important process parameters for the industrial processing of boron steel are:

- *Heat treatment condition*. The austenitization conditions in the heating furnace are chosen with respect to the metallurgical transformation and, in the case of the USIBOR coated steel, the intermetallic alloying reaction between the iron substrate and the Al–Si coating, which is unique to that steel. The maximum heating rate is 12°C/s so as to allow the alloying reaction and preserve the integrity of the layer.[1] The austenitization temperature ranges from 880°C to 940°C and the soaking time from 4 to 10 min, so that a homogeneous austenite phase is achieved, and abnormal or excessive grain growth is avoided.
- *Transfer from heating furnace to press*. The transfer operation is usually fast in order to minimize heat loss and to maintain the elevated temperature of the

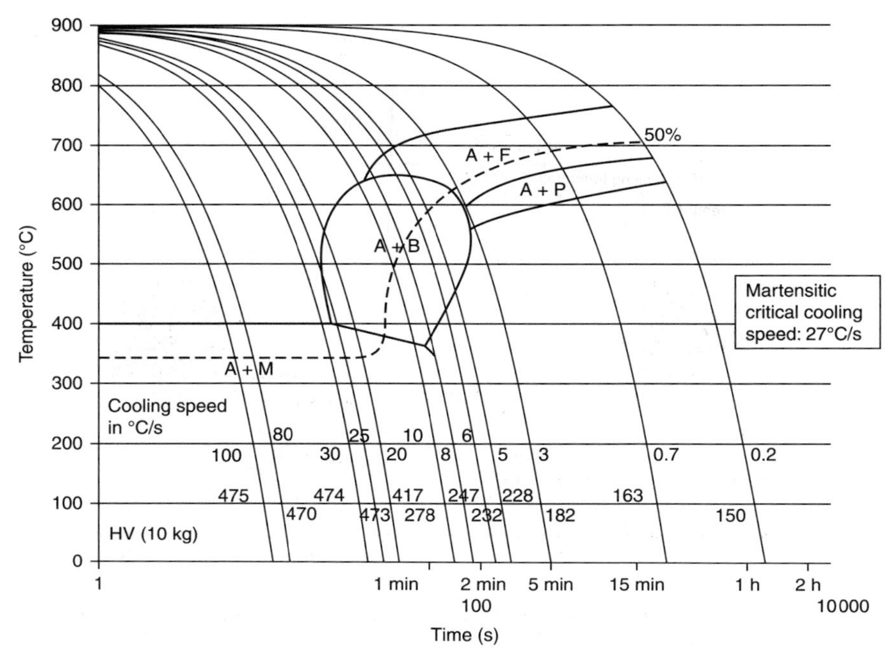

8.2 CCT diagram of USIBOR 1500 P (from ArcelorMittal).

Table 8.1 Chemical composition of USIBOR 1500 P (maximum %)

C	Mn	Si	Sr	Ti	B
0.25	1.40	0.35	0.30	0.05	0.005

heated blank, which avoids a reduction in formability and the occurrence of local phase transformation. In practice, the transfer time is shorter than 7 s.

- *Forming speed.* A fast closing speed of the tool is necessary to reduce heat exchange between the blank and the die during deformation, so that the formability of the steel is high and its flow stress is low.
- *Quenching.* The quenching speed in the cold die has to be sufficiently high (minimum 27°C/s for boron steel) that single-phase martensite can be obtained. Therefore, for mass production, hot-stamping dies are water cooled. The formed component is removed from the cold die when its temperature is sufficiently low, and hence thermal distortion can be minimized and no phase transformation occurs after removal.
- *Trimming.* Owing to the high strength of the material, laser trimming is employed. Alternatively, die-trimming with a low clearance can be used to shorten the process sequence.

8.1.2 Applications in the automotive industry

Owing to increasing concerns about safety, it is required that motor vehicles be designed for crashworthiness to protect the occupants. One of the most straightforward ways is to increase the strength of safety-critical components. Thus, exploring and developing new types of high-strength materials has become popular worldwide. In general, the trend in car body manufacturing is to use improved or new materials which can reduce car body weight, provide better fuel economy, allow smoother surfaces and more complex shapes, and possess high formability and excellent crashworthiness. Hot stamping and cold die quenching is thus being developed to meet both environmental and safety requirements.

The main advantages of the hot stamping and cold die quenching of boron steel are a significant increase in the strength of automobile safety-critical components and, consequently, an increase in the strength/weight ratio; and an improvement in the geometrical accuracy of formed parts. By use of this process, car body weights can be reduced, and thus less gasoline is consumed and less tail exhaust is emitted, which is more environmentally friendly; in addition, the safety of automobiles is enhanced.

The range of potential applications of hot-stamped boron steel is even wider than that of conventional high-strength steels (HSS) and advanced high-strength steels (AHSS), especially when combined with tailored blanks, and covers the whole range of reinforcement structures in vehicles, as shown in Fig. 8.3.[2] This is because the use of tailor welded blanks (TWBs) can significantly improve the crashworthiness, as the material thickness, strength and elongation are locally adjusted.

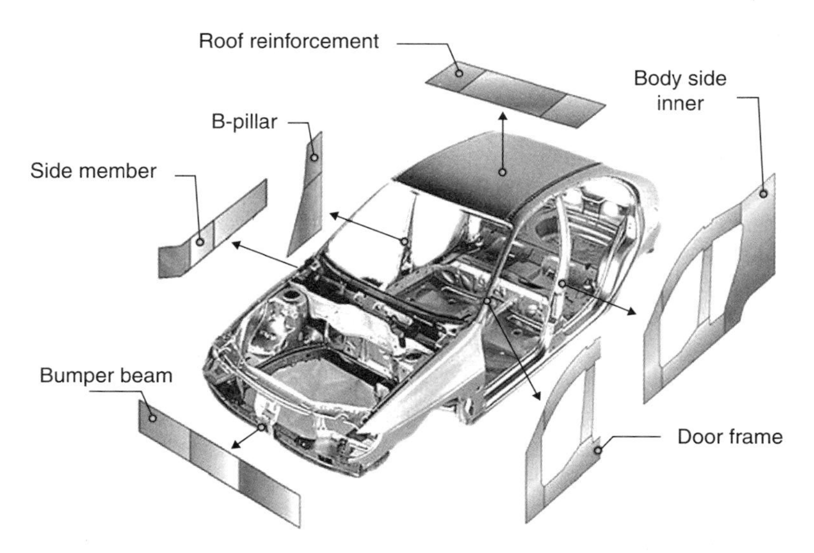

8.3 Potential applications of hot-stamped laser-welded blanks combining USIBOR 1500P and DUCTIBOR 500P.[3]

8.2 Phase transformations on heating: experimentation and modelling

One of the key features of this process is austenitization during the heating and soaking periods, for the purpose of forming a single austenite phase with a homogeneous microstructure and small grain size. Control of the austenitization is vital for obtaining the required high strength of formed boron steel parts. The characteristics of the austenite affect the microstructure of the component and hence the mechanical properties resulting from the forming/quenching operation. The heating and soaking need to be carefully designed so that the original phase can be fully austenitized and the optimum mechanical properties of the austenite can be achieved. A thorough understanding of the austenitization process is fundamental, and the use of a material model can aid in this. Fick's second law is often used to describe the austenitization process; however, it is difficult to solve and the formulation is usually complex. Factors such as the incubation period and various characteristics of the heating process are normally ignored. A mechanism-based model which is easy to solve and to implement would provide a more convenient method.

8.2.1 Austenitization mechanism

Typically, the austenitization process can be divided into three steps: nucleation, growth and homogenization. The formation of austenite depends greatly on the parent phase prior to heating. Grain growth and homogenization is a carbon-diffusion-controlled process, which depends significantly on the microstructure, chemical composition, soaking temperature and soaking time.[3]

Nucleation

From the relationship between the amount of austenite formed and the austenitizing time for a eutectoid steel, it has been shown that there is an incubation period for the formation of the first austenite nucleus, after which more nuclei develop and grow at a much higher rate. The incubation time is required in order to satisfy the thermodynamic conditions for nucleation. The eutectoid reaction of ferrite + carbide austenite occurs when austenite nucleates. Therefore, austernite is expected to nucleate at carbide-ferrite interfaces. As the driving force for heterogeneous grain nucleation is determined by thermodynamics and geometry, the surface energy at the interfaces of the parent and product phases has to be taken into account. When the temperature is increased to that at which austenitization occurs, the free energy change increases with increasing temperature, resulting in an increasing rate of austenite nucleation. Therefore, the start and completion of austenitization will occur earlier at a higher temperature than at a lower one in the isothermal case, as shown in Fig. 8.4, where γ, θ and α represent the austernite,

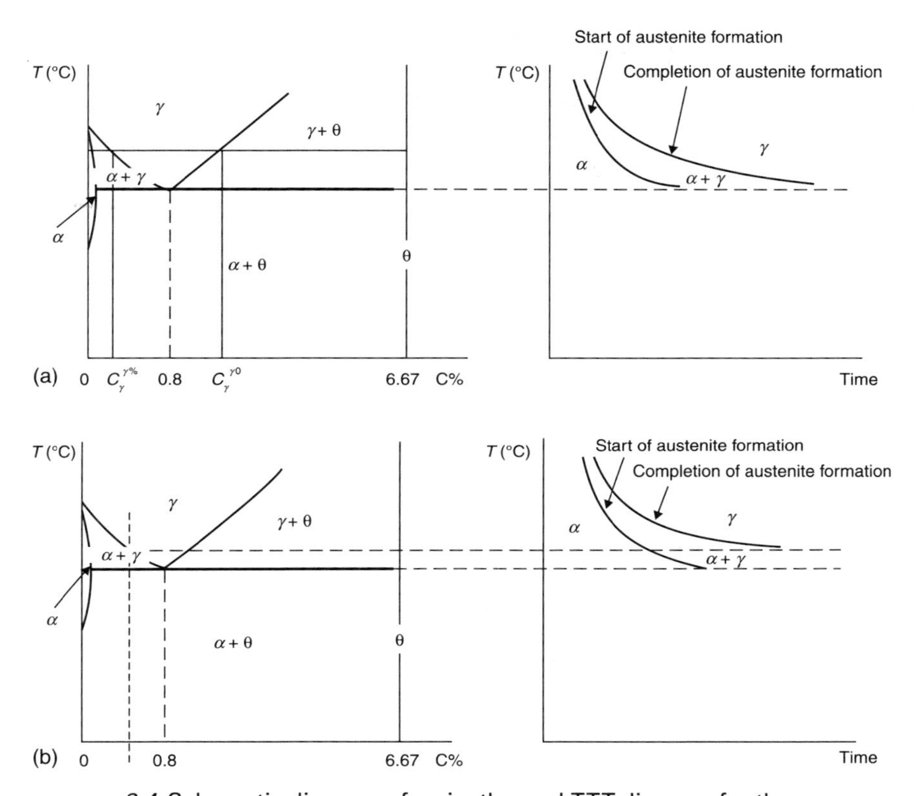

8.4 Schematic diagram of an isothermal TTT diagram for the formation of austenite. (a) Eutectoid steel (C% = 0.8); (b) hypoeutectic steel (C% < 0.8).[5]

cementite and ferrite regions, respectively; C% is the average carbon concentration; $C_\gamma^{\gamma\theta}(\%)$ is the carbon concentration in the austenite at the θ/γ interface; and $C_\gamma^{\gamma\alpha}(\%)$ is the carbon concentration in the austenite at the γ/α interface.[4]

Grain growth of austenite

After the nucleation of austenite, further austenitization will be determined by grain growth. For a Fe–C steel, grain growth of austenite is controlled by diffusion of carbon at the θ/γ and γ/α interfaces, and thus it is greatly affected by the factors of temperature, material composition and the grain size of the parent phases. The carbon distribution is uneven in the growing austenite. The evolution of the carbon concentration and carbon distribution profile in each phase is schematically shown in Fig. 8.5.[5] Here, $r_{\gamma\theta}$ and $r_{\gamma\alpha}$ are the positions of the θ/γ interface and γ/α interface, respectively; r_0 is the initial position of the θ/α interface, which is equal to the initial radius of an assumed spherical cementite particle.

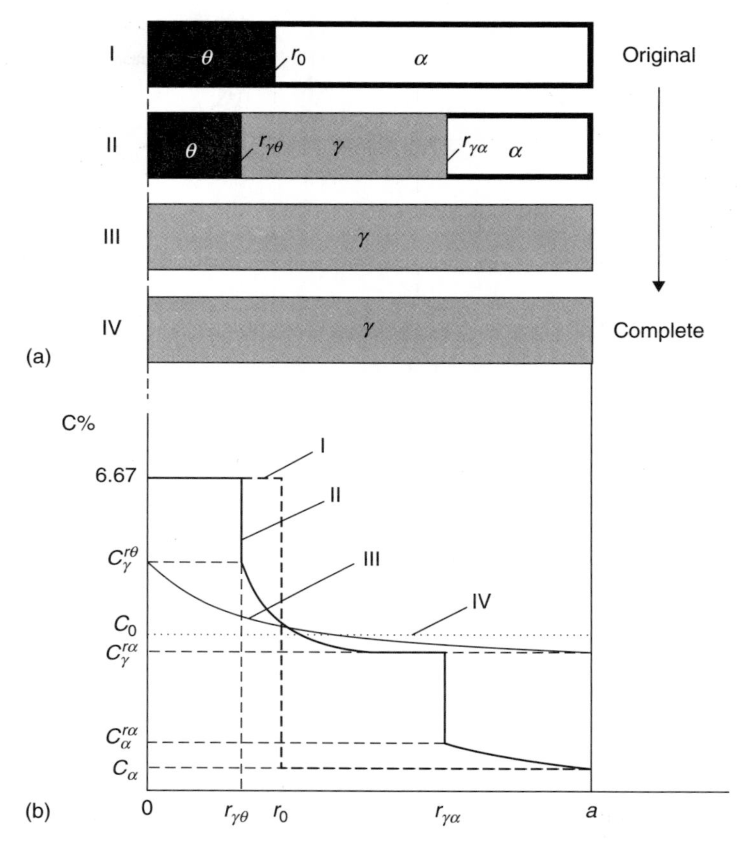

8.5 (a) Illustration of austenite nucleation, grain growth and homogenization in a two-phase (θ–α) steel; (b) carbon distribution profile in each phase at various stages of austenitization.[6]

Homogenization of austenite

The homogenization of austenite includes the processes of dissolution of retained cementite and chemical homogenization. Behind the moving interface between the austenite and parent phases, the austenite may contain undissolved carbide in the form of spheroidized particles,[6] termed retained cementite. Immediately after the parent phase disappears and the steel is fully transformed into austenite, the chemical composition is still not uniform. The development of the austenite grain size and the carbon concentration profile during the whole austenitization process is illustrated in Fig. 8.5 for a low carbon steel whose average carbon content is C_0. Here, r is the distance from the zero point.

Stage I represents the initial state, of parent phases composed of cementite and ferrite; Stage II represents the growth of austenite; Stage III represents the homogenization process; and Stage IV is when the final homogenized austenite is obtained. The corresponding carbon distribution profile is shown in Fig. 8.5(b). A significant carbon concentration gradient exists in the austenite region. Curve III corresponds to the Stage III in Fig. 8.5(a). To produce an acceptable homogenized microstructure, the steel has to be reheated or held in the austenite temperature region for a length of time sufficient to dissolve the retained cementite and eliminate the uneven chemical composition in the austenite. A uniform carbon distribution in the austenite, equal to the average carbon content in the virgin material, will be the final result (Curve IV in Fig. 8.5(b)). This process is termed homogenization, and is a carbon-diffusion-controlled process for Fe–C steels. Compared with austenite nucleation and grain growth, homogenization is usually far slower. The carbon content in the cementite (θ) is 6.67%.

8.2.2 Experimental study of austenitization in hot stamping and cold die quenching

The effect of the heating rate on austenitization can be determined by performing heat treatment tests, for example using a Gleeble simulator. The test conditions are shown in Fig. 8.6. The specimen is heated at different heating rates, followed by soaking at different temperatures and for different times. The volume fraction of martensite in the final phase, which can be measured using scanning electron microscopy (SEM), is equal to the volume fraction of

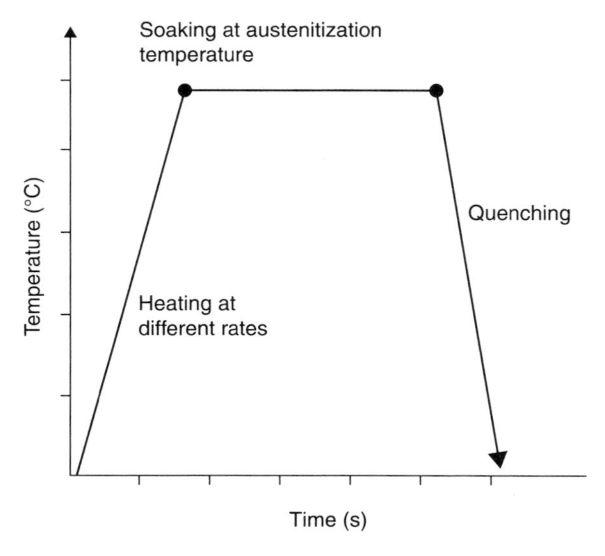

8.6 Test programme for a phase transformation study.

austenite immediately before quenching. By such tests, the start and finish temperatures of the austenitization process can be determined, as well as the growing volume fraction of austenite at particular temperature levels during continuous heating.

8.2.3 Constitutive modelling of austenitization of boron steel

A set of mechanism-based unified constitutive equations for modelling the evolution of austenite during continuous heating is described in this section. To develop these equations, the effects of temperature and heating rate on the incubation time were rationalized, and the effects of temperature and heating rate on the development of the austenite volume fraction were analysed as well. The evolution of strain with the increasing volume fraction of austenite at different heating rates was evaluated. Optimization techniques and procedures were developed to determine the model from the results of dilatometry experiments.

Modelling of incubation time

The incubation time, which is a function of temperature and heating rate, is an important parameter in determining the starting temperature of austenitization. A higher heating rate leads to a higher start temperature. A variable x, which varies from 0 to 1, was introduced to describe the incubation period. As soon as the temperature rises to the critical austenitization temperature of the steel, usually 727°C, incubation starts. At the beginning of the incubation period, $x = 0.0$. When the incubation period approaches its end, the value of x is close to 1.0, and the austenitization process occurs. At a higher heating rate, the incubation covers a wider temperature range above 727°C. To model the incubation process, the following rate equation was introduced:

$$\dot{x} = A(1-x)\dot{T}^{-\gamma_1} \cdot \left(\frac{T}{T_c} - 1\right) \qquad [8.1]$$

where A is a material constant related to the incubation period, γ_1 is a constant to regulate the heating rate during the incubation period, T is the absolute temperature, and T_c is the theoretical critical austenitization temperature of the steel; $T_c = 1000\,\text{K} = 727°\text{C}$. If the temperature is lower than T_c, then $(T/T_c - 1) < 0$, and $\dot{x} = 0$.

Modelling of the evolution of the austenite fraction

As the austenitization of steel is a diffusion process, it is necessary to include a diffusion factor in the model to evaluate the volume fraction of austenite. The diffusion coefficient, which is temperature-dependent for a thermal austenitization process, can be described as follows:

$$D = D_0 \exp\left(-\frac{Q_D}{RT}\right) \qquad [8.2]$$

where D_0 is a material constant, Q_D is the activation energy, T is the absolute temperature and R is the universal gas constant. The introduction of D represents the effect of temperature on the diffusion rate of carbon or alloying elements in steels. It is also known from experimental results that austenitization takes place after incubation and that the phase transformation is fast at the beginning. With the accumulation of austenite, the austenitization process slows down. The growth rate of the fraction of austenite is a function of the accumulated volume fraction of austenite. Therefore, the evolution rate of the volume fraction of austenite is controlled by several parameters:

$$\dot{v}_a = f(D, T, \dot{T}, v_a) \qquad [8.3]$$

Both temperature and heating rate have to be included, because the austenitization of steel is temperature-dominated; the heating process plays an important role. To model the evolution of the volume fraction of austenite, the following rate equation is introduced:

$$\dot{v}_a = D\left(x\frac{T}{T_c} - 1\right)^{m_a} \cdot (1 - v_a)^{n_a} \cdot \dot{T}^{\gamma_2} \qquad [8.4]$$

where m_a and n_a are material constants used to regulate the effects of temperature and of accumulation of austenite, and γ_2 is a constant to regulate the effect of heating rate during the phase transformation. The term $(xT/T_c - 1)$ is introduced to determine the start temperature of the austenitization process. If $(xT/T_c - 1) < 0$, then $\dot{v}_a = 0$. Austenitization only occurs when $(xT/T_c - 1) \geq 0$.

Evaluation of dilatometry strain

Throughout the process of austenitization, the volume change of the material is characterized to evaluate the amount of austenite that has been transformed. The thermal strain can be summarized by the phase transformation part of the dilatometric strain curve, which is given by

$$\dot{\varepsilon}_\theta = \left[\alpha_\alpha - (\alpha_\alpha - \beta_\gamma) \cdot v_\alpha\right]\dot{T} - \theta_v \dot{v}_\alpha \qquad [8.5]$$

where α_α (K^{-1}) and β_γ (K^{-1}) are the thermal expansion coefficients of ferrite and austenite, respectively, which can be measured directly from the dilatometric strain curve. For boron steel, $\alpha_\alpha = 15.251 \times 10^{-6}$/K and $\beta_\gamma = 21.6305 \times 10^{-6}$/K. θ_v is a material constant related to the volume difference between the original material lattice (ferrite) and the austenite lattice.

The unified constitutive equations

By introducing the incubation variable and inserting the austenite volume fraction evolution equation (Eq. 8.4) into the thermal strain equation (Eq. 8.5), we can formulate a set of unified constitutive equations to model the austenitization process as follows:

$$\dot{x} = A(1-x)\dot{T}^{-\gamma_1} \cdot \left(\frac{T}{T_c} - 1\right)$$

$$\dot{v}_a = D\left(x\frac{T}{T_c} - 1\right)^{m_\alpha} \cdot (1-v_\alpha)^{n_\alpha} \cdot \dot{T}^{\gamma} \tag{8.6}$$

$$\dot{\varepsilon}_\theta = [\alpha_\alpha - (\alpha_\alpha - \beta_\gamma) \cdot v_\alpha]\dot{T} - \theta_v \dot{v}_\alpha$$

$$T = T + \dot{T}\,\Delta t$$

where A, γ_2, γ_1, m_a, n_a and θ_v are material constants; α_α and β_γ are the thermal expansion coefficients of ferrite and austenite, respectively; D is the diffusion coefficient and is temperature-dependent; the temperature T is in kelvin; and T_c is the theoretical critical austenitization temperature of the steel. As before, T_c = 727°C = 1000 K. If $(T/T_c - 1) < 0$, then $\dot{x} = 0$; if $(xT/T_c - 1) < 0$, then $\dot{v}_\alpha = 0$.

Determination and validation of the unified material model of austenitization

The determination of the constants in the equation set in Eq. 8.6 is based on experimental dilatometry results obtained from heat treatment tests, as described in Section 8.2.2. To highlight the phase transformation process, enlarged strain–temperature curves predicted by Eq. 8.6, in the temperature range from 727°C (1000 K) to 950°C (1223 K), are presented in Fig. 8.7, where the x-axis is the temperature and the y-axis is the thermal strain. The model is calibrated via an optimization method by fitting the predicted dilation curves obtained from the model to the experimental data; the material constants A, D_0, n_a, θ_v, m_a, Q_D, γ_1 and γ_2, are determined by the fitting process. As shown in the figure, at a particular temperature, the thermal strains are smaller at lower heating rates, representing a larger volume fraction of austenite. The reason is that lower heating rates allow a longer time for the phase transformation. As a consequence, a larger amount of austenite is formed. The results of the analysis of some dilatation curves obtained from heat treatment tests are summarized in Fig. 8.8.[7] The triangle symbols represent experimentally measured temperatures.

The measured fractions of austenite (measured as martensite) are also presented in Fig. 8.8. As shown in the figure, for a heating rate of 5°C/s, when the percentage of austenite increases from 0 to 21%, the temperature increase is around 24°C; when the percentage of austenite increases from 77% to 95%, the temperature increase is around 60°C. The austenitization rate becomes smaller as the absolute

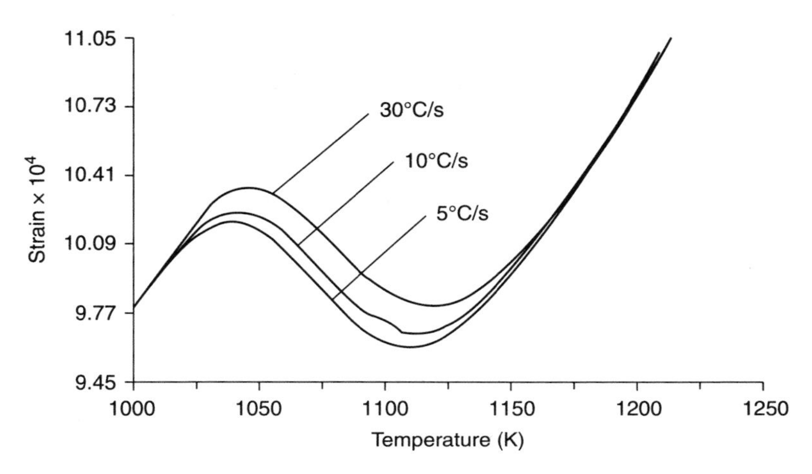

8.7 Predicted strain–temperature relationships for the phase transformation parts of dilatometry curves. The heating rate varies from 5°C/s to 30°C/s.

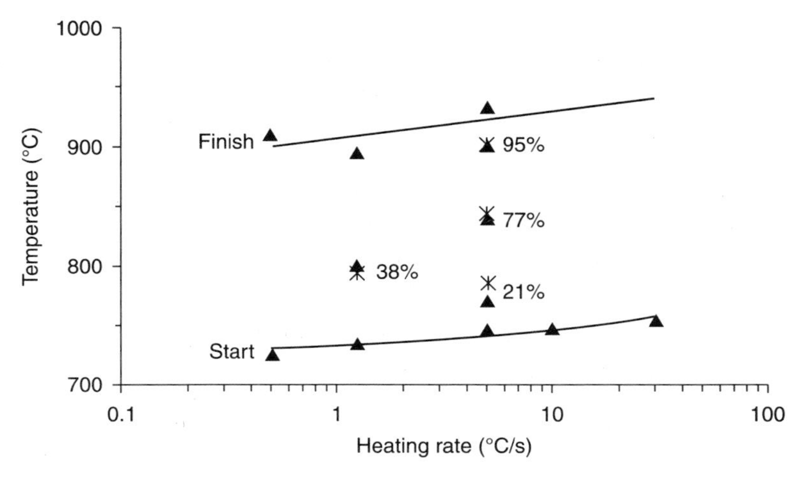

8.8 Comparison of experimental (▲) and computed (solid lines and ✳) starting and finishing temperatures of the austenitization process, and the evolution of the fraction of austenite at heating rates of 1.25°C/s and 5°C/s.[8]

amount of austenite increases. This shows that the austenite evolution rate is also a function of the volume fraction of austenite itself, and decreases with the growth of the austenite fraction. The evolution of austenite is multi-controlled by the diffusion coefficient, the temperature, the heating rate and the accumulation of austenite.

The start and finish temperatures of austenitization at heating rates from 0.5°C/s to 30°C/s are predicted by the model that was determined, as well as the evolution

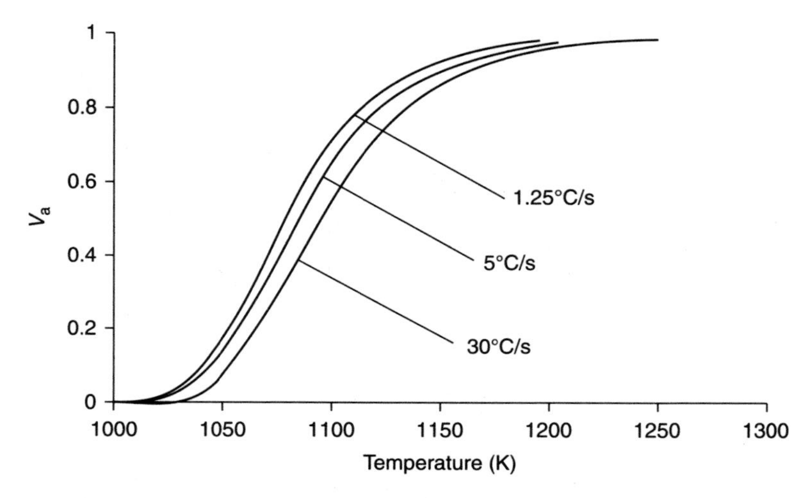

8.9 Evolution of computed fraction of austenite vs temperature. The heating rates are 1.25°C/s, 5°C/s and 30°C/s.

of the volume fraction of austenite at heating rates of 0.5°C/s and 1.25°C/s. The results are summarized in Fig. 8.8 by the solid lines. The predicted volume fractions of austenite at certain temperature levels are indicated by the star symbols. The predicted volume fractions of austenite as a function of temperature at different heating rates are presented in Fig. 8.9.

Since the model is physical-mechanism based and does not depend on the parent or austenite microstructure, it can potentially be used for various types of steels, with the material constants redetermined for the specific steel.

8.3 Phase transformations on cooling: experimentation and modelling

Bainite and martensite transformations are observed during cooling in the process of hot stamping and cold die quenching. The martensite transformation is a typical non-diffusive process, which is fast and difficult to model. The bainite transformation in the process is studied and modelled in this section. The fraction of bainite transformation is predicted and, consequently, the fraction of martensite transformation is determined.

8.3.1 Mechanism of bainite transformation

The formation of bainite in steels requires redistribution of carbon and is thus diffusional, but not necessarily diffusion controlled. Many measurements have shown that the growth rate of bainite is too rapid to be attributed to carbon diffusion. There have been many arguments about whether the bainite transformation is a diffusional process or a martensitic transformation.[8–10] As early as the 1950s, it was observed that bainite and Widmanstätten ferrite in a hypereutectoid steel grew simultaneously and produced a martensite-like upheaval on the free surface of specimens, but grain boundary ferrite allotriomorphs did not. Bainite forms by shear, as does martensite, but the process is accompanied by diffusion of carbon in the parent phase. Further experimental results have shown that the formation of bainitic ferrite at a free surface gives rise to surface relief, and it has been well demonstrated that the shape change has the characteristics of an invariant plane strain.[11] It has thus been proved that the transformation to bainitic ferrite is displacive and that the lattice deformation within the transformed volume has a significant component of shear. Taking account of the fact that significant diffusion of carbon involved, the bainite transformation is therefore widely accepted as being diffusional-displacive.[12, 13]

8.3.2 Experimental study of bainite transformation in hot stamping and cold die quenching

It is well known that the bainite and martensite transformations are related to the cooling rate. However, in the hot stamping and cold die quenching process, it is important to know the effect of the strain due to hot stamping on the phase transformation. Figure 8.10 shows a test programme to characterize this effect.

An experimental investigation of the effect of strain on the bainite transformation was performed by introducing deformation during the cooling process shown in Fig. 8.10. The start and finish temperatures of the bainite and martensite transformations were deduced from dilatation curves, and the result is summarized in Fig. 8.11. The dashed curves are CCT diagrams of the boron steel without

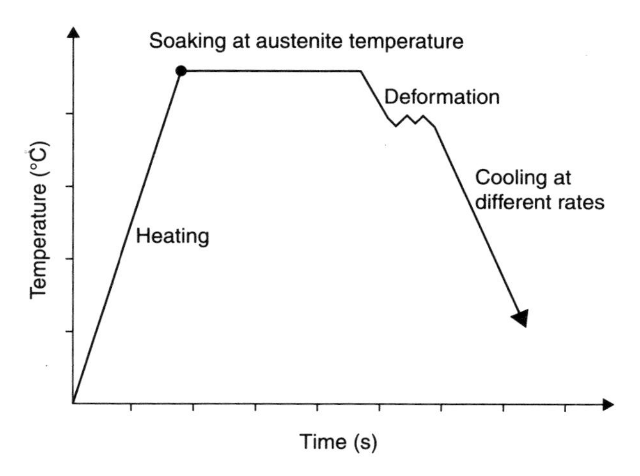

8.10 Test programme and temperature profile for phase transformation study.

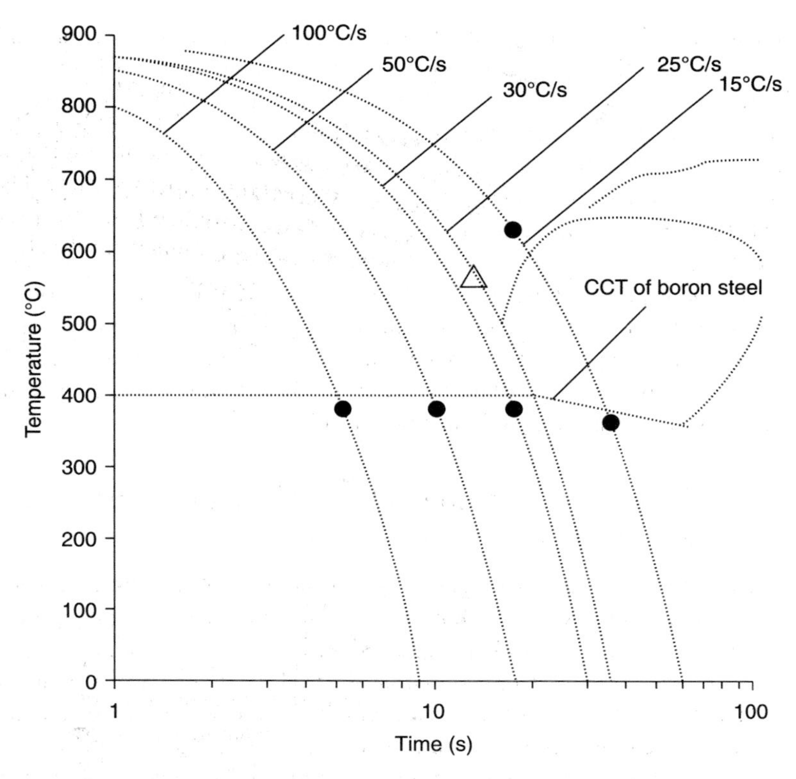

8.11 Critical temperatures of bainite and martensite transformations when pre-deformation is introduced.

deformation. The symbols show the measured critical temperatures of the bainite and martensite transformations when the steel was pre-deformed at 800 and 600°C. Compared with the CCT diagrams of the steel where no pre-deformation was introduced, the start temperature of the bainite transformation increased for cooling rates of 15°C/s and 25°C/s, and the start temperature of the martensite transformation decreased by less than 15°C. The microstructure of the steel formed at 800°C and quenched at 15°C/s was examined, and also that of the steel formed at 600°C and quenched at 25°C/s. Mixed phases of bainite and martensite were obtained in these two cases. For the same cooling rate, a larger amount of bainite was observed compared with that in steel that had not been pre-deformed.

8.3.2 Constitutive modelling of bainite transformation during cooling processes

In this section, mechanism-based unified constitutive equations to predict the evolution of bainite during quenching are introduced. The critical radius of newly formed bainite is deduced, and the evolution rate of the newly formed phase at different cooling rates is evaluated. Optimization procedures to calibrate the unified constitutive equations from heat treatment tests and relevant microstructure observations are described.

Nucleation and growth

Theoretically, there is an equilibrium temperature T_{eq} between the austenite and bainite phases. The two phases can coexist at this temperature. When the temperature is higher than T_{eq}, bainite will automatically transform into austenite to maintain the lowest free energy of the system, and vice versa. In order to achieve the transformation from austenite to bainite, a small degree of undercooling is required. If the austenite is undercooled before transformation occurs by ΔT, where $\Delta T = T_{eq} - T$, the bainite transformation will be accompanied by a decrease in free energy ΔG_V, which provides the driving force for the transformation, as shown in Fig. 8.12. In the figure, T_{eq} is the equilibrium temperature of austenite and bainite, and G_V^A and G_V^B are the free energies of the initial (austenite) and final (bainite) states, respectively. When the undercooling is ΔT, the corresponding decrease in free energy ΔG_V is

$$\Delta G_V = G_V^A - G_V^B = \frac{\Delta T}{T_{eq}} \cdot Q_V \qquad [8.7]$$

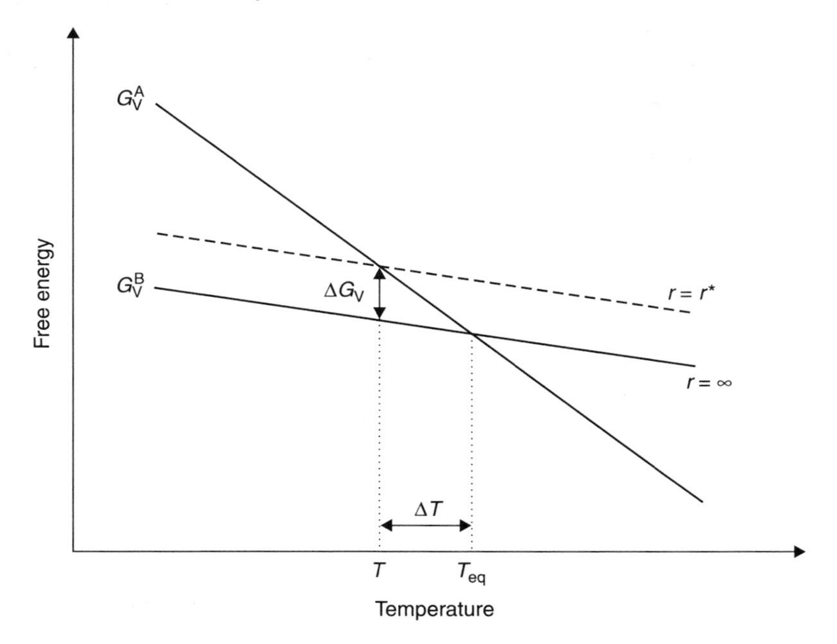

8.12 Changes in the volume free energy of the bainite/austenite transformation as a function of temperature.

where Q_V is the activation energy for nucleation of bainite, which is proportional to the driving force for the transformation.

According to nucleation theory based on homogeneous nucleation, the change in free energy when a particle of a new phase is formed can be ascribed to two contributions: the decreased free energy per unit volume (ΔG_V) and the increased interfacial energy (S_{AB}) of small particles. The magnitude of the volume free energy ΔG_V depends on the degree of undercooling. Assuming that the particle of the new phase is a sphere with radius r, the change in free energy can be written as

$$\Delta G = -\frac{4}{3}\pi r^3 \Delta G_V + 4\pi r^2 S_{AB} \quad [8.8]$$

The interfacial energy S_{AB} is usually difficult to determine, but can be described as being proportional to a temperature ratio:[14]

$$S_{AB} \propto \frac{\sqrt{\Delta T}}{T} \quad [8.9]$$

It can be seen from Eq. 8.8 that the interfacial-energy term increases as r^2, whereas the released volume free energy increases as r^3. Therefore the creation of small particles of solid will always lead to a free energy increase when the radius of the newly formed particle is in a certain range. As a consequence, austenite can be maintained in a metastable state even if the temperature is below T_{eq}. If the radius r is large, the r^3 term will play the key role and the volume free energy term will be dominant in Eq. 8.8. In this case, an increase in the value of r will decrease the free energy of the entire system, as shown in Fig. 8.13, and the austenite will be decomposed.

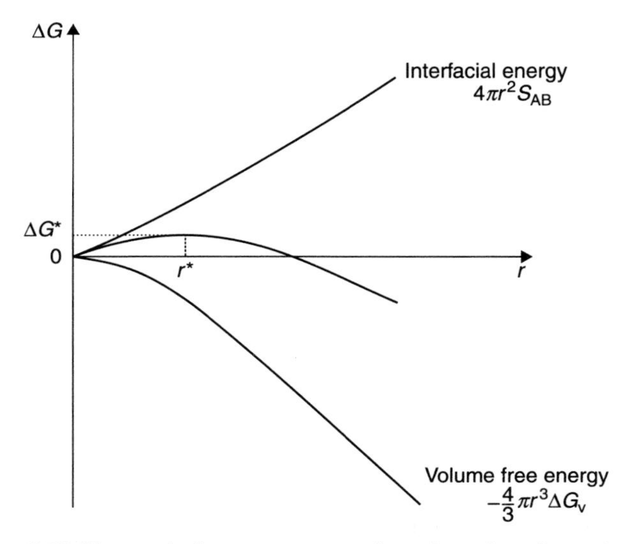

8.13 Change in free energy as a function of nucleus size.

Based on molecular dynamics and kinetic theory, it can be seen from Fig. 8.13 that a critical radius r^* exists which is associated with a maximum excess free energy, the activation energy barrier ΔG^*. Thus, r^* can be determined by differentiating the change in free energy with respect to the radius r and equating the result to zero, as follows:

$$\frac{\mathrm{d}\Delta G}{\mathrm{d}r} = -4\pi r^2\,\Delta G_V + 8\pi r S_{AB} = 0 \qquad [8.10]$$

Hence,

$$r^* = \frac{2S_{AB}}{\Delta G_V} \qquad [8.11]$$

Substituting Eqs 8.7 and 8.9 into Eq. 8.11 gives

$$r^* = \frac{A_1 T_{eq}}{T\sqrt{\Delta T Q_V}} \qquad [8.12]$$

where A_1 is a material constant.

For nucleated particles of the new phase, only those with radii greater than r^* can grow. The growth rate of the particles of the new phase is temperature-dependent and follows a Gaussian function, as shown in Fig. 8.14. When the temperature is lower than T_{eq}, a large undercooling will result in a significant decrease in the free energy change. Meanwhile, the carbon concentration in the austenite increases owing to the drop in temperature; hence the carbon diffusion distance is reduced. These factors will assist the bainite transformation and

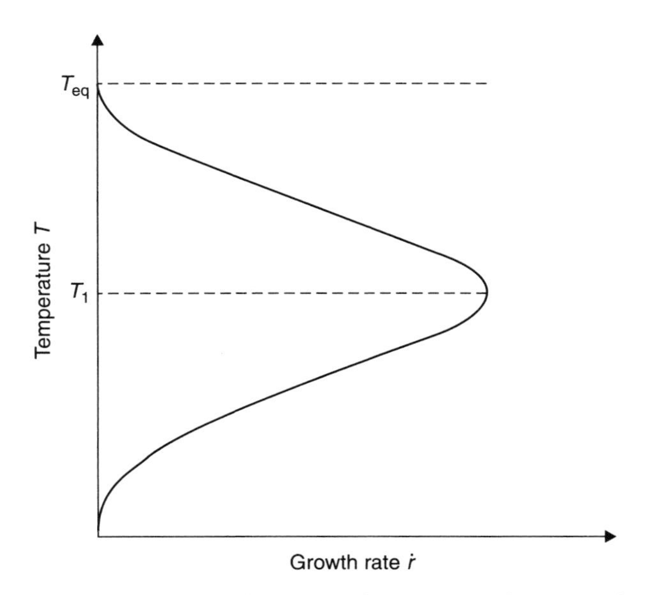

8.14 Growth rate of the size of bainite as a function of temperature.

increase the growth rate. With continuous cooling, the transformation temperature becomes lower and sharply reduces the activation energy of the phase transformation, which leads to a subsequent decrease in the growth rate of the new phase. Therefore, a modified Gaussian function is proposed to describe the growth rate of the bainite particle size:

$$\dot{r} = A_2 \exp\left(-\frac{|T - T_1|^{n_1}}{B_1} \right) \qquad [8.13]$$

where A_2, B_1, n_1 and T_1 are material constants. The highest growth rate of r is achieved at an intermediate temperature T_1. When the radius r of the nucleated particle reaches the critical value r^*, the particle can grow larger, and the start of a new phase arises. The time for the radius to reach r^* is termed the incubation time. Before r reaches r^*, the steel is in the incubation period.

Evolution of volume fraction

The start of formation of a new phase is defined as the time when the radius r of the new-phase particle is greater than r^*. The fraction of newly formed phase increases with increasing size of the new-phase particle. Therefore, a term $(r - r^*)/r$ is introduced to take into account the effect of the radius of the new-phase particle, and normalize the term into the range 0 to 1. During the incubation period, if $r < r^*$, the growth rate of the volume fraction $\dot{X} = 0$.

When bainite begins to form, the nucleation rate, which is temperature-dependent, becomes one of the major factors dominating the amount of transformed phase. The relation between nucleation rate and temperature can be considered as a modified Gaussian distribution as shown in Fig. 8.15. When the undercooling is less than ΔT, nucleation is negligible, as the driving force is too small for nuclei to grow. The nucleation rate increases with increasing undercooling in the temperature range where $T > T_2$. With very large undercooling, when $T < T_2$, the nucleation rate will decrease because of slow diffusion due to the low temperature. A maximum nucleation rate is obtained at an intermediate temperature T_2. The effect of temperature on the amount of phase transformation can be introduced by means of the nucleation rate, in the form of a modified Gaussian function.

Considering the effects of new-phase particle size and nucleation rate, the growth rate of the bainite fraction X during the transformation from austenite to bainite can be presented as

$$\dot{X} = A_3 \dot{r} \left(\frac{r - r^*}{r} \right)_+^{n_2} \exp\left(-\left(\frac{|T - T_2|}{B_2} \right)^{n_3} \right) (1 - X)^{n_4} \qquad [8.14]$$

where A_3, B_2 and T_2 are material constants; the powers n_2, n_3, n_4 give higher regulation of the equation. The term $(1 - X)^{n_4}$ is introduced to ensure that the volume fraction of the phase transformation X varies from 0.0 to 1.0, as well as to

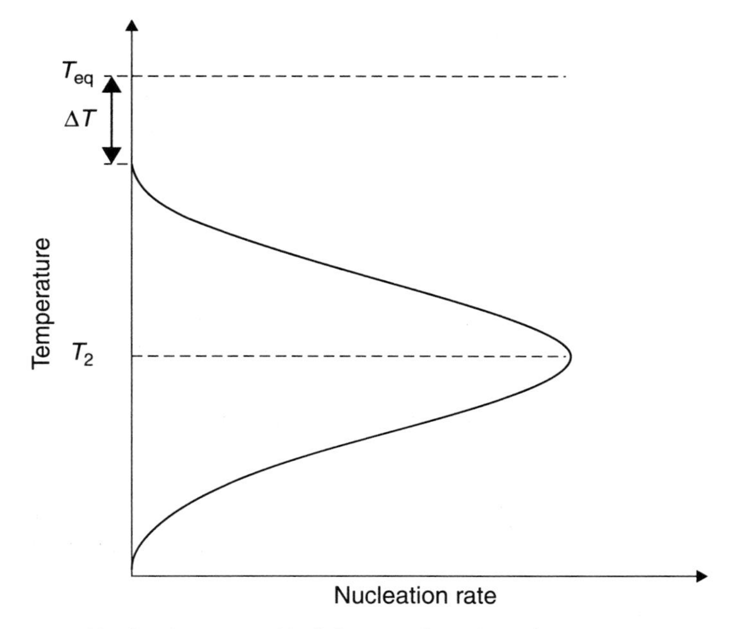

8.15 Nucleation rate of bainite as a function of temperature.

control the effect of the transformed phase on the transforming phase, because the growth rate $\dot{X}$ will decrease with the accumulation of transformed phase. For the term $((r - r^*)/r)^{n_2}$, only positive values are considered: if $r - r^* \leq 0$, $\dot{X} = 0$. Since only two phases, bainite and martensite, exist in hot-stamped parts in the range of cooling investigated, the fraction of martensite X_M can thus be obtained from

$$X_M = 1 - X_B \tag{8.15}$$

By introducing the critical radius of a new-phase particle, the grain size growth rate and the growth rate of the volume fraction of bainite, we can formulate a set of unified constitutive equations predicting the volume fraction of bainite transformation during the quenching of boron steel as follows:

$$\begin{cases} r^* = \dfrac{A_1 T_{eq}}{T\sqrt{T_{eq} - TQ_V}} \\ \dot{r} = A_2 \exp\left(-\dfrac{|T - T_1|^{n_1}}{B_1}\right) \\ \dot{X} = A_3 \dot{r} \left(\dfrac{r - r^*}{r}\right)^{n_2} \exp\left(-\left(\dfrac{|T - T_2|}{B_2}\right)^{n_3}\right)(1 - X)^{n_4} \end{cases} \tag{8.16}$$

where $A_1, A_2, A_3, T_1, T_2, B_1, B_2, n_1, n_2, n_3, n_4$ and Q_V are material constants, and need to be determined from experimental results and CCT diagrams. The temperature is in kelvin. T_{eq} is the equilibrium temperature, which is given by the CCT diagram. For boron steel, T_{eq} is equal to 983 K when no pre-deformation is introduced.

Calibration of the model

In the equation set in Eq. 8.16, the equations for r^* and $\dot{r}$ predict the start of the bainite transformation, as the phase transformation starts when $r \geq r^*$. Thus, the material constants in the equations for r^* and $\dot{r}$, which are $T_1, n_1, B_1, A_1, A_2, Q_V$, and T_{eq}, can be determined by fitting the start temperature and time of the bainite transformation to the CCT diagram.

A comparison of the best fit of the computed starting temperature of the bainite transformation and the CCT diagram of boron steel is shown in Fig. 8.16. Here, the cooling rate varies from 10°C/s to 25°C/s; these values are in the range of cooling rates where a mixed phase of bainite and martensite can be obtained. The triangle symbols show the computed temperatures of the bainite transformation for cooling rates of 10, 15, 20 and 25°C/s. It can be seen that the predicted result agrees well with the CCT diagram.

To determine the constants in the equation for $\dot{X}$, the model has to be fitted to the fraction of transformed bainite. The fraction of transformed bainite was measured from the microstructure of heat-treated steel. A comparison of the

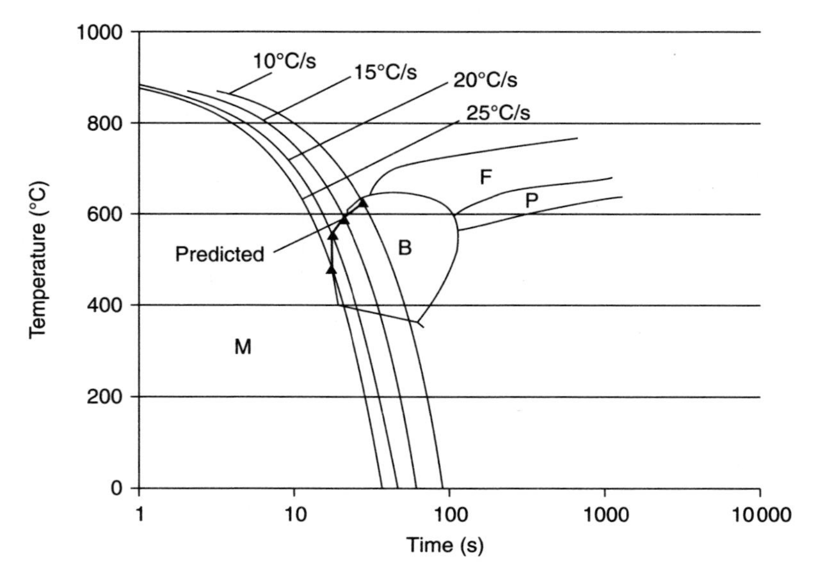

8.16 Comparison of the predicted starting temperature of the bainite transformation and the CCT diagram of boron steel.

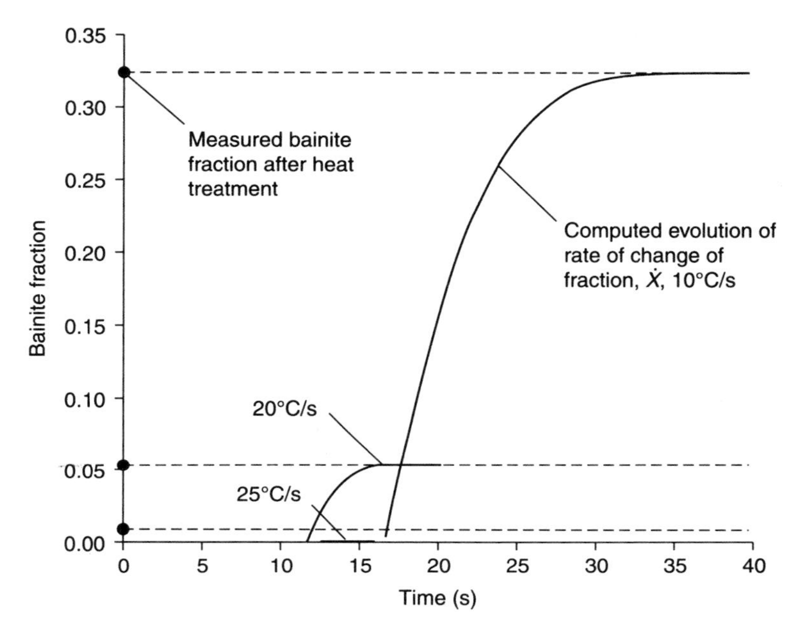

8.17 Comparison of predicted and experimentally-observed bainite fractions.

predicted bainite fraction and the experimentally-observed bainite fraction is shown in Fig. 8.17. Good agreement is achieved.

Modelling bainite transformation of pre-deformed material

Pre-deformation has been reported to affect the start temperature and the amount of bainite transformation[15–17] in a complex manner, and thus is of importance when high-strength steels are being formed. Plastic deformation introduced before the bainite transformation changes the amount of energy required to drive the phase transformation, and it promotes the bainite transformation, as observed in tests. It is therefore necessary for this effect to be quantified. In this section, an energy-based material model which can be used to predict the start temperature of the bainite transformation and the fraction of transformation as a function of pre-strain is described, in which the effect of strain is accounted for by introducing an energy factor f_ρ into Eq. 8.7:

$$\Delta G_{\mathrm{V}} = G_{\mathrm{V}}^{\mathrm{A}} - G_{\mathrm{V}}^{\mathrm{B}} = \frac{\Delta T}{T_{\mathrm{eq}}} \cdot Q_{\mathrm{V}} + f_\rho \tag{8.17}$$

Since the promotion of the bainite transformation by pre-deformation is believed to be due to the accumulation of dislocations, the energy factor is thus defined as

$$f_\rho = \alpha_B \cdot \bar{\rho} \quad [8.18]$$

Consequently,

$$\Delta G_V = G_V^A - G_V^B = \frac{\Delta T}{T_{eq}} \cdot Q_V + \alpha_B \cdot \bar{\rho} \quad [8.19]$$

where α_B is a material constant having the same units as f_ρ. The units are J/mol. Therefore,

$$r^* = \frac{2S_{AB}}{\Delta G_V} = \frac{A_1 T_{eq}}{T\sqrt{\Delta T} Q_V + \frac{TT_{eq}}{\sqrt{\Delta T}} \alpha_B \bar{\rho}} \quad [8.20]$$

In this approach, the effect of strain on the bainite transformation is represented by the dislocation density instead of the external force or the strain. For example, during relaxation, although the external force has been removed, the density of dislocations that exist gradually decreases to zero if the time is sufficiently long. Thus the effect of strain on the bainite transformation decreases with decreasing dislocation density according to this model, which accurately describes the real situation. The effect of strain rate on the bainite transformation can also be described by including a term f_ρ in the equation for ΔG_V. A higher strain rate results in a higher dislocation density and, consequently, a higher volume fraction of bainite in the final phase; a lower strain rate allows a longer time for relaxation, which results in a lower dislocation density; therefore, a small rate of deformation results in less pronounced promotion of the bainite transformation.

The equations for $\dot{r}$ and $\dot{X}$ keep the same form as that shown in Eq. 8.16. The new equation set is shown below in Eq. 8.21. The constants need to be determined by fitting to experimental data of the kind shown in Fig. 8.11. A comparison of experimentally measured and predicted CCT diagrams is shown in Fig. 8.18. The solid curves show the predicted CCT diagrams when the steel is pre-deformed at 800°C and 600°C to a strain of 0.1 at 1.0/s, and the dots show the measured results. The thick dashed curve in Fig. 8.18(a) shows the predicted CCT diagram when the steel is pre-deformed at 800°C to a strain of 0.05 at 1.0/s. It can be seen from the figure that the predicted start temperatures of the bainite transformation agree well with the experimentally observed values when pre-deformation is introduced.

$$\begin{cases} r^* = \dfrac{A_1 T_{eq}}{TQ_V\sqrt{T_{eq} - T} + \left(TT_{eq} / \sqrt{T_{eq} - T}\right)\alpha_B \bar{\rho}} \\ \dot{r} = A_2 \exp\left(-\dfrac{|T - T_1|^{n_1}}{B_1}\right) \\ \dot{X} = A_3 \dot{r} \left(\dfrac{r - r^*}{r}\right)^{n_2} \exp\left(-\left(\dfrac{|T - T_2|}{B_2}\right)^{n_3}\right)(1 - X)^{n_4} \end{cases} \quad [8.21]$$

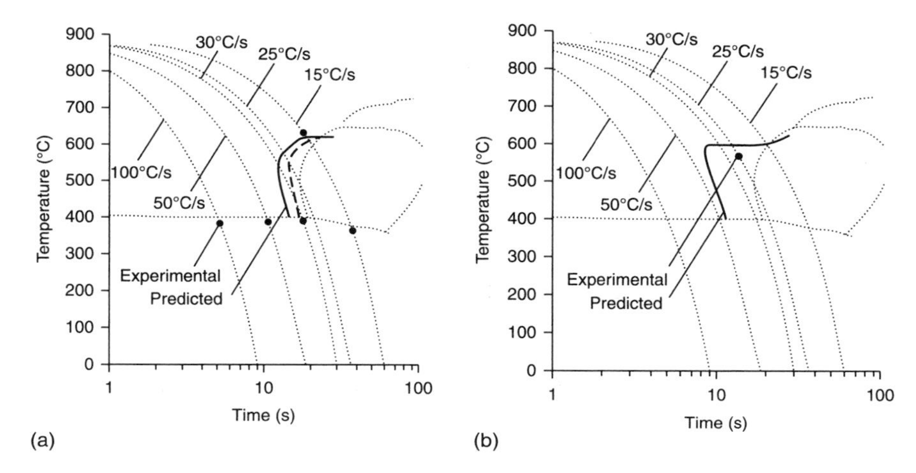

8.18 Comparison of experimentally obtained CCT diagrams for boron steel including the effect of strain (dots), and predicted data (solid curves and thick dashed curve) obtained with the proposed model. (a) Pre-deformation at 800°C, (b) pre-deformation at 600°C.

In order to determine the constants in the equation for $\dot{X}$, the fractions of bainite were measured using SEM images for samples of the steel that had experienced pre-deformation at 800°C and 600°C to a strain of 0.1 at 1.0/s. In both cases, a mixed microstructure of martensite and bainite was observed. The constants were determined by comparing the experimentally-measured bainite fractions with the predicted values. In Fig. 8.19, the solid curves show the predicted evolution of the bainite fraction (B%) and the open circles show the

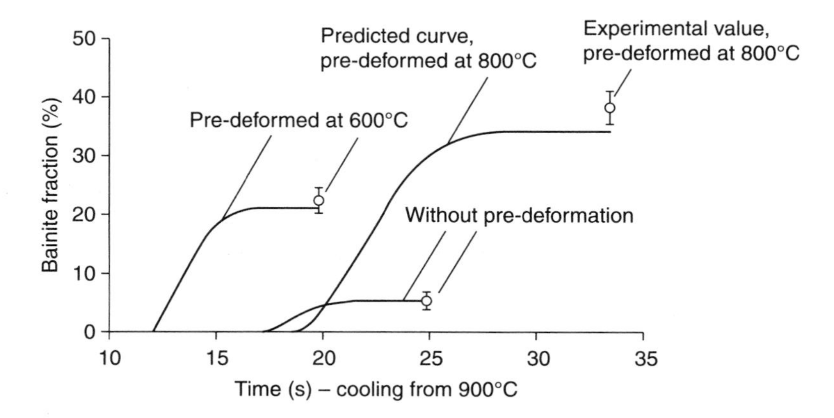

8.19 Comparison of predicted and experimentally observed bainite fractions including the effect of strain. The temperatures of pre-deformation were 600°C and 800°C. The strain rate was 1/s and the strain was 0.1.

experimentally measured values. The pre-deformation temperatures were 600°C and 800°C. Both test pieces were tensioned to a strain of 0.1 at a rate of 1.0/s. Good agreement between the predicted and experimental results was achieved. The proposed bainite transformation model can successfully predict the evolution of the bainite transformation in the boron steel investigated; as a consequence, the fraction of martensite transformation in the hot stamping process can be deduced by use of Eq. 8.15.

8.4 Conclusion and future trends

Thanks to the advantages of the hot stamping and cold die quenching process, the market for hot-stamped structures in the automotive industry has expanded rapidly. In Europe, the estimated total consumption of flat boron steels for hot stamping and cold die quenching was around 60 000 to 80 000 t/year in 2004, and more than doubled, to 300 000 t/year, in 2009.[1, 2, 18] The components targeted cover most of the crash-relevant parts of car structures, for example bumpers, side impact reinforcements, A- and B-pillars, tunnels, and front and rear members.

The constitutive models developed in this chapter for the hot stamping and cold die quenching process show great promise for future work. For example, the modelling of the bainite transformation has been described in this chapter, and the predicted results agree well with the experimental results. This modelling is sufficient to describe the phase transformations in boron steel in a cold die quenching process; however, other phase transformations may occur in slow cooling situations, and these should also be modelled. Also, the effects of strain on other phase transformations, for example the ferrite and pearlite transformations, need to be determined. Furthermore, a knowledge of the effect of strain at large levels (e.g. greater than 0.1) is important for describing phase transformations after large deformations.

To maximize the advantages of the hot stamping and cold die quenching process and enable further weight reduction of formed parts, it is desirable to produce panel parts having different mechanical properties in different regions.[19–21] This could be done by using different controlled cooling rates in different regions of the workpiece. Thus, a tailored microstructure distribution within one component could be obtained. This concept has great potential in automotive applications for producing advanced safety-critical components. A technique that would enable panel parts to be formed with a tailored microstructure distribution, and hence a tailored mechanical-property distribution, would be of great interest to passenger car manufacturers and steel companies. Constitutive models which can predict microstructure distributions, as well as the localized mechanical properties due to these microstructure distributions, would be valuable for improving the conventional hot stamping and cold die quenching process.

Constitutive material models based on physical mechanisms will continue to be an important method of describing complex material behaviour for the foreseeable future. This chapter is intended to enhance the understanding of phase transformations in boron steel in both heating and cooling situations. The methodology and results are transferable to other forming techniques and other materials. The research described here can be extended to other types of boron steels, as well as other types of quenchable steels.

8.5 References

1. Wilsius J, Hein P, Kefferstein R. Status and future trends of hot stamping of USIBOR 1500P®, in *Tagungsband zum 1. Erlanger Workshop Warmblechunformung*, 2006.
2. Hein P, Wilsius J. Status and innovation trends in hot stamping of USIBOR 1500 P. *Steel Research International* 2008;79:85.
3. Brooks CR. *Principles of the Austenitization of Steels*. London: Elsevier Applied Science, 1992.
4. Cai J, Lin J, Wilsius J, Dean TA. Austenitization mechanisms and modelling methods for steels. *World Journal of Engineering* 2008;5:10.
5. Cai J, Lin J, Wilsius J, Dean TA. Theoretical and experimental investigations on austenitization of a boron steel. *Steel Research International* 2008;2:183.
6. Roberts GA, Mehl RF. The mechanism and the rate of formation of austenite from ferrite–cementite aggregates. *Transactions of the ASM* 1943;31:613.
7. Cai J, Lin J, Wilsius J, Dean TA. On austenitizaton of boron steel: modelling investigation. *Steel Research International* 2008;1:155.
8. Aaronson HI, Revnolds WT, Shiflet GJ, Spanos G. Bainite viewed three different ways. *Metallurgical Transactions A* 1990;21:1343.
9. Bhadeshia HKDH. The bainite transformation: unresolved issues. *Materials Science and Engineering A* 1999;273–275:58.
10. Ohmori Y, Maki T. Bainitic transformation in view of displacive mechanism. *Materials Transactions, JIM* 1991;32:631.
11. Muddle BC, Nie JF. Formation of bainite as a diffusional–displacive phase transformation. *Scripta Materialia* 2002;47:187.
12. Muddle BC, Nie JF, Hugo GR. Application of the theory of martensite crystallography to displacive phase transformations in substitutional nonferrous alloys. *Metallurgical and Materials Transactions A* 1994;25:1841.
13. Christian JW. Lattice correspondence, atomic site correspondence and shape change in 'diffusional–displacive' phase transformations. *Progress in Materials Science* 1997;42:101.
14. Garrett RP, Xu S, Lin J, Dean TA. A model for prediction austenite to bainite phase transformation in producing dual phase steels. *International Journal of Machine Tools and Manufacture* 2004;44:831.
15. Chang LC, Bhadeshia H. Stress-affected transformation to lower bainite. *Journal of Materials Science* 1996;31:2145.
16. Shipway PH, Bhadeshia H. The effect of small stresses on the kinetics of the bainite transformation. *Materials Science and Engineering A* 1995;201:143.
17. Shipway P, Bhadeshia HKDH. Mechanical stabilisation of bainite. *Materials Science and Technology* 1995;11:1116.
18. Hein P. A global approach of the finite element simulation of hot stamping. *Advanced Materials Research* 2005;6:763.

19. Cumming DM, Gabbianelli G, Mellis JJ, Uchiyama H. *Multiple Material Bumper Beam*. Royal Oak, MI: Magna International Inc., 2005.
20. Thomas D, Detwiler DT. *Microstructural Optimization of Automotive Structures*. Willoughby, OH: Honda Motor Co., Ltd, 2009.
21. Bodin H. *Method of Hot Stamping and Hardening a Metal Sheet*. Luleå, Sweden: Gestamp Hardtech AB, 2010.

9
Modelling microstructure evolution and work hardening in conventional and ultrafine-grained microalloyed steels

J. MAJTA and K. MUSZKA, AGH – University of Science and Technology, Poland

Abstract: Aspects of the mechanical behaviour of microalloyed steels characterised by grain refinement due to large plastic deformation are discussed. The studies presented in this chapter are intended to promote understanding of the mechanical behaviour of multiphase and ultrafine-grained materials related to the refinement of the microstructure, particularly taking into account the effects of microalloying. Several levels of grain refinement are identified with respect to changes in the strengthening mechanisms. Experimental results have been used to calibrate constitutive equations describing plastic flow mechanics. These equations have been modified and systematically ordered for ultrafine-grained microalloyed steels. Experimental results supported by multiscale numerical analyses have allowed the development of a rheological model for ultrafine-grained microalloyed steels.

Key words: microalloyed steel, ultrafine-grained materials, mechanical properties, computer modelling, work hardening.

9.1 Introduction

Microalloyed (MA) steels have been available for structural applications for many years. However, there is a growing market for these attractive structural materials at the present time, and the factors that are encouraging their expanded usage are numerous and complex. For example, as recently as 2009, over 10% of the world's production of structural plate and long products contained microalloying additions of Nb (Jansto, 2010). Undoubtedly, the evolutionary improvement of MA steels will be continued well into the future (Gladman, 1997; DeArdo, 1998; Matlock *et al.*, 2005; Jansto, 2010). This progress is accompanied by the necessity to use modelling processes, to understand both microstructure development and the mechanisms responsible for the final mechanical properties. At present, several different branches of industry are interested in ultrafine or submicron-grained steels (less than one micrometre). For example, the automotive industry requires a suitable combination of material ductility and strength, as well as good high-strain-rate performance with a sufficient level of crash energy absorption.

Refining of the grain size into the submicron region in MA steels shows promise for the development of new steel types with both ultrahigh strength and superductility. Grain size control is one of the most attractive methods for the improvement of mechanical properties. The strength, hardness, toughness and ductility can be significantly enhanced by use of this strengthening mechanism. Moreover, the method is especially attractive because it does not require expensive alloying additions and it allows the use of an existing production technology.

The combined effect of the complex chemical composition and grain refinement determines the formation of the microstructure, which controls the mechanical properties in a new way. Cross-application of similar MA steel grade systems is being specified for different end user requirements. The important feature of MA steels is that a wide range of microstructures can be obtained by changing the steel composition and varying the thermomechanical treatment. In Nb-containing MA steels, a very fine ferrite grain size is produced by nucleation in hardened, unrecrystallised austenite.

The evolution of the microstructure is connected with physical processes taking place in the deformed steel such as recrystallisation, precipitation and phase transformation, depending on the particular parameters of the deformation process, i.e. the strain, strain rate and temperature. A large volume fraction and a fine dispersion of a second phase effectively increases the work-hardening rate by promoting the accumulation of dislocations around interphase boundaries, which results finally in better ductility. The application of thermomechanical treatment leads to a grain size of a few micrometres.

Conventional MA steels typically have a grain size of about 8–12 μm. Nowadays, ultrafine-grained steels can be obtained by using various fabrication routes, both in the laboratory and on the industrial scale. The routes can be divided into two major groups, namely:

- advanced thermomechanical processing (ATP);
- severe plastic deformation (SPD) methods.

Various kinds of nanostructures can be obtained by using SPD processing. In these processes, metallic materials are subjected to significant plastic deformation, up to an equivalent strain exceeding 4 or 5 (Altan *et al.*, 2006; Zehtbauer and Zhu, 2009). The ultrafine-grained (UFG) materials produced in this way are well suited for many modern applications owing to their outstanding properties, i.e. high strength, toughness and work hardening. The ATP route has been successfully applied to various steel grades (C–Mn, interstitial-free and microalloyed), resulting in grain refinement in the range of 1 to 3 μm and improved mechanical properties. The ultrafine-grained ferrite–pearlite microstructure obtained provides benefits from both grain refinement and dual-phase strengthening. It results in an attractive strength/ductility ratio owing to the presence of a hard second phase similar to that observed in dual-phase steels.

This chapter contains a brief overview of the techniques that lead to ultrafine-grained MA steels. Furthermore, we describe the determination of the impact of the processing conditions on grain refinement and on the mechanical properties, and employ these phenomena in a rheological model of UFG MA steels. The support of the computational finite element method (FEM) allowed us to simulate the different stress–strain relations in different locations in the deformed material. The levels of grain refinement where changes in the deformation mechanisms are observed, are identified and related to the resulting changes in the strengthening mechanisms in the materials discussed. Differences in the description of the deformation and strengthening mechanisms operating in UFG and conventional MA steels are also mentioned.

9.2 Thermomechanical and severe plastic deformation processing of ultrafine-grained microalloyed (MA) steels

During the hot working of plain carbon steels, the development of the microstructure is not as pronounced as it is in MA steels. The addition of microalloying elements to low carbon steels is used to facilitate the austenite conditioning process, and to influence the phase transformation kinetics as well as provide further precipitation hardening of the ferrite. The microalloying elements of primary interest are Nb, Ti, V, Al and in some case B and Mo, singly or in combination. First of all, the metallurgy of MA steels should be considered in terms of quantitative relationships between the microstructural parameters and the mechanical properties. Effective control of the mechanical properties can be obtained and the quality of products can be improved by understanding the effects of processing conditions and the microstructural changes that occur during processing. A proper understanding of the specific mechanisms by which plastic deformation leads to a refined grain size is of paramount importance. However, in the case of MA steels, the synergetic effects of microstructure evolution, precipitation processes and the influence of microalloying elements in solution makes the analysis of this problem especially complex.

Qualitative and, especially, quantitative evaluation of the microstructure of these steels is extremely difficult. The mechanical properties of ultrafine-grained MA steels are affected by a combination of strengthening due to grain boundaries, dislocation structures (the dislocation arrangement and cell structure), precipitation and solid solution hardening. The manufacture of MA steels must carefully integrate not only strength but also toughness and ductility. This necessity starts to be the main problem in the case of UFG steels, which require a grain size in the range of 0.5–2 μm. Microalloying elements can interact with the microstructure evolution in different ways depending on the reheating conditions, the deformation history and parameters, and the cooling conditions. As a result of these interactions, evolution of the dislocation structure occurs and, finally, various levels of refinement of the microstructure are achieved. The 'condition' of the grain boundaries and dislocation structures directly determines the mechanical properties of the final product.

In order to produce UFG steels, advanced thermomechanical processing with a dynamic austenite-to-ferrite transformation (deformation-induced ferrite transformation (DIFT) (Yang and Wang, 2003)) has been extensively investigated, at least during the last thirty years (Essadiqi and Jonas, 1988; Beynon *et al.*, 1992; Mintz and Jonas, 1994; Yada *et al.*, 2000; Hickson *et al.*, 2002). DIFT is a kind of solid state transformation induced by deformation, which can be applied effectively to produce fine or ultrafine ferrite grains. Another example of the ATP route is based on strain-induced dynamic transformation (SIDT) (Li *et al.*, 1998; Choi *et al.*, 2003), also called dynamic strain-induced ferrite transformation (DSIFT) (Beladi *et al.*, 2004). In this case the deformation is mostly applied not before, as in DIFT-controlled process, but during the austenite-to-ferrite transformation. SIDT can be used very effectively for grain refinement in low carbon steels, with approximately 2 μm ferrite grains obtained by dynamic transformation in the austenite–ferrite two-phase region. The potential for grain refinement of MA steels during the annealing of cold-rolled strip was also investigated by Lesch *et al.* (2007). This work addressed how fine-grained structures can be achieved during rapid transformation annealing (RTA) using the effect of transformation-induced grain refinement during both heating and cooling. It was found that an increased amount of microalloying elements in a cold-deformed MA steel resulted in a more intensive interaction of recrystallisation and transformation during annealing in the intercritical (austenite plus ferrite) region and led to a more refined microstructure.

Thermomechanical processing cannot refine grains to the submicrometre or nanometre range. SPD techniques have made it possible to produce submicron structures and nanostructures in ultrafine-grained steels. There have been several recent developments in improving processing technologies that use SPD techniques (Valiev *et al.*, 2006, 2000; Valiev and Langdon, 2010). Materials produced by SPD usually have a high initial dislocation density, which is introduced during the deformation process. Therefore, saturation of the dislocation density results in a very low work-hardening rate during subsequent deformation, and hence leads to poor ductility (Wang and Ma, 2004). Microstructures that contain microalloying elements in solution as well as in hard dispersed particles within a ductile matrix exhibit a higher dislocation density at a given strain. These obstacles serve as a source for geometrically necessary dislocations, which, in turn, increase the work-hardening rate. However, it is also observed that ordered dislocation structures, represented by cell boundaries, are more difficult to achieve in MA than in plain carbon steels because of the much lower mobility of the dislocations and grain boundaries (Majta *et al.*, 2006). Recently published results support earlier observations that cells formed in interstitial-free (IF) steels are more sharply defined but larger than cells formed in MA steels (Stefańska-Kądziela *et al.*, 2007; Muszka and Majta, 2010).

9.3 The principles of deformation-induced grain refinement

It is difficult to produce ultrafine-grained ferrite–pearlite steels with a ferrite grain size below 2 μm, since small-sized austenite does not transform to pearlite but decomposes into ferrite and dispersed cementite particles (Tsuchida *et al.*, 2008). Therefore, cementite has been also widely used as a second-phase component in UFG steels (Song *et al.*, 2006). In the case of advanced thermomechanical processing of MA steels, the most effective method used to refine the ferrite microstructure involves the previously mentioned DIFT (or SIDT) process, which is, in turn, promoted by the effect of niobium precipitates, provided that recrystallisation of austenite does not occur (Majta *et al.*, 1999; Majta, 2000; Dong and Sun, 2005; Muszka *et al.*, 2009).

Further grain refinement (below 1 μm) can be obtained when SPD routes are applied. From the microstructural point of view, the large deformation applied in SPD processes introduces a high volume fraction of low-angle grain boundaries (LABs), which can subsequently be transformed into high-angle grain boundaries (HABs) with the aid of annealing or dynamic recovery. This leads to a much higher refinement level compared with ATP routes. Hence, the main factor in the process of grain refinement is an evolution of the dislocation substructure, consisting of the generation of dislocations, their self-organisation into structures inside the grains, evolution of the structure into LABs, characterised by subgrains, and transformation of LABs into HABs. During the evolution of the dislocation structure, concurrent effects of its various elements are observed, namely interaction of dislocations with LABs and HABs, the effects of boundaries, dispersed particles and solute atoms on hardening, and so on. The warm and cold deformation applied in the ATP approach also causes high refinement of the microstructure and leads to subdivision of grains (into dislocation cells and subgrains), whereas hot working develops larger but more homogeneous and polygonised grains or subgrains with strain-induced precipitation of Nb(C,N). SPD processes facilitate a transformation of traditional plate-like pearlite into spheroidised cementite particles. During plastic deformation, pearlitic cementite lamellae disintegrate into short fragments. During large-strain deformation and annealing, these fragments spheroidise into discrete cementite particles. All the aforementioned factors and, first and foremost, decreasing grain size, influence the physical phenomena and involve different deformation and strengthening mechanisms compared with coarse-grained microstructures. These, in turn, affect the strength, ductility and toughness of the final products.

As previously mentioned, the application of severe plastic deformation results in strong subdivision of grains into subgrains and cells, which subsequently transform into a stable microstructure with high-angle grain boundaries. This process is called continuous recrystallisation (Oscarsson *et al.*, 1994; Belyakov *et al.*, 2000) or recrystallisation *in situ* (Tsuji *et al.*, 2000); and was reported by Wang *et al.* (1993) as '*in situ* grain boundary evolution from dislocation tangles into high angle near-equilibrium state'. It can be observed that the principal effect of annealing on the dislocation structure is the development of a more pronounced delineation of the cell walls. Additionally, as was stated by Keh and Weissman (1963), during the early stages of annealing the free dislocations inside the cells are attracted by dislocations located in the tangled cell walls. In the meantime, the dislocations in the cell walls probably undergo some slight rearrangement. It is believed that these effects involve a relaxation of the long-range stress field and are ultimately responsible for the disturbances in the flow stress which have been observed. The dislocation structure is also dependent on the strain rate (Majta *et al.*, 2005; Mukai *et al.*, 1995). The strain rate affects both the short-range and the long-range stress (Uenishi and Teodosiu, 2004). However, it has been found (Wang and Ma, 2004) that additional deformation applied after annealing significantly decreases these effects and, additionally, an ultrafine inhomogeneous deformation microstructure is observed in this case.

Many published observations have shown that the number of HABs formed across a grain depends on the crystal orientation. When the effective strain increases, the misorientation angle across the deformation-induced dislocation boundaries increases and the typical cell block structures start to reorient into lamellar structures. However, after annealing, these dislocation structures undergo some rearrangement, and a more stable deformation microstructure is observed. The presence of second-phase particles can greatly accelerate the grain-refining process by promoting heterogeneous deformation. The inhomogeneity of the strain distribution disrupts the planarity of lamellae and ribbon grain structures and encourages them to break up at lower strains. As a result, a higher fraction of HABs are formed at the same strain in an alloy containing a significant volume fraction of second-phase particles compared with a single-phase alloy.

One common feature of nanocrystalline and UFG materials is the high driving force for grain growth. In the case of MA steels, as presented by Chen *et al.* (2009) using a model which incorporates grain boundary thermodynamics into a kinetic model for grain growth, solute drag can retard grain growth, and stabilisation of the UFG microstructure is achieved through a reduction in grain boundary energy. As presented by Michels *et al.* (1999), equilibrium segregation to grain boundaries can be attained at lower temperatures not by diffusion of solute atoms to the boundaries but rather by the entrapment of solute at the boundaries as they move. In general, it can be expected that the inhibition of grain growth in UFG MA steels at lower annealing temperatures is due to solute drag on the grain boundaries, but at higher temperatures it is due to pinning of the boundaries by nanoscale intermetallic precipitates (Perez *et al.*, 1998; Koch *et al.*, 2008).

While plastic deformation of b.c.c. metals occurs during the early stages by dislocation glide on the most stressed $\{110\}\langle 111\rangle$ slip system, the dislocation density on secondary slip systems increases continuously during deformation, leading to accelerated processes of substructure formation. In contrast to f.c.c. metals, dislocation movement in b.c.c. metals can be easily retarded by existing obstacles, which results directly in an increased dislocation density and supports the possibility of substructure rearrangement. Also, the accumulated strain leads to the activation of slip on systems other than $\{110\}$. As a result, the dislocation substructure can be developed more, and an increase in the work-hardening rate

($d\sigma/d\varepsilon$, where σ is the flow stress and ε is the true strain) is observed. Changes in the strain path also play a significant role. Unfortunately, the evolution of deformation structures to very high plastic strains with cyclic changes in the strain path, has still not been clearly described. It can be said in summary that in the case of MA steels, the formation of stabilised UFG structures needs a higher accumulated energy, first of all because of alloying elements that have a strong effect of inhibiting recovery, but the effect of microstructural refinement is much more pronounced than in low carbon steels.

9.4 Effects of microstructure evolution on mechanical properties of ultrafine-grained microalloyed steel

It is already well known that UFG materials have unique and very attractive mechanical properties compared with materials with coarse-grained microstructures. In both the structural and the functional sense, ultrafine-scale structures represent a frontier area for the exploration of new structure–property regimes in materials science (Embury and Sinclair, 2001). A proper description of the constitutive laws governing the mechanical behaviour of UFG materials that is based on the actual physical phenomena occurring in these microstructures, can significantly improve the accuracy of the computer modelling process.

There are a number of well-known constitutive laws that describe the mechanical response of conventional structures (with a mean grain size down to 1 μm). The situation becomes much more complicated in the case of ultrafine-grained and nanostructured materials. As was presented earlier, such structures are obtained mainly by a large amount of cold deformation followed by an annealing process. This processing route produces dislocation structures with a significant fraction of LABs. These materials represent an excellent combination of high yield strength and low brittle–ductile transition (BDT) temperature. It was deduced (Tanaka *et al.*, 2010) that the decrease in the BDT temperature produced by grain refining was not due to an increase in the dislocation mobility controlled by short-range barriers, as in a dislocation pile-up model, but rather by a decrease in the spacing of dislocation sources. Molecular dynamics simulations revealed that moving dislocations impinged against grain boundaries and were then re-emitted with increasing strain. This indicates that grain boundaries can be new sources in ultrafine-grained materials, which increases toughness at low temperatures.

The impact properties of ultrafine-grained MA and low/medium carbon steels have been investigated by Tsuji *et al.* (2004), Hanamura *et al.* (2004), Song *et al.* (2005a,b) and Murty and Torizuka (2010). It was confirmed that decreasing the grain size limits the propagation of initiated cleavage cracks and raises the fracture toughness in the transition region. HABs are also more efficient in improving the toughness of steels because when an intragranular cleavage crack moves across a HAB, the crack front usually branches according to the change of preferred fracture plane (Furuya *et al.*, 2007). Such branching results in an additional work of fracture. In contrast, when a crack meets a LAB, it can typically penetrate such an interface without a substantial change in the propagation direction and without branching. That is why it is necessary to clearly identify and quantitatively characterise the character of grain boundaries together with an analysis of the grain size.

Unfortunately, grain refinement significantly decreases the ductility, represented most often by the uniform elongation in tensile tests. In the case of MA and low carbon steels, the b.c.c. structure is expected to lead to poor results compared with f.c.c. metals. In the nanocrystalline range, iron samples show a lack of work hardening and even some softening, which is related to unstable deformation and leads to decreased ductility (Wei *et al.*, 2004; Ma, 2005; Benito *et al.*, 2010). The reduction in area in tensile tests is also an important measure of the ductility of UFG materials, especially in the case of MA steels, because it is significantly affected by second phases and inclusions. However, measurement of the reduction in area requires a standard tensile specimen, which, in turn, requires bulk material for the fabrication of the specimen. The balance between the reduction in area and the tensile strength for UFG steels is far better than for conventional ferrite–pearlite steels and bainitic steels, and is even superior to that for tempered martensitic steels (Murty and Torizuka, 2010). Recently published work has suggested several ideas to improve the uniform elongation. These include the use of dispersed oxide, cementite or martensite (Sakai *et al.*, 2001; Tsuji *et al.*, 2002; Hayashi and Nagai, 2002); the use of a bimodal microstructure or distribution of grain size (Benito *et al.*, 2010); and deforming the UFG material at a very low temperature and/or high strain rate, thereby increasing the high-strain-rate sensitivity or introducing heavy twinning into the material during processing (Murty and Torizuka, 2010).

It has been already proven that the physical phenomena governing the deformation and strengthening mechanisms of UFG and nanostructured materials are different from those of their conventional counterparts; these phenomena are still poorly understood (Song *et al.*, 2006; Ponge *et al.*, 2007; Muszka *et al.*, 2007a,b). For example, the decrease in the true stress–strain slope from a positive to a zero value, and at later stages even a negative value, may not in fact be related to conventional work hardening, but should rather be seen as a result of an evolving grain boundary network, accommodating the applied stress by both intergrain and intragrain deformation mechanisms (Van Swygenhoven *et al.*, 2003). Recently, interesting results were obtained in a nanostructured IF steel produced by accumulative roll bonding (Huang *et al.*, 2010). By applying cold rolling to annealed samples of a nanostructured IF steel, both the flow stress and the elongation can be significantly changed and an optimal combination of mechanical properties can be obtained.

Also, there is still clearly a lack of systematic investigations concerning the effect of grain refinement on the mechanical properties of such materials measured at high strain rates. It is observed that in the case of coarse-grained microstructures, the uniform and total elongations increase with increasing strain rate, while in the case of UFG materials, the uniform elongation does not change. This indicates that for conventional microstructures, the increase in the total elongation at high strain rates is mainly due to the increase in the uniform elongation, while for UFG materials it can be attributed to an increase in the post-uniform elongation (Tanaka *et al.*, 2008; Tsuchida *et al.*, 2002).

The dynamic behaviour of UFG IF and MA steels seems to be especially important for many reasons, for instance because of their use in the automotive industry as materials that are characterised by improved crashworthiness (high energy absorption capability). As shown by Hodgson and Beladi (2004), smaller-grain-size b.c.c. metals are much stronger at low strain rates, but are characterised by less relative strengthening at high strain rates. The total elongation decreases with decreasing grain size and, in contrast, has a tendency to increase with increasing strain rate (da Silva and Ramesh, 1997). In the case of conventional b.c.c. metals, the work-hardening rate increases with decreasing temperature and is almost independent of the ferrite grain size, while in the case of UFG microstructures the effective work hardening decreases, becoming negligible for grain sizes less than 300 nm (for both low and high strain rates). Also, the strain rate sensitivity decreases with decreasing grain size, in contrast to f.c.c. nanocrystalline metals (Jia *et al.*, 2003; Wei *et al.*, 2004). The effect of grain size on mechanical behaviour at various strain rates has been studied (e.g. Zurek *et al.*, 2009).

9.5 Application, results and discussion

This section presents some aspects of the deformation-induced microstructure evolution of ultrafine-grained MA steels. An investigation was performed on two commonly applied steels: a high-strength Nb microalloyed steel (Y-MA, X65) and an IF steel. The chemical compositions of these steels are listed in Table 9.1. Nb is a key microalloying element in MA steels, and Ti is an important component that may affect properties of all of the materials investigated here. The data needed for a successful and complete interpretation of all of the mechanical tests that were performed to define the correlation between microstructure and mechanical properties were obtained from the results of transmission electron microscopy (TEM) and scanning electron microscopy (SEM) analyses. A range of different

Table 9.1 Chemical composition of the steels investigated

Steel	C	Mn	Al	Si	Ti	Nb
Y-MA	0.07	1.36	0.02	0.27	0.031	0.067
X65	0.07	1.10	0.041	0.25	0.02	0.04
IF	0.0022	0.112	0.037	0.009	0.073	–

grain sizes was developed using multiaxial compression tests (performed with a MaxStrain system) (Majta *et al.*, 2006; Majta and Muszka, 2007; Muszka *et al.*, 2010) and single-pass hot rolling experiments (Muszka *et al.*, 2007a; Muszka, 2008). The deformation schedules were designed to involve various phenomena (continuous recrystallisation *in situ*, dynamic recrystallisation and dynamic strain-induced transformation), using thermomechanical treatment to finally obtain a range of grain sizes from approximately 12 to 0.6 μm. The principles behind the deformation schedules for the single-pass hot rolling experiments (using the X65 steel) with various thermomechanical treatments were presented by Muszka *et al.* (2007a). In the case of the multiaxial compression tests performed on the IF and Y-Ma steels, the schedules are shown in Table 9.2, and the results of these experiments will be discussed in detail below.

To understand the effect of the ferrite refinement produced by the rolling tests on the mechanical behaviour of the microalloyed steels, the flow stress curves for the X65 steel obtained under quasi-static and dynamic loading conditions are summarised in Fig. 9.1(a). The mechanical properties of particular specimens suggest that besides the different levels of grain size strengthening, the various volumes of precipitates and the microalloying elements in solution control the strain rate sensitivity. It is also evident that there is a lack of a grain size effect on the flow stress when dynamic loading conditions are applied. The mechanical properties of the specimens after multiaxial compression were also characterised by tensile testing. The stress–strain curves for the Y-MA and IF steels are shown in Fig. 9.1(b). Both of these steels show the typical behaviour of nanostructured metals, i.e. a high yield strength and limited uniform elongation followed by a relatively large post-necking elongation.

TEM microstructures of the ultrafine-grained IF steels produced by the multiaxial compression tests are presented in Fig. 9.2. It can be observed that the application of severe plastic deformation (total strain = 20, schedule 3 in Table 9.2) enabled strong subdivision of the grains into cells and subgrains that were subsequently transformed into a stable microstructure (with the aid of annealing

Table 9.2 Schedules of the multiaxial compression tests

Schedule				First cycle			Second cycle		
	T_{D1} (°C)	t_R (s)	ε_1	T_A (°C)	t_A (s)	CR_1 (°C/s)	T_{D2} (°C)	ε_2	CR_2 (°C/s)
1	20	–	5	–	–	–	–	–	–
2	20	–	5	500	1200	4	–	–	–
3	20	–	20	500	1200	4	–	–	–
4	500	300	4	–	–	4	–	–	–
5	20	–	5	500	1200	–	500	5	4

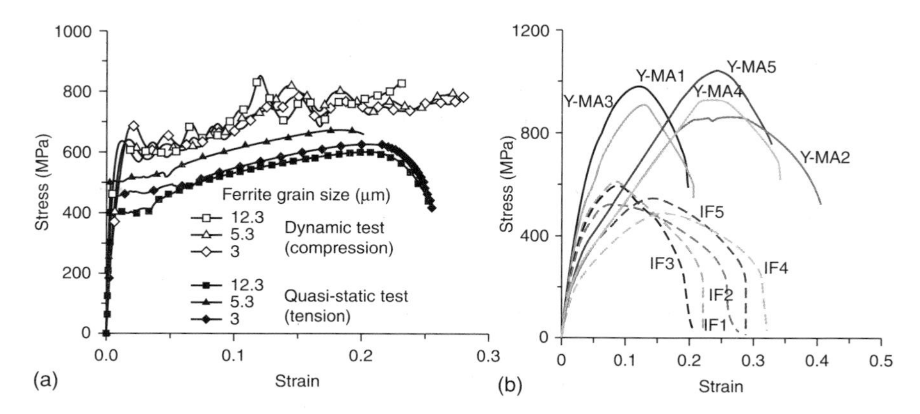

9.1 (a) Flow stress curves (for X65 steel) obtained under quasi-static and dynamic loading conditions. (b) Stress–strain curves for the Y-MA and IF steels.

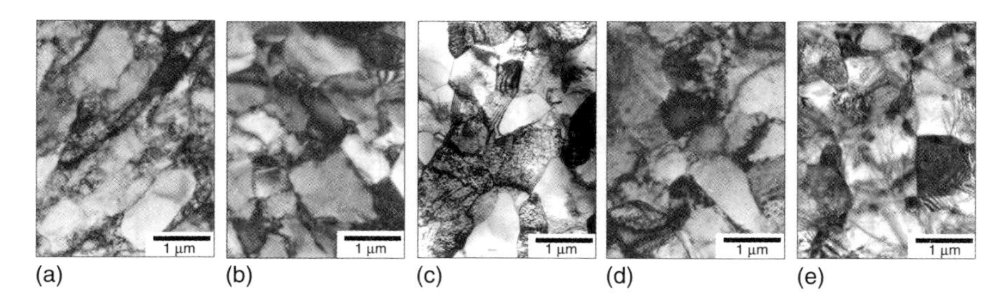

9.2 TEM micrographs of the ultrafine-grained IF steels produced by the multiaxial compression tests. (a)–(e) Microstructures after the deformation schedules 1–5, respectively, in Table 9.2.

applied at 500°C). The volume fraction of grains with high-angle boundaries, measured using electron backscattered diffraction (EBSD), was about 50% in the case of the IF steel, and 70% in the case of the Y-MA steel. It can be inferred that the greater number of stable HABs, the higher level of precipitation and the presence of solid solution strengthening led to the higher strength observed in this steel. However, we can also expect that the presence of precipitates and the higher volume fraction of grains with HABs in the case of the Y-MA steel will be the main sources of increased ductility, i.e. its ability to continue to show an increase in work hardening during tension (Hayashi and Nagai, 2002; Majta *et al.*, 2006). The increase in the ductility of the severely deformed steel can be also attributed to the finely dispersed cementite particles, which effectively accelerate the work-hardening rate. The condition for plastic instability, described by the Considère criterion ($\sigma > d\sigma/d\varepsilon$) (Considère, 1886), shows that since the flow stress increases with grain refinement according to the Hall–Petch relation, a higher work-hardening rate is required to avoid plastic instability and improve the ductility. Also, it has been demonstrated that precipitation strengthening due to cementite particles improved the fatigue strength of ultrafine-grained steels without lowering the fatigue limit ratios (Furuya *et al.*, 2007).

As presented in Figs 9.2 and 9.3, our TEM examinations clearly revealed a decrease in the dislocation density inside grains as a result of annealing. However, single dislocations were still present in the volumes between the boundaries. This observation leads to the suggestion that the observed hardening may be caused by enhanced dislocation recovery due to interaction between dislocations. The HABs, which are more important for strengthening than LABs, were developed especially effectively during deformation to a strain of 20 with subsequent annealing.

The nanostructured metals and alloys produced by SPD are in an extremely non-equilibrium state. The dispersed precipitates of (Nb,Ti)(C,N) effectively pin dislocations so that the effects of deformation in severe multiaxial compression accumulate. Comparing Figs 9.2 and 9.3, it can be seen that when the total strain increases, a higher volume fraction of LABs is generated in the Y-MA steel. This suggests a much stronger tendency for this steel to create dislocation structures in the form of cells and subgrains. Such a substructure can be transformed into HABs under certain favourable conditions (e.g. after applying subsequent annealing). Besides the volume fraction of the precipitates, their size is also important in terms of their interactions with dislocations. Only fine precipitates are active in the trapping of dislocations. The larger precipitates pin only a limited number of dislocation walls. In UFG materials, hard precipitates also initiate, drag and pin dislocations, and hence dynamic recovery is reduced. As a result, significant dislocation storage is observed, which in turn is required for compatible plastic strains, allowing a high work-hardening rate that leads to larger uniform strains and at the same time increases the tensile strength (Copreaux *et al.*, 1993).

As mentioned earlier, the subdivision of grains into cell boundaries is more difficult in MA steels than in, for example, IF steels, owing to the much lower

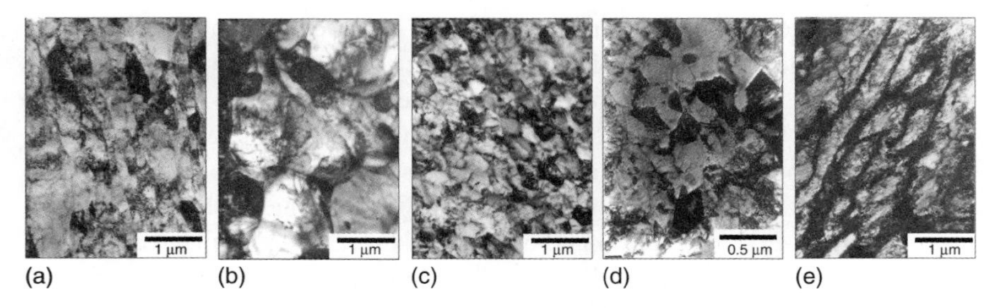

9.3 TEM micrographs of Y-MA steel. (a)–(e) Microstructures after the deformation schedules 1–5, respectively, in Table 2.9.

mobility of the dislocations and grain boundaries (Muszka *et al.*, 2007b; Majta and Muszka, 2007). The significantly different microstructures presented in Figs 9.2 and 9.3 that were obtained in the tests support this thesis. However, the much stronger grain refinement, together with precipitation and solid solution strengthening, make UFG MA steels more attractive from the point of view of optimisation of mechanical properties.

In the case of f.c.c. metals, it is well known that the cell size evolves as a function of the precipitate volume fraction. The mechanism governing this behaviour can be described by an Orowan-type formalism. A dislocation may avoid particles or obstacles by leaving the slip plane in the vicinity of each particle, or it may avoid the particles by the Orowan mechanism. In the case of b.c.c. metals, only the secondary wall spacing seems to depend on the precipitates. It has been suggested that the formation of cells is controlled mainly by the movements of primary screw dislocations, and this is mostly related to the lattice friction rather than to dislocation–dislocation or dislocation–particle interactions. The lattice friction is still important and seems to control the development of the planar dislocation substructure, which is thus independent of the volume fraction of precipitates; the closed-cell structure results either from multiple slip activity or from interaction between dislocations and particles.

Also, from a recent study by the authors (Muszka *et al.*, 2007b), it is believed that the refined dislocation structure of MA steels results from interaction with fine precipitates, which forces dislocations to form fine local cell structures, in contrast to the situation where uniform dislocation lines propagate owing to a lower number of precipitates. As the volume fraction of precipitates increases, the average size of the closed cells decreases (Muszka *et al.*, 2007b; Stefańska-Kądziela *et al.*, 2007). It has also been found that the volume fraction of precipitated carbides has an effect on the proportions of the different types of cells. The higher the volume fraction, the lower the ratio of closed cells to open cells; the dislocation substructure is similar to those obtained after low-temperature deformation (Kocks and Canova, 1981; Copreaux *et al.*, 1993).

9.6 Multiscale modelling of the flow stress of conventional and ultrafine-grained microalloyed steels

The strain–stress data obtained in the experimental work described above and the TEM and EBSD observations in that work were used to calibrate a model of the flow stress, which will be introduced in this section.

From a microscopic viewpoint, the plastic deformation and work hardening of a crystal are caused by dislocation motion and accumulation. In order to model the mechanical behaviour of UFG materials, there is a need, first of all, to understand the deformation and strengthening mechanisms governing the plastic deformation of such materials. To capture these phenomena in a sufficient way, a multiscale modelling approach needs to be utilised, since conventional material models are not able to bridge the gap arising from the different scales on which those phenomena take place. Thus, hierarchical material models involving different scales are required in order to predict some of the interactions between microscopic dislocation fields and macroscopic deformation fields computationally. Also, recent progress in material characterisation methods, such as EBSD, focused ion beam (FIB) methods, and the most recent synchrotron radiation X-ray imaging and diffraction (3-DXRD) techniques, offers new possibilities for *in situ* observation of materials on different scales, and provides new information about the deformation and damage mechanisms operating in polycrystalline materials, as well as about their anisotropy.

In order to create robust numerical models that take these data into consideration, a combination of several different discrete and continuous modelling techniques (the FEM, cellular automata (CA), molecular dynamics (MD) and Monte Carlo (MC) methods) on various scales is needed. In general, multiscale modelling methodologies can be classified into two groups (De Borst 2008; Madej, 2010): upscaling methods and concurrent multiscale computing. In the upscaling class of methods, constitutive models on higher scales are constructed from observations and models on lower, more elementary scales. In concurrent multiscale computing, one strives to solve the problem simultaneously on several scales by an a priori decomposition of the domain.

Another example of a multiscale modelling approach is a combination of crystal plasticity and the FEM (CPFEM) (Rothers *et al.*, 2010; Svyetlichnyy *et al.*, 2010). This makes use of the basic principles of mechanics to achieve a transition from the macroscale to the microscale and is more often used in the recently developed 'digital material representation' (DMR) approach (Madej, 2010), which can represent material microstructure in an explicit manner. Its main conceptual advantage is that it can combine a variety of mechanical effects, which are direction-dependent owing to the underlying crystalline structure, i.e. slip planes, crystallographic orientation and anisotropy.

In order to use the modelling techniques presented above to simulate the plastic deformation of UFG materials, there is still a need to properly understand the mechanisms governing the plastic deformation and strength of these materials and to formulate a constitutive description. For polycrystalline materials deformed at low temperatures, the flow stress can be considered as the sum of three main components: (i) the stress required to activate the mobile dislocations, (ii) the stress resistance due to the creation of new dislocations and the interactions between them, and (iii) the stress resistance due to the grain boundaries. Difficulties in the measurement of dislocation densities and the lack of clarity in the assessment of such measurements have led to solutions where the dislocation strengthening effect is divided into two components: (i) dislocation strengthening due to free, statistically stored dislocations, and (ii) substructure strengthening due to dislocation cells. There are a number of theories that describe the correlation between the flow stress and the dislocation cell size, mostly taking into consideration the distance between dislocation cell walls or, equivalently, the dislocation cell diameter D_B. Generally, these models can be considered as one of two types: static, where the relation between the cell size and flow stress is derived based on an energy minimisation condition (i.e. Holt's model) (Holt, 1970), and dynamic, where the evolution of the dislocation structure during deformation is assumed to be dynamic (i.e. Edward's model) (Edward *et al.*, 1982; Embury and Sinclair, 2001). Hence, it can be stated that the sizes of the cells and subgrains and the evolution of the misorientation angles of LABs in MA steels are directly correlated with the chemical composition and the accumulated plastic strain.

The main difference between the plastic deformation mechanisms in conventional coarse-grained and nanostructured materials arises from the number of dislocations, which is much smaller within the nanostructured grains than in coarse-grained materials. Dislocations in UFG materials are usually generated from grain boundaries, in coarse-grained materials, where dislocations appear as a result of intragranular sources such as Frank–Read sources (Hirth and Lothe, 1968). This leads to different distributions of 'dislocation charges' in UFG/nanostructured materials compared with typical materials and causes certain peculiarities in the strain-hardening behaviour. The significance of grain boundaries increased when it was recognised that they can also serve as sources of dislocations (Ashby, 1970; Li and Liu, 1963). Ashby proposed statistically stored and geometrically necessary dislocations, the latter being required to accommodate the incompatibility stresses at the grain boundaries due to anisotropy. Other researchers (Gifkins, 1976; Meyers *et al.*, 2006) also postulated that when the volume fraction of grain boundaries becomes significant, each grain can be represented schematically by two regions: a central 'core' and a grain boundary region, called the 'mantle'. The mechanical responses of these two regions are different, with work hardening being more pronounced in the mantle owing to several factors being present in this region that contribute to the increased hardening: grain boundary dislocation sources, a change in the orientation of the plane of maximum shear, and elastic and plastic incompatibility. The ratio between the volumes of these two regions is dependent on the grain size.

In order to propose new (or modify existing) physically based models relating deformation and strengthening mechanisms, there is a need to fully understand the above-mentioned phenomena. So far, for conventional materials, the relationship between their strength and grain size has been accurately explained by the concept of pile-ups (Ashby, 1970) and is consistent with the well-known Hall–Petch (H–P) relationship (Hall, 1951; Petch, 1953),

$$\sigma_p = \sigma_0 + k_y d^{-1/2} \quad [9.1]$$

where σ_p is the yield stress; σ_0 is the internal friction; k_y is the H–P slope, which is a coefficient characterising the contribution of grain boundaries to hardening; and d is the grain size.

This traditional approach gives a linear relationship between the grain size and the yield strength, with a constant value of k_y. However, as the grain size is reduced, the pile-up mechanism breaks down, and a change in the H–P slope has been reported by many researchers in deformed structures with a grain size less than 1 μm (Conrad, 2003; Hansen, 2004). Besides the above-mentioned mechanisms identified in UFG materials, the Coble creep mechanism has been found to play a major role during deformation at ambient temperature, leading to inversion of the H–P slope (Carlton and Ferreira, 2007). Additionally, a lack of strain hardening in UFG materials and nanostructures, resulting from low dislocation densities within the grains, has been identified to be responsible for the low ductility of these materials. As has already been mentioned, one of the ways to avoid this problem is to increase the work-hardening rate (in accordance with the Considère criterion). The work-hardening rate in the case of MA steels is also affected by the interaction between disperse second-phase particles and dislocations. A correlation between the work-hardening rate and the dispersion of hard second-phase particles was proposed in the well-known work of Ashby (1966, 1970).

This idea was adopted in several studies to improve the strength–ductility balance of UFG steels (Tsuji *et al.*, 2002; Hayashi and Nagai, 2002; Ohmori *et al.*, 2004). For example, the problem of modelling the work hardening observed in a UFG low carbon steel due to dispersed cementite was studied by Ohmori *et al.* (2004), who presented the following equations:

$$\frac{\mathrm{d}\sigma}{\mathrm{d}\varepsilon} = 8\alpha G b^{1/2}\left[\frac{C}{D} + 16\beta\left(\frac{f}{d}\right)\varepsilon\right]^{-1/2}\beta\left(\frac{f}{d}\right) \qquad [9.2]$$

and

$$\beta = (D - \bar{x})^3 / D^3 \qquad [9.3]$$

where β is the volume fraction of the intragranular region, $\bar{x}$ is the mean thickness of the grain boundary region, D is the grain size, f is the volume fraction of the second phase, d is the mean diameter of the particles, ε is the strain, b is the magnitude of the Burgers vector, G is the shear modulus, and α and C are constants. This model predicts that the work-hardening rate is simply proportional to $(\beta f/d)^{1/2}$ when the grain size hardly affects the geometrically necessary dislocation density. Such a rough assumption can also explain the effects of precipitation strengthening in UFG MA steels.

In another physical model, based on dislocation bowing (Nes *et al.*, 2004), yielding is closely related to the critical shear stress required to make a semicircle configuration of a Frank–Read source. Theoretically, the critical stress for such a condition can be approximated by the following expression:

$$\tau_{\mathrm{cr}} = \frac{Gb}{2\pi L(1-\nu)}\left[\left(1 - \frac{3}{2}\nu\right)\ln\left(\frac{L}{b}\right) - 1 + \frac{\nu}{2}\right] \qquad [9.4]$$

where v is Poisson's ratio and L is the average dislocation length. For grain sizes larger than 100 nm, it can be assumed that L is equal to $\rho^{-1/2}$.

Because the spacing of mobile dislocations in UFG materials and nanomaterials is comparable to the grain size, and because, when a stress is applied, these dislocations will be forced to bow out of the boundaries, stable pile-up configurations do not form, and so this mechanism cannot be used to explain the Hall–Petch relationship for such microstructures. For large deformations at room temperature, it was proposed that the stress required for dislocation migration in a substructure consisting of a mixture of LABs and HABs with a Frank network of dislocations inside the subgrains or grains is represented by

$$\tau = \tau_{t} + \tau_{p} + \alpha_{1} Gb\sqrt{\rho_{i}} + \alpha_{2} Gb\left(\frac{1}{\delta} + \frac{1}{D}\right) \qquad [9.5]$$

where $\hat{\tau}_p$ is the contribution to the flow stress caused by non-deformable particles (the Orowan by-pass stress $\hat{\tau}_p = Gb/\lambda$, where λ is the particle spacing), ρ_i is the dislocation density in the subgrain interior and δ is the separation of the LABs. D is either the grain size in the case of undeformed polycrystalline metals or the separation of HABs in the case of heavily deformed metals.

The contribution from LABs and HABs to the strengthening mechanism can be taken into consideration not only by determination of their separation but also by specifying their density. In the case of severely deformed structures that are characterised both by dislocation substructures and by HABs, the contribution of LABs and HABs to the strength can be expressed in an additive form (Hansen, 2004):

$$\sigma = \sigma_{0} + M\alpha Gb\sqrt{1.5bS_{V}\theta_{LAB}(1-f)} + k\sqrt{\frac{S_{V}}{2}f} \qquad [9.6]$$

where σ_0 is the frictional stress, S_V is the area of boundary per unit volume, f is the density of HABs and θ_{LAB} is the average misorientation angle of the LABs.

Good predictions have been found using this equation in an analysis of the experimental flow stresses of cold-rolled Al, Ni and IF steel. In this equation, S_V can be replaced with $2/D_B$, where D_B is the average boundary spacing. On the basis of the above analysis, it can be concluded that the main differences in the description of the constitutive laws for the strengthening mechanisms of ultrafine-grained and nanostructured structures are a result of the presence of significantly increased volumes of both LABs and HABs, which act as obstacles to dislocations. Hence, it can be said in summary that the ability to predict the mechanical properties of such microstructures depends on properly established representations of these phenomena in the constitutive equations for the flow stress. The problem is especially complicated in the case of MA steels, where the dislocations responsible for creating the cells and subgrains interact with other microstructural features, such as precipitates, impurity atoms and grain boundaries, and can be beneficial to the development of an ultrafine-grained microstructure. Unfortunately, in the case of MA steels, no clear methodology exists for transferring this knowledge into a proper macroscopic model.

In order to use computer simulation to model the mechanical behaviour and assess the uniform elongation of UFG materials, it is necessary to use a proper rheological model that takes into consideration the contribution from substructure strengthening. Based on the above-mentioned study of substructure strengthening, a modification of the Khan–Huang–Liang (KHL) flow stress model was proposed so that it relates the strain hardening to the strain, strain rate and dislocation cell size (Khan *et al.*, 2006):

$$\sigma_{\mathrm{p}} = \sigma_{\mathrm{gb}}\left[1+B^{*}\left(1-\frac{\ln\dot{\varepsilon}}{\ln D_{\mathrm{p0}}}\right)^{n_1}(\varepsilon^{\mathrm{p}})^{B^{*}n_0}\right]\left(\frac{\dot{\varepsilon}}{\dot{\varepsilon}_{\mathrm{r}}}\right)^{C}\left(\frac{T_{\mathrm{m}}-T}{T_{\mathrm{m}}-T_{\mathrm{r}}}\right)^{m} \quad [9.7]$$

where σ_p is the flow stress; σ_{gb} is the grain boundary strengthening; ε_p is the equivalent plastic strain; $\dot{\varepsilon}_r$ is the reference strain rate; $\dot{\varepsilon}$ is the equivalent plastic strain rate; T, T_m and T_r are the deformation, melting and reference temperatures, respectively; and D_{p0}, B, n_0, n_1, C and m are model parameters.

Recently, a further modification of the KHL model has been proposed, where the parameter B is replaced by $B^* = B/\sigma_{gb}$. At the same time, the parameter B^* is a function of the strain to some power. This modification leads to better agreement with the experimental results. Based on the above-mentioned discussion and the assumption that severe plastic deformation leads to the development of a dislocation structure, it was assumed that the grain boundary strengthening can be treated as a sum of effects of low- and high-angle boundaries, in the same way as in Eq. 9.7:

$$\sigma_{\mathrm{gb}} = \sigma_0 + M\alpha Gb\sqrt{1.5bS_{\mathrm{V}}\theta_{\mathrm{LAB}}(1-f_{\mathrm{HABs}})} + k\sqrt{\frac{S_{\mathrm{V}}}{2}f_{\mathrm{HABs}}} \quad [9.8]$$

The values of the parameters d_b, θ_{LABs} and f_{HABs} were measured using EBSD analysis, and the values of M, α, G and b were taken from the literature. The rest of the parameters were identified using inverse methods based on the results of tensile tests (load vs displacement).

This model was successfully applied in FEM multiscale modelling of the mechanical behaviour of UFG MA steels obtained using severe plastic deformation. Submodelling, which is a concurrent multiscale computing method that is an effective finite element technique, was used here to bridge different scales. In the submodelling technique, in general, the first step is to solve a problem with one mesh describing the global domain. Then another, generally finer mesh is used to reanalyse a part of the global domain which is of particular interest. Thus, either the displacement field or the stress field of a global domain is interpolated at the boundary of the submodel, which has a fine mesh along the edges. In this study, a global model with a coarse mesh was prepared in order to model 3D tensile tests of flat specimens. Then a submodel was built which involved the local features of interest, i.e. the microstructures developed during severe plastic deformation. The microstructures were designed using a DMR approach (Fig. 9.4). Digital microstructures representing different levels of grain refinement and substructure strengthening were chosen and attached to the central location in the global model. The parameters of the identified KHL material model applied in the submodels were diversified using a Gaussian distribution function to reflect differences in the crystallographic orientations.

Examples of experimental results are shown in Figs 9.5–9.7. The distributions of the equivalent plastic strain for both the global model and the submodel, recorded at the time of the onset of necking, are presented in Fig. 9.5 for samples that were deformed at room temperature to total strains of 5 and 20. Additionally, the sample deformed to a strain of 20 was subsequently annealed in order to allow LABs to rearrange into more stable HABs, and thus was characterised by a much higher volume fraction of HABs. The average cell size in this case was much smaller than that of the sample that experienced less deformation. In all cases it can be seen that the results given by the submodel show inhomogeneity in the strain at the level of the microstructure, which has not been captured by the global model. As can be seen in Fig. 9.5, strain localisation starts to develop in the microstructure along the grain boundaries and has a non-uniform character. Areas with narrow strain localisation bands are visible between the grains. These results describe the material behaviour more precisely than the global model.

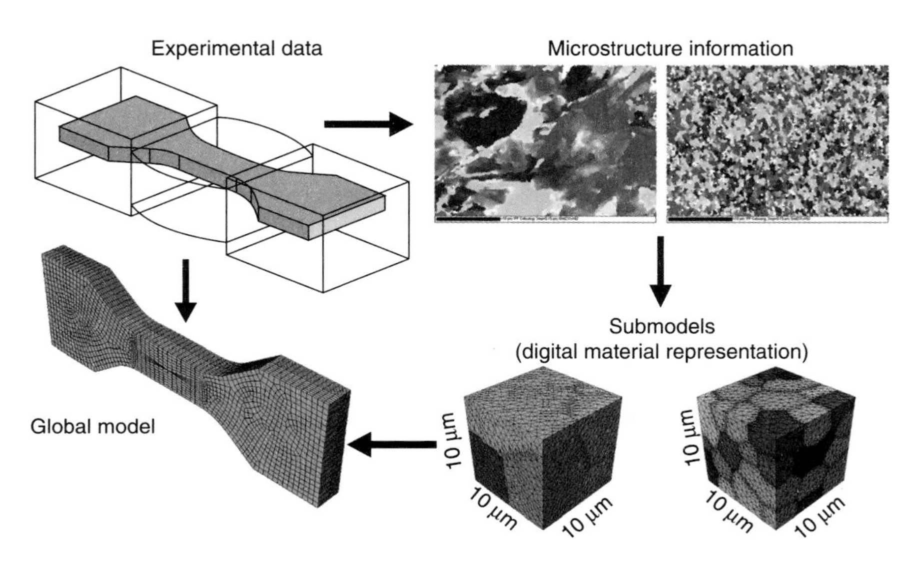

9.4 Microstructures designed using a digital microstructure representation (DMR) approach.

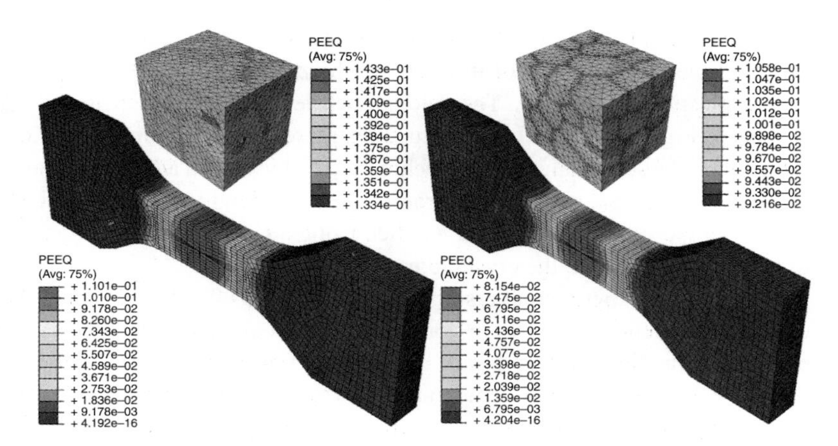

9.5 Distributions of equivalent plastic strain for both the global model and the submodel recorded at the time of the onset of necking.

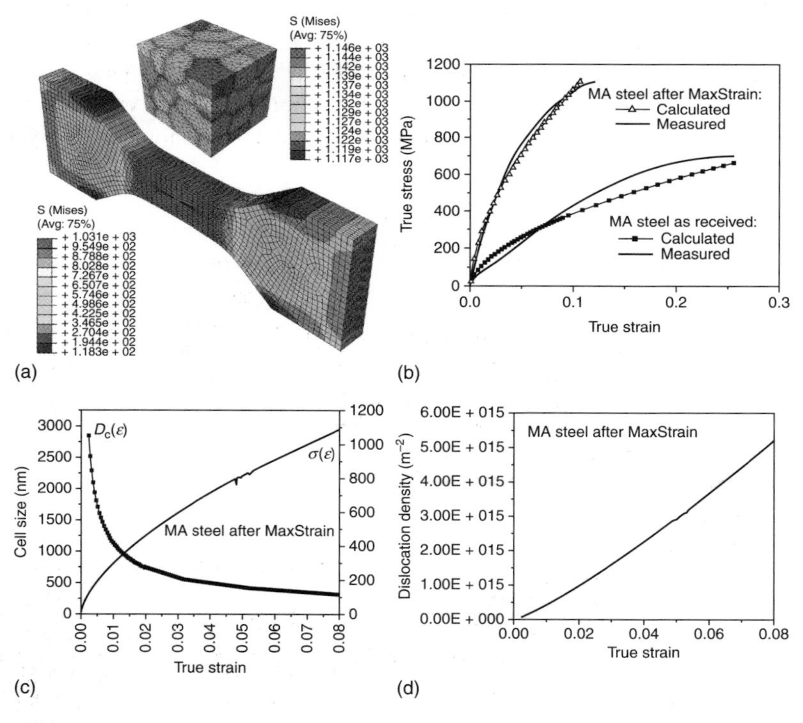

9.6 Example results of multiscale modelling: (a) distribution of von Mises stress in a sample after a deformation of 20 (annealed), (b) comparison of the calculated and measured results, and corresponding calculations of (c) the changes in cell size and (d) dislocation density during the tensile test.

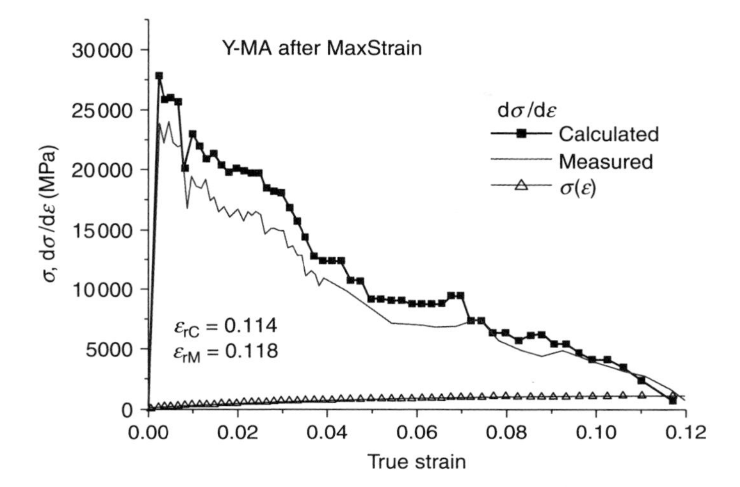

(a)

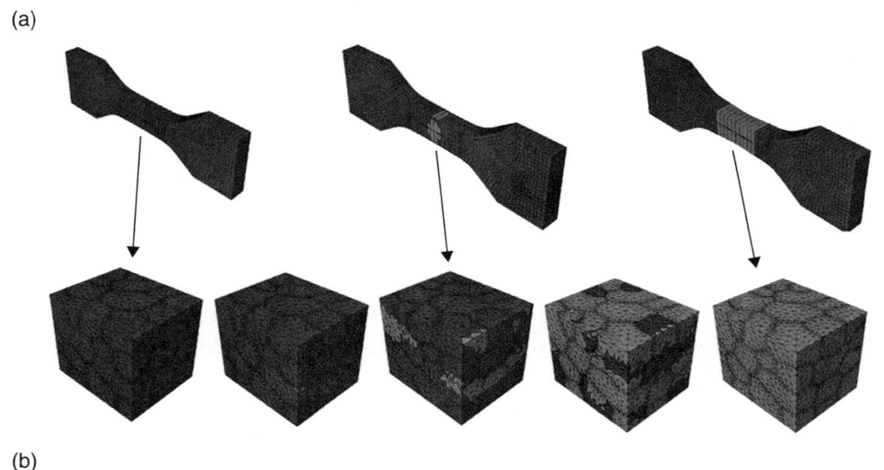

(b)

9.7 Examples of determination of uniform elongation using multiscale simulation. (a) Comparison of calculated and measured results; (b) onset of necking in subsequent simulation steps.

It was also found (Fig. 9.6) that for the specimen characterised by the smaller average cell size, both the von Mises stress and the dislocation density are higher. If, however, the dislocation density measurements for the structure characterised by a high volume fraction of high-angle grain boundaries can be treated reasonably (strain of 20, annealed), the results for the material deformed at room temperature without subsequent annealing are in contradiction to the experimental observations. Figure 9.7 shows an example in which, by using this approach, it is also possible to determine in an accurate way the uniform elongation of UFG structures and the beginning of plastic instability (onset of necking). Values of the work-hardening rate vs true strain calculated from global models using Abaqus Standard are plotted and compared with the experimental data. Additionally, the data from the submodel show more precise information about the onset of the plastic instability. Uniform plastic deformation occurs as long as the true stress is below the value of the work-hardening rate ($d\sigma/d\varepsilon < \sigma$) (shown by dark mesh elements). When the two quantities are equal, uniform deformation stops and necking begins (shown as light grey mesh elements).

In general, the results presented above show that a combination of continuum mechanics and dislocation theory can be used successfully to map dislocation strengthening and to build a rheological model of ultrafine-grained microalloyed steels.

9.7 Conclusion and future trends

We have shown in this chapter that deformation-induced microstructure evolution and work hardening can be obtained in conventional and ultrafine-grained microalloyed steels by several different methods. In general, these methods can be defined as advanced thermomechanical processing, together with others that can be defined as severe plastic deformation techniques (ISUGS, 2005; Weng, 2009). ATP can be applied in existing, conventional industrial plants. UFG materials obtained using SPD techniques are characterised by high strength; however, their application on an industrial scale is limited, not only because it is difficult to produce final products in bulk form but also because of poor ductility (Zhao and Liao, 2010). The large density of grain boundaries significantly restricts or prohibits the activation of processes traditionally associated with plastic deformation, such as dislocation multiplication at Frank–Read sources and dislocation pile-ups at grain boundaries. This is one of the barriers to advanced structural applications. Because of these concerns, producing ultrafine-grained MA steels for practical structural applications is currently a focus of research investigations.

The second-phase nanoparticles present in non-equilibrium grain boundaries impede grain boundary sliding, increase the resistance of nanostructured materials to localisation of plastic strain and improve the ductility of these materials (Naydenkin and Grabovetskaya, 2010). In the case of UFG MA steels, the problem of grain growth and the resulting loss of the unique properties of these materials is effectively avoided owing to a reduction in the grain boundary energy. Atoms of microalloying elements segregate to vacancies or other distorted areas on the grain boundaries, and hence the driving force for coarsening is eliminated, resulting in a metastable equilibrium state (Kirchheim, 2002; Rajgarhia *et al.*, 2010). A proper understanding of the deformation and strengthening mechanisms that govern the mechanical response of UFG materials can provide a way to propose guidelines to improve their ductility. It is a challenge to future research to quantify and model the correlations between the microstructural evolution, chemical composition and ductility of UFG MA steels.

As mentioned above, both the SPD process and the resulting evolution of the microstructure involve inhomogeneity of the mechanical properties of UFG materials. Therefore, an interpretation of the results of mechanical testing of these materials presents some difficulties, which can be avoided when computer simulation is incorporated. Recently published studies show that by combining the TEM technique, EBSD and the Considère criterion it is possible to develop models capable of predicting the plastic instability of necking during tension tests and to make real progress in this field (Majta *et al.*, 2010). Results of calculations show very good accuracy with respect to data obtained from experiments. The approach presented here can be successfully used for prediction of the mechanical response of microalloyed steels subjected to deformation by means of SPD, with high microstructural and mechanical inhomogeneity, where the dislocation substructure, precipitation and solid solution strengthening are significant.

The application of the FEM in calculations and modelling of the mechanical response of UFG materials also needs properly built rheological models. In the case of severely deformed microstructures, existing flow stress models need to be modified and justified for new conditions (Muszka *et al.*, 2007b, 2010; Majta *et al.*, 2009). Also, it has been shown that it is an unacceptable simplification to transfer the analysis from the microscale to the nanoscale. The combination of microstructure observations and FEM simulations has significantly contributed to our current understanding of the physical mechanisms involved in the deformation and strengthening of UFG materials. The problems discussed in this chapter are relevant to analysing SPD and ATP processes, where multiscale modelling of both microstructural evolution and the strengthening mechanisms is necessary.

9.8 References

Altan B S, Miskioglu I, Purcek G, Mulyukov R R and Artan R (2006), *Severe Plastic Deformation: Towards Bulk Production of Nanostructured Materials*, New York: Nova Science Publishers.

Ashby M F (1966), 'Work hardening of dispersion-hardened crystals', *Philos Mag*, 14, 1157–1178.

Ashby M F (1970), 'The deformation of plastically non-homogeneous materials', *Philos Mag*, 21, 399–424.

Beladi H, Kelly G L, Shokouhi A and Hodgson P D (2004), 'The evolution of ultrafine ferrite formation through dynamic strain-induced transformation', *Mater Sci Eng A*, 371, 343–352.

Belyakov A, Sakai T and Miura H (2000), 'Fine-grained structure formation in austenitic stainless steel under multiple deformation at 0.5 T_m', *Mater Trans JIM*, 41, 476–484.

Benito J A, Tejedor R, Rodrguez-Baracaldo R, Cabrera J M and Prado J M (2010), 'Ductility of bulk nanocrystalline and ultrafine grain iron and steel', *Mater Sci Forum*, 633–634, 197–203.

Beynon J H, Gloss R and Hodgson P D (1992), 'The production of ultrafine equiaxed ferrite in a low carbon microalloyed steel by thermomechanical treatment', *Mater Forum*, 16, 37–42.

Carlton C E and Ferreira P J (2007), 'What is behind the inverse Hall–Petch behavior in nanocrystalline materials', in *Size Effects in the Deformation of Materials: Experiments and Modeling*, Lilleodden E, Besser P, Levine L and Needleman A (eds), Materials Research Society Symposium Proceedings 976E, Warrendale, PA: Materials Research Society, p. 0976-EE01-04.

Chen Z, Liu F, Wang H F, Yang W, Yang G C and Zhou Y H (2009), 'A thermokinetic description for grain growth in nanocrystalline materials', *Acta Mater*, 57, 1466–1475.

Choi J K, Seo D H, Lee J S, Um K K and Choo W Y (2003), 'Formation of ultrafine ferrite by strain-induced dynamic transformation in plain low carbon steel', *ISIJ Int*, 43, 746–754.

Conrad H (2003), 'Grain size dependence of the plastic deformation kinetics in Cu', *Mater Sci Eng A*, 341, 216–228.

Considère A (1886), *Mèmoire sur l'emploi du fer et de l'acier dans les constructions*. Paris.

Copreaux J, Lanteri S and Schmitt J H (1993), 'Effect of precipitation on the development of dislocation substructure in low carbon steels during cold deformation', *Mater Sci Eng A*, 164, 201–205.

da Silva M G and Ramesh K T (1997), 'The rate-dependent deformations of porous pure iron', *Int J Plast*, 13, 587–610.

DeArdo A J (1998), 'Microalloyed strip steels for the 21st century', *Mater Sci Forum*, 284–286, 15–26.

De Borst R (2008), 'Challenges in computational materials science', *Comput Mater Sci*, 43, 1–15.

Dong H and Sun X (2005), 'Deformation induced ferrite transformation in low carbon steels', *Curr Opin Solid State Mater Sci*, 9, 269–276.

Edward G H, Etheridge M A and Hobbs B E (1982), 'On the stress dependence of subgrain size', *Textures Microstruct*, 5, 127–152.

Embury J D and Sinclair C W (2001), 'The mechanical properties of fine-scale two-phase materials', *Mater Sci Eng A*, 319–321, 37–45.

Essadiqi E and Jonas J J (1988), 'Effect of deformation on the austenite-to-ferrite transformation in a plain carbon and two microalloyed steels', *Metall Trans A*, 19, 417–426.

Furuya Y, Matsuoka S, Shimakura S, Hanamura T and Torizuka S (2007), 'Fatigue strength of ultrafine ferrite–cementite steels and effects of strengthening mechanisms', *Metall Trans A*, 38, 2984–2991.

Gifkins R C (1976), 'Grain-boundary sliding and its accommodation during creep and superplasticity', *Metall Trans A*, 7, 1225–1232.

Gladman T (1997), *The Physical Metallurgy of Microalloyed Steels*, London: Institute of Materials.

Hall E O (1951), 'The deformation and ageing of mild steel: III. Discussion of results', *Proc Phys Soc London B*, 64, 747–753.

Hanamura T, Yin F and Nagai K (2004), 'Ductile–brittle transition temperature of ultrafine ferrite/cementite microstructure in a low carbon steel controlled by effective grain size', *ISIJ Int*, 44, 610–617.

Hansen N (2004), 'Hall–Petch relation and boundary strengthening', *Scr Mater*, 51, 801–806.

Hayashi T and Nagai K (2002), 'Improvement of strength–ductility balance for low carbon ultrafine-grained steels through strain hardening design', *Trans Jpn Soc Mech Eng*, 68 A, 1553–1558.
Hickson M R, Hurley P J, Gibbs R K, Kelly G L and Hodgson P D (2002), 'The production of ultrafine ferrite in low-carbon steel by strain-induced transformation', *Metall Mater Trans*, 33A, 1019–1026.
Hirth J P and Lothe J (1968), *Theory of Dislocations*, New York: McGraw-Hill.
Hodgson P D and Beladi H (2004), 'The formation of ultrafine grained steel microstructures through thermomechanical processing', *Steel-Grips*, 2, Suppl., 45–51.
Holt D L (1970), 'Dislocation cell formation in metals', *J Appl Phys*, 41, 3197–3201.
Huang X, Kamikawa N and Hansen N (2010), 'Strengthening mechanisms and optimization of structure and properties in a nanostructured IF steel', *J Mater Sci*, 45, 4761–4769.
ISUGS (2005), *International Symposium on Ultrafine Grained Structures*, Iron and Steel Supplement, 40, Beijing: Metallurgical Industry Press.
Jansto S G (2010), 'Process and physical metallurgy, applications and sustainability of niobium bearing steels' in *Materials Science and Technology*, Houston, TX, 1627–1638.
Jia D, Ramesh K T and Ma E (2003), 'Effects of nanocrystalline and ultrafine grain sizes on constitutive behavior and shear bands in iron', *Acta Mater*, 51, 3495–3509.
Keh A S and Weissman S (1963), *Electron Microscopy and Strength of Crystals*, New York: Wiley, 231–300.
Khan A S, Suh Y S, Chen X, Takacs L and Zhang N (2006), 'Nanocrystalline aluminum and iron: mechanical behavior at quasi-static and high strain rates, and constitutive modeling', *Int J Plast*, 22, 195–209.
Kirchheim R (2002), 'Grain coarsening inhibited by solute segregation', *Acta Mater*, 50, 413–419.
Koch C C, Scattergood R O, Darling K A and Semones J E (2008), 'Stabilization of nanocrystalline grain size by solute additions', *J Mater Sci*, 43, 7264–7272.
Kocks U F and Canova G R (1981), 'Mechanisms and microstructures', in Hansen N, Leffers T and Lilholt H (eds), *Deformation of Polycrystals*, Roskilde: Risø National Laboratory, 185.
Lesch C, Alvarez P, Bleck W and Gil Sevillano J (2007), 'Rapid transformation annealing: a novel method for grain refinement of cold-rolled low-carbon steels', *Metall Mater Trans*, 38A, 1882–1890.
Li J C M and Liu G C T (1963), 'Energy of elliptical dislocation loops', *Philos Mag*, 14, 413–414.
Li J C M, Yada H and Yamagata H (1998), 'In situ observation of $\gamma \rightarrow \alpha$ transformation during hot deformation in an Fe–Ni alloy by an X-ray diffraction method', *Scr Mater*, 39, 963–967.
Ma E (2005), 'Four approaches to improve the tensile ductility of high-strength nanocrystalline metals', *Eng Perf*, 14, 430–434.
Madej Ł (2010) *Development of the Modelling Strategy for the Strain Localization Simulation Based on the Digital Material Representation*, Krakow: AGH University of Science and Technology Press.
Majta J (2000), 'Complete model for niobium-microalloyed steels deformed under hot working conditions', in *Thermomechanical Processing of Steels*, IOM Communications, London: Chameleon Press, 1, 322–331.
Majta J and Muszka K (2007), 'Mechanical properties of ultra fine-grained HSLA and Ti-IF steels', *Mater Sci Eng A*, 464, 186–191.

Majta J, Zurek A K and Pietrzyk M (1999), 'Modeling of developing inhomogeneities in the ferrite microstructure and resulting mechanical properties induced by deformation in the two-phase region', in *4th International Conference on Recrystallization and Related Phenomena*, Sakai T, Suzuki H G (eds), Tsukuba City, Japan Institute of Metals, 691–696.

Majta J, Stefanska-Kadziela M and Muszka K (2005), 'Modeling of strain rate effects on microstructure evolution and mechanical properties of HSLA and IF-Ti steels', in *Fifth International Conference on HSLA Steels, HSLA Steels 2005, Iron & Steel*, Suppl 40, Hainan, China 513–517.

Majta J, Muszka K and Stefańska-Kądziela M (2006), 'Study of mechanical properties of ultrafine grained HSLA and Ti-IF steels', in *Proceedings of the 5th International Conference on Mechanics and Materials in Design*, Silva Gomes J F and Shaker Meguid A (eds), Porto, 441–447.

Majta J, Doniec K, Muszka K (2010) 'On the utilization of plastic instability criterion in ductility assessment of ultrafine-grained microalloyed steel', *Mater Sci Forum*, 638–642, 1977–1982.

Majta J, Muszka K, Doniec K and Svyetlichnyy D (2009) 'Mapping the dislocation strengthening in rheological model of BCC ultrafine-grained structures', in *X International Conference on Computational Plasticity Fundamentals and Applications: COMPLAS*, Onate E, Owen D R J and Suárez B (eds), CIMNE, Barcelona.

Matlock D K, Krauss G and Speer J G (2005), 'New microalloyed steel applications for the automotive sector', *Mater Sci Forum*, 500–501, 87–96.

Meyers M A, Mishra A and Benson D J (2006), 'Mechanical properties of nanocrystalline materials', *Prog Mater Sci*, 51, 427.

Michels A, Kril C E, Ehrhardt H, Birringer R and Wu D T (1999), 'Modelling the influence of grain-size-dependent solute drag on the kinetics of grain growth in nanocrystalline materials', *Acta Mater*, 47, 2143–2152.

Mintz B and Jonas J J (1994), 'Influence of strain rate on production of deformation induced ferrite and hot ductility of steels', *Mater Sci Technol*, 10, 721–727.

Mukai T, Ishikawa K and Higashi K (1995), 'Influence of strain rate on the mechanical properties in fine-grained aluminum alloys', *Mater Sci Eng A*, 204, 12–18.

Murty N S V S and Torizuka S (2010), 'Mechanical properties of ultrafine grained steels processed by large strain–high Z deformation – a review', *Mater Sci Forum*, 633–634, 211–221.

Muszka K (2008), 'An effect of grain refinement on the strengthening mechanisms of low carbon steels subjected to plastic deformation' (in Polish), PhD thesis, AGH, Krakow, Poland.

Muszka K and Majta J (2010), 'Study of the effect of ultrafine-grained microstructure of mechanical behavior of microalloyed steels', in *Polish Metallurgy 2006–2010 in Time of the Worldwide Economic Crisis*, Swiatkowski K, Blacha L, *et al.* (eds), Committee of Metallurgy of the Polish Academy of Sciences, Krakow: Publishing House AKAPIT, 243–263.

Muszka K, Majta J and Hodgson P D (2007a), 'Modeling of the mechanical behavior of nanostructured HSLA steels', *ISIJ Int*, 47, 1221–1227.

Muszka K, Majta J and Hodgson P D (2007b), 'Study of the grain size effect on the deformation behavior of microalloyed steels', in *Proceedings of Materials Science and Technology*, September 16–20, Detroit, MI, 493–504.

Muszka K, Hodgson P D and Majta J (2009), 'Study of the effect of grain size on the dynamic mechanical properties of microalloyed steels', *Mater Sci Eng A*, 500, 25–33.

Muszka K, Dymek S, Majta J and Hodgson P (2010), 'Microstructure and properties of a C–Mn steel subjected to heavy plastic deformation', *Arch Metall Mater*, 55, 641–645.

Naydenkin E V and Grabovetskaya G P (2010), 'Deformation behavior and plastic strain localization of nanostructured materials produced by severe plastic deformation', *Mater Sci Forum*, 633–634, 107–119.

Nes E, Marthinsen K and Holmedal B (2004), 'The effect of boundary spacing on substructure strengthening', *Mater Sci Technol B*, 20, 1377–1382.

Ohmori A, Torizuka S and Nagai K (2004), 'Strain-hardening due to dispersed cementite for low carbon ultrafine-grained steels', *ISIJ Int*, 44(6), 1063–1071.

Oscarsson A, Hutchinson B, Nicol B, Bate P and Ekström H E (1994), 'Misorientation distribution and the transition to continuous recrystallization in strip cast aluminium alloys', *Mater Sci Forum*, 157–162, 1271–1276.

Perez R J, Jiang H G, Dogan C P and Lavernia E J (1998), 'Grain growth of nanocrystalline cryomilled Fe–Al powders', *Metall Mater Trans A*, 29, 2469–2475.

Petch N J (1953), 'The cleavage strength of polycrystals', *J Iron Steel Inst*, 174, 25–28.

Ponge D, Song R and Raabe D (2007), 'The formation of ultrafine grained microstructure in a plain C–Mn steel', in *International Symposium on Ultrafine Grained Steels 2007*, Kitakyushu, Japan, October 24–26, 2.

Rajgarhia R K, Spearot D E and Saxena A (2010), 'Behavior of dopant-modified interfaces in metallic nanocrystalline materials', *JOM*, 62(12), 70–74.

Rothers F, Eisenlohr P, Bielere T R and Raabe D (2010), *Crystal Plasticity Finite Element Methods in Materials Science and Engineering*, Weinheim: Wiley-VCH.

Sakai Y, Ohtaguchi M, Belyakov A, Kimura Y, Hara T and Tsuzaki K (2001), in *Proceedings of the International Symposium on Ultrafine Grained Steels (ISUGS-2001)*, Takaki S and Maki T (eds), ISIJ, Tokyo, 290–296.

Song R, Ponge D and Raabe D (2005a), 'Mechanical properties of an ultrafine grained C–Mn steel processed by warm deformation and annealing', *Acta Mater*, 53, 4881–4892.

Song R, Ponge D, Raabe D and Kaspar R (2005b), 'Microstructure and crystallographic texture of an ultrafine grained C–Mn steel and their evolution during warm deformation and annealing', *Acta Mater*, 53, 845–858.

Song R, Ponge D, Raabe D, Speer J G and Matlock D K (2006), 'Overview of processing, microstructure and mechanical properties of ultrafine grained bcc steels', *Mater Sci Eng A*, 441, 1–17.

Stefańska-Kądziela M, Majta J, Dymek S and Muszka K (2007), 'Effect of high strain rate on the dislocation structure of microalloyed and IF steels', *Arch Metall Mater*, 52, 223–229.

Svyetlichnyy D, Majta J, Muszka K and Łach Ł (2010), 'Modelling of microstructure evolution subjected to severe plastic deformation', in *Advances in Materials and Processing Technologies:* Paris AMPT 2010.

Tanaka M, Higashida K and Shimokawa T (2010), 'The effect of severe plastic deformation on the brittle–ductile transition in low carbon steel', *Mater Sci Forum*, 833–634, 471–480.

Tanaka N, Okitsu Y and Tsuji N (2008), 'Dynamic deformation behavior or ultrafine grained aluminium produced by ARB and subsequent annealing', *J Mater Sci*, 43, 7385–7390.

Tsuchida N, Tomota Y and Nagai K (2002), 'High-speed deformation for an ultrafine-grained ferrite–pearlite steel', *ISIJ Int*, 42, 1594–1596.

Tsuchida N, Masuda H, Harada Y, Fukaura K, Tomota Y and Nagai K (2008), 'Effect of ferrite grain size on tensile deformation behavior of a ferrite–cementite low carbon steel', *Mater Sci Eng A*, 488, 446–452.

Tsuji N, Ueji R, Ito Y and Saito Y (2000), 'In-situ recrystallization of ultra-fine grains in highly strained metallic materials', in *Proceedings of the 21st Risø International Symposium on Materials Science*, Hansen N, Huang S, Jensen D J, Lauridsen E M, Leffers T, *et al.* (eds), Risø National Laboratory, Roskilde, 607–616.

Tsuji N, Ueji R, Minamino Y and Saito Y (2002), 'A new and simple process to obtain nano-structured bulk low-carbon steel with superior mechanical property', *Scr Mater*, 46, 305–310.

Tsuji N, Okuno S, Koizumi Y and Minamino Y (2004), 'Toughness of ultrafine grained ferritic steels fabricated by ARB and annealing process', *Mater Trans*, 45, 2272–2281.

Uenishi A and Teodosiu C (2004), 'Constitutive modeling of the high strain-rate behaviour of interstitial-free steel', *Int J Plast*, 20, 915–936.

Valiev R Z and Langdon T G (2010), 'Achieving exceptional grain refinement through severe plastic deformation: new approaches for improving the processing technology', *Metall Mater Trans*, DOI:10.1007/s11661-010-0556-0.

Valiev R Z, Islamgaliev R K and Alexandrov I V (2000), 'Bulk nanostructured materials from severe plastic deformation', *Prog Mater Sci*, 45, 103–189.

Valiev R Z, Estrin Y, Horita Z, Langdon T G, Zehetbauer M J and Zhu Y T (2006), 'Producing bulk ultrafine-grained materials by severe plastic deformation', *JOM*, 58, 33–39.

Van Swygenhoven H, Budrovich Z, Derlet P M and Hanaoui A (2003), 'Are deformation mechanisms different in nanocrystalline metals? Experiments and atomistic computer simulations', in *Processing and Properties of Structural Nanomaterials*, Shaw L L, Suryanarayana C and Mishra R S (eds), Chicago: TMS, 3–10.

Wang J T, Holita Z, Furukawa M, Nemoto M, Tseney N, *et al.* (1993), 'An investigation of ductility and microstructural evolution in an Al–3%Mg with submicron grain size', *J Mater Res*, 8, 2810–2818.

Wang Y M and Ma E (2004), 'Three strategies to achieve uniform tensile deformation in a nanostructured metal', *Acta Mater*, 52, 1699–1709.

Wei Q, Kecskes L, Jiao T, Hartwig K T, Ramesh K T and Ma E (2004), 'Adiabatic shear banding in ultrafine-grained Fe processed by severe plastic deformation', *Acta Mater*, 52, 1859–1869.

Weng Y (2009), *Ultra-Fine Grained Steels*, Beijing: Metallurgical Industry Press, and Berlin: Springer.

Yada H, Li C M and Yamagata H (2000), 'Dynamic $\gamma \rightarrow \alpha$ transformation during hot deformation in iron–nickel–carbon alloys', *ISIJ Int*, 40, 200–206.

Yang Z M and Wang R Z (2003), 'Formation of ultra-fine grain structure of plain low carbon steel through deformation induced ferrite transformation', *ISIJ Int*, 43, 761–766.

Zehtbauer M J, Zhu Y T (2009), *Bulk Nanostructured Materials*, Weinheim: Wiley-VCH.

Zhao Y and Liao X (2010), 'Ductility of bulk nanostructured materials', *Mater Sci Forum*, 633–634.

Zurek A K, Muszka K, Majta J and Wielgus M (2009), 'Multiscale analysis of the effect of grain size on the dynamic behavior of microalloyed steels', in *9th International DYMAT Conference on the Mechanical and Physical Behaviour of Materials under Dynamic Loading*, 7–11 September, Brussels.

Part III

Microstructure evolution in the processing of other metals

10

Aging behavior and microstructure evolution in the processing of aluminum alloys

D. SHAN and L. ZHEN, Harbin Institute of Technology, China

Abstract: The chapter begins by reviewing the characteristics and applications of aluminum alloys. It then focuses on microstructure evolution during plastic processing, and on aging behavior and age hardening in aluminum alloys. The characterization and test methods related to this research are also discussed. Finally, the chapter includes two case studies of aging behavior and microstructure evolution in the processing of 7000 series aluminum alloys and discusses future trends in the field of aluminum alloys.

Key words: microstructure evolution, aging behavior, age hardening, plastic processing, aluminum alloys.

10.1 Introduction

Compared with other materials, aluminum alloys possess a higher specific modulus, higher specific strength, higher corrosion resistance and better workability. Owing to this superior combination of properties, aluminum alloys are widely applied in a number of industries, especially in the fields of aerospace, automobile and high-speed train engineering. In the past few decades, almost all advanced industrial countries have conducted projects to plan, guide and support research and development on aluminum alloys (Sato, 2010; Liu and Lin, 2010; Hirsch and Laukli, 2010; Williams and Starke Jr, 2003; Zhong, 2002).

Aluminum alloys have been so widely used because they have good plasticity and can easily be formed into complex-shaped parts according to requirements. Aluminum has high plasticity at room temperature because of its face-centered cubic (fcc) crystal structure and 12 slip systems. The plasticity of aluminum increases with increasing temperature. When the temperature reaches 400°C, the elongation is double that at room temperature, owing to the formation of new slip planes. Plastic forming technology is a processing method for obtaining workpieces with a particular shape, microstructure and properties. The technology for the plastic processing of aluminum alloys includes mainly extrusion, die forging, semisolid forming and sheet forming (Fang *et al.*, 2009; Browne and Battikha, 1995). Some typical parts formed by these processes are shown in Fig. 10.1.

During the forming process, aluminum alloys undergo various thermal, stress and strain conditions; a complex evolution of the microstructure thus takes place. This evolution determines the final microstructure and has a dominant effect on

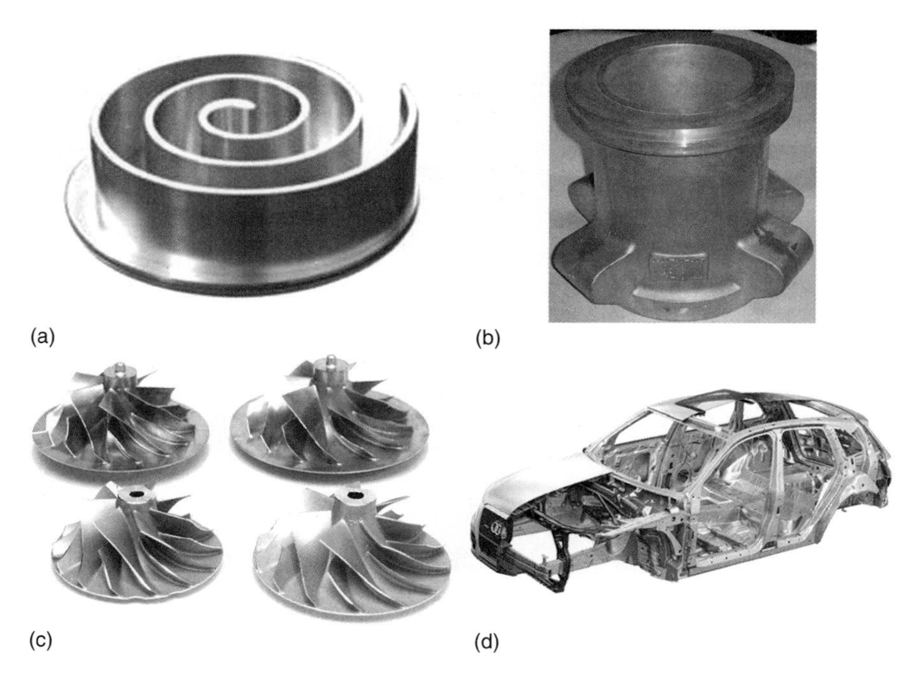

10.1 Aluminum alloy parts formed by various plastic forming methods: (a) scroll formed from 4032 aluminum alloy; (b) cylindrical housing for helicopter formed from 7075 aluminum alloy; (c) turbocharger impeller; (d) automotive body panel for Audi Q7.

the service performance of the formed parts. Microstructural evolution can be observed on multiple scales. Elongated grains, dynamic recovery (DRV) and dynamic recrystallization (DRX) may occur in aluminum alloys during processing. These processes are all on the scale of grains. On the other hand, precipitate particles of various types, shapes and sizes may be formed during aging treatment. Additionally, features related to structural defects, such as the accumulation of dislocations and their resulting distribution, vary greatly during processing and aging treatments.

Deformed aluminum alloys can be classified into two categories: age-hardenable alloys and strain-hardenable alloys. Aluminum alloys in the first category can be strengthened by heat treatment, such as in the case of the 2000 series (Al–Cu–Mg alloys), 6000 series (Al–Mg–Si alloys) and 7000 series (Al–Zn–Mg alloys). Alloys in the second category can be strengthened only by cold working, such as commercially pure aluminum (1000 series), Al–Mn alloys (3000 series) and Al–Mg alloys (5000 series).

10.2 Microstructure evolution during plastic processing: the effects of hot working on microstructure and properties

Plastic forming operations for aluminum alloys can be classified according to the forming temperature into hot working, warm working and cold working. Plastic deformation at temperatures above the recrystallization temperature and below the melting temperature (T_m) is called hot working. The upper hot-working temperature is limited by the melting point of the most fusible constituent. The recrystallization temperature for aluminum alloys is about $0.5T_m$. The melting temperature of aluminum is 660°C. Therefore, the hot-working temperature of aluminum alloys falls in the range of 330–550°C. Cold working is a metal forming process performed around room temperature, and no softening occurs. Plastic forming processes at temperatures between those of cold and hot working are called warm working. The evolution of the microstructure during these processes differs greatly between the different types of processes.

10.2.1 Effects of hot working on cast structure and properties

Most alloying elements and impurities in aluminum alloys may combine with Al atoms to form hard, brittle intermetallic compounds. Accompanying the formation of intermetallic compounds, regional segregation, dendritic segregation and intercrystalline segregation may occur during the cooling process after casting, resulting in a severe decrease in plasticity. In addition, casting defects such as porosity and shrinkage cavities are inevitable, which may also lead to a decrease in plasticity and strength. The microstructure and properties of aluminum casting alloys can be improved by plastic operations. Figure 10.2 shows the effect of

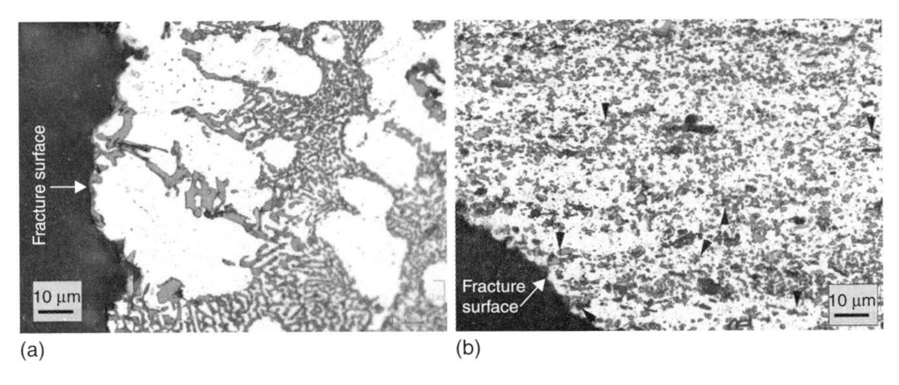

10.2 Effects of plastic deformation on the microstructure of a cast Al–11 mass % Si alloy (Ma *et al.*, 2005): (a) as-cast state; (b) microstructure after processing by RD-ECAP with 32 passes.

rotary-die equal-channel angular pressing (RD-ECAP) on the microstructure of a cast Al–11 mass % Si alloy. As can be seen in the figure, the silicon corals were broken up by RD-ECAP and the dispersion of silicon particles became more homogeneous after plastic deformation.

Hot deformation leads to an increase in the strength of aluminum casting alloys, which may be attributed to the following reasons:

- The coarse dendritic crystal structure is broken down to form a refined structure with small, equiaxed grains via recrystallization (Kaibyshev *et al.*, 2003).
- The hard, brittle intermetallic particles accumulated at the grain boundaries are broken up and distributed more evenly.
- The density increases because of void closure during plastic deformation.
- The high temperature due to heating before forging and the plastic deformation lead to an increase in the ability of atoms to diffuse, resulting in a uniform chemical composition.

Therefore, in order to improve the microstructure and properties of aluminum casting alloys, sufficient deformation must be ensured. Otherwise, the as-cast structure may be retained in forgings. In practice, cast aluminum alloys cannot be used directly as billets for die forging, because the deformation within the billet during die forging is inhomogeneous and some stagnant zones or zones of small deformation exist. Therefore, before die forging, the casting must first be extruded or free-forged to eliminate the as-cast structure in the billet.

10.2.2 Effects of hot working on coarse grain structures in aluminum alloys

The deformation behavior of grains depends mainly on the degree and temperature of hot deformation. During the hot working of aluminum alloys, the grains become finer with increasing degrees of deformation. However, unsuitable deformation conditions always lead to the occurrence of a coarse grain structure (Fig. 10.3) during the processing of aluminum alloys. The strength limit and yield limit will decrease significantly owing to the formation of this coarse grain structure. For example, a coarse grain structure may occur during hot working of aluminum alloys when the degree of deformation is small.

Coarse grains are easily formed in deformed aluminum alloys, wrought aluminum and hard aluminum. The coarse grains are distributed mainly in thick zones with a small degree of deformation and in shear zones with a large deformation. In addition, the surfaces of forgings are usually covered with a layer of coarse grains. A coarse grain structure may occur for the following reasons:

- The extent of deformation is small and falls in the critical deformation region (12–15%).

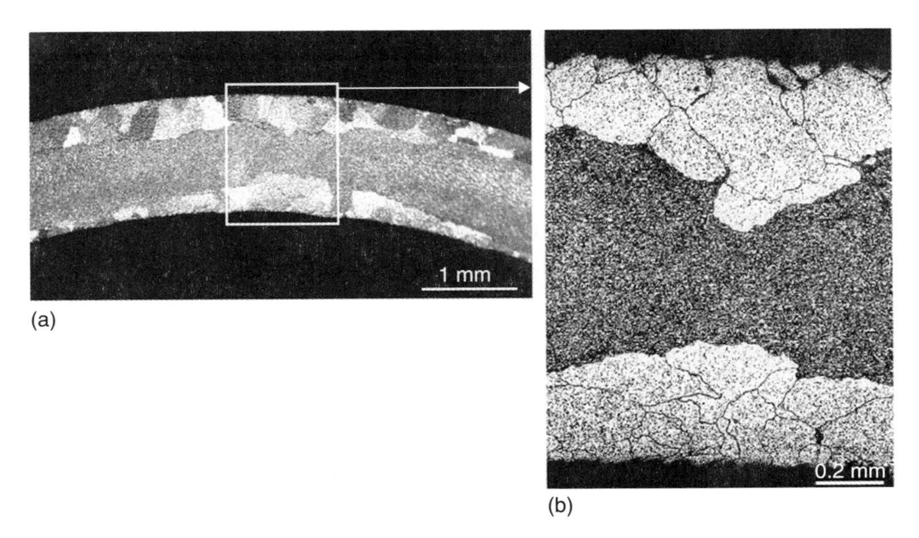

10.3 Coarse surface grain structure in 6005A tube (Birol, 2010): (a) structure of transverse section; (b) magnified image.

- The extent of deformation is large, accompanied by severe shear deformation, which is especially obvious at low deformation temperatures. For instance, a coarse-grained peripheral zone usually appears after heat treatment in extruded aluminum alloy parts.
- A coarse grain structure may be caused by overheating or the inheritance of a coarse-grained peripheral zone from the billet.

Since no allotropic transformation occurs in aluminum, a coarse grain structure in aluminum alloy forgings cannot be refined by heat treatment. Therefore, in order to avoid the formation of a coarse grain structure, the proper technique must be adopted during processing, i.e., the final reduction in the forging process is required to be either greater or lesser than the critical strain, the forging should not be held for a long time at a high temperature, and an appropriate final forging temperature should be ensured.

10.2.3 Effects of hot working on dynamic recovery (DRV) and dynamic recrystallization (RRX)

For a long time, many researchers have suggested that DRV is the only softening mechanism in aluminum and its alloys, owing to the high stacking fault energy and small self-diffusion energy, resulting in the ready occurrence of dislocation slip and climb. However, some researchers have found that, under certain conditions, DRX may occur during hot working of aluminum alloys. DRX and

DRV always proceed simultaneously owing to the easy occurrence of DRV, which leads to a conversion between them. The relative content of substructure formation and DRX varies according to the deformation conditions. This characteristic of aluminum alloys does not occur in low-stacking-fault-energy materials.

Dynamic recovery

DRV is the only softening mechanism in commercial aluminum and its alloys when they are hot worked at strain rates of 10^{-2}–10^{2} s^{-1} at temperatures higher than 350°C. Microscopically, the presence of grains elongated parallel to the deformation direction, with fine, equiaxed, steady subgrains 1–10 μm in size (Fig. 10.4), is the main microstructural morphology. Macroscopically, the flow stress increases rapidly to a steady state and the actual work-hardening rate almost equals zero. The stress in the steady state corresponds to a change in the substructure, and the final mechanical properties of deformed aluminum alloys depend on the size of the substructure. The bigger the substructure, the lower the steady-state stress. The substructure size increases with increasing temperature and decreasing strain rate, but the corresponding steady-state stress exhibits the opposite trend (McQueen and Hockett, 1970). Therefore, the required subgrain microstructure and the corresponding properties can be obtained by controlling the processing conditions.

Dynamic recrystallization

Aluminum and its alloys exhibit very high rates of DRV owing to the high stacking fault energy, which is generally expected to completely inhibit the occurrence of DRX. However, under certain conditions, such as very high temperature or low strain rate, the formation of new grains during hot deformation of aluminum has

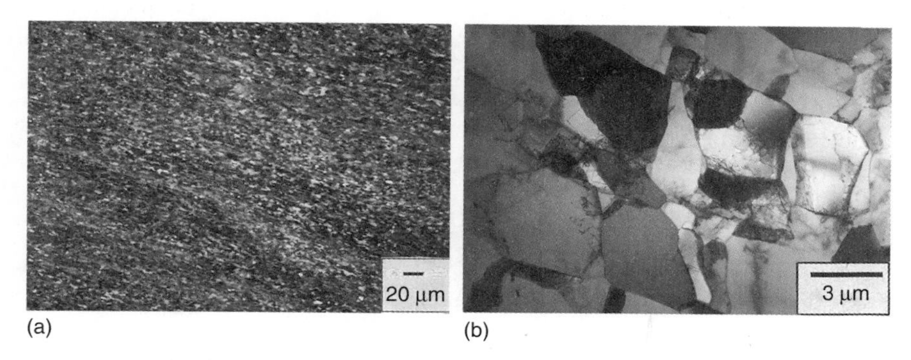

10.4 Subgrains in Al after a torsional strain of 3.2 at 300°C, 1 s^{-1} (McQueen and Blum, 2000): (a) SEM-EBSI, ×200; (b) TEM; the mean subgrain size is 1.61 μm.

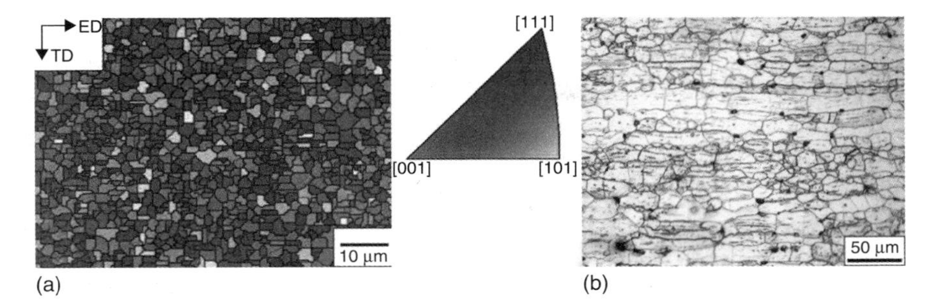

10.5 Grains produced by dynamic recrystallization in aluminum alloys during hot working: (a) 1421 aluminum alloy, 300°C, 10 s^{-1}(Kaibyshev *et al.*, 2005); (b) 7050 aluminum alloy, 460°C, 10^{-4} s^{-1} (Hu *et al.*, 2008).

been frequently reported, as shown in Fig. 10.5. Several mechanisms have been proposed for DRX processes in aluminum alloys, such as discontinuous DRX (DDRX), continuous DRX (CDRX) and geometrical DRX (GDRX) (Yamagata *et al.*, 2001; Doughertya *et al.*, 2003; Blum *et al.*, 1996). CDRX and GDRX are commonly observed in aluminum and its alloys. DDRX rarely occurs in aluminum and its alloys except in two specific cases: in high-purity aluminum (99.999% purity) (Yamagata, 1995) and in aluminum alloys containing large particles. A high purity favors DDRX by increasing the grain boundary mobility. Large particles (>1 μm) promote particle-stimulated nucleation, for example in Al–Mg–Mn alloys (Gourdet and Montheillet, 2000).

DRX includes nucleation and grain growth, which depends mainly on the stored deformation energy (thermodynamics) and the diffusion coefficient (dynamics). The material itself and the processing conditions, for example the original microstructural state, the stacking fault energy, the deformation conditions and the deformation mode, are the main factors that influence the stored deformation energy and the diffusion coefficient (Lin *et al.*, 1988). The following text discusses these influences:

- the original microstructural state;
- the stacking fault energy;
- the deformation conditions;
- the deformation mode.

Factors connected with the original microstructural state, such as the grain size, the substructure, the presence of impurities, and the size and distribution of second phases, have a significant influence on DRX. For example, superfine grains in aluminum alloys change from grains with a fibrous structure into equiaxed grains during superplastic deformation. The effects of substructure and second phases on DRX depend on the size of these. Second-phase particles with a large size (> 1 μm) can improve DRX because of dislocation accumulation around them, leading to the formation of DRX nuclei. However, second-phase particles of small size will inhibit the occurrence of DRX. The DRX temperature depends on the impurity concentration. The higher the impurity concentration, the lower the DRX temperature. For example, the DRX temperatures of aluminum with impurity contents of 99.999%, 99.995%, 99.94% and 99.5% are 375, 450, 550 and 600°C, respectively (Ravichandran and Prasad, 1991).

As noted above, a second influencing factor is the stacking fault energy. Many researchers have found that DRX occurs in Al–Mg and Al–Zn alloys during hot deformation. Experimental results show that the stacking fault energy decreases rapidly when a large amount of Mg, Zn or Cu is added, and thus the energy for dislocation cross-slip increases, which makes it difficult for DRX to occur.

A third influencing factor is the deformation conditions. Complete DRX of the material can be obtained by a suitable choice of the deformation temperature, the strain rate and the extent of deformation. Many publications indicate that increasing the deformation temperature can improve DRX because the diffusion of atoms is improved. However, high temperatures also improve DRV, and the stored deformation energy is too low for the nucleation of DRX grains. DRX does not occur at low temperatures less than $0.5T_m$. It is interesting that DRX is found in aluminum alloys during drawing at room temperature. Nearly complete DRX can be achieved and very fine equiaxed grains can be obtained for low-constituent aluminum alloys during drawing to a true strain of 4–5 at room temperature at high tensile rates. This indicates that at low temperatures, DRX can also happen when other deformation parameters are changed. The effects of strain rate on the softening mechanism are very complex. Some researchers have proposed that decreasing the strain rate increases the deformation time, and then DRX can proceed sufficiently far. However, some other researchers have opposite opinions.

A fourth factor is the deformation mode. Large differences exist in the strain distributions produced by different deformation modes. At present, the research on DRX in aluminum alloys is focused mostly on hot extrusion, hot compression and hot torsion, and it has been shown that these deformation modes improve the occurrence of DRX. Even more DRX can be observed in pure aluminum during hot torsion.

Therefore, adjusting the deformation parameters and deformation modes can inhibit the process of DRV and improve the occurrence of DRX.

10.2.4 Effects of hot working on the formation of fibrous structure (flow lines)

A large number of impurities and intermetallic compounds (e.g., Cu_2Al and Mg_2Si) exist in aluminum alloys. These impurities and intermetallics are distributed with a strip or chain morphology parallel to the main deformation direction, and the resulting fibrous structure results in anisotropic mechanical properties. The strength parallel to the fiber direction is high, and is low in the

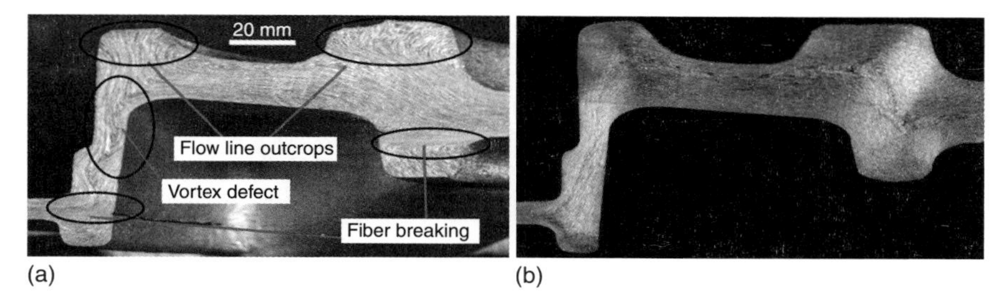

10.6 Flow lines in a rotating seat made from 7075 aluminum alloy. (a) Flow line defects in the ring seat forging; (b) flow lines formed by an optimized forging process.

transverse direction. This anisotropy is related to the difficulty of obtaining DRX in aluminum alloys. The anisotropy becomes obvious for alloys in which DRX is extremely hard to obtain; in such cases the elongation in the longitudinal direction can be twice as high as that in the transverse direction.

The distribution of the fibrous structure has important effects on the properties of aluminum alloy parts. Discontinuous flow lines, vortices and fiber breaking all decrease the plasticity, fatigue strength and corrosion resistance. For example, flow line outcrops, vortices and fiber breaking have been observed in a rotating seat made from 7075 aluminum alloy (Fig. 10.6(a)). With a reasonable hot-working process, however, the flow lines are distributed parallel to the forging shape and the direction of maximum stress (Fig. 10.6(b)), and good properties are obtained (Zhang *et al.*, 2009).

10.3 Microstructure evolution during plastic processing: the effects of cold working on microstructure and properties

The range of application of cold-formed parts made from aluminum and its alloys has been significantly increased owing to the good formability of these materials and the high strength that may be obtained in the final product. Cold-formed parts composed of aluminum alloys include tubes, pump housings, bicycle components, high-pressure cylinders, automotive transmission and power steering components, light reflectors, and so on (Bay, 1997). Besides unalloyed aluminum, a large variety of aluminum alloys can be cold forged. These are grouped into the following series: AA1000, 2000, 3000, 5000, 6000 and 7000.

There is a close relationship between the cold-worked microstructure and the extent of deformation. When the extent of deformation is small, the original grains are elongated parallel to the deformation direction and the shape of the grains can still be observed. With an increase in the extent of deformation, the profile of the original grains gradually breaks down and disappears.

10.3.1 Changes in grain shape

After cold deformation, as a result of the changes in grain profile, the grains are either elongated or reduced and flattened parallel to the main deformation direction. As the grains are elongated, the impurity particles or second-phase particles between grains are also elongated, and a fibrous structure may be formed after cold deformation.

Substructure

Subgrains exist in the grains of sufficiently cold-worked metals; this is known as substructure. The extent to which a substructure is formed, the size of the substructure, and the orientation difference within it depend on the purity, the extent of deformation and the deformation temperature (Rosen *et al.*, 1995; Hurley and Humphreys, 2003). For materials consisting of a pure metal and a second phase, the subgrains are small, and they have large orientation differences and a large crystal distortion for large deformations at low temperature. The substructure has important effects on work hardening during cold working because it inhibits dislocation slip, owing to the different crystal orientations of cells and the presence of dislocation tangles at cell walls. Therefore, refinement of the substructure increases the strength of aluminum alloys.

Deformation texture

During the cold working of aluminum alloys, the operative slip systems tend to rotate toward the direction of the axis of the external force, which has the result that all the grains have a tendency to have the same orientation. The orientation of the grains transforms from a disordered state to an ordered state; this is called a preferred orientation. The fibrous structure formed has a strict orientation relationship, known as a deformation texture. Figure 10.7 shows an image of the texture of an AA5086 aluminum alloy deformed to a large strain (Roy *et al.*, 2011). The occurrence of texture leads to anisotropy in aluminum alloys. For example, 'ears' always occur during the deep drawing of aluminum alloy sheet, which is evidence of texture-caused anisotropy. In actual production, it is important to control the deformation conditions, to use the favorable aspects of the deformation texture, and to avoid its adverse aspects.

10.4 Aging behavior and age hardening

Aging is the most important process for heat-treatable aluminum alloys and the most effective way to achieve their full potential. The final procedure for fabricating parts from heat-treatable aluminum alloys is usually an aging treatment, which then greatly influences their performance in service. Since age-

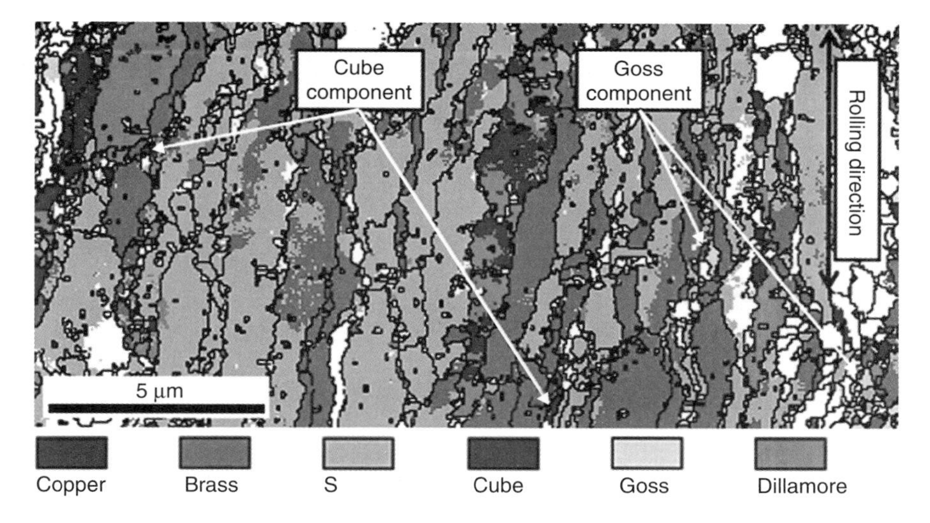

10.7 Map generated by electron backscattering diffraction (EBSD) showing the distribution of various texture components in the microstructure of an AA5086 aluminium alloy processed by accumulative roll bonding (Roy *et al.*, 2011).

hardening behavior was first observed in 1904 (Wilm, 1906), research on the aging process has continued for more than 100 years, as has the application of the process. Innovations in the use of the aging process have contributed greatly to the optimization of the performance of aluminum alloys, and this research field is long-lasting and ever-renewing.

The characteristics of precipitates, including the type, size, shape and distribution, are the dominant factors that affect the performance of aluminum alloys. These characteristics are established during the aging process and are sensitive to precipitation behavior. Control of the aging behavior is thus a key element in tailoring the microstructure and optimizing performance. Obtaining thermodynamic data and reliable microstructure–performance correlations is a basic step in designing aging processes.

10.4.1 Heat-treatable Al alloys

Solution strengthening is rather weak in most aluminum alloys, and thus other strengthening methods are necessary to obtain high strength. Of the various strengthening methods, precipitation strengthening is the most effective. It is also referred to as age hardening, since precipitates are obtained during the aging process. Aluminum alloys that can be strengthened by precipitation are classified as 'heat-treatable aluminum alloys'.

The basic precondition for precipitation hardening is the formation of precipitates. The presence of alloying elements with a solubility that varies greatly is the basic precondition for forming precipitates. When aluminum alloys of this class are heated to a high temperature (the solid solution temperature) and held there for a sufficient period of time, the alloying elements dissolve into the matrix in high concentration. When the alloy is then quenched to a low temperature, the atoms of alloying elements may be 'frozen' into the Al matrix to form a supersaturated solid solution (SSSS). When this SSSS is heated to a moderate temperature, the supersaturated alloying elements separate out and combine with other elements to form particles of intermetallic compounds in the Al matrix; this is called precipitation.

The alloying elements that are frequently used to generate precipitates in aluminum alloys include Cu, Zn, Mg, Si, and Li. To obtain precipitates with an intense strengthening effect, two or more elements are usually used simultaneously to form precipitates that have an obvious mismatch with the Al matrix. In fact, most of the commercial alloys contain two to four alloying elements. The elements added in high concentration are called master alloying elements, and those added in low concentration trace elements. For instance, Mg, Cu, Si and Zn are usually applied as master alloying elements, and Zr, Ag, Li, Sc and others as trace elements.

Based on the type of master alloying elements, heat-treatable aluminum alloys can be classified into a few series, for instance the Al–Cu series (including the Al–Cu–Mg and Al–Cu–Li series), the Al–Mg–Si series, the Al–Zn–Mg–Cu series and the Al–Li series. According to the naming convention established by Alcoa, these series are named the 2000 series, 6000 series, 7000 series and 9000 series, respectively. When different alloying elements are used, the precipitates are accordingly different. The master alloying elements and main strengthening precipitates in typical heat-treatable aluminum alloys are listed in Table 10.1.

The aluminum alloys in different series show different characteristics of their performance because of the different properties of the precipitates in them. For instance, most aluminum alloys in the 2000 series possess medium strength, high toughness and excellent damage tolerance; most 7000 series Al alloys possess

Table 10.1 Master alloying elements and precipitates in the main aluminum alloys

Series	Master alloying elements	Precipitates
2000	Al, Cu, (Mg)	GP $\rightarrow \theta' \rightarrow \theta'$ ($CuAl_2$)
6000	Al, Mg, Si	GP $\rightarrow \beta' \rightarrow \beta''$ (Mg_2Si)
7000	Al, Zn, Mg	GP $\rightarrow \eta' \rightarrow \eta$ ($MgZn_2$)
9000	Al, Li	δ' (Al_3Li) $\rightarrow \delta$ ($AlLi$)

higher strength, but lower toughness, lower damage tolerance and higher stress corrosion cracking (SCC) sensitivity.

10.4.2 Aging behavior

Generally, the aging behavior of aluminum alloys includes four stages, namely

- formation of atom clusters;
- formation of Guinier–Preston (GP) zones;
- formation/growth of metastable phases;
- formation/coarsening of the equilibrium phase.

The aging behavior of 7055 aluminum alloy is demonstrated in Figs 10.8–10.10 to reveal the precipitation sequence (Chen *et al.*, 2009). It is clear that no precipitates can be observed in the solid-solution-treated specimen. After aging for 5 min, nanometer-scale particles are observed within the grains, as shown in Fig. 10.8(b). Selected-area electron diffraction (SAED) patterns indicate that the particles at this stage are mainly GP zones. After aging for 30 minutes, the GP zones grow larger and some of them transform to the η' phase, as Figs 10.8(c) and 10.9(a) suggest. High-resolution transmission electron microscopy (HRTEM) images and the corresponding images filtered by fast Fourier transform (FFT) shown in Fig. 10.10(a) and (b) indicate that most of the η' phase particles and the GP zones are highly coherent with the Al matrix. After aging for 5 h, most of the precipitates observed in Fig. 10.8(d) are still the η′ phase, as the SAED pattern in Fig. 10.9(b) suggests. The η' phase particles at this stage are much larger than those observed in the early stages and most of them have lost coherency with the Al matrix, as demonstrated in Fig. 10.10(c) and (d). After aging for 44 h, most of the η' phase transforms to incoherent, coarse particles of the equilibrium η phase, as shown in Figs 10.8(e), 10.9(c) and 10.10(e).

The aging of 7055 aluminum alloy exhibits the typical SSSS → GP → η' → η sequence when it is aged at 160°C, as mentioned above. The sequence of phase transitions may vary slightly as the composition or aging process changes. For instance, GP zones can hardly be observed in 9000 series aluminum alloys, and the formation of GP zones in 7000 series alloys can be greatly suppressed when

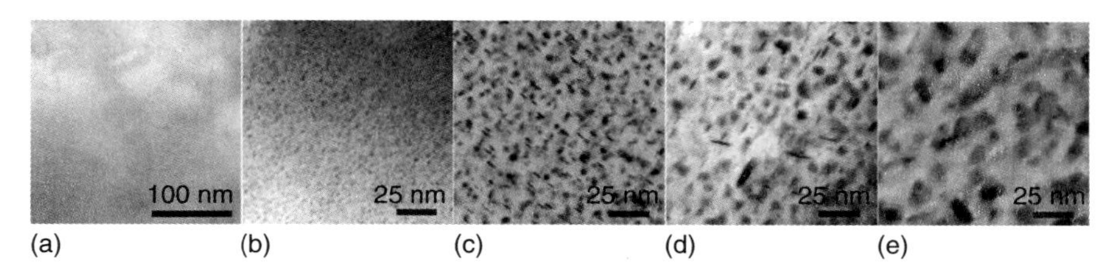

10.8 TEM bright-field images of precipitates in an AA7055 Al alloy aged at 160°C for different times: (a) solid solution treated; (b) aged for 5 min; (c) aged for 30 min; (d) aged for 5 h; (e) aged for 48 h.

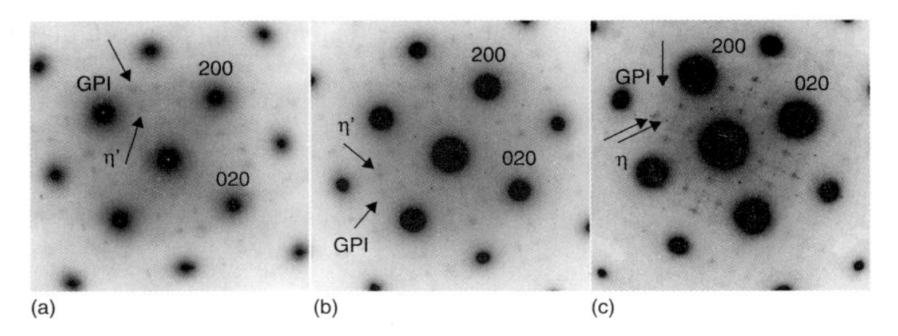

10.9 SAED patterns of AA7055 Al alloy aged at 160°C for different times: (a) 30 min <001> (b)5 h <001> (c) 48 h <001>.

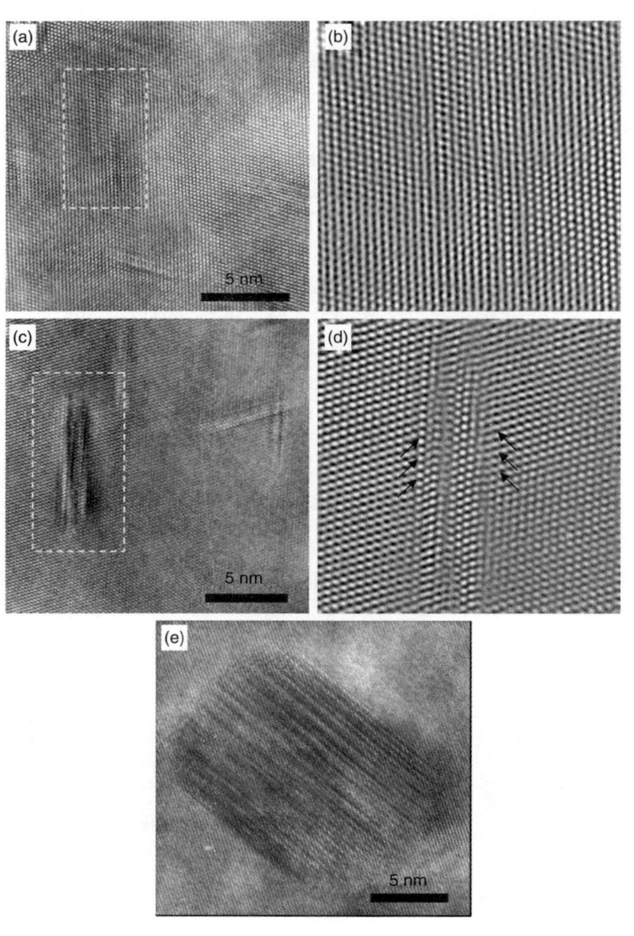

10.10 HRTEM and corresponding FFT-filtered images of precipitates in AA7055 alloy aged at 160°C for different times: (a), (b) 30 min, GPI; (c), (d) 5 h, η'; (e) 48 h, η.

the alloys are aged at elevated temperatures. Also, variations in composition may affect the thermodynamic characteristics of the aging behavior. For instance, when SSSSs of 7000 series aluminum alloys are exposed to room temperature, aging behavior can hardly be detected; in the case of 2000 series aluminum alloys, however, the natural aging is quite intense.

10.4.3 Effect of precipitation on mechanical properties

The mechanical properties of aluminum alloys are highly dependent on the characteristics of the precipitates in them. GP zones and coherent or semicoherent precipitates dispersed in the Al matrix contribute to a high yield strength. The strain field induced by the mismatch between the Al matrix and the precipitates can enhance the resistance to dislocation sliding and thus strengthen the alloy. In contrast, coarsened, and usually noncoherent, particles of equilibrium phases contribute little to the resistance to dislocation sliding, since they are sparse. It is easy to understand why aluminum alloys subjected to different aging conditions possess disparate mechanical performance.

The effect of precipitates on the strength of aluminum alloys is quite illustrative. In the SSSS state, aluminum alloys are quite soft, since the solution-strengthening effect of the alloying elements is rather weak, as shown in the starting segment of the aging–strength curve of 7055 aluminum alloy in Fig. 10.11. Once aging behavior starts, atom clusters and GP zones form quickly in the Al matrix, as shown in Fig. 10.8(b) and (c). These particles are very fine and are coherent with the Al matrix, and thus cause an intense strengthening effect. Accordingly, the strength of aluminum alloys increases sharply during the initial stage, as observed

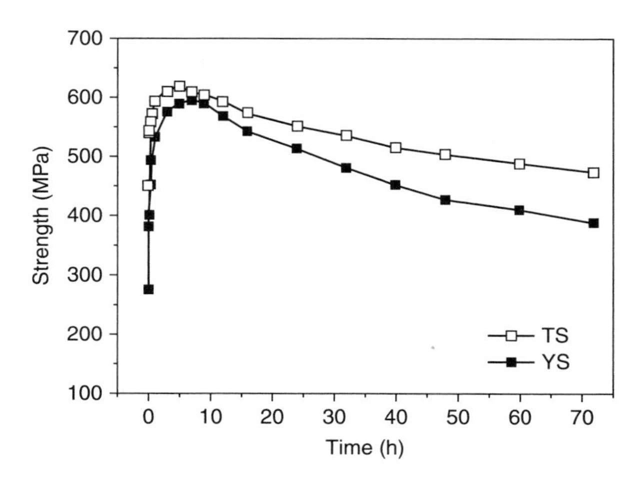

10.11 Evolution of the strength of AA7055 Al alloy aged at 160°C. (TS, tensile strength; YS, yield strength.)

in Fig. 10.11. As the number of particles increases, the strength of the aluminum alloy increases continuously until most of the alloying elements have separated out from the matrix, when a peak-aged alloy is obtained. Specifically, a yield strength of 595 MPa was obtained for 7055 aluminum alloy after aging at 160°C for 8 h. When these alloys are aged for longer times, metastable precipitates grow quickly and then transform gradually to equilibrium phases, and the coherency between the precipitates and the matrix gradually vanishes, as shown in Fig. 10.10(d). Accordingly, the strengthening effect of these precipitates weakens significantly, and the aluminum alloy is then in an over-aged state. Although the strength is much lower than that of peak-aged alloys, however, over-aged aluminum alloys possess better SCC resistance, which is quite important for practical applications.

Although investigations of aging behavior at certain temperatures provide important information for process design, good combinations of properties can hardly be obtained through just the single-step aging process discussed above. A few two-step or even three-step processes, known as T73, T74, T76, T77 and so on, have been developed to meet the requirements of applications. Pre-deformation has also been introduced to tailor the precipitation behavior. The application of these processes has significantly improved the performance of aluminum alloys. For instance, Alcoa applied T77, a three-step process, to treat 7055 alloy and successfully obtained a good combination of properties, including high strength, high toughness, high tolerance and high corrosion resistance (Chen *et al.*, 2010; Lee *et al.*, 1995; Dixit *et al.*, 2008).

TEM bright-field images of a 7055-T77 aluminum alloy are shown in Fig. 10.12. Fine precipitates, most of which were identified as η' phase particles by

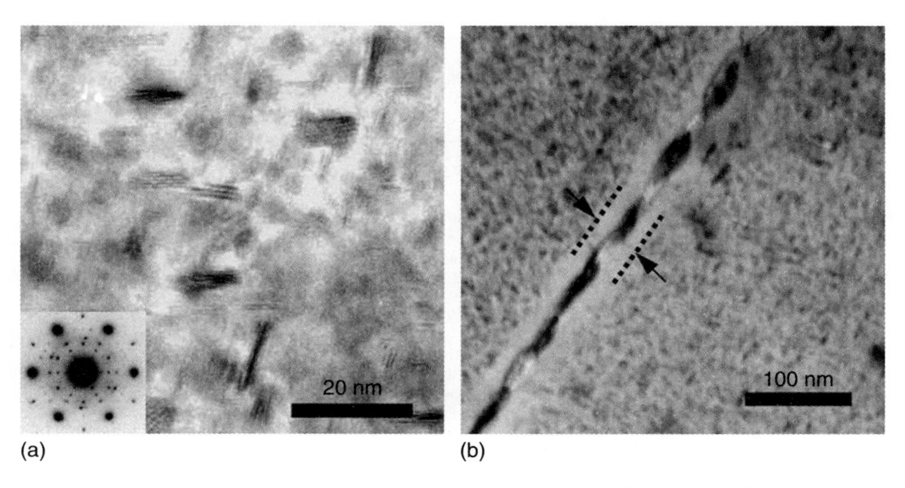

10.12 Precipitates in grains and on grain boundaries in 7055-T77 aluminum alloy: (a) in grains; (b) on grain boundaries.

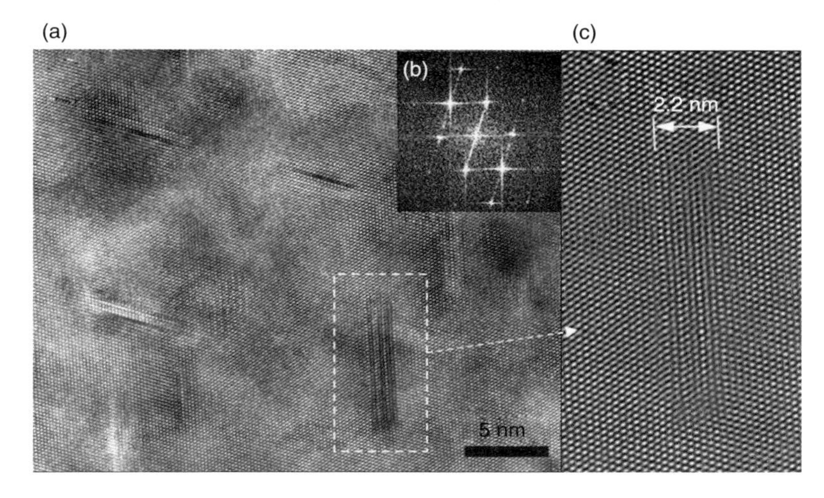

10.13 (a–c)HRTEM and corresponding FFT-filtered images of η' phase particles in grains in a 7055-T77 aluminum alloy.

SAED analysis, about 10 nm in length and 1–3 nm in width, can be observed in the grains. HRTEM images indicate that these precipitates are coherent with the Al matrix, as shown in Fig. 10.13. On the other hand, coarse η phase particles are observed to be distributed discontinuously on the grain boundaries in the alloy, as shown in Fig. 10.12(b); this is believed to be favorable for SCC resistance. It can then be concluded that the interior of the grains is in a peak-aged condition and the grain boundaries are in an over-aged condition when 7055 aluminum alloy is treated by the T77 process. This 'composite' precipitation induces an excellent performance combination in this alloy, including high strength, high roughness, high corrosion resistance, high damage tolerance and so on. Detailed information about the performance of 7055-T77 alloys is listed in Table 10.2.

Table 10.2 Typical mechanical performance of 7055-T77 aluminum alloys

Strength		K_{1c} (MPa m$^{1/2}$)	δ (%)	Resistance to exfoliation corrosion	FCGR (da/dn)
$\sigma_{0.2}$ (MPa)	σ_b (MPa)				
≥614	≥634	≥26	≥11	EA	4.10 × 10^{-4} (ΔK = 10 MPa/√m, R = 0.5)

10.5 Characterization and test methods

10.5.1 Methods of microstructure characterization

Various kinds of technologies are used to observe and characterize the microstructure of aluminum alloys on different scales. Optical microscopes are widely used to observe polished-and-etched specimens to obtain qualitative information about the size, shape and orientation of grains. Scanning electron microscopy (SEM) is another powerful tool for observing the grain structure and precipitates in aluminum alloys. Additionally, the large depth of field of SEM enables it to observe rough surfaces, such as fracture sections.

Electron backscattering diffraction (EBSD) is a newly developed method for acquiring large amounts of information about grain structures (Cabibbo *et al.*, 2005; Kamaya *et al.*, 2005; Xun and Tan, 2004). Classifying grain structures and obtaining quantitative statistics about them, which is quite important for the study of microstructure evolution, thus becomes feasible.

TEM is widely used to characterize microstructures, including precipitates, grain boundary structure and dislocations, which then promotes the study of deformation behavior, damage behavior and strengthening mechanisms (Chen *et al.*, 2009; Chen *et al.*, 2010). Lattice images can also be obtained by using HRTEM, which may provide basic information about interface structures.

Small-angle X-ray scattering (SAXS) technology is another method for the characterization of precipitation processes (Guyot and Cottignies, 1996; Luzzati *et al.*, 1961). In combination with careful data processing, SAXS provides statistical information about shape and size distributions over a relatively large range.

3D atom probe tomography (3D APT) is a newly developed technology for structure characterization. It is now widely used for semiquantitative compositional analyses of solute clusters and nanosized precipitates in aluminum alloys. The resolution of 3D APT is well below 50 pm in the z-direction and 100–150 pm in the x–y plane. A typical result obtained from 3D-APT is shown in Fig. 10.14 (Moody *et al.*, 2011; Marceau *et al.*, 2010; Gault *et al.*, 2009; Marlaud *et al.*, 2010).

10.5.2 Performance-testing methods for aluminum alloys

The mechanical properties of aluminum alloys are usually evaluated by hardness measurement, tensile testing and thermal simulation. Hardness measurement is the most simple and effective way to evaluate the mechanical properties of aluminum alloys, and it can provide an intuitive estimate of the strength of the specimens measured. Among the various methods of hardness measurement, measurements of the Brinell hardness and Vickers hardness are the most widely used methods for aluminum alloys.

Tensile testing is a method widely used to evaluate the mechanical performance of aluminum alloys. The elastic limit, elongation, elastic modulus, yield and

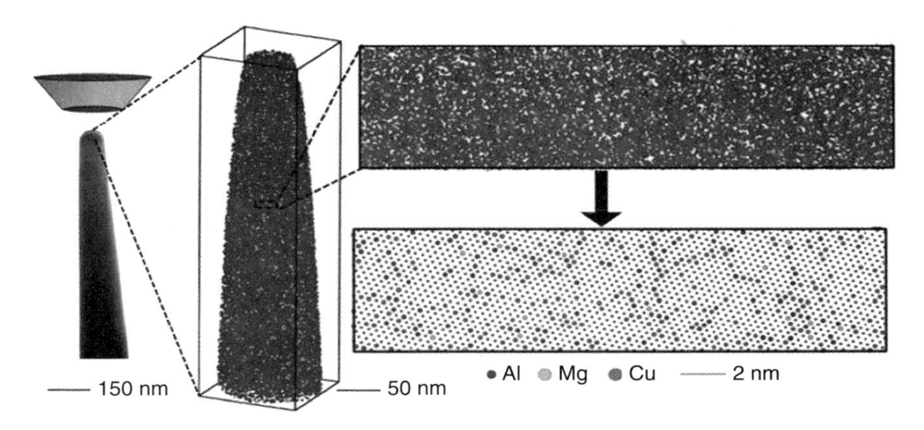

10.14 Lattice rectification in atom probe microscopy. The true atomic positions have been rectified from the raw data (Moody *et al.*, 2011).

tensile strengths, and many other parameters can be obtained from tensile testing. Rod or flake specimens are usually used in tensile tests.

The processability of aluminum alloys is usually evaluated using thermal-simulation systems. Dynamic thermal forming processes, such as forging, torsion and rolling, under various thermal and stress conditions, can be performed in such a system. The mechanical properties, microstructure evolution and deformation behavior can be systematically characterized and analyzed, and this can provide reliable data for designing processes.

10.5.3 Modeling technology

Computation and simulation are now attracting more and more attention because of their unique contribution to the description of microstructure and prediction of performance. Various modeling techniques have been developed to solve problems on different scales and of different types.

Molecular dynamics modeling is an effective method for simulating the motion of systems of molecules based on the equations of classical mechanics. In the field of materials science, this method is now being applied to simulate the evolution of microstructure and defects in aluminum alloys, including precipitation, dislocation reactions and so on (Du *et al.*, 2009; Stegailov *et al.*, 2009).

The Monte Carlo method is an important statistical simulation method based on probability and statistics. It is now being applied widely in the field of materials science to simulate microstructure evolution on the scale of grains. The newly established kinetic Monte Carlo method has also been used to simulate precipitation behavior in the initial stages of aging on the atomic scale (Cerezo *et al.*, 2005; Malerba *et al.*, 2005). This method is simple, quick and flexible.

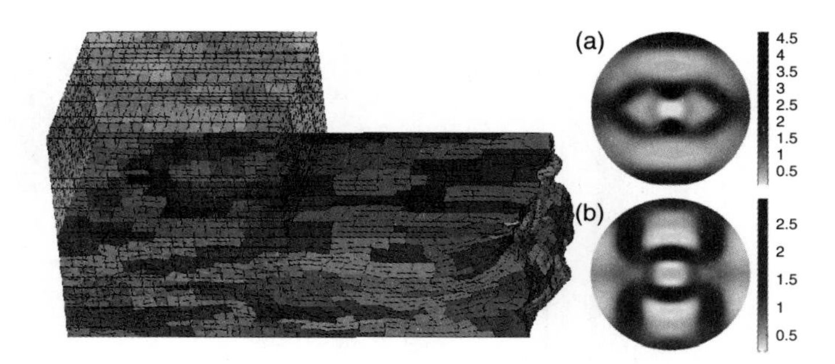

10.15 Grain deformation (left) and the corresponding evolution of texture during compression (right). (a) the [111] pole figure, (b) the [100] pole figure.

The phase field method is another method of numerical simulation. It is now being applied to modeling and predicting mesoscale morphological and microstructural evolution in materials (Wang *et al.*, 2007; Vaithyanathan *et al.*, 2007). The phase field method is quite powerful and the related algorithm is quite simple. In recent years, phase field modeling has been used to simulate the evolution of grain structures, precipitates and defects in aluminum alloys.

The evolution of microstructure on the scale of grains and precipitates, and stress and temperature fields in parts of samples, can be described by the method of finite element analysis (FEA). Some core problems in the field of aluminum alloys, including dislocation dynamics and the deformation of polycrystals, have been partially described by FEA models. The images in Fig. 10.15 demonstrate the deformation of grains and the corresponding evolution of the texture during a compression process (Quey *et al.*, 2011; Eidel, 2011).

10.6 Case studies and applications

10.6.1 Aging behavior and microstructure evolution in the processing of 7075 aluminum alloy

The evolution of the microstructure during the isothermal precision forging and subsequent aging treatment of 7075 aluminum alloy to form a rotating socket sleeve (Fig. 10.16), is chosen as an example. The 7075 alloy is a high-strength aluminum alloy belonging to the Al–Zn–Mg–Cu series that can be strengthened by heat treatment. The isothermal precision forging was carried out on a hydraulic press with a nominal working pressure of 50 000 kN, and the deformation temperature was 450°C. The forging was carried out by a three-step forming process, namely preforming, preforging and finish-forging. The chemical composition of the 7075 aluminum alloy is listed in Table 10.3, and the original microstructure is shown in Fig. 10.17.

Table 10.3 Chemical components of aluminum alloy 7075 (wt%)

Elements	Cr	Mn	Si	Cu	Zn	Mg	Ti	Fe	Al
Percentage by weight	0.23	0.081	0.063	1.49	5.8	2.8	0.024	0.45	Balance

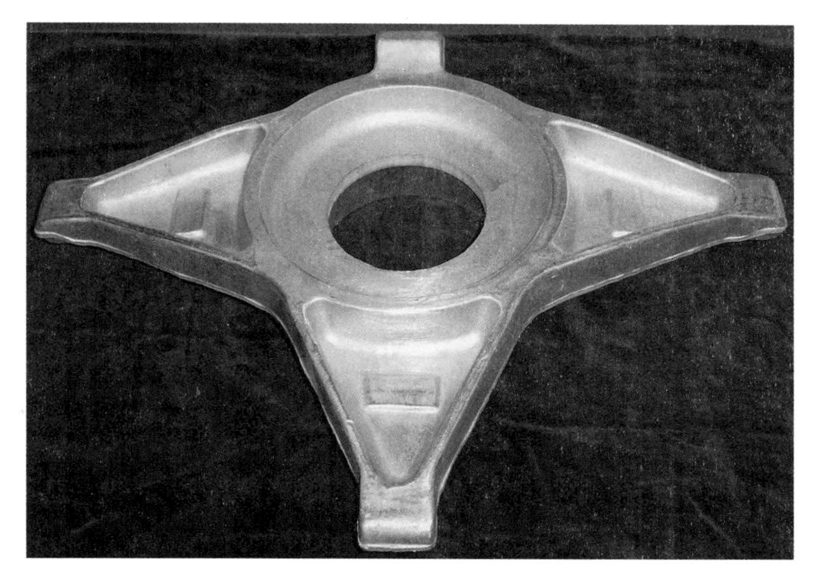

10.16 Photograph of 7075 alloy forging for a rotating socket sleeve.

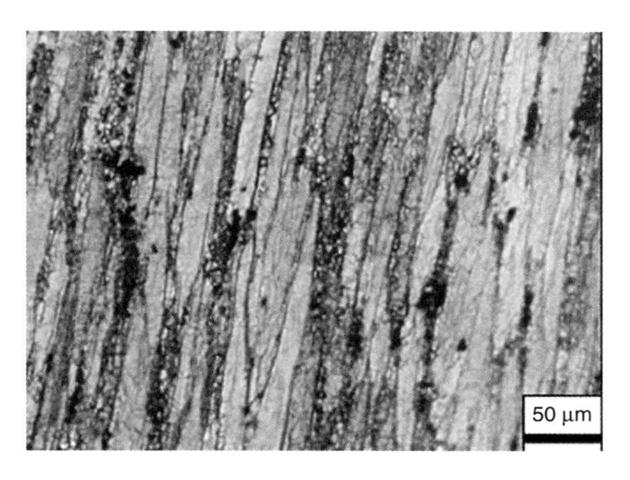

10.17 Original microstructure of 7075 alloy.

Optical and TEM microstructures of the final forgings obtained with different deformations are shown in Figs 10.18 and 10.19. As can be seen from these figures, the radial flow of metal in the forging without preforming is very small. The grains are coarse, and no obvious directivity is observed, owing to the small deformation, resulting in low tensile properties in the radial direction. With increasing deformation, the grains are elongated parallel to the deformation direction. In addition, the subgrains become finer and incomplete DRX occurs. The extent of DRX increases with increasing deformation. A large amount of small, approximately equiaxed subgrains occurs at a deformation of 80%. The volume fraction of the second phase also increases with increasing deformation.

In order to obtain good mechanical properties and stress corrosion resistance, an aging treatment following a solution treatment is essential for the precipitation of a strengthening phase in ultra-strength aluminum alloys. TEM micrographs of a forging alloy after different aging treatments following solution treatment for 45 min at 465°C are shown in Fig. 10.20, and the room temperature properties are

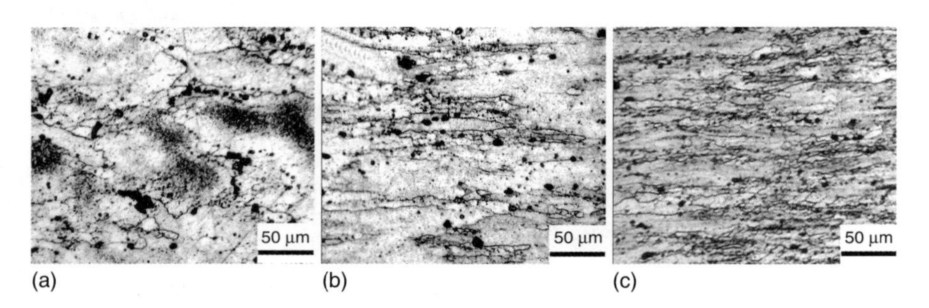

10.18 Optical microstructure of final forgings with different deformations: (a) deformation of 0%; (b) deformation of 50%; (c) deformation of 80%.

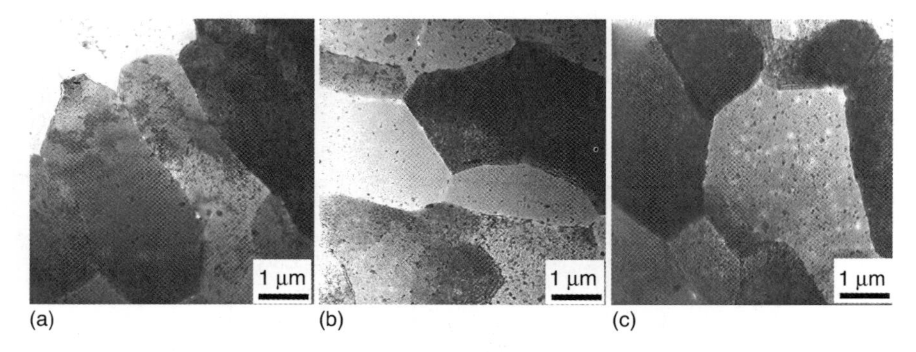

10.19 TEM microstructure of final forgings with different deformations: (a) deformation of 0%; (b) deformation of 50%; (c) deformation of 80%.

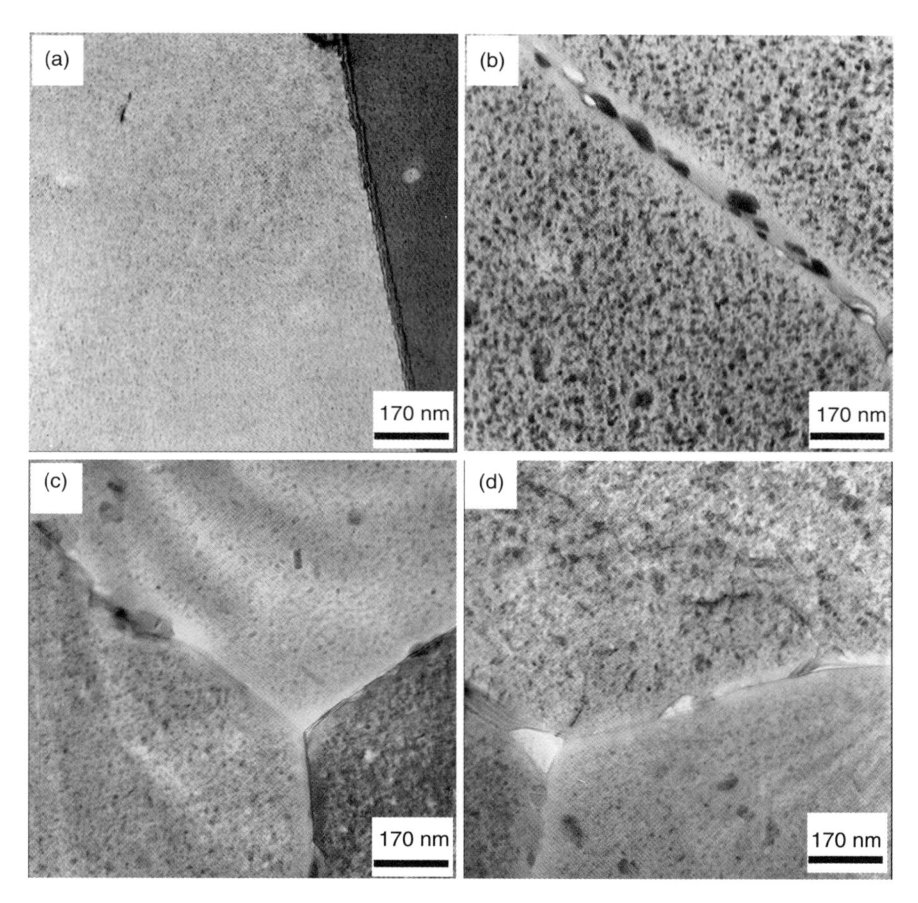

10.20 TEM micrographs of a forging alloy after different aging treatments: (a) peak aging (T6); (b) over-aging treatment (T73); (c) double aging treatment; (d) retrogression and re-aging treatment (RRA).

listed in Table 10.4. After aging under T6 conditions, the precipitate phase in the grains has the finest structure, leading to high strength and hardness. However, the stress corrosion resistance is very low owing to the fine precipitates at grain boundaries. The main precipitate phase in the matrix is the η phase ($MgZn_2$) after the first step of low-temperature pre-aging and the following second step of high-temperature final aging. The size of the precipitates is relatively large, and their volume fraction is small. The strength is low because of the small inhibition effects of the precipitates on dislocation motion. The alloy after double aging at 120°C for 6 h + 165°C for 12 h or after RRA aging at 120°C for 24 h + 200°C for 1 h + 120°C for 24 h, not only has better plasticity but also has better strength, hardness and stress corrosion resistance than the alloy after normal aging.

Table 10.4 Room temperature properties of 7075 aluminum alloy rotating-socket sleeve after different aging treatments

Sample number		Aging system	σ_b (MPa)	σ_s (MPa)	δ (%)	HBS	ICAS (%)
1	T6	120°C × 120 h	587.4	520.3	4.21	184.8	18.3
2	T73	110°C × 7 h + 177°C × 9 h	526.6	418.7	6.9	154.2	21.873
3	Double aging	120°C × 6 h + 165°C × 12 h	533.6	461.2	5.33	163.2	21.932
4		120°C × 6 h + 165°C × 16 h	505.5	430.3	3.98	160.2	21.5
5		120°C × 6 h + 165°C × 20 h	500.8	405.3	5.5	158.2	21.673
6		120°C × 6 h + 165°C × 24 h	452.1	366	4.49	145.6	22.753
7	RRA	120°C × 24 h + 160°C × 1 h + 120°C × 24 h	603.1	544.3	4.06	192.8	18.136
8		120°C × 24 h + 180°C × 1 h + 120°C × 24 h	593.9	537	4.31	190.2	19.897
9		120°C × 24 h + 200°C × 1 h + 120°C × 24 h	544.6	461.7	5.02	164.8	21.962
10		120°C × 24 h + 220°C × 1 h + 120°C × 24 h	414.7	302	5.86	129.6	22.169

10.6.2 Prediction of precipitation behavior and strength for AA7055 aluminum alloy

The AA7055 aluminum alloy has attracted great attention because it provides the excellent performance that is desired in the aircraft industry. This alloy has been thoroughly studied since it was developed by the Alcoa Company in 1990. A few aging processes, including T74 and T77, have been developed for this alloy, as mentioned above. However, investigation of the precipitation behavior during a single-step process is still helpful in obtaining further insight into this alloy. Here, we describe an investigation of the precipitation behavior and the related hardening effect by TEM, SAXS and computational modeling.

Figure 10.21 shows bright-field TEM images near a <011> zone axis of the Al matrix of an AA7055 specimen aged at 120°C for different aging times. HRTEM images of precipitates obtained under various conditions are shown in Fig. 10.22.

After aging for 5 min, darkly imaging precipitates in the Al matrix could be observed, as shown in Fig. 10.21(a). It can be seen that the contrast of most of these precipitates has a round appearance, but some appear slightly elongated. The uniform distribution of these precipitates in the matrix indicates that they are predominantly formed by homogeneous nucleation in the highly supersaturated solid solution. The HRTEM images in Fig. 10.22(a) and (b) indicate that these precipitates are fully coherent with the Al matrix lattice and that their average size is about 1–2 nm. The corresponding selected-area diffraction patterns (SADPs) indicated that these precipitates are mainly GPI zones.

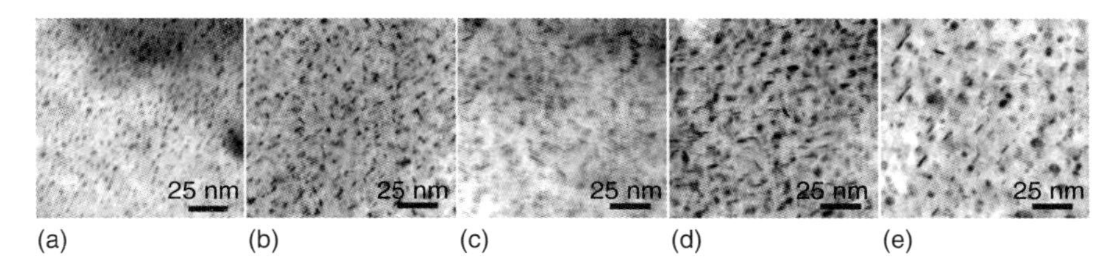

10.21 TEM bright-field images of precipitates in AA7055 Al alloy aged at 120°C for (a) 5 min, (b) 1 h, (c) 5 h, (d) 24 h and (e) 48 h.

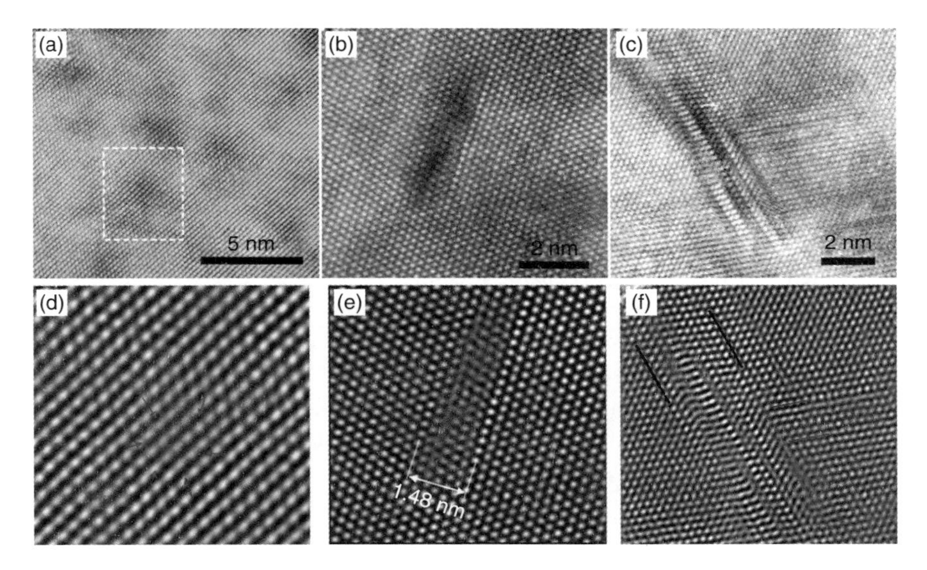

10.22 HRTEM and corresponding FFT-filtered images of precipitates in AA7055 Al alloy aged at 120°C for different times: (a), (d) 5 min, GPI; (b), (e) 5 h, η'; (c), (f) 48 h, η'.

After aging for 60 min, a lot of rod-like precipitates, as well as nearly round ones, could be observed in bright-field TEM images, as shown in Fig. 10.21(b). The rod-like precipitates are about 1 nm in width and 3 nm in length. They are also uniformly distributed in the Al matrix. The corresponding SADPs indicate that most of these precipitates are still GPI zones but that some metastable precipitates with a platelet morphology are beginning to appear.

After aging for 300, 1440 and 2880 min, rod-like and nearly round precipitates can still be observed, as shown in Fig. 10.21(c)–(e), respectively. The number of small nearly round GPI zones decreases with increasing aging time, while the numbers of large rod-like precipitates and large nearly round ones increase.

HRTEM observations show that the η' phase precipitates are still coherent with the Al matrix lattice after aging for 300 min, as shown in Fig. 10.22(c) and (d). They then become semicoherent with the Al matrix lattice after aging for 2880 min, as shown in Fig. 10.22(e) and (f). From the corresponding SADPs, it can be seen that after 300 min of aging at 120°C, the GPI density has decreased and the volume fraction of η' phase has increased. After aging for 1440 min, the volume fraction of η' phase increases further and some η precipitates have already formed. After aging for 2880 min, the GPI diffraction spots become weaker and it becomes hard to separate the spots arising from the η' and η phases.

Quantitative statistical data on the size and volume fraction of the precipitates in the AA7055 aluminum alloy was acquired using the SAXS method. As shown in Fig. 10.23(a), during aging at 120°C the precipitates grow quickly to about 3.2 nm in less than 5 h and then remain unchanged as aging proceeds. However, during aging at 160°C, the precipitates grow continuously to larger than 11 nm. It can be seen from Fig. 10.22(b) that when the sample is aged at 160°C, the volume fraction of precipitates increases sharply to 5% in less than 12 hours and then remains at this level. In contrast, the volume fraction of precipitates increases continuously for more than 60 h when the sample is aged at 120°C. It thus can be concluded that the coarsening of the precipitates is rather intense at 160°C but rather weak at 120°C. On the other hand, the volume fraction of precipitates changes slightly with aging temperature; the final level is about 5–6%.

To predict the yield strength of AA7055 aluminum alloy, a model was constructed. A few strengthening mechanisms were considered synthetically to improve the accuracy of the predictions. The model can be described briefly as follows:

$$\sigma_y = \sigma_i + \Delta\sigma_{GB} + \Delta\sigma_{ss} + (1-H)\Delta\sigma_{cutting,p} + H\Delta\sigma_{passing,p} \qquad [10.1]$$

Here, σ_i is the intrinsic strength of pure aluminum, $\Delta\sigma_{GB}$ is the strength increment related to grain boundaries, $\Delta\sigma_{ss}$ is the strength increment related to solid solution

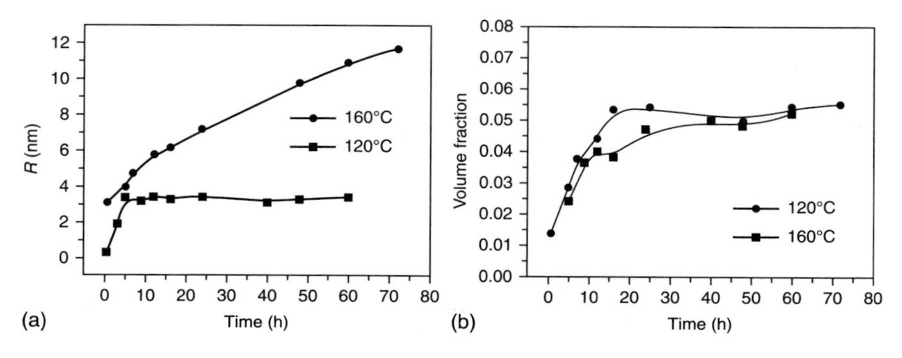

10.23 Evolution of (a) size and (b) volume fraction of precipitates in AA7055 Al alloy aged at 120°C and 160°C for different times.

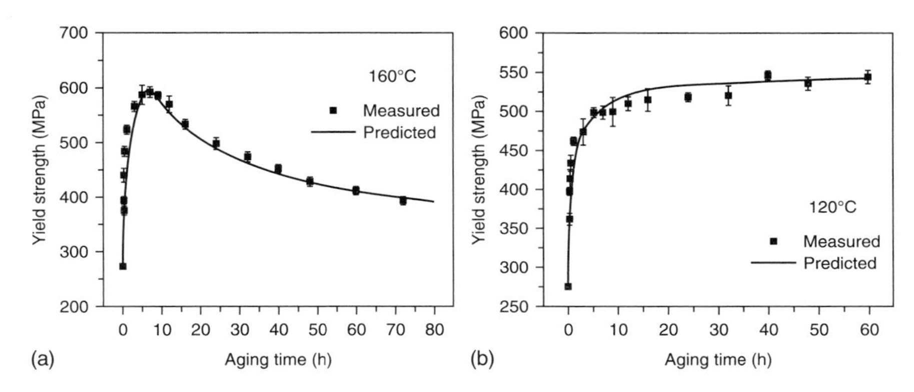

10.24 Yield strength measured by experiment compared with model predictions for AA7055 Al alloy aged at (a) 160°C and (b)120°C.

hardening, $\Delta\sigma_{\text{cutting, p}}$ is the strength increment related to the behavior of dislocations as they cut the sites of GPI zones and fine η' phase particles, $\Delta\sigma_{\text{passing, p}}$ is the strength increment related to the behavior of dislocations as they pass the sites of η phase and η' phase particles of larger size, and H is a parameter to adjust the relative intensity of $\Delta\sigma_{\text{cutting, p}}$ and $\Delta\sigma_{\text{passing, p}}$. Here, $\Delta\sigma_{\text{cutting, p}}$ and $\Delta\sigma_{\text{passing, p}}$ are calculated based on the parameters of the precipitates discussed above, including the type, size, volume fraction, shape, density and so on.

The results shown in Fig. 10.24 suggest that the yield strength of 7055 aluminum alloy in various states can be predicted accurately all through the aging process by using the model described above.

10.7 Conclusion and future trends

The future trends in aluminum alloys appear to be twofold. One aspect is that new aluminum alloys with high overall performance are being developed for the aerospace and aviation industries. The other aspect is that the overall performance and specific properties of aluminum alloys will be improved by using new processing methods and equipment. It is extremely important that the properties of aluminum alloys should be improved by means of tailoring and controlling the microstructure during casting and processing.

The effect of the microstructure on the properties of aluminum alloys is usually determined either on the scale of the grain structure or on the scale of precipitates. When high overall performance is desired, microstructure control must be carried out on multiple scales. Here, the control of grain shape, grain size, orientation, precipitation, grain boundary structure, dislocation structure and so on should be considered integrally (Sepehrband *et al.*, 2010). However, the effects of different scales of the microstructure on the properties are usually interconnected, and are sometimes consistent and sometimes contradictory. For instance, small grains and fine, coherent precipitates within grains are favorable for high strength; however, large grains are believed to be beneficial for damage tolerance, and coarsened precipitates on grain boundaries contribute to low SCC sensitivity. Thus, in order to understand the mechanisms that control the properties of aluminum alloys and to tailor them, the development of multiscale coupled modeling is unavoidable; it has been applied to understand the behavior with respect to fatigue, fracture and even corrosion.

Additionally, combining forming and heat treatment together may provide simultaneous control of the shape, microstructure and performance of parts. Combined processing is expected to be more efficient and simple, since it greatly shortens the procedure. Some studies have been carried out to develop combined processing. For instance, creep forming and aging have been combined to develop the technology of creep–age forming (CAF) for the fabrication of large parts with large radii of curvature (Pitcher and Styles, 2000; Zhu and Starke Jr, 2001a,b; Bakavos and Prangnell, 2004). In the CAF process, a part is elastically loaded into a mold and held at an artificial-aging temperature for a certain period of time. Stress relaxation occurs by creep, and thus permanent deformation occurs. By means of the CAF process, parts are formed and heat-treated simultaneously and the fabricated parts possess a precise shape, high mechanical performance and low residual stress, which satisfies the requirements of large size, long life and high reliability demanded by the aircraft industry. Some studies are being done on merging solid solution treatment and hot forming so that the parts can be used directly after being formed. Although the feasibility of doing this has been preliminarily confirmed, there is still a long way to go in developing reliable merged processes.

10.8 Acknowledgments

The authors gratefully acknowledge the contributions made to some sections and the proofreading of the text done by Dr Y. Y. Zong and Dr J. T. Jiang. We acknowledge Elsevier and Cambridge University Press for permission to reproduce some figures from their journals.

10.9 References

Bakavos D and Prangnell P B (2004), 'A comparison of the effects of age forming on the precipitation behaviour in 2xxx, 6xxx and 7xxx aerospace alloys', in Muddle B C, Morton A J and Nie J F, (eds), *9th International Conference on Aluminum Alloys*, 2–5 August 2004, Brisbane, Australia. North Melbourne: Institute of Materials Engineering Australasia, 124–131.

Bay N (1997), 'Cold forming of aluminium – state of the art', *Journal of Materials Processing Technology*, 71, 76–90.

Birol Y (2010), 'Impact of partial recrystallization on the performance of 6005A tube extrusions', *Engineering Failure Analysis*, 17, 1110–1116.

Blum W, Zhu Q, Merkel R and McQueen H J (1996), 'Geometric dynamic recrystallization in hot torsion of Al–5Mg–0.6Mn (AA5083)', *Materials Science and Engineering A*, 205, 23–30.

Browne D J and Battikha E (1995), 'Optimisation of aluminium sheet forming using a flexible die', *Journal of Materials Processing Technology*, 55, 218–223.

Cabibbo M, Evangelista E and Scalabroni C (2005), 'EBSD FEG-SEM, TEM and XRD techniques applied to grain study of a commercially pure 1200 aluminum subjected to equal-channel angular-pressing', *Micron*, 36, 401–414.

Cerezo A, Hirosawa S, Sha G and Smith G D W (2005), '3-D atomic-scale experimental and modelling studies of the early stages of precipitation', in Laughlin D E, Howe J M, Lee J K, Dahmen U and Soffa W A, (eds), *International Conference on Solid–Solid Phase Transformations in Inorganic Materials 2005: Diffusional Transformations*, 29 May–3 June, 2005, Phoenix, AZ. Warrendale, PA: The Minerals, Metals & Materials Society, 252–256.

Chen J Z, Zhen L, Yang S J and Dai S L (2010), 'Effects of precipitates on fatigue crack growth rate of AA7055 aluminum alloy', *Transactions of the Nonferrous Metals Society of China*, 20, 2209–2214.

Chen J Z, Zhen L, Yang S J, Shao, W Z and Dai S L (2009), 'Investigation of precipitation behavior and related hardening in AA 7055 aluminum alloy', *Materials Science and Engineering A*, 500, 34–42.

Dixit M, Mishra R S and Sankaran K K (2008), 'Structure–property correlations in Al 7050 and Al 7055 high-strength aluminum alloys', *Materials Science and Engineering A*, 478, 163–172.

Doughertya L M, Robertson I M and Vetrano J S (2003), 'Direct observation of the behavior of grain boundaries during continuous dynamic recrystallization in an Al–4Mg–0.3Sc alloy', *Acta Materialia*, 51, 4367–4378.

Du N, Qi Y, Krajewski P E and Bower A F (2009), 'The effect of solute atoms on aluminum grain boundary sliding at elevated temperature', *Metallurgical and Materials Transactions A*, 42, 651–659.

Eidel B (2011), 'Crystal plasticity finite-element analysis versus experimental results of pyramidal indentation into (001) fcc single crystal', *Acta Materialia*, 59, 1761–1771.

Fang G, Zhou J and Duszczyk J (2009), 'Extrusion of 7075 aluminium alloy through double-pocket dies to manufacture a complex profile', *Journal of Materials Processing Technology*, 209, 3050–3059.

Gault B, Moody M P, de Geuser F, Haley D, Stephenson L T and Ringer S P (2009), 'Origin of the spatial resolution in atom probe microscopy', *Applied Physics Letters*, 95, 034103.

Gourdet S and Montheillet F (2000), 'An experimental study of the recrystallization mechanism during hot deformation of aluminium', *Materials Science and Engineering A*, 283, 274–288.

Guyot P and Cottignies L (1996), 'Precipitation kinetics, mechanical strength and electrical conductivity of AlZnMgCu alloys', *Acta Materialia*, 44, 4161–4167.

Hirsch J and Laukli H I (2010), 'Aluminum in innovative light-weight car design', in Kumai S, Umezawa O, Takayama Y, Tsuchida T. and Sato T, (eds), *12th International Conference on Aluminum Alloys*, 5–9 September 2010, Yokohama, Japan. Tokyo: Japan Institute of Light Metals, 46–53.

Hu H E, Zhen L, Yang L, Shao W Z and Zhang B Y (2008), 'Deformation behavior and microstructure evolution of 7050 aluminum alloy during high temperature deformation', *Materials Science and Engineering A*, 488, 64–71.

Hurley P J and Humphreys F J (2003), 'The application of EBSD to the study of substructural development in a cold rolled single-phase aluminium alloy', *Acta Materialia*, 51, 1087–1102.

Kaibyshev R, Sitdikov O, Goloborodko A and Sakai T (2003), 'Grain refinement in as-cast 7475 aluminum alloy under hot deformation', *Materials Science and Engineering A*, 344, 348–356.

Kaibyshev R, Shipilova K, Musin F and Motohashi Y (2005), 'Continuous dynamic recrystallization in an Al–Li–Mg–Sc alloy during equal-channel angular extrusion', *Materials Science and Engineering A*, 396, 341–351.

Kamaya M, Wilkinson A J and Titchmarsh J M (2005), 'Measurement of plastic strain of polycrystalline material by electron backscatter diffraction', *Nuclear Engineering and Design*, 235, 713–725.

Lee E W, Kim N J, Jata K V and Frazier W E (1995), 'Corrosion behavior of 7055-T7751: a new high strength aluminum alloy for aircraft application', in Lee E W, Kim N J, Jata K V and Frazier W E, (eds), *3rd Symposium on Light Weight Alloys for Aerospace Applications*, 13–16 February 1995, Las Vegas. Warrendale, PA: The Minerals, Metals & Materials Society, 29–41.

Lin J P, An X Y and Lei T Q (1988), 'Dynamic recrystallization in aluminum alloys', *Metal Science & Technology*, 7,107–116.

Liu Q and Lin L (2010), 'Current status of research and industries of Al sheets in China', in Kumai S, Umezawa O, Takayama Y, Tsuchida T and Sato T, (eds), *12th International Conference on Aluminum Alloys*, 5–9 September 2010, Yokohama, Japan. Tokyo: Japan Institute of Light Metals, 20–29.

Luzzati V, Witz J and Nicolaieff A (1961), 'Détermination de la masse et des dimensions des protéines en solution par la diffusion centrale des rayons X mesurée à l'échelle absolue: Exemple du lysozyme', *Journal of Molecular Biology*, 3, 367–378.

Ma A, Saito N, Takagi M, Nishida Y, Iwata H, *et al.* (2005), 'Effect of severe plastic deformation on tensile properties of a cast Al–11mass% Si alloy', *Materials Science and Engineering A*, 395, 70–76.

Malerba L, Becquart C S, Hou M and Domain C (2005), 'Comparison of algorithms for multiscale modelling of radiation damage in Fe–Cu alloys', *Philosophical Magazine*, 85, 417–428.

Marceau R K W, Sha G, Ferragut R, Dupasquier A and Ringer S P (2010), 'Solute clustering in Al–Cu–Mg alloys during the early stages of elevated temperature ageing', *Acta Materialia*, 58, 4923–4939.

Marlaud T, Deschamps A, Bley F, Lefebvre W and Baroux B (2010), 'Evolution of precipitate microstructures during the retrogression and re-ageing heat treatment of an Al–Zn–Mg–Cu alloy', *Acta Materialia*, 58, 4814–4826.

McQueen H J and Blum W (2000), 'Dynamic recovery: sufficient mechanism in the hot deformation of Al (<99.99)', *Materials Science and Engineering A*, 290, 95–107.

McQueen H J and Hockett J E (1970), 'Microstructures of aluminum compressed at various rates and temperatures', *Metallurgical Transactions*, 1, 2997–3004.

Moody M P, Gault B, Stephenson L T, Marceau R K W, Powles R C, *et al.* (2011), 'Lattice rectification in atom probe tomography: toward true three-dimensional atomic microscopy', *Microscopy and Microanalysis*, 17, 226–239.

Pitcher P D and Styles C M (2000), 'Creep age forming of 2024A, 8090 and 7449 alloys', *Materials Science Forum*, 331–337, 455–460.

Quey R, Dawson P R and Barbe F (2011), 'Large-scale 3D random polycrystals for the finite element method: generation, meshing and remeshing', *Computer Methods in Applied Mechanics and Engineering*, 200, 1729–1745.

Ravichandran N and Prasad Y V R K (1991), 'Dynamic recrystallization during hot deformation of aluminum: a study using processing maps', *Metallurgical and Materials Transactions*, 22A, 2339–2348.

Rosen G I, Jensen D J, Hughes D A and Hansen N (1995), 'Microstructure and local crystallography of cold rolled aluminium', *Acta Metallurgica Materialia*, 43, 2563–2579.

Roy S, Singh D S, Suwas S, Kumar S and Chattopadhyay K (2011), 'Microstructure and texture evolution during accumulative roll bonding of aluminium alloy AA5086', *Materials Science and Engineering A*, 528, 8469–8478.

Sato T (2010), 'Innovative development of aluminum research and technologies in Japan', in Kumai S, Umezawa O, Takayama Y, Tsuchida T. and Sato T, (eds), *12th International Conference on Aluminum Alloys*, 5–9 September 2010, Yokohama, Japan. Tokyo: Japan Institute of Light Metals, 1–9.

Sepehrband P, Wang X, Jin H O and Esmaeili S (2010), 'Interactions between precipitation and annealing phenomena during non-isothermal processing of an AA6xxx alloy', in Kumai S, Umezawa O, Takayama Y, Tsuchida T. and Sato T, (eds), *12th International Conference on Aluminum Alloys*, 5–9 September 2010, Yokohama, Japan. Tokyo: Japan Institute of Light Metals, 308–313.

Stegailov V V, Kuksin A Y, Norman G E and Yanilkin A V (2009), 'Molecular dynamic modeling of plasticity of Al and Al–Cu alloys under dynamic loading', *AIP Conference Proceedings*, 1195, 781–784.

Vaithyanathan V, Wolverton C and Chen L Q (2007), 'Multiscale modeling of θ' precipitation in Al–Cu binary alloys', *Acta Materialia*, 52, 2973–2987.

Wang W, Murray J L, Hu S Y, Chen L Q and Weiland H (2007), 'Modeling of plate-like precipitates in aluminum alloys – comparison between phase field and cellular automaton methods', *Journal of Phase Equilibria and Diffusion*, 28, 258–264.

Williams J C and Starke Jr E A (2003), 'Progress in structural materials for aerospace systems', *Acta Materialia*, 51, 5775–5799.

Wilm A (1906), German patent DRP244554.

Xun Y and Tan M J (2004), 'EBSD characterization of 8090 Al–Li alloy during dynamic and static recrystallization', *Materials Characterization*, 52, 187–193.

Yamagata H (1995), 'Dynamic recrystallization and dynamic recovery in pure aluminum at 583 K', *Acta Metallurgica Materialia*, 43, 723–729.

Yamagata H, Ohuchida Y, Saito N and Otsuka M (2001), 'Nucleation of new grains during discontinuous dynamic recrystallization of 99.998 mass% aluminum at 453 K', *Scripta Materialia*, 45, 1055–1061.

Zhang Y Q, Shan D B and Xu F C (2009), 'Flow lines control of disk structure with complex shape in isothermal precision forging', *Journal of Materials Processing Technology*, 209, 745–753.

Zhong J (2002), 'Progress in the basic research of improving aluminum materials quality', *Light Alloy Fabrication Technology*, 30, 1–10.

Zhu A W and Starke Jr E A (2001a), 'Materials aspects of age-forming of Al–xCu alloys', *Acta Materialia*, 117, 354–358.

Zhu A W and Starke Jr E A (2001b), 'Stress aging of Al–xCu alloys: experiments', *Acta Materialia*, 49, 2285–2295.

11 Microstructure control in creep–age forming of aluminium panels

L. ZHAN, Central South University, China and
J. LIN and D. BALINT, Imperial College London, UK

Abstract: The chapter begins with an introduction to the creep–age forming (CAF) process and its importance in the aircraft and aerospace fields because of its obvious advantages for producing components comprising large high-strength aluminium alloy panels with aerofoil sections and complex curvatures. The importance of precipitation control in CAF is then presented. Testing methods for stress/strain ageing, and modelling methods for precipitation hardening are described. The applications of CAF are introduced and future trends in this field are discussed. A summary is given at the end of the chapter.

Key words: creep–age forming, precipitation control, stress ageing, precipitation hardening, modelling, aluminium alloys.

11.1 Introduction to the creep–age forming (CAF) process and its importance

By the middle of the twentieth century, traditional mechanical metal forming methods were showing themselves to be inadequate for producing components comprising large high-strength aluminium alloy panels with aerofoil sections and complex curvatures, such as those used in modern aircraft and aerospace metal structures. To deal with this problem, a new forming method was developed by Textron Aerostructures, which has been proven to be very useful for forming components with these shape characteristics and good mechanical properties. This method is called 'creep–age forming' (CAF) or 'autoclave age forming'. The fundamental mechanism on which it is based is the stress relaxation and/or creep phenomena that can arise when a metal alloy is artificially aged. This chapter is concerned exclusively with aluminium alloys, as, to date, they are virtually the only alloys for which CAF has been used.

Stress relaxation or creep takes place as a metal with internal stress is exposed to a high-temperature environment for a period of time. The degree of stress reduction or creep deformation depends on the initial stress level, the temperature, the ageing time and the metallurgical/mechanical behaviour of the material. Another phenomenon that can occur in some metal alloys is age hardening. This increases their strength when they are held at an elevated temperature in a supersaturated solid solution state for a period of time. Several series of aluminium alloys are artificially age-hardenable. The concept of creep–age forming is based upon an understanding of the above two phenomena and their simultaneous utilization. Ageing is a process that can increase the strength of the metal, while creep and/or stress relaxation during ageing is the mechanism used to form and retain the shape of the part. Significant research and development on CAF have been carried out over the last 20 years. This includes forming-equipment design (Holman, 1989; Brewer *et al.*, 1992; Levers, 2005), precipitate formation and growth (Lin *et al.*, 2006; Chen *et al.*, 2004), the microstructural mechanism of the forming process (Starink *et al.*, 2006; Zhu and Starke, 2001a), springback prediction (Jeunechamps *et al.*, 2006; Huang *et al.*, 2007; Jackson *et al.*, 2005), the evolution of mechanical properties during CAF (Zhu and Starke, 2001b), and other factors with a bearing on added value.

Recently, the popularity of creep–age forming as a method for forming large integrally stiffened lightweight structures for use in the aircraft and aerospace industries has been increasing significantly. Three stages are usually included in such a process: Firstly, a machined but undeformed workpiece is placed on a forming tool and a vacuum is created between the two so that the workpiece is forced to conform to the shape of the tool by atmospheric pressure acting on its upper surface. Owing to the small curvatures normally required in an aerospace component, the strain introduced in the workpiece is largely elastic. If the workpiece is strong, either a pressurized autoclave or mechanical clamping can be used to supplement the vacuum and force the workpiece into close contact with the tool surface. Secondly, the workpiece, still held against the tool by vacuum, is placed in an oven. Both the workpiece and the tool are heated to the ageing temperature of the alloy and held in the furnace for a period of time, typically between 24 and 48 h. Finally, the temperature is reduced to room temperature, the pressure is released and the workpiece is allowed to spring back to a shape somewhere between its undeformed shape and the tool shape (Holman, 1989). A typical creep–age forming process is shown schematically in Fig. 11.1. The loading and unloading phases are usually characterized by a linearly elastic response from the alloy, while the ageing phase is characterized by nonlinear stress relaxation and/or creep deformation (Sallah *et al.*, 1991).

Compared with conventional metal forming processes, CAF has the following advantages:

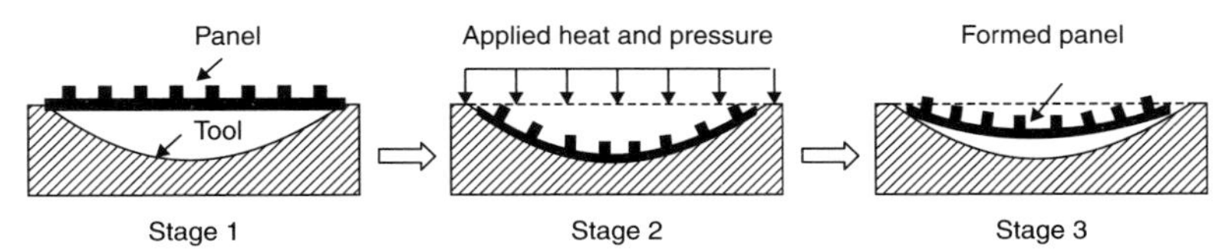

11.1 Creep–age forming process.

- Low residual stresses in formed parts, which enhance resistance to fatigue and stress corrosion cracking, and also component shape stability (Holman, 1989; Zhu and Starke, 2001b).
- Low forming stress. Usually, the part is stressed below the material yield stress, which decreases the possibility of forming processing cracks or the occurrence of plastic instability or rupture, compared with traditional plastic forming processes (Holman, 1989).
- Low forming equipment costs for manufacturing extra-large panel components. A large-capacity, expensive press would be required for forming large panel parts conventionally, which would be prohibitively costly, particularly for small batch-quantity production. Also, the furnace used for ageing incurs no significant extra cost, as an ageing furnace would be required to improve the mechanical properties of the alloy in any type of processing route. A pressurized autoclave would add an unknown cost, though.
- Age hardening and stress relaxation occur synchronously in the forming process, causing changes in the microstructure of the alloy and increasing its strength. This simplifies the part production procedure by combining two processes (age hardening and forming) into one, and thus the manufacturing cost is greatly reduced, with significant energy savings (Holman, 1989; Zeng *et al.*, 2008).
- Stress-oriented precipitation occurs during creep–age hardening, which offers an additional means for controlling the mechanical properties of high-strength aluminium alloys containing semi-coherent second-phase particles. That is to say, stress-ageing may provide a way to control anisotropy in high-strength aluminium alloys by balancing the effects of strong crystallographic texture (Hargarter *et al.*, 1998).

However, CAF also has some disadvantages, which currently constrain the range of component types that it can be used to manufacture. More research is required to overcome its disadvantages and expand its range of application. Identified disadvantages include:

- At present, it is recognized that CAF can be used only for metals which are age-hardening, such as the 2xxx, 6xxx and 7xxx series aluminium alloys (Bakavos *et al.*, 2004). However, it is believed that CAF could also be used for some other metals in which significant creep and/or stress relaxation could occur at their artificial-ageing or tempering temperature, if a heat treatment process is normally needed to improve their mechanical properties.
- A long processing time is needed to manufacture one component, and thus CAF can be used only to produce small batch quantities or single products. Process developments are required to enable many components to be produced in one heat treatment cycle.
- Springback of components is unavoidable and, currently, is too complicated to be accurately predicted, which makes tool design for net-shape manufacture difficult.

It has been found that stress-orientated precipitation decreases the yield strength of some metal alloys compared with those aged without an external force being applied under the same ageing conditions. For example, Zhu and Starke (2001b) pointed out that when stress–age forming of Al–Cu alloys was done, the yield strength of the stress-aged specimens was lower than that of the stress-free-aged specimens, for the same peak ageing conditions. This was true when the test direction was either parallel or perpendicular to the stress direction during stress-ageing. Similar results were obtained in tensile stress-ageing treatments for cube-textured polycrystalline Al–4Cu and Al–5Cu–0.8Mg–0.6Mn–0.5Ag alloys (Hargarter *et al.*, 1998).

Based on the present understanding of the CAF process and its forming mechanism, a definition of its application domains can be given as follows:

- Forming of panel parts with small curvatures and small plastic strain, such as aircraft wing panels. The mechanism of the forming of a panel part is high-temperature creep, or stress relaxation under initial elastic deformation/ loading conditions. It is common practice to constrain the initial material deformation to be within the elastic region in this forming process, to avoid any possible microdamage to the material due to plastic deformation. Thus the total strain that occurs during forming is relatively small.
- For metals, significant creep or stress relaxation can take place below the yield strength at the ageing or heat treatment temperature. Creep and stress relaxation can take place at stresses below the yield stress in many metals, such as heat-treatable aluminium alloys, nickel-based superalloys and titanium alloys, at their ageing temperature.

Compared with conventional metal forming methods, CAF is a low-cost forming technique, which utilizes lower forming forces at the ageing temperature while producing dimensionally stable contoured panels containing low levels of residual stress.

11.2 The importance of precipitation control in CAF

11.2.1 Precipitation hardening in CAF

CAF is a combined process of artificial ageing and stress-induced deformation of the workpiece. During the artificial-ageing period in CAF, constituents of the metal precipitate out, altering the microstructure of the material, which improves its mechanical properties, for instance by an increase in the yield strength and tensile strength. This mechanism is called 'age hardening' or 'precipitation hardening'. Because of the simultaneous increase in strength with relaxation of the metal during CAF, the actual deformation is different from the conventional creep/stress relation behaviour. Ho (2004) compared conventional creep deformation with creep in CAF. That author pointed out that in conventional creep studies, lifetime and creep damage are investigated for engineering components working at high temperatures, which have been artificially aged to their highest strength for creep resistance. Thus the initial creep rate is low. But in CAF, the material is initially in a solution-treated and quenched state, and is relatively soft. Thus the primary creep rate is higher in CAF than in the conventional creep deformation of structural components. Consequently, more creep strain is accumulated in a CAF process, which is a favourable aspect of the forming process which tends to reduce springback. As the material undergoes age hardening and also stress relaxation, the creep rate decreases gradually and becomes constant when the secondary creep stage is reached.

11.2.2 Stress/strain age (precipitation) hardening

The age hardening and stress-induced creep deformation that arise in CAF are simultaneous with, and are affected by, hardening due to stress/strain precipitation, in two ways. First, the increase in strength of the workpiece material inhibits creep and stress relaxation. Second, there is an effect on the stress level (varying levels can arise when different curvatures on bending radii are formed) and direction on precipitate nucleation and growth, which contributes to the mechanical properties of age-formed components. This effect has been investigated extensively for many age-hardening alloy systems. Zhu and Starke (2001a) investigated the effect of the stress level on precipitation hardening and the resulting yield strength of Al–*x*Cu alloys using the experimental fixture shown in Fig. 11.2.

In this study, a series of Al–*x*Cu samples of different sizes were mounted in the fixture, under a compressive spring load, and put into a furnace at a particular ageing temperature for each alloy. The creep deformation of materials aged at

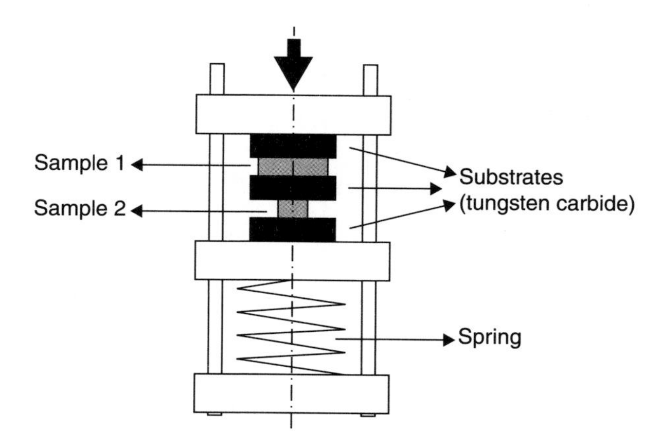

11.2 Schematic representation of the stress-ageing sample fixture used by Zhu and Starke (2001a).

different temperatures (160, 180, 201 and 220°C) and compressive stress levels (0, 9, 12, 31, 60, 96 and 146 MPa) was investigated. It was found that a relatively high stress, for example greater than 100 MPa for the Al–4Cu alloy, was required for creep to take place during the ageing period.

Other research on stress-ageing conditions has shown that the stress direction affects the plate-like θ'' or θ' precipitates in Al–4Cu alloys (Zhu and Starke, 2001b). Figure 11.3(a) shows the effect of compressive stress on the precipitate distribution. It was observed that (1) ageing under an applied stress generates an aligned θ''/θ' precipitate structure with a higher number density and larger diameter of favoured plates than of unfavoured plates; (2) compressive stress leads to a structure dominated by perpendicular θ''/θ' plate precipitates, whereas tensile stress favours the formation of parallel θ''/θ' plates; and (3) the yield strength of a stress-aged specimen was lower than that of a stress-free-aged specimen under the same peak ageing conditions, whether the test direction was parallel or perpendicular to the ageing-stress direction (see Fig. 11.3(b)). It was also found that the alignment of the plate-shaped precipitates led to lower yield strength values in the stress-aged specimens compared with the stress-free-aged specimens, where randomly oriented plates (on their habit planes) were formed.

Hargarter *et al.* (1998) investigated the effect of preferential alignment of plate-shaped precipitates on the yield strength anisotropy of Al–Cu–Mg–Ag and Al–Cu alloys. It was found that externally applied stresses during ageing result in preferential precipitation of θ' and Ω on those variants of the $\{100\}$ and $\{111\}$ habit planes which form the smallest angle with the load. An experimental comparison of yield strength between conventionally aged and tensile-stress-aged materials was carried out, and the results are shown in Fig. 11.3(b). It can be seen that the yield strength of the conventionally aged material is higher than that of the material that was aged while being stressed in tension.

Bakavos *et al.* (2004, 2006) carried out experimental research, using industrial-scale age-forming tools, to evaluate the through-thickness microstructure variation of AA7475 and AA2022. Plates of the two alloys with a thickness of 20 mm were age-formed on a single-curvature tool with a radius of 3180 mm.

The outer surface of the test piece was subjected to a tensile stress and the inner surface to a compressive stress. The maximum tensile and compressive principal stresses at the outer and inner surfaces were ±185 MPa at the beginning of the CAF process. The results showed that for AA2022, a very strong preferential alignment of the θ' phase was formed parallel to the direction of the tensile principal stress and perpendicular to the direction of the compressive principal stress at the outer and inner surfaces; no alignment of the Ω phase was detected. Also, it was suggested that the neutral axis shifted towards the outer (tensile-stress) surface of the plate. For AA7457, no precipitate alignment could be detected in the samples, and significant particle coarsening occurred in the most highly stressed regions (at the surfaces). The average precipitate size was measured at five positions from the top to the bottom surface and the results are

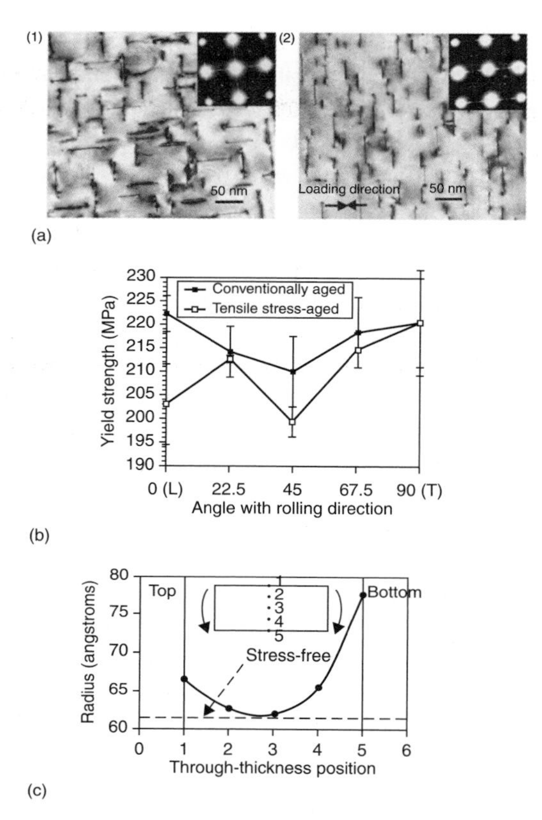

11.3 Effect of stress on age-precipitation hardening: (a) TEM bright-field images and diffraction patterns of (1) conventionally aged and (2) stress-aged (40 MPa) Al–4Cu single crystals aged at 201°C for 11 h (Zhu and Starke, 2001b). (b) Yield strength under slightly over-aged conditions without (conventionally aged) and with an applied tensile stress of 50 MPa (stress-aged) (Hargarter *et al.*, 1998). (c) Mean particle radius of η' precipitates in AA7475 (Eberl *et al.*, 2008).

shown in Fig. 11.3(c). The top surface was subjected to tensile stresses and the precipitate size there was bigger than in the unstressed material, but not as big as that at the bottom surface, which was in compression. This indicates that the precipitates grew more quickly under compressive stress in AA7475. At the neutral surface (location 3 in Fig. 11.3(c)), where the stress and strain were zero, the precipitate size was similar to that in the material aged under stress-free conditions, which is denoted by the dashed line in Fig. 11.3(c).

Related work was undertaken by Eberl *et al.* (2008), who compared the evolution of precipitation hardening between stress-aged and conventionally aged control samples of AA6056, AA2022 and AA7475 alloys, using transmission electron microscopy and small-angle scattering techniques. Some of the results obtained were as follows. (1) No differences could be detected for the AA6056 alloy. (2) The AA2022 alloy showed precipitate alignment for the stress-aged samples, but this had no impact on the static properties, and neither could any anisotropy be detected in the alloy. (3) In the AA7475 alloy, some evolution of the precipitate alignment at intermediate stages of the ageing treatment could be detected. Alignment was not evident in the fully aged metal, but the stress-aged sample (aged at a stress of 245 MPa) had slightly poorer static mechanical properties (by about 6%), and an average particle size of 68 Å was measured, compared with 62 Å for the conventionally aged sample.

The effects of ageing time and stress level on the precipitate size in AA7055 were investigated by Zhan *et al.* (2011a). Some important results were obtained. With increasing ageing time, the size of platelet-shaped precipitates (known as Guinier–Preston or GP zones) and spherical-shaped η' precipitates increased, and the number of platelet-shaped precipitates increased too, while the number of η' precipitates decreased. Precipitates of almost the same size were distributed continuously and uniformly on the grain boundaries, with quite narrow precipitate-free zones (PFZs), at 5 h of ageing time. When the ageing time was increased to 8 h, the distribution of the precipitates on the grain boundaries became discontinuous and the PFZs became wider, and with a further increase in the ageing time, the precipitates became distributed continuously on the grain boundaries again and the PFZs became wider, as shown in Fig. 11.4. There was no evident alignment of the precipitates at the end of the stress-ageing period.

The evolution of precipitate size with stress level and ageing time was studied, as shown in Table 11.1 (Zhan *et al.*, 2011a). From comparison of the stress-aged alloy with the stress-free-aged alloy, it was found that after 20 h of ageing, the precipitates in the stress-aged alloy were slightly coarser and the PFZs were wider; the density of retained η' was lower, with the development of η phase, as shown in Fig. 11.5 (Zhan *et al.*, 2011a).

The evolution of the yield strength under different stress/creep–age conditions ($\sigma = 0$, 252.2 and 308.9 MPa) was also studied, as shown in Fig. 11.6 (Zhan *et al.*, 2011b). It can be seen that significant hardening takes place during thermal

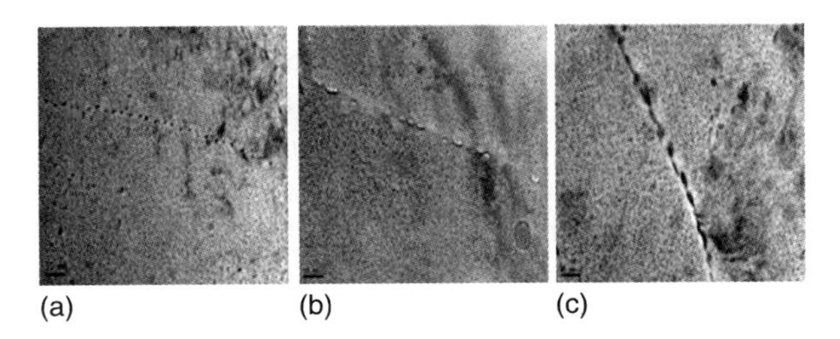

11.4 TEM microstructure of AA7055 creep-aged under a stress of 250 MPa at 120°C for (a) 5 h; (b) 8 h; (c) 20 h.

Table 11.1 Average precipitate size after ageing at different stress levels and for different ageing times at 120°C

Stress level (MPa)	Time (h)	Average precipitate size (nm)
0	20	4.5
190	20	6.2
	5	5.8
250	8	6.8
	20	9.0

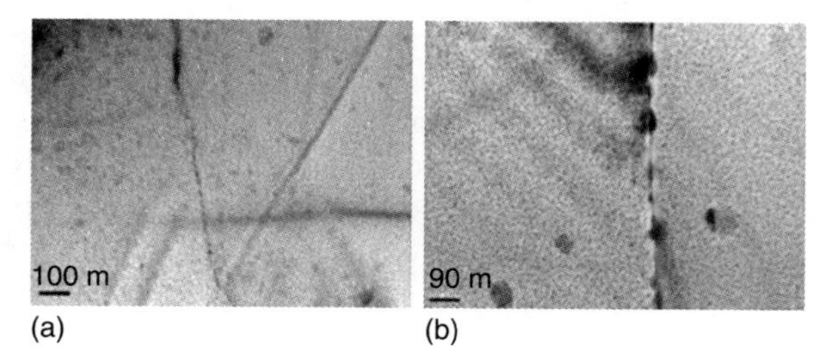

11.5 TEM micrographs of 7055 alloy creep-aged under different stresses at 120°C for 20h at (a) 0 MPa; (b) 190 MPa.

exposure at a constant stress. The hardening can be attributed to two mechanisms: age hardening due to precipitation from supersaturated solid solution in the material matrix, and work hardening due to an increase in dislocation density (i.e. creep deformation). Owing to these hardening mechanisms, initially the yield strength of the material increases with time and reaches a peak value; after that, over-ageing occurs and the yield strength gradually decreases. Two important phenomena can be perceived from Fig. 11.6. First, the yield strength increases more quickly under stress-ageing conditions than under stress-free conditions (σ = 0 MPa) in the initial ageing stage. The yield strength varies from 513 to 528 MPa when the external tensile stress changes from 252.2 to 308.9 MPa after 5 h of stress-ageing, while the yield strength is only 470 MPa for stress-free ageing at the same ageing time. Second, the ageing time required to reach peak strength decreases in an obvious way with increasing external stress. The peak ageing time is 27.5 h for stress-free conditions and 21.9 h and 19.5 h for stress-ageing at 252.2 and 308.9 MPa, respectively. Thus it can be concluded that creep-ageing with stress promotes the hardening (precipitation) process, and the ageing time needed for the material to reach its peak strength decreases accordingly. The higher the stress level, the shorter the ageing time for reaching peak strength.

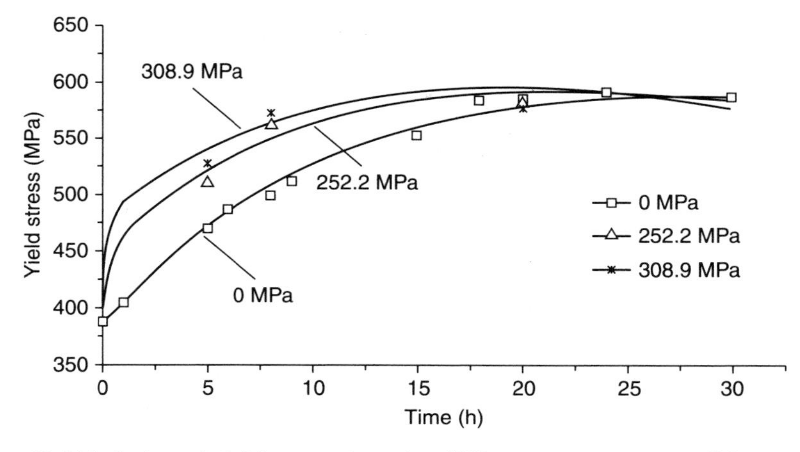

11.6 Variation of yield strength under different creep–age conditions.

In summary, experimental evidence has revealed that stress and creep deformation do affect the mechanical properties of aged aluminium alloys. The precipitate orientations are related to the stress direction (tension or compression) for some alloys, particularly AA2022. The creep deformation and the stresses enable precipitates to grow more quickly. Thus over-ageing may take place in regions with high stress levels.

11.3 Testing methods for stress/strain ageing

In age forming, the temperature of the workpiece is elevated and artificial ageing occurs *concurrently* with the deformation strains that arise as the workpiece conforms to the shape of the die with which it is deformed (Holman, 1989). Stress relaxation takes place by thermally activated plastic flow and/or creep, but the total strain (and the workpiece shape) remains virtually constant during the age-forming process. In this process, the applied loading is usually lower than 60% of the tensile strength. As mentioned before, both creep and stress relaxation happen during the CAF process. Thus constant-stress creep-ageing tests and constant-strain stress relaxation ageing tests are two major methods that are relevant to CAF. Jambu *et al.* (2002) studied creep–age forming of non-age-hardenable alloys such as AlMgSc alloys and 5xxx alloys, used for fabrication of fuselage shells. Both creep tests and stress relaxation tests were carried out at elevated temperatures to study the thermal stability and mechanical properties of AlMgSc alloys during CAF. The experimental results showed that the duration of the process can be reduced to less than half of that of the original CAF process without deterioration of the properties of the alloy.

11.3.1 Constant-stress creep-ageing test

The constant-stress creep-ageing test is a method designed to investigate both the creep and the ageing behaviour of metal alloys under constant stress for a controlled amount of time at the peak ageing temperature. This test is very similar to the conventional creep test, apart from the fact that the material used is not artificially aged; it is only solution heat-treated, and then quenched. Therefore, the material is expected to be less strong initially but to exhibit a lot of hardening during the test. The hardening can be attributed to ageing due to thermal exposure and to creep deformation. At the end of the test, the material's yield strength is expected to increase owing to the hardening mechanisms that operate. The experimental results are usually used as data to develop a new set of physically based, unified creep-ageing constitutive equations, capable of modelling both the physical and the mechanical behaviour of the alloy during creep–age forming, which will be discussed in the following section. Ho (2004) conducted creep-ageing tests on AA7010 using an ESH constant-stress creep machine. Similar constant-stress creep-ageing tests were carried out on AA7055 by Zhan *et al.* (2011b) using a Mayes 100 kN testing machine and the VCP software package.

The most popular experimental procedure, as shown in Fig. 11.7 (Zhan *et al.*, 2011b), can be described briefly as follows:

- First, the specimen is fitted and aligned in the middle of a furnace, and a thermocouple is wired in the middle of the specimen gauge length.
- The furnace is closed and the heating is switched on. Thermal cotton is used to cover the top and bottom of the furnace to reduce heat loss. The closed

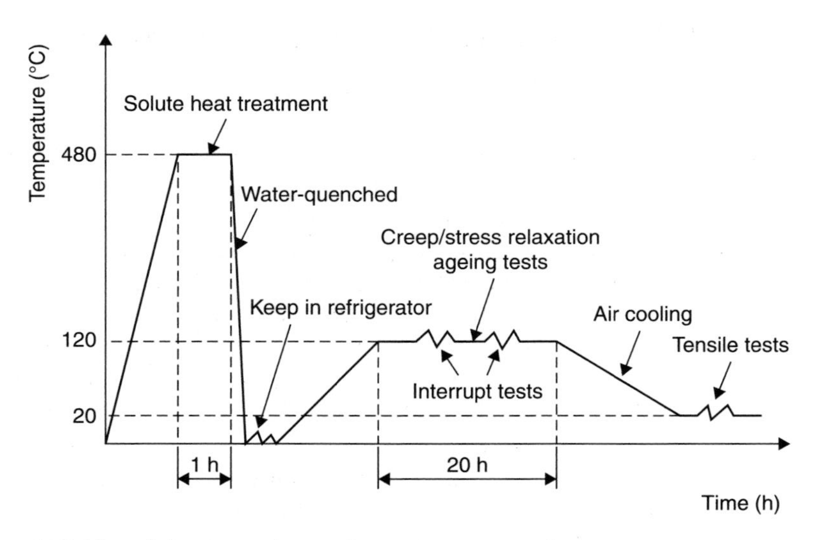

11.7 Material preparation and test programme for constant-stress creep-ageing tests (Zhan *et al.*, 2011b).

furnace takes up to 1–2 h to rise from room temperature to a steady peak ageing temperature (for example, 120°C for AA7055).
- When the temperature becomes steady at the peak ageing temperature of the alloy (120°C for AA7055), a load is applied and the elongation of the specimen is measured.
- The extension is measured every 10 s for the first 30 min. The time interval is then increased to 60 s for the rest of the experimental period.
- The data logger is stopped when the time reaches the alloy's peak ageing time (20 h for AA7055). The heating is switched off, the furnace is opened and the load is removed.

The creep-ageing strains for AA7055 specimens at 120°C are plotted in Fig. 11.8 (Zhan *et al.*, 2011b). All of the creep-ageing strain curves have a similar shape. The strain increases with time but at a decreasing rate throughout the test. This kind of pattern is typically observed in the primary stage of conventional creep deformation.

11.3.2 Constant-strain stress relaxation ageing test

The stress induced during the forming of large-curvature components is sometimes very high, and the stress level at or near the sheet surface may exceed the material's yield strength. In a CAF process, stresses will be relieved by thermal exposure. The higher the initial stress level, the higher the stress relaxation rate. As time goes on, the stress relaxation rate decreases according to the pattern of a power law or a sinh law. Therefore, the stress relaxation behaviour is dominant during the early period of CAF, where the time is too short for a significant amount of time-dependent creep deformation to take place. Thus constant-strain stress

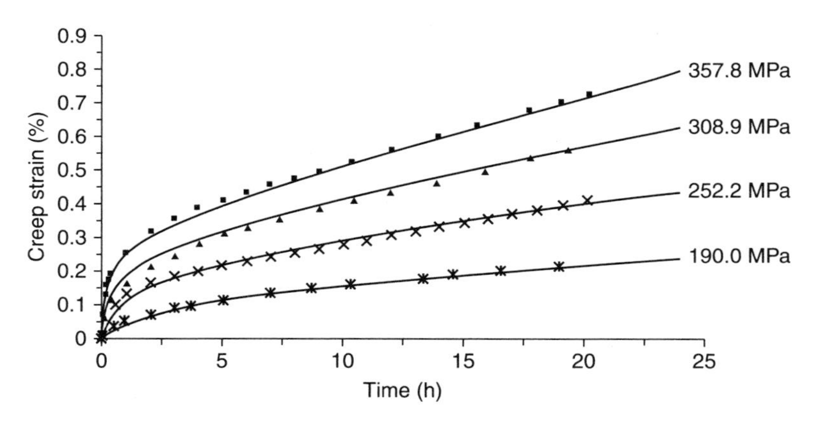

11.8 Curves for creep-ageing under different stress levels (Zhan *et al.*, 2011b).

relaxation ageing tests are also a very important part of the investigation of the stress relaxation behaviour in CAF.

Ho (2004) carried out uniaxial stress relaxation ageing tests using a Zwick 1484 machine. The machine was equipped with a set of twin recirculating balls and had a maximum loading capacity of ±200 kN. A closed, indirect resistance-heated air furnace, connected to a Eurotherm heat generator, was used to provide constant heating to the specimen. During the test, a thermocouple was fitted close to the centre of the specimen gauge length to monitor the temperature. The load, displacement and force data were logged to the computer attached, and a chart recorder was used to record the force history throughout the experiment. The experimental procedure is described below:

- The specimen was fitted onto the grips. The crosshead was adjusted slowly and carefully, so that the specimen was just tight in place and any slack was removed.
- The furnace was closed and the heating was switched on. Thermal cotton was used to cover the top and bottom of the furnace to reduce heat loss. Heating for up to 1 h was necessary for the furnace temperature to become steady at 150°C.
- When the temperature became steady, a preload of 100 N was applied to make sure that no slipping would occur during the test, which would have affected the tension-loading results later.
- Once the preload was achieved, loading of the specimen was continued at a constant crosshead stroke rate of 0.5 mm/min until the target stroke was achieved.
- When the target stroke was achieved, the crosshead was kept at the same position, while the force relaxed. The holding time was 24 h.
- Throughout the experiment, the force history was recorded using the chart recorder connected to the Zwick machine through the computer.
- After 24 h, the recording was stopped and the results were processed.

In stress relaxation ageing tests, the stress level of the material drops sharply from the maximum stress during the initial period of the test. The relaxation rate decreases as the stress level decreases. This kind of relaxation pattern can be modelled using a power law or sinh law. Constant-strain stress relaxation ageing curves for AA7055 specimens at 120°C are plotted in Fig. 11.9 (Zhan *et al.*, 2011c).

11.3.3 Comparison between constant-stress and constant-strain ageing

Until now, most work has focused on constant-stress creep-ageing tests to build a set of constitutive equations for CAF. A comparative study of constant-stress and constant-strain ageing was carried out for AA7055 by Zhan *et al.* (2011c).

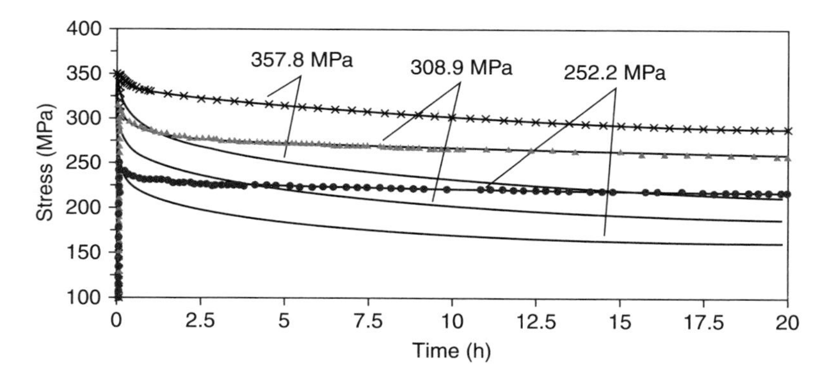

11.9 Constant-stress strain relaxation ageing curves under different initial stress levels (Zhan *et al.*, 2011c).

In order to study the similarities and dissimilarities between the creep behaviour and the stress relaxation behaviour of age-formed aluminium alloys, both creep and stress relaxation experiments were conducted with plate-shaped 7055 aluminium alloy specimens on a 100 kN tensile testing machine at 120°C for 20 h, under stress levels from 190 to 350 MPa. Similar trends were observed in the variation of the creep and the stress relaxation behaviour. Both creep and stress relaxation curves can be divided into two stages. During the first stage, a higher creep rate and stress relaxation rate occur, which increase with stress level but decrease with ageing time. During the second stage, both the creep rate and the stress relaxation rate reach their lowest values and then remain constant. A detailed description has been given by Zhan *et al.* (2011c).

It is widely believed that both thermally activated stress relaxation and creep take place during the ageing period. Figures 11.10(a)–(c) show corresponding stress relaxation and creep curves and the stress–strain relationship during age forming. The stress relaxation (as shown in Fig. 11.10(a)) may be due to thermally activated diffusion, dislocation recovery and/or creep. During the age-forming process, the stress level (below the yield stress) in the workpiece reduces from σ_1 to σ_2, even though the total strain ε_T remains constant throughout the period of ageing. The residual elastic strain ε_e springs back after unloading, while the amount of inelastic strain ε_{in} is responsible for shaping the part. Heat, stress and time are the three main factors in creep. Under age-forming conditions, creep deformation occurs (as shown in Fig. 11.10(b)), which transforms part of the initial elastic strain into plastic strain and accumulates continuously as time passes at higher stresses and ageing temperatures.

Most of the previous work regards creep as the only reason for stress relaxation. Thus constant-stress creep-ageing tests are usually used to describe deformation behaviour during the age-forming process and to predict the springback of age-formed panels. However, as mentioned before, during age forming, it is the total

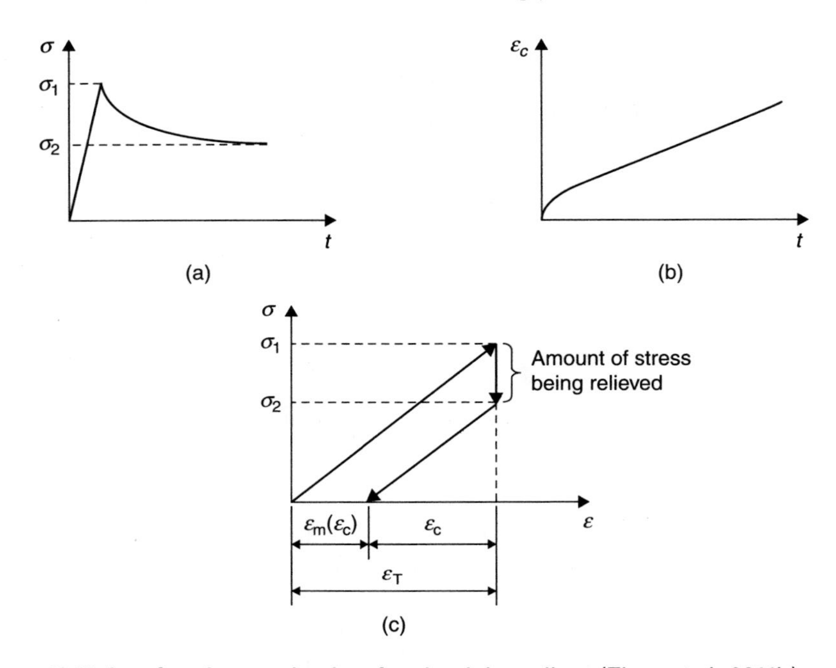

11.10 Age-forming mechanism for aluminium alloys (Zhan *et al.*, 2011b). (a) Stress relaxation; (b) creep deformation; (c) the stress relaxation that occurs during age forming.

strain and not the stress that remains constant throughout the period of ageing. That is to say, constant-strain stress relaxation ageing should be more appropriate than constant-stress creep ageing to describe age-forming processes.

In order to study the correlation between creep and stress relaxation under age-forming conditions, a new set of unified physically based creep-ageing constitutive equations has been formulated based on an understanding of creep forming and the hardening mechanisms of age forming. A multi-step fitting process using experimental data from creep-ageing tests was then used to obtain the material constants. As can be seen from Fig. 11.8, good fitting results (solid curves) were achieved between the constitutive equations constructed and the experimental results from creep-ageing tests. Figure 11.8 shows a comparison of the predicted (solid curves) and experimental (symbols) creep strains for four stress levels. It can be observed that very close agreement is achieved. The accumulated errors are less than 2%.

The constitutive equations were then used to predict the stress relaxation behaviour under three initial stress levels, as shown in Fig. 11.9 (solid curves). Some disagreements were found between the predicted and experimental stress relaxation data. The predicted stress relaxation results are much lower than the experimental results, which shows that there is not a one-to-one correspondence between creep and stress relaxation: creep deformation is the major but not the only reason for stress relaxation under age-forming conditions. Thus, in order to study the deformation and related springback behaviour in the age-forming process, not only creep-ageing tests but also stress relaxation tests should be carried out at the same time.

11.4 Modelling of precipitation hardening

11.4.1 Mechanisms of precipitation hardening

To model precipitation hardening, springback and the evolution of the mechanical properties of materials during CAF, it is important to have fully determined, physically based, unified creep-ageing constitutive equations. Thus an understanding of the relevant forming and precipitation-hardening mechanisms is important.

Precipitation hardening

Precipitation hardening, or age hardening, refers to a mechanism that takes place at the artificial-ageing stage of a heat treatment process, which is commonly used to improve the mechanical properties of commercial aluminium alloys. It is only applicable to heat-treatable aluminium alloys.

Heat-treatable alloys

Heat-treatable alloys are those that can achieve higher strength by heat treatment. Alloys in this group contain one or more elements chosen to give higher strength by precipitation hardening. The heat-treatable alloys include aluminium alloys in the 2000 (Al–Cu–Mg), 6000 (Al–Mg–Si) and 7000 (Al–Zn–Mg–Cu) series, which are commonly used for aerospace and automotive applications.

Heat treatment

The purpose of heat treatment is to increase the strength and hardness of an alloy. It is designed to alter the mode of occurrence of the soluble alloying elements, particularly copper, magnesium, silicon and zinc, which can combine with one another to form intermetallic compounds. Figure 11.11(a) shows a binary phase diagram of a heat-treatable aluminium alloy. A typical heat treatment involves the following stages (Ho, 2004):

1. Solution heat treatment (SHT) at a relatively high temperature T_0 within the single-phase (α) region in Fig. 11.11(a), to dissolve the alloying elements.
2. Rapid cooling, or quenching, across the solvus line, usually to room temperature. This leads to a metastable supersaturated solid solution (SSSS) at T_1. The equilibrium structure should be $\alpha + \beta$, but limited diffusion does not allow the β phase to form.

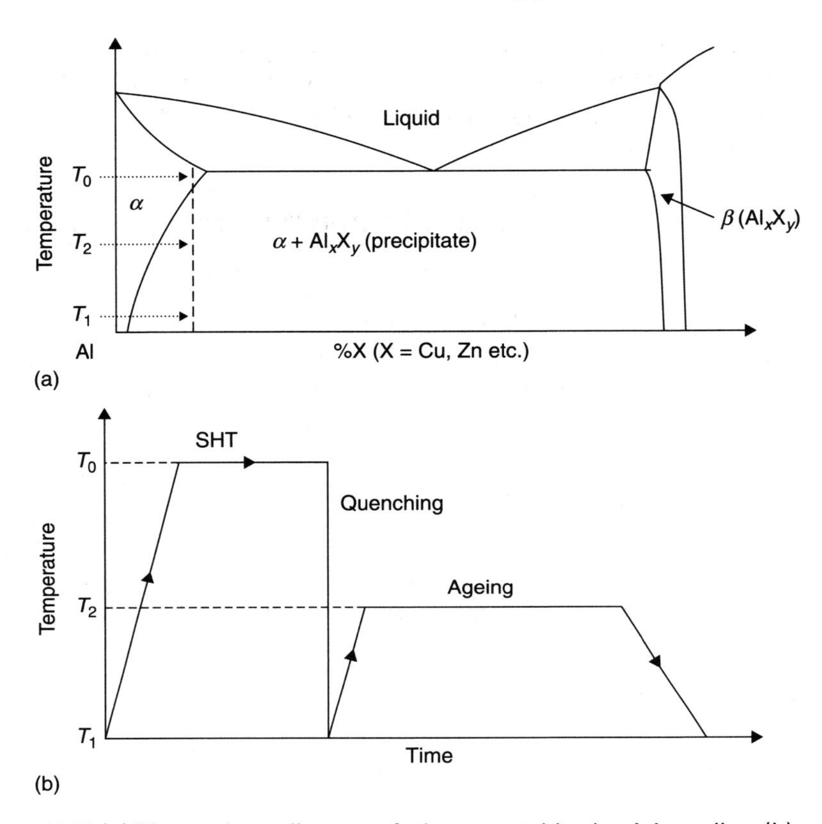

11.11 (a) Binary phase diagram of a heat-treatable aluminium alloy. (b) Typical temperature profile for the heat treatment.

3. Ageing heat treatment, where the SSSS is heated to T_2. The SSSS decomposes to form finely dispersed precipitates. This stage is usually achieved by artificial ageing for a convenient time at one, or sometimes two, intermediate temperatures.

Figure 11.11(b) shows a typical temperature profile during a full heat treatment process. Because of the complex interactions that take place during the processes of heat treatment, different alloys have different characteristics that require careful selection and control of the heating operations and specific combinations of temperature and time if the required properties are to be achieved in the heat-treated product.

Microstructure evolution in precipitation-hardening mechanism

The ageing behaviour of 7xxx aluminium alloys under isothermal heat treatment has been examined in detail by several investigators (Poole *et al.*, 1997; Ferragut

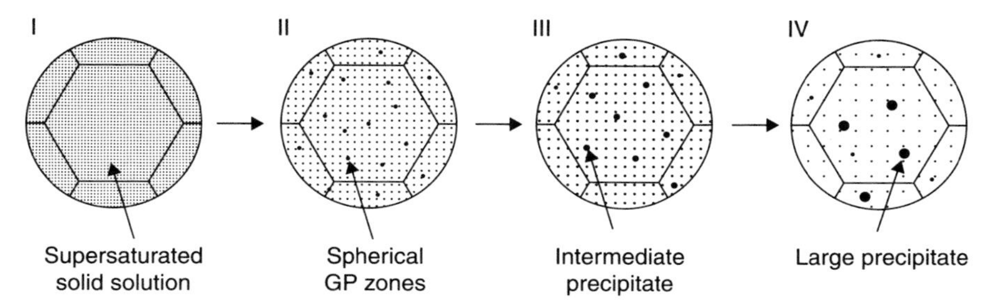

(I) Supersaturated solid solution, (II) nucleation of GP zones (coherent), (III) precipitate coarsening (semi-coherent), (IV) over-aged condition (incoherent).

11.12 Schematic representation of AA7xxx at different ageing stages.

et al., 1999; Kashyap *et al.*, 2000; Poole *et al.*, 2000). The generally accepted ageing sequence is

$$\alpha_{\text{solid solution}} \rightarrow \alpha + \text{spherical GP zones} \rightarrow \alpha + \eta' \rightarrow \alpha + \eta$$

where α is the aluminium matrix, η' is a transition phase and η is the equilibrium phase ($MgZn_2$). Figure 11.12 shows a schematic representation of the evolution of the microstructure of a 7xxx aluminium alloy undergoing an ageing (precipitation-hardening) process.

The age-hardening mechanism can be divided into several stages, as shown in Fig. 11.12. The quenching process retains the supersaturated solid solution in the aluminium matrix (stage I), as there is insufficient time for precipitates to nucleate in a water-quenched aluminium alloy. The initial yield strength reflects contributions from the intrinsic strength and solute hardening.

In the early stages of ageing, ordered solute-rich clusters, called GP zones, form (stage II). GP zones are only one or two atom planes in thickness; they have a similar crystal structure to that of the Al matrix and are coherent. Because of this, their interfacial energy is low, making their nucleation easy. The yield strength starts to increase as precipitates nucleate and coarsen, but because the GP zones are small and coherent, they can be cut by dislocations. During this stage, the number of precipitates increases drastically with time. The difference in atomic size between zinc and aluminium strains the lattice. The hardening that occurs is therefore due to the increased work required to move dislocations through the strained lattice and the work required for dislocations to pass through the GP zones. As the coarsening process proceeds, the supersaturation of the SSSS in the matrix phase decreases. This decrease in the concentration of solute atoms in the matrix results in a decrease in the solute hardening. However, the decrease in solute hardening is less than the increase due to precipitation and dislocation hardening. Therefore, the overall strength of the material continues to increase with time as the precipitates continue to increase in size at constant volume fraction.

Eventually, the GP zones themselves are replaced by the more stable η' phase, i.e. stage III in Fig. 11.12. The typical size of the precipitates is between 1 and 10 nm (Ferragut *et al.*, 1999), and they are semi-coherent with the Al matrix. During this stage, the size of the precipitates increases and the amount of their constituents in the matrix phase steadily decreases, approaching the equilibrium value. As the precipitates become larger, the coherent particles lose their coherency and the dislocation–precipitate elastic interaction is diminished, while the precipitate spacing becomes larger. There is no further increase in strength due to a decrease in the concentration of solute in the matrix, as this concentration has reached its equilibrium value. The precipitation hardening reaches its maximum value at this stage, when the precipitates reach the 'peak ageing' size (the best match of precipitate radius and spacing), and thus the peak strength is obtained.

During stage IV, the number of precipitates decreases and the precipitate spacing increases further; here, the equilibrium η phase forms as large incoherent particles. These particles are large enough to be bypassed by dislocations by the Orowan bowing process without being cut. The strength of the material begins to decrease. The material is now in an over-aged condition.

Figure 11.13 shows precipitates in three 7000 series aluminium alloys in the peak aged condition after ageing at 150°C (Caraher *et al.*, 1998). The dark and more or less spherical particles in each alloy are η' precipitates. The diameter of the largest precipitate is less than 50 nm. Figure 11.14 shows the evolution of the precipitate radius in AA7010 and a ternary alloy (a modified AA7010 alloy) during ageing at 160°C (Deschamps *et al.*, 1999). From the figure, it can be seen that the precipitate radius grows with ageing time at a temperature of 160°C.

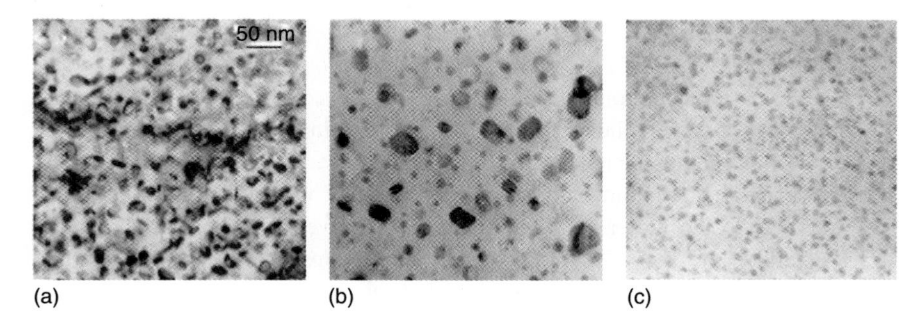

11.13 Microstructure of 7000 series aluminium alloys for (a) Al–1.72Zn–3.4Mg–0.1Ag, (b) Al–1.72Zn–3.4Mg–0.37Cu and (c) Al–1.72Zn–3.4Mg–0.1Ag–0.37Cu in peak aged condition after ageing at 150°C (Caraher *et al.*, 1998).

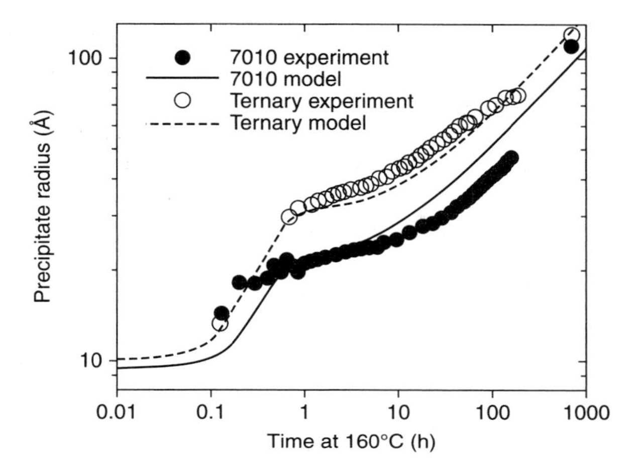

11.14 Evolution of precipitate radius during ageing at 160°C (Deschamps *et al.*, 1999).

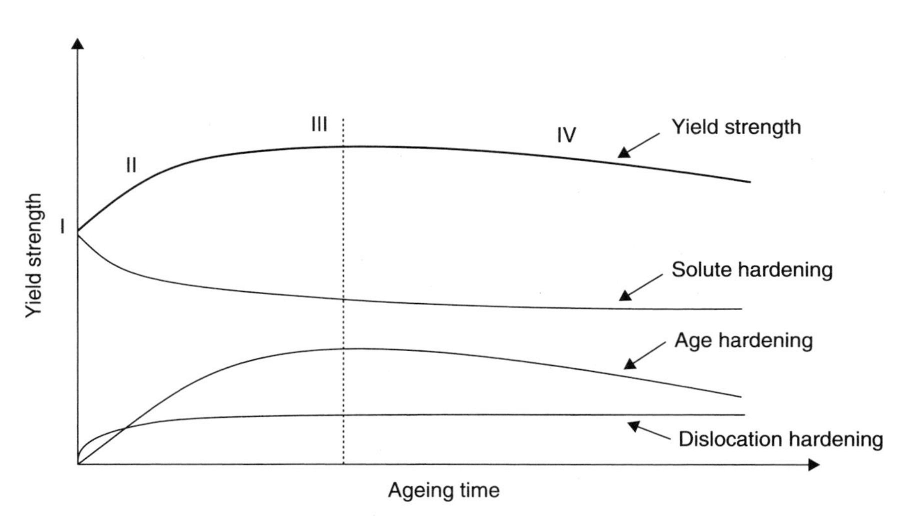

11.15 Contributions of solute hardening, age hardening and dislocation hardening to the yield strength of AA7055 as a function of ageing time.

Contributions to yield strength

The precipitation or ageing mechanism improves the yield strength of the material. Figure 11.15 shows a schematic diagram of the relative contributions of solute hardening, age hardening and dislocation hardening to the yield strength of a 7xxx aluminium alloy, and the corresponding ageing stages.

At $t = 0$, there has been insufficient time for precipitates to nucleate in the quenched aluminium alloy. The initial yield stress reflects contributions from the intrinsic stress and solute hardening.

The yield strength of the material starts to increase as precipitates start to nucleate and coarsen. Another mechanism contributing to hardening is due to dislocations, which are generated by creep deformation. At the ageing temperature, annealing takes place, which reduces the dislocation density. Thus the dislocation density reaches a saturated or balanced condition at the end of primary creep. This implies that dislocation hardening increases quickly in the initial stage of the creep-ageing process and reaches a fixed value when secondary creep starts, as shown in Fig. 11.15. Dislocations also accelerate the process of precipitate formation and growth.

As the precipitate-coarsening process proceeds, the overall strength of the material increases with time, as explained earlier. Eventually, however, there is no further increase in strength, because the concentration of solute in the matrix has reached its equilibrium value.

As the particles (precipitates) become larger, the coherent precipitates lose their coherency and the dislocation–precipitate elastic interaction is diminished. In addition, the decreasing number of precipitates and the increase in mean intercentre spacing between precipitates cause a decrease in the particle strengthening and an overall decrease in the strength of the material. Thus, the strength of the material reaches a maximum value and begins to decrease. As coarsening continues, the material's strength continues to decrease (see stage IV in Fig. 11.15 for illustration). Using a 7000 aluminium alloy as a model material, the research described below concentrates on the early stages of the ageing mechanism, where the material is proceeding towards its peak strength at stage III in Fig. 11.15.

11.4.2 Constitutive modelling of precipitation hardening

Over the years, the modelling of precipitation hardening in aluminium alloys has been widely studied by a number of researchers (Shercliff and Ashby, 1990a,b; Poole *et al.*, 1997; Deschamps *et al.*, 1999; Ferragut *et al.*, 1999; Myhr *et al.*, 2001). The models presented by these authors incorporate physical models for the evolution of the precipitation and for the relationship between the microstructure and the strength of the alloy. In summary, the components of these models include expressions for particle coarsening, solute hardening, dispersion hardening due to precipitate shearing, and precipitate bypassing.

Modelling of precipitate radius growth

During the ageing period, coarsening is described by the growth of the mean precipitate radius r_p with time. To date, no attempt has been made to include changes in particle shape or in the distribution of precipitate size during coarsening, although it is acknowledged that these can be important. It is assumed that a single precipitate type dominates the mechanical behaviour throughout the whole precipitation or ageing period, where the precipitates coarsen at constant volume fraction, losing coherency and changing structure as they do so. For a single peak ageing curve, the coarsening kinetics can be approximated by a single kinetic equation, and, using a cubic coarsening law, the evolution of the mean precipitate radius r_p with ageing time is given by Wagner (1961) as below:

$$r_p^3 - r_{p0}^3 = c_1 \left[\frac{1}{T} \exp\left(-\frac{Q_A}{\kappa T} \right) \right] t \qquad [11.1]$$

where r_{p0} is the initial mean precipitate radius at the ageing time $t = 0$, c_1 is a material constant, T is the temperature, κ is the universal gas constant and Q_A is an activation energy for ageing. The term $[(1/T) \exp(-Q_A/\kappa T)]t$ is called the kinetic strength P, of the thermal process. For practical ageing times, $r_p^3 >> r_{p0}^3$, and Eq. 11.1 can be rewritten as

$$r_p = c_1 P^{1/3} \qquad [11.2]$$

For isothermal ageing, the term [. . .] in Eq. 11.1 is constant. The growth rate of the precipitate radius can be obtained by differentiating Eq. 11.1 with respect to time, which gives

$$\dot{r}_p = c_2 r_p^{-2} \qquad [11.3]$$

where c_2 combines all the material constants. Equation 11.3 summarizes the growth rate of the mean precipitate radius during isothermal, static (stress-free) ageing. This equation shows that as the mean precipitate radius increases, the growth rate of the precipitate radius decreases.

Modelling of material strength increment

As discussed before, the strength of the workpiece material increases during the artificial-ageing process. The main contribution to the increment of material strength is from solute hardening and dispersion hardening. The contribution to the strength due to solute hardening, discussed by Shercliff and Ashby (1990), is given by

$$\Delta\sigma_{SS} = c_3 \bar{c}^{2/3} \qquad [11.4]$$

where $\bar{c}$ is the mean concentration of solute atoms in the aluminium matrix and c_3 is related to the size, modulus and electronic mismatch of the solute atoms.

The contribution to the strength due to a volume fraction f of precipitates of mean radius r_p depends on the type of dislocation interaction. For shearing of precipitates by dislocations, it is described by

$$\Delta\sigma_{ps} = c_4 f^{1/2} r_p^{1/2} \qquad [11.5]$$

while for precipitates which are bypassed by dislocations, it is described by

$$\Delta\sigma_{pb} = c_5 \frac{f^{1/2}}{r_p} \qquad [11.6]$$

where c_4 and c_5 are constants for a given alloy system. By substituting Eq. 11.2 into both Eq. 11.5 and Eq. 11.6, and introducing P_p for normalization (where $P/P_p = 1$ at peak strength), for precipitate growth at constant volume fraction, Eqs 11.5 and 11.6 can be rewritten as

$$\Delta\sigma_{ps} = 2S_0\left(\frac{P}{P_p}\right)^{1/6} \text{ and } \Delta\sigma_{pb} = 2S_0\left(\frac{P}{P_p}\right)^{-1/3} \qquad [11.7]$$

Here, all the unknown constants are combined into a single parameter S_0, with dimensions of strength. A convenient mathematical form for defining the net contribution of precipitation to the strength is given by

$$\Delta\sigma_{ppt} = \left[\frac{1}{\Delta\sigma_{ps}} + \frac{1}{\Delta\sigma_{pb}}\right]^{-1} \qquad [11.8]$$

Substituting from Eq. 11.7,

$$\Delta\sigma_{ppt} = \frac{2S_0(P/P_p)^{1/6}}{1+(P/P_p)^{1/2}} \qquad [11.9]$$

At the peak strength, $P/P_p = 1$ and the precipitate strength is S_0.

The net contribution to the yield strength σ_Y of a material subjected to artificial ageing is given by

$$\sigma_Y = \sigma_{SS} + \sigma_{ppt} \qquad [11.10]$$

In the modelling approach discussed above, the thermodynamics and kinetics of solution heat treatment, precipitation and coarsening have been established and the interaction between dislocations and the strength-giving precipitates has been elucidated. However, all proposed models are capable only of modelling the static ageing effect (stress-free ageing); there is as yet no overall model available that can predict the dynamic ageing behaviour (i.e. age hardening related to the creep strain rate) during CAF. During CAF, as the creep deformation increases, the dislocation density increases, and this will enhance or accelerate the age-hardening mechanism (Poole *et al.*, 1997). Thus the development of unified constitutive equations that can model both static and dynamic ageing effects (i.e. so-called 'creep-ageing' behaviour) is a must for better prediction of material behaviour during CAF.

11.4.3 Development of physically based unified creep-ageing constitutive equations

Unified creep-ageing constitutive equations

Many mechanism-based constitutive equations for creep damage have been developed over the past decades, but their derivation and use have been aimed mainly at conventional creep damage situations in structures exposed to high temperature (Kowalewski *et al.*, 1994). They are not appropriate for modelling the CAF process, since different information is required in the two different situations. Constitutive relations established for conventional sheet-forming processes, such as those described by Asnafi (2001) and Xue *et al.* (2001a,b), are not suitable for CAF either, as they contain no consideration of the influence of precipitation hardening on mechanical properties. To meet the specific needs of CAF technology, research has been carried out to investigate and establish new constitutive equations for materials under creep–age forming conditions.

Owing to the complications and limitations of classical plasticity theory, a different approach called 'unified theories' has been proposed and practised (Chaboche, 1989; Lin and Yang, 1999; Li *et al.*, 2002). In this approach, the classical separation of the strain into a time-independent plastic strain and a time-dependent creep strain is replaced by a total inelastic strain. All aspects of inelastic deformation, such as plasticity, hardening, creep and recovery, are treated by a set of equations. These theories often make use of internal state variables subjected to evolution rules and are suitable for a broad range of applications.

Ho *et al.* (2004) proposed a set of mechanism-based unified creep-ageing constitutive equations. This was the first attempt to model the interacting effects of creep and precipitate nucleation, growth and hardening. But this equation set is only suitable for 7xxx aluminium alloys with spherical precipitates.

Based on the 'unified theories' approach and ageing kinetics, and assuming isothermal ageing and particle growth at constant volume fraction, a new set of physically based, unified creep-ageing constitutive equations may take the form shown below. For ease of description, the equation set is listed first (Zhan *et al.*, 2011b):

$$\dot{\varepsilon}_{\mathrm{c}} = A_1 \sinh\{B_1[|\sigma|(1-\bar{\rho}) - k_0\sigma_{\mathrm{y}}]\}\mathrm{sign}\{\sigma\} \qquad [11.11]$$

$$\dot{\sigma}_{\mathrm{A}} = C_{\mathrm{A}}\dot{\bar{r}}^{m_1}(1-\bar{r}) \qquad [11.12]$$

$$\dot{\sigma}_{\mathrm{SS}} = C_{\mathrm{SS}}\dot{\bar{r}}^{m_2}(\bar{r}-1) \qquad [11.13]$$

$$\dot{\sigma}_{\mathrm{dis}} = A_2 \cdot n \cdot \bar{\rho}^{n-1}\dot{\bar{\rho}} \qquad [11.14]$$

$$\sigma_{\mathrm{y}} = \sigma_{\mathrm{SS}} + \sqrt{\sigma_{\mathrm{A}}^2 + \sigma_{\mathrm{dis}}^2} \qquad [11.15]$$

$$\dot{\bar{r}} = C_r(Q-\bar{r})^{m_3}(1+\gamma_0\bar{\rho}^{m_4}) \qquad [11.16]$$

$$\dot{\bar{\rho}} = A_3(1-\bar{\rho})|\dot{\varepsilon}_c| - C_p\bar{\rho}^{m_5} \quad [11.17]$$

where A_1, B_1, k_0, C_A, m_1, C_{SS}, m_2, A_2, n, C_r, Q, m_3, γ_0, m_4, A_3, C_p and m_5 are material constants. Equation 11.11 describes the evolution of creep strain. The creep rate is not only a function of the stress σ and the dislocation density $\bar{\rho}$, but also a function of the age (precipitation) hardening σ_A, the solute hardening σ_{SS} and the dislocation hardening σ_{dis}, which together contribute to the material's yield strength σ_Y, which varies during a CAF process.

In the present model, the yield strength σ_Y derives from three sources, σ_{dis}, σ_A and σ_{SS} (see also Fig. 11.15). Deschamps *et al.* (2005) used the precipitate radius to simplify the modelling of the ageing mechanism as the precipitates evolve or grow monotonically during isothermal ageing. However, two problems arise owing to this simplification. First, the initial precipitate radius is difficult to determine and varies with different aluminium alloys. Second, according to classical ageing mechanisms (Ringer and Hono, 2000), in addition to the precipitate radius, the space between the precipitates is another important factor that influences precipitation-hardening behaviour, and this should also be considered in the hardening equations. Thus a more general parameter, the normalized precipitate size, is introduced here:

$$\bar{r} = \frac{r}{r_c} \quad [11.18]$$

where r_c is the precipitate size in the peak ageing state, which considers the best match of precipitate size and spacing for the alloy. When $0 \le \bar{r} < 1$, under-ageing occurs; $\bar{r} = 1$ represents peak ageing, and $\bar{r} > 1$ over-ageing. This approach simplifies the modelling process significantly. The evolution of the normalized precipitate size is given in Eq. 11.16, which will be detailed later.

The term $(1 - \bar{\rho})$ in Eq. 11.11 is used to model the primary creep, which is the contribution of the dislocation density to the creep rate. This is the same as the effect of strain hardening on creep. But the variation of the dislocation density is related to dynamic and static recovery, in addition to creep deformation. A detailed description of the normalized dislocation density will be given later.

Equations 11.12 and 11.13 represent the evolution of age hardening and solute hardening, which are described in terms of the normalized precipitate size $\bar{r}$ and its evolution rate $\dot{\bar{r}}$ (Eq. 11.16). The contribution to strengthening from shearable precipitates can arise from a variety of mechanisms, such as chemical hardening and coherency strain hardening. However, the overall contribution to strengthening from various mechanisms is summarized in Eq. 11.12, where C_A describes the interaction between dislocations and shearable precipitates. Equation 11.13 approximates the contribution from solid solution strengthening (solute hardening), where resistance is caused by solute atoms that obstruct dislocation motion. In Eq. 11.13, C_{SS} is a constant related to the size, modulus and electronic mismatch of the solute, and m_2 describes the depletion of solute into precipitates. As the concentration of the solute atoms decreases, the solid solution strengthening decreases, acting more or less like a 'softening' mechanism. Equation 11.14 describes the evolution of dislocation hardening, which is a function of the normalized dislocation density $\bar{\rho}$, which is defined by (Lin *et al.*, 2005)

$$\bar{\rho} = \frac{\rho - \rho_i}{\rho_m} \qquad [11.19]$$

where ρ_i is the dislocation density in the virgin material (the initial state), and ρ_m is the maximum (saturated) dislocation density that the material could have. Thus ρ varies from ρ_i to ρ_m. This has the result that the normalized dislocation density $\bar{\rho}$ varies from 0 (the initial state) to 1 (the saturated state) under the condition that $\rho_i << \rho_m$. The evolution of the normalized dislocation density is given in Eq. 11.17. The first term in Eq. 11.17 represents the development of a dislocation density due to creep deformation and dynamic recovery. The second term gives the effect of static recovery on the dislocation density at higher temperatures (Lin *et al.*, 2005).

Combining Eqs 11.12, 11.13 and 11.14 gives Eq. 11.15, which describes the overall contribution to the yield strength for a material undergoing an ageing process. The key feature of Eq. 11.15 is that the yield stress σ_Y changes dynamically during the creep-ageing period owing to age and dislocation hardening. This kind of behaviour could not be modelled using the conventional viscoplastic constitutive equations (Lin and Yang, 1999).

Equation 11.16 describes the evolution of the precipitate radius under isothermal ageing conditions. As precipitates evolve or grow monotonically during isothermal ageing, the coarsening kinetics of the ageing mechanism can be modelled using an equation for the growth of the normalized precipitate size, Eq. 11.16, in which the nucleation and growth of precipitates are related to the dislocation density given in Eq. 11.17. Because CAF is a combination of creep deformation and age hardening, Eq. 11.16 includes the dislocation density effect to describe the dynamic ageing behaviour. That is, as the normalized dislocation density (or creep deformation) increases, the age-hardening effect increases. The effect of the dislocation density on precipitate nucleation and growth is controlled by the parameters γ_0 and m_4 in the equation. For stress-free ageing conditions (no creep), the normalized dislocation density is zero, and thus Eq. 11.16 models only the static ageing behaviour. Comparing Eq. 11.16 with the classical equation for the growth rate of the mean precipitate radius given by Wagner (1961), it can be seen that they are of different forms but describe a similar mechanism, where the normalized precipitate size grows at a decreasing rate.

The form of Eq. 11.16 was chosen because it has several advantages. First, the normalized precipitate size $\bar{r}$ is used to substitute for the conventional precipitate radius r_p, which can provide a synthetic consideration of the effect of both precipitate radius and precipitate spacing if an optimal precipitate size, referring to the peak ageing strength of different aluminium alloys, can be provided. Second, to avoid numerical difficulties, the normalized precipitate size has been placed in the numerator instead of in the denominator. Third, the term Q, a material constant for a given alloy, is used to represent the saturation limit for depletion of zinc solute atoms in the aluminium matrix. In a practical ageing mechanism, when the depletion of zinc solute atoms eventually reaches its saturation value, precipitation will stop; hence the precipitates will stop growing. Fourth, the power term m_3 gives flexibility to the equation, as different alloys behave differently. Most importantly, Eq. 11.16 is capable of describing both static and dynamic ageing behaviour, while conventional precipitate growth equations can model only static ageing behaviour.

Determination of material constants

The proposed creep-ageing constitutive equations consist of a set of nonlinear ordinary differential equations. They cannot be solved analytically. Hence, a numerical integration method was used to solve the equations (Cao *et al.*, 2008). The determination of material constants in unified constitutive equations is not an easy job, and significant efforts have been made over the years (Lin and Yang, 1999; Cao and Lin, 2008; Li *et al.*, 2002). However, to simplify the determination process and speed up convergence, the whole process was divided into a number of stages (Zhou and Dunne, 1996) for individual problems. A multiple-step fitting process was used to obtain the material constants efficiently.

Step 1: Determine the material constants related to the normalized precipitate size using experimental results on the evolution of the yield strength with time under stress-free conditions.

Step 2: Determine the material constants related to the yield strength using experimental results on the evolution of the yield strength with time under different stress levels.

Step 3: Determine the material constants related to the creep strain rate according to the initial and minimum creep rates at different stress levels.

Step 4: Re-evaluate all previously determined material parameters according to experimental creep data and yield strength data, and optimize the previously determined material parameters in steps 1, 2 and 3 to obtain the best-fit results.

11.4.4 Finite element (FE) modelling of precipitation hardening

Precipitation hardening is a key feature of CAF. The evolution of precipitates affects not only the creep/stress relaxation behaviour, but also the achievement of the material's final strength. As discussed before, CAF is based on a complicated combination of stress relaxation, creep and age hardening. Creep deformation takes place at low stress levels, and the amount of plastic deformation is directly related to the ageing temperature and time. This is significantly different from other metal forming processes, where elastic–plastic deformation of the material is dominant. For industrial applications, numerical studies on springback and precipitation hardening in CAF need to be available.

For the past 20 years, most of the finite element (FE) methods for simulation or numerical calculation have taken only macroscopic parameters, such as stress relaxation, creep strain, springback and so on, into consideration. The research carried out has concentrated mainly on the prediction of springback in CAF processes using simple creep or stress relaxation models. The coupling of ageing or precipitation hardening and creep or stress relaxation has not been considered.

Stress-ageing and creep-ageing phenomena in aluminium alloys have been understood further in recent years, and industry needs to be able to know and predict the mechanical-property distribution of creep–age-formed aircraft wing parts. Thus Ho *et al.* (2004) proposed a set of mechanism-based unified creep-ageing constitutive equations for AA7010, which were implemented into ABAQUS through the user-defined subroutine CREEP. In addition to creep deformation and stress relaxation, the precipitate growth during ageing and the evolution of the yield stress increment during CAF were predicted. Similar work has also been done by Zhan *et al.* (2011b) for AA7055 at 120°C, which will be discussed below.

11.4.5 Case study: verification of modelling and discussion

Multiaxial constitutive equations

In a manner similar to that for creep deformation (Lin *et al.*, 1993), the uniaxial sinh-law equations can be generalized by consideration of a dissipation potential function. First, consider Eq. 11.1 for the creep strain rate without the hardening and other state variables, which then reduces to

$$\dot{\varepsilon}_c = A_1 \sinh(B_1\sigma) \qquad [11.20]$$

Equation 11.20 can be generalized to multiaxial conditions by assuming an energy dissipation potential of the form

$$\psi = \frac{A_1}{B_1}\cosh(B_1\sigma_e) \qquad [11.21]$$

where $\sigma_e = (3S_{ij} \cdot S_{ij}/2)^{1/2}$ is the effective stress and the $S_{ij} = \sigma_{ij} - \delta_{ij}\sigma_{kk}/3$ are stress deviators. Assuming normality and the associated flow rule, the multiaxial relationship is given by

$$\frac{d\varepsilon^p_{ij}}{dt} = \dot{\lambda}\frac{\partial\psi}{\partial S_{ij}} = \frac{3A_1}{2}\left(\frac{S_{ij}}{\sigma_e}\right)\sinh(B_1\sigma_e) \qquad [11.22]$$

On reintroduction of the hardening and grain growth variables, the effective plastic strain rate $\dot{p}$ for the sinh-law material model can be written as

$$\dot{p} = A_1 \sinh\{B_1[\sigma_e(1-\bar{\rho}) - k_0\sigma_y]\} \qquad [11.23]$$

and then the set of multiaxial viscoplastic constitutive equations, integrated within a large-strain formulation, may be written as follows:

$$D_{ij}^{\mathrm{P}} = (3S_{ij} / 2\sigma_{\mathrm{e}})\dot{p} \qquad [11.24]$$

$$\dot{\sigma}_{\mathrm{A}} = C_{\mathrm{A}}\dot{\bar{r}}^{m_1}(1-\bar{r}) \qquad [11.25]$$

$$\dot{\sigma}_{\mathrm{SS}} = C_{\mathrm{SS}}\dot{\bar{r}}^{m_2}(\bar{r}-1) \qquad [11.26]$$

$$\dot{\sigma}_{\mathrm{dis}} = A_2 \cdot n \cdot \bar{\rho}^{n-1}\dot{\bar{\rho}} \qquad [11.27]$$

$$\sigma_{\mathrm{y}} = \sigma_{\mathrm{SS}} + \sqrt{\sigma_{\mathrm{A}}^2 + \sigma_{\mathrm{dis}}^2} \qquad [11.28]$$

$$\dot{\bar{r}} = C_{\mathrm{r}}(Q-\bar{r})^{m_3}(1+\gamma_0\bar{\rho}^{m_4}) \qquad [11.29]$$

$$\dot{\bar{\rho}} = A_3(1-\bar{\rho})\dot{p} - C_p\bar{\rho}^{m_5} \qquad [11.30]$$

$$\hat{\sigma}_{ij} = GD_{ij}^{\mathrm{e}} + 2\lambda D_{kk}^{\mathrm{e}} \qquad [11.31]$$

where D_{ij}^{P} is the rate of plastic deformation, $D_{ij}^{\mathrm{e}} = D_{ij}^{\mathrm{T}} - D_{ij}^{\mathrm{P}}$ is the rate of elastic deformation, D_{ij}^{T} is the rate of total deformation and $\hat{\sigma}_{ij}$ is the Jaumann rate of the Cauchy stress. G and λ are the Lamé elasticity constants. The multiaxial constitutive equations have been implemented into the large-strain finite element solver MSC.MARC through a user-defined subroutine and used to simulation creep–age forming of a 3D structural component.

FE model and numerical procedures

FE simulation of a model panel was carried out using a doubly curved tool shape. The material was AA7055 and was creep–age formed at a constant temperature of 120°C. In the CAF simulation, a flat metal sheet was located against the tool, the surface of which formed a cavity in the shape required. For convenience of locating the workpiece on the tool surface in the forming simulations, four soft springs, which could be compressed to zero volume, were used to support the weight of the workpiece. The FE model is shown in Fig. 11.16.

The FE model shown in Fig. 11.16 consisted of a square aluminium sheet and a rigid spherical tool surface connected by springs. The aluminium sheet had dimensions of 300 × 300 × 8 mm. A four-node quadrangular thin-shell element was used for the analysis. The initial rigid tool surface had a radius of 1000 mm in the x direction and 2000 mm in the z direction, and a maximum depth of 40.8 mm in the y direction. Friction, as a surface interaction property, was related to the contact pair by specifying a friction coefficient of 0.3 to simulate a non-lubricated condition. The equation set established was implemented in the commercial FE code MSC.MARC via the user-defined subroutine CRPLAW.

Multi-step numerical procedures were developed to simulate CAF and to evaluate the springback using MSC.MARC, as shown in Fig. 11.17. The procedures can be described as follows:

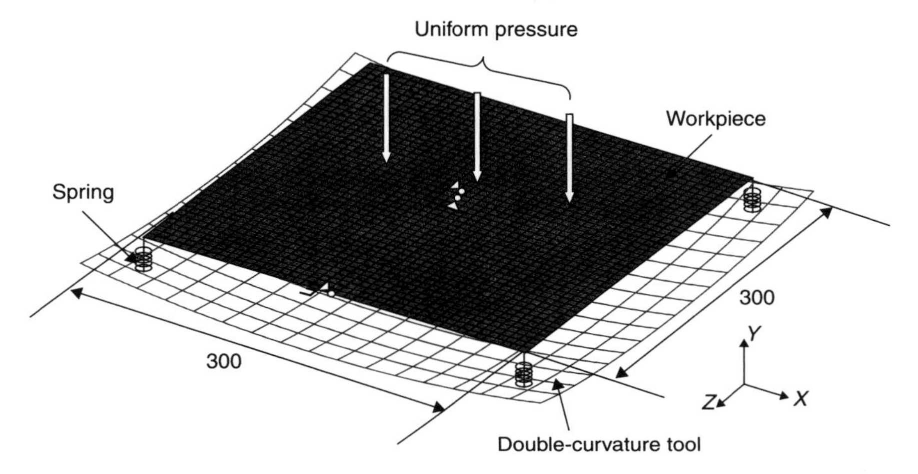

11.16 FE model with boundary and loading conditions.

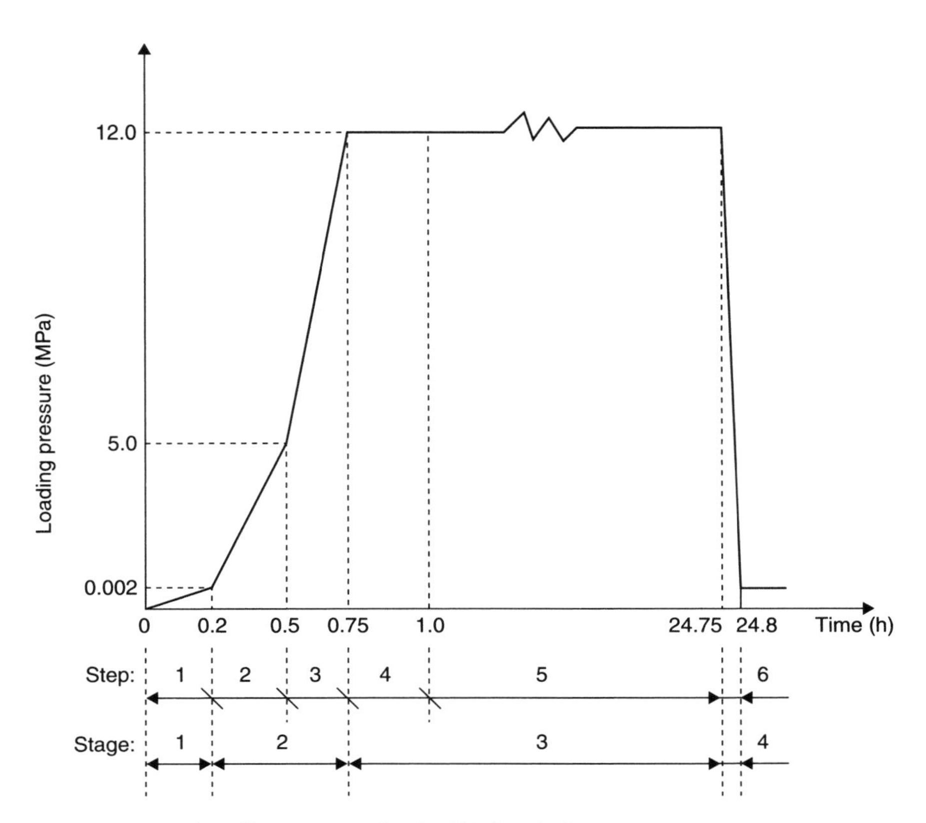

11.17 Loading process in the FE simulation.

Stage 1: Apply a small but sufficient uniform load on top of the aluminium panel to overcome the stiffness of the four springs and make sure that the four corners of the panel touch the tool surface.

Stage 2: Continue to apply a uniform pressure load to deform the panel until it is completely in contact with the tool surface.

Stage 3: Maintain the loading pressure to hold the aluminium sheet on the tool surface for a certain period, for example 24 h. This allows ageing, creep and stress relaxation to take place, which introduces plastic deformation and forms the panel part.

Stage 4: Remove the pressure incrementally and allow the aluminium panel to spring back. Data points for the panel after springback were collected.

Prediction of springback

Figure 11.18 shows the distribution of the surface von Mises stresses arising in the creep–age forming of an 8 mm thick double-curvature aluminium panel part. The total forming time was 24 h. Figure 11.18(a) shows the initial loading stage, where a lower pressure was applied to deform the whole workpiece into the tool shape. At this stage, the surface stress reached about 347.2 MPa. At this stress level, creep and stress relaxation took place very quickly initially. With the workpiece held to the tool shape for 24 h, significant creep took place, which reduced the stresses in the panel. The maximum stress was reduced to approximately 200.1 MPa from 347.2 MPa, as shown in Fig. 11.18(b), at the end of the creep-ageing period. Springback occurred upon release of the forming pressure, since the stresses were not fully relaxed by creep. Figure 11.18(c) shows the residual stress distribution after springback. The majority of the residual stress distribution was less than 42.1 MPa.

Modelling of creep strain, evolution of precipitate radius and yield strength

Figure 11.19 shows the distribution of the creep strain on the upper surface of the aluminium sheet during various stages of ageing. As shown in Fig. 11.19(a), after 1 h of creep ageing, the maximum creep strain of the material sheet reached approximately 0.167%. The creep strain was distributed evenly in the central part of the plate, where the stress level was expected to be high; the creep strain had a high value of approximately 0.16%. After 10 h of ageing (see Fig. 11.19(b)), the creep strain increased further and the high-creep-strain area slowly decreased. As ageing continued, the creep deformation became almost steady-state, and the increment of creep strain was not significant. At the end of the ageing period (t = 24 h), the major part of the aluminium sheet had a creep strain of around 0.348% and the stresses were reduced to close to the threshold stress of the material, $0.2\sigma_Y$ (see Fig. 11.18(b)).

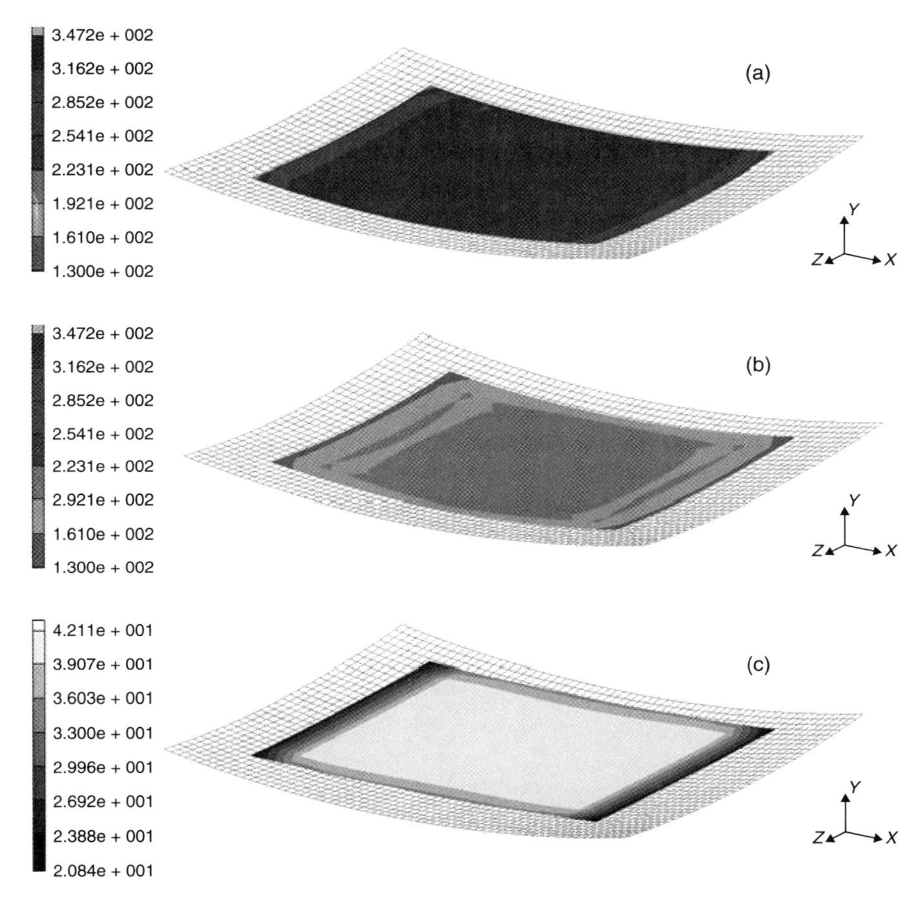

11.18 Distribution of von Mises stress (MPa) at different stages of the CAF process. (a) Panel deformed to its target shape at stage 2 (t = 0.75 h); (b) held for 24 h at stage 3 (t = 24.75 h); (c) after unloading at stage 4 (t = 24.8 h).

The effects of creep deformation on the evolution of the normalized precipitate size and yield strength of the material can also be predicted. The modelling results are shown in Fig. 11.20. The left-hand part of Fig. 11.20 shows the evolution of the normalized precipitate size during CAF. At an ageing time of t = 1 h, the normalized precipitate size is very small, and varies from 0.063 to 0.07. Upon thermal exposure, the normalized precipitate size increases with ageing time as creep deformation and age hardening take place simultaneously. For example, the increase in creep strain from t = 10 h to t = 24 h caused growth of the normalized precipitate size from 0.51 to 1.06.

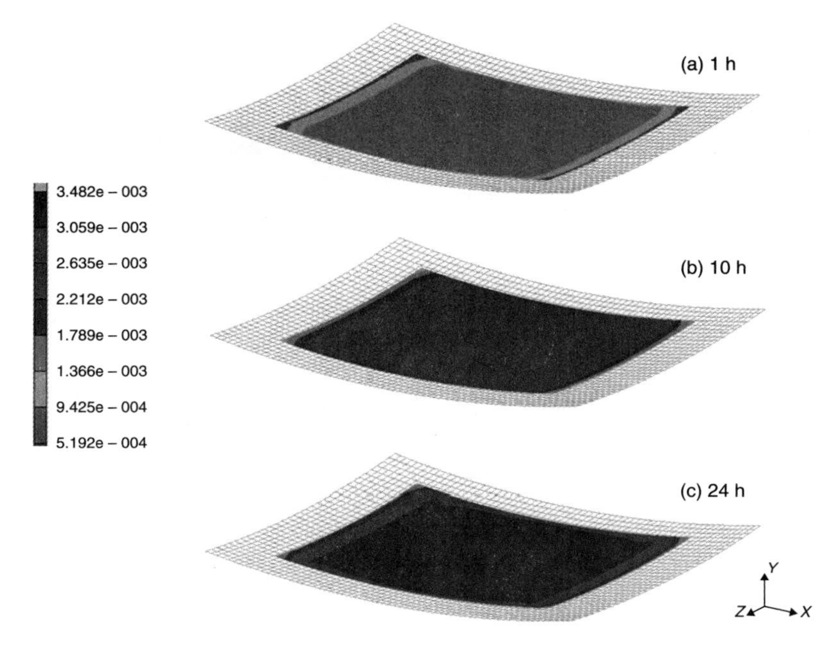

11.19 Distribution of creep strain during the ageing period at (a) 1 h, (b) 10 h and (c) 24 h.

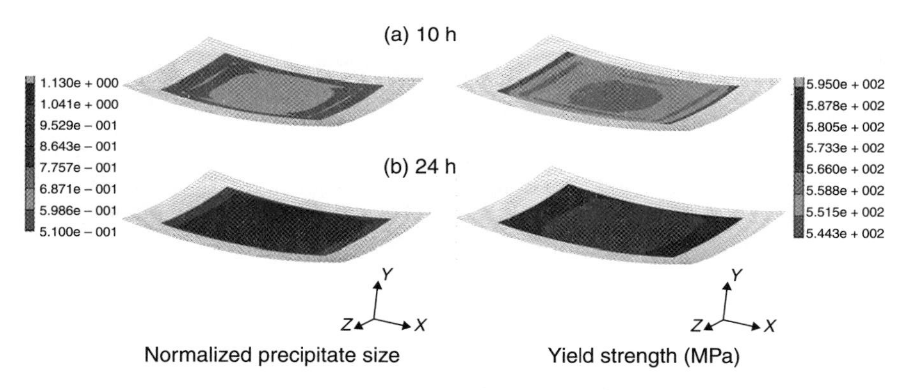

11.20 Variation and evolution of normalized precipitate size (left) and yield stress (right) during the ageing process at (a) 10 h and (b) 24 h.

The evolution of the yield strength during CAF is shown in the right-hand part of Fig. 11.20. The distribution of the yield strength has a pattern that is more or less similar to the distribution of the precipitate radius at the various stages. This is because, according to the constitutive equations, the increase in the yield strength of the alloy is mainly the result of the growth of the normalized precipitate size, although dislocation hardening increases the yield strength also, mainly at the primary creep stage. At the end of the CAF ($t = 24$ h), the yield strength of the material, which is now in the area of maximum normalized precipitate size, is 585.3 MPa, which indicates a slightly over-aged condition.

11.5 Applications and future trends

11.5.1 Current applications of CAF

Currently, the CAF process is used mainly for manufacturing large aircraft wing panels. In general, 7xxx series alloys are used for the upper panels of wings and 2xxx series alloys for the lower panels. Some applications are summarized below:

- Both the upper and the lower wing skins of the B-1B′ Long Range Combat Aircraft used by the USA are produced using CAF. Textron Aerostructures fabricates the wing skins from machined aluminium panels 15.2 m long and 2.7 m wide at the inboard end, tapering to 0.9 m wide at the outboard end. Panel thicknesses range from 63.5 mm at the inboard end to 2.5 mm at the outboard end.
- Textron Aerostructures also manufactures the upper wing skins of the Gulfstream G-IV aircraft and the Airbus A330/340 using the CAF process.
- BAE Systems manufactures the upper wing panels of the Hawk aircraft using the alloy AA7475.
- Bennetts Associates have worked with Airbus on the development of CAF to produce the wing skin for the latest A380 aircraft, using AA7055 alloy. The length of the wing skin is 33 m, its widest section is 2.8 m and it has a double-curvature aerodynamic surface. Its thickness varies from 3 to 28 mm. Figure 11.21 shows a panel part.

11.21 Wing panel for the world's largest commercial aircraft, the A380.

11.5.2 Future trends

Currently, most components manufactured using the CAF technique are aircraft wing skin panels, but the process has potential applications in other fields of manufacture. For example, it has already been used to produce wide-chord fan blades in Rolls Royce engines (Mavromihales *et al.*, 2003).

In the glass industry, CAF has been used to manufacture flat and hollow glass products (Lochegnies *et al.*, 1995; Lochegnies *et al.*, 1996). In order to obtain the final form expected by the designer, the glass, after gradual heating in a furnace, is placed on a stainless steel skeleton packed with a stainless steel grid. The skeleton is a support ring on which the glass goes out of shape. Then, under its own weight, the sheet goes out of shape until it reaches the designer's desired geometry.

CAF could possibly be used in the world of space launch vehicles, which incorporate isogrid panel construction (Holman, 1989), and in the marine industry for the production of ship hull panels.

In the development of CAF technology, significant research has been carried out on materials testing, process testing, materials modelling and process modelling. Owing to the specific requirements of the process, both experimental and modelling techniques have been advanced. The current state of the art is the following:

- CAF is suitable for the production of extra-large panel components which are made of heat-treatable materials, and in which stress relaxation/creep deformation occurs at their artificial ageing temperature, although it is currently used mainly for the production of aircraft wing panels.
- The equipment cost is relatively low for CAF compared with other conventional metal forming processes. Usually, only one tool is required if a pressure differential sufficiently high to deform the workpiece can be obtained. The process combines heat treatment and forming of materials into one operation and saves energy in production.
- The amount of springback is large in CAF. Thus most of the process testing and modelling has been concentrated on springback evaluation, in order to design appropriate tool shapes to compensate for springback. However, one of the most important features of CAF is the generation of good mechanical properties. Less research, particularly using materials and process modelling, has been carried out on this aspect.
- To model precipitation hardening and the evolution of the mechanical properties of materials during CAF, it is important to have fully determined, physically based unified creep-ageing constitutive equations. Currently, there is a lack of research in this area, although an initial attempt has been made (Ho *et al.*, 2004; Zhan *et al.*, 2011b).
- Most experimental and process simulation research has concentrated on single-curvature tool surfaces. This is not suitable for the modelling of real

cases of CAF, since this process has been developed mainly for the production of double-curvature panel parts. The springback has features that are different between the forming of single- and double-curvature panel parts.

- Precipitate formation and growth in different stress states have not been studied in depth, but it has been shown that stress and strain do affect the evolution, orientation and mechanical properties of precipitates. Further research needs to be carried out to generate sufficient data for formulating and determining physically based creep-ageing constitutive equations for process simulation.
- Significant tests need to be carried out to characterize the effects of stress and strain and their directions on the formation, growth, shape, size, spacing, orientation and anisotropic mechanical properties of precipitates in age-hardening materials. A set of unified creep-ageing constitutive equations should be developed and calibrated using experimental data to enable the evolution of physical and mechanical properties during CAF processes to be modelled. This set of constitutive equations could then be implemented into commercial FE process simulation packages to optimize processes and to design tools for springback compensation.

11.6 References

Asnafi N (2001), 'On springback of double-curved auto body panels', *Int J Mech Sci*, 43, 5–37.

Bakavos D, Prangnell P B, Dif R (2004), 'A comparison of the effect of age forming on the precipitation behaviour in 2xxx, 6xxx and 7xxx aerospace alloys', in: *Proceedings of the 9th International Conference on Aluminium Alloy (ICAA9)*, Brisbane, Australia, 124–131.

Bakavos D, Prangnell P B, Eberl F, Gardiner S (2006), 'Through thickness microstructural gradients in 7475 and 2022 age formed bend coupons', *Mater Sci Forum*, 519–521 (Part 1), 407–412.

Brewer H M Jr, Holman M C (1992), *Method of Tool Development*, US Patent 5168169.

Cao J, Lin J (2008), 'A study on formulation of objective functions for determining material models', *Int J Mech Sci*, 50, 193–204.

Cao J, Lin J, Dean T A (2008), 'An implicit unitless error and step-size control method in integrating unified viscoplastic/creep ODE-type constitutive equations', *Int J Numer Methods Eng*, 73, 1094–1112.

Caraher S K, Polmear I J, Ringer S P (1998), 'Effects of Cu and Ag on precipitation in Al–4Zn–3Mg (wt. %)', *Proceedings of the 6th International Conference on Aluminium Alloys (ICAA6)*, Sato T, Kumai S, Kobayashi T., Murakami Y, (eds.), Toyohashi, Japan. Tokyo: Japan Institute for Light Metals, 2, 739–744.

Chaboche J L (1989), Constitutive equations for cyclic plasticity and cyclic viscoplasticity, *Int J Plast*, 5, 31–43.

Chen D Q, Zheng Z Q, Li S C, Chen Z G, Liu Z Y (2004), 'Effect of external stress on the growth of precipitates in Al–Cu and Al–Cu–Mg–Ag alloys', *Acta Metall Sinica*, 40(8), 799.

Deschamps A, Solas D, Bréchet Y (1999), 'Modeling of microstructure evolution and mechanical properties in age-hardening aluminium alloys', *Proceedings of EUROMAT 99*, Munich, 3, 121–132.

Deschamps A, Solas D, Bréchet Y (2005). *Modeling of Microstructure Evolution and Mechanical Properties in Age-hardening Aluminium Alloys in Microstructures, Mechanical Properties and Processes: Computer Simulation and Modelling*, vol. 3. Weinheim: Wiley-VCH.

Eberl F, Gardiner S, Campanile G, Surdon G, Venmans M, Prangnell P (2008), 'Age formable panels for commercial aircraft', *Proceedings of IMechE 222 (Part G: J Aerospace Eng)*, 873–886.

Ferragut R, Somoza A, Tolley A (1999), 'Microstructural evolution of 7012 alloy during the early stages of artificial ageing', *Acta Mater*, 47, 4355–4364.

Hargarter H., Lyttle M T, Starke E A Jr (1998), 'Effects of preferentially aligned precipitates on plastic anisotropy in Al–Cu–Mg–Ag and Al–Cu alloys', *Mater Sci Eng A*, 257, 87–89.

Ho K C (2004), 'Modelling of age-hardening and springback in creep age-forming', PhD thesis, University of Birmingham.

Ho K C, Lin J, Dean T A (2004), 'Constitutive modelling of primary creep for age forming an aluminium alloy', *J Mater Process Technol*, 153–154, 122–127.

Holman M C (1989), 'Autoclave age forming large aluminium aircraft panels', *J Mech Work Technol*, 20, 477–488.

Huang L, Wan M, Chi C L, Ji X S (2007), 'FEM analysis of spring-backs in age forming of aluminum alloy plates', *Chin J Aeronaut*, 20, 564–569.

Jackson M J, Peddieson J, Foroudastan S (2005), 'Age-forming of beam structures – analysis of springback using a unified viscoplastic model', *Proc ImechE Part L: J Mater: Des Appl*, 219, 17–24.

Jambu S, Lenczowski B, Rauh R (2002), 'Creep forming of AlMgSc alloys for aeronautic and space applications', *Proceedings of CAS 2002 Congress*, Vol. 632, Bremen, Germany.

Jeunechamps P P, Ho K C, Lin J, Ponthot J P, Dean T A (2006), 'A closed form technique to predict springback in creep age-forming', *Int J Mech Sci*, 48, 621–629.

Kashyap K T, Ramachandra C, Chatterji B, Lele S (2000), 'A model for two-step ageing', *Bull Mater Sci*, 23(5), 405–411.

Kowalewski Z L, Hayhurst D R, Dyson B F (1994), 'Mechanism-based creep constitutive equations for an aluminium alloy', *J Strain Anal*, 29, 309–316.

Levers A (2005). *Aircraft Component Manufacturing Tool and Method*, European Patent EP1581357.

Li B, Lin J, Yao X (2002), 'A novel evolutionary algorithm for determining unified creep damage constitutive equations', *Int J Mech Sci*, 44(5), 987–1002.

Lin J, Yang J (1999), 'GA-based multiple objective optimization for determining viscoplastic constitutive equations for superplastic alloys', *Int J Plast*, 15, 1181–1196.

Lin J, Hayhurst D R, Dyson B F (1993), 'A new design of uniaxial testpiece with slit extensometer ridges for improved accuracy of strain measurement', *Int J Mech Sci*, 35(1), 63–78.

Lin J, Liu Y, Farrugia D C J, Zhou M (2005), 'Development of dislocation based-unified material model for simulating microstructure evolution in multipass hot rolling', *Philos Mag A*, 85(18), 1967–1987.

Lin J, Ho K C, Dean T A (2006), 'An integrated process for modelling of precipitation hardening and springback in creep age-forming', *Int J Mach Tools Manuf*, 46, 1266–1270.

Lochegnies D, Francois E, Oudin J (1995), 'New computer-aided-design, finite-element modelling and 3-dimensional measuring of rear screen creep forming', *Proc Inst Mech. Eng Part E: J Process Mech Eng*, 209(E2), 137–143.

Lochegnies D, Marion C, Carpentier E, Oudin J (1996), 'Finite element contributions to glass manufacturing control and optimization: 1. Creep forming of flat volumes', *Glass Technol*, 37(4), 128–132.

Mavromihales M, Mason J, Weston W (2003), 'A case of reverse engineering for the manufacture of wide chord fan blades (WCFB) used in Rolls Royce aero engines', *J Mater Process Technol*, 134(3), 279–286.

Myhr O R, Grong O, Andersen SJ (2001), 'Modelling of the age hardening of Al–Mg–Si alloys', *Acta Mater*, 49, 65–75.

Poole W J, Shercliff H R, Castillo T (1997), 'Process model for two step age hardening of 7475 aluminium alloy', *Mater Sci Technol*, 13(11), 897–904.

Poole W J, Sæter J A, Skjervold S, Waterloo G (2000), 'A model for predicting the effect of deformation after solution treatment on the subsequent artificial aging behavior of AA7030 and AA7108 alloys', *Metall Mater Trans A*, 31, 2327–2338.

Ringer S P, Hono K (2000), 'Microstructural evolution and age hardening in aluminium alloys: atom probe field-ion microscopy and transmission electron microscopy studies', *Mater Charact*, 44(1–2), 101–131.

Sallah M, Peddieson J, Foroudastan S (1991), 'A mathematical model of autoclave age forming', *J Mater Process Technol*, 28, 211–219.

Shercliff H R, Ashby M F (1990a), 'A process model for age-hardening of aluminium alloys – I: the model', *Acta Metall Mater*, 38(10), 1789–1802.

Shercliff H R, Ashby M F (1990b), 'A process model for age-hardening of aluminium alloys – II: Applications of the model', *Acta Metall Mater*, 38(10), 1803–1812.

Starink M J, Gao N, Kamp N, Wang S C, Pitcher P D, Sinclair I (2006), 'Relations between microstructure, precipitation, age-formability and damage tolerance of Al–Cu–Mg–Li (Mn, Zr, Sc) alloys for age forming', *Mater Sci Eng A*, 418, 241–249.

Wagner C Z (1961), *Electrochemistry*, 65, 581–591.

Xue P, Yu T X, Chu E (2001a), 'An energy approach for predicting springback of metal sheets after double-curvature forming', Part 1: axisymmetric stamping, *Int J Mech Sci*, 43, 1983–1914.

Xue P, Yu T X, Chu E (2001b), 'An energy approach for predicting springback of metal sheets after double-curvature forming', Part 2: unequal double-curvature forming, *Int J Mech Sci*, 43, 1915–1924.

Zeng Y S, Huang X, Huang S (2008), 'The research situation and the developing tendency of creep age forming technology', *J Plast Eng*, 15(3), 1–8.

Zhan L, Li Y, Huang M (2011a), 'Effect of process parameters on microstructures of 7055 aluminum alloy in creep age forming', *2011 International Conference on Information Engineering for Mechanics and Materials (ICIMM 2011)*, August 13–14, Shanghai, China.

Zhan L, Lin J, Dean Tr A, Huang M (2011b), 'Experimental studies and constitutive modelling of the hardening of aluminium alloy 7055 under creep age forming conditions', *Int J Mech Sci*, 53, 595–605.

Zhan L, Li Y, Huang M, Lin J (2011c), 'Comparative study of creep and stress relaxation behaviour for 7055 aluminium alloy', *Adv Mater Res*, 314–316, 772–777.

Zhou M, Dunne F P (1996), 'Mechanism-based constitutive equations for the superplastic behaviour of a titanium alloy', *J Strain Anal*, 31(3), 187–196.
Zhu A W, Starke E A Jr (2001a), 'Stress ageing of Al–*x*Cu alloys: experiments', *Acta Mater*, 49, 2285–2295.
Zhu A W, Starke E A Jr (2001b), 'Materials aspects of age-forming of Al–*x*Cu alloys', *J Mater Process Technol*, 117, 354–335.

12

Microstructure control in processing nickel, titanium and other special alloys

C. SOMMITSCH, R. RADIS and A. KRUMPHALS, Graz University of Technology, Austria and M. STOCKINGER and D. HUBER, Böhler Schmiedetechnik GmbH & Co KG, Austria

Abstract: This chapter gives an overview of materials modelling and microstructure control for some special alloys, namely nickel-based superalloys, titanium alloys and intermetallics. After a description of the applications and production processes of these alloys, the resulting microstructures and related mechanical properties are presented. Also, the possibility of process and materials optimization is shown in two case studies. The chapter closes with future trends and some sources of additional information.

Key words: special alloys, materials processing, microstructure control, materials modelling, process simulation.

12.1 Introduction

Innovative materials and processes to produce them are enabling sophisticated technologies. Materials that are multi-functional and smart, and possess physical and engineering properties superior to existing materials, are constantly needed for continued technical advances in a variety of fields, such as that of special materials. In modern times, the development, processing and characterization of new materials have been greatly aided by novel approaches to materials design and synthesis that are based on a fundamental and unified understanding of the processing–structure–properties–performance relationships for a wide range of materials (Asthana *et al.*, 2006).

The intrinsic properties of materials depend on the microstructure, and the microstructure alters during processing. Microstructure is not accessible to observation during processing, so controlling it requires the ability to predict how a given process step will cause it to evolve or change. This can be guided by the use of modelling and process simulation. Historically, the control of microstructure, for example during the hot forging of titanium- and nickel-based alloys, has advanced since the introduction of high-performance workstations and finite element software coupled with semi-empirical materials laws in order to simulate complex process chains and the evolution of structure. Nowadays, producers of aircraft components, for example, have to prove that they are able to simulate their production processes and to predict the final microstructure and hence the mechanical properties of their products.

There have been many examples of control of the microstructure of special alloys, such as iron-, nickel- and cobalt-based superalloys, and titanium alloys and intermetallics. The historical developments in the processing of special alloys have brought about considerable advances, for example increases in the operating temperatures of superalloys. Superalloys were originally iron-based and cold wrought prior to the 1940s. In the 1940s, investment casting of cobalt-based alloys significantly raised operating temperatures. The development of vacuum melting in the 1950s allowed very fine control of the chemical composition of superalloys and reduction in contamination, and this in turn led to a revolution in processing techniques, such as the introduction of directional solidification of alloys and single-crystal superalloys (Sims *et al.*, 1987; Kracke, 2010). Many forms of superalloys are present in gas turbine engines. Polycrystalline nickel-based superalloys are used for the discs of low- and high-pressure gas turbines, where the material can be produced by powder metallurgy via the classical vacuum melting and remelting technology. Turbine blades can be polycrystalline, have a columnar grain structure, or be single-crystal. Polycrystalline blades are formed using casting technology in a ceramic mould. Columnar-grain-structured blades are created using directional solidification techniques and have grains parallel to the major stress axes. Single-crystal superalloys are formed as single crystals using a modified version of the directional solidification technique, so there are no grain boundaries in the material.

There is an ongoing search for materials that can be applied at even higher temperatures than nickel-based superalloys. The intermetallic titanium aluminides, with their low density, could be a choice for the future, and they find use in several applications, including automobiles and aircraft. The development of TiAl-based alloys began about 1970; however, these alloys have only been used in these applications since about 2000. Titanium aluminide contains three major intermetallic compounds: γ-TiAl, α_2-Ti_3Al and $TiAl_3$. Of these three, γ-TiAl has received the most interest and applications. γ-TiAl has excellent mechanical properties and oxidation and corrosion resistance at temperatures above 600°C, which makes it a possible replacement for traditional nickel-based superalloy components in aircraft turbine engines.

Titanium-based special alloys have been developed since the 1950s for lightweight high-strength applications below 500°C. Since the early 1960s, their use has shifted significantly from military applications to commercial ventures. Although pure titanium was valued for its blend of high strength, low weight and excellent durability, even stronger materials were needed for aerospace use. In the 1950s, a high-strength alloy called Ti–6Al–4V was developed, and found immediate use in engine and airframe parts. A leaner alloy called Ti–3Al–2.5V with higher ductility was created, which could be processed by special tube-making equipment. Today, virtually all the titanium tubing in aircraft and aerospace equipment consists of Ti–3Al–2.5V alloy. Its use spread in the 1970s to sports products such as golf club shafts, and in the 1980s to wheelchairs, ski poles, pool cues and tennis rackets.

In the following sections, three main representatives of special materials are discussed, namely nickel-based superalloys, titanium alloys and titanium aluminides. After a description of their applications, the production processes are outlined, followed by the resulting microstructures and related mechanical properties. Special focus is put on modelling of the evolution of the materials during processing and methods for the simulation of the manufacturing processes. Two case studies are given to show the possibility of optimization of both the processes and the materials. The chapter closes with future trends and some sources of further information and advice.

12.2 Application of special alloys such as nickel-based alloys, titanium alloys and titanium aluminides

12.2.1 Nickel-based superalloys

Nickel-based superalloys usually combine high strength and corrosion resistance during service at elevated temperatures. Initially, their development was encouraged and driven by the insight that the efficiency of thermal power generation machines can be increased by increasing the combustion temperature and/or pressure. Consequently, they have been widely used in high-performance combustion engines, such as gas turbines in aircraft, and for power generation in thermal, nuclear and fossil fuel power plants. The typical nickel-based components in this energy sector are rotors, turbine discs, blades, shafts, bearings, spindles and bolts, as well as casings for stationary gas and steam turbines. In the aircraft industry, most of the rotating turbine parts and also the casings, links and some of the engine mounts are typically made of high-performance nickel-based superalloys. Some similar applications where nickel-based superalloys are also used are in turbocharger discs for large diesel engines and in high-performance racing car engines.

Furthermore, the chemical industry uses these alloys for applications in highly corrosive environments, containing brines, carbonates, phosphates, sulphates, chlorides, nitrates or just seawater. In this respect, the oil and gas industry should be especially mentioned, where these materials are used in gas and oil exploration, and in refining and transport. The typical components are downhole equipment, wellheads, pipes, valves and pump wheels. Owing to their excellent high-temperature corrosion resistance, nickel-based alloys are also often used for valves in large ship diesel engines (see Fig. 12.1).

The metal processing industry uses nickel-based superalloys in several components of extrusion and forging machines, especially tools for forging, shaping and deep drawing, where wear resistance plays an important role.

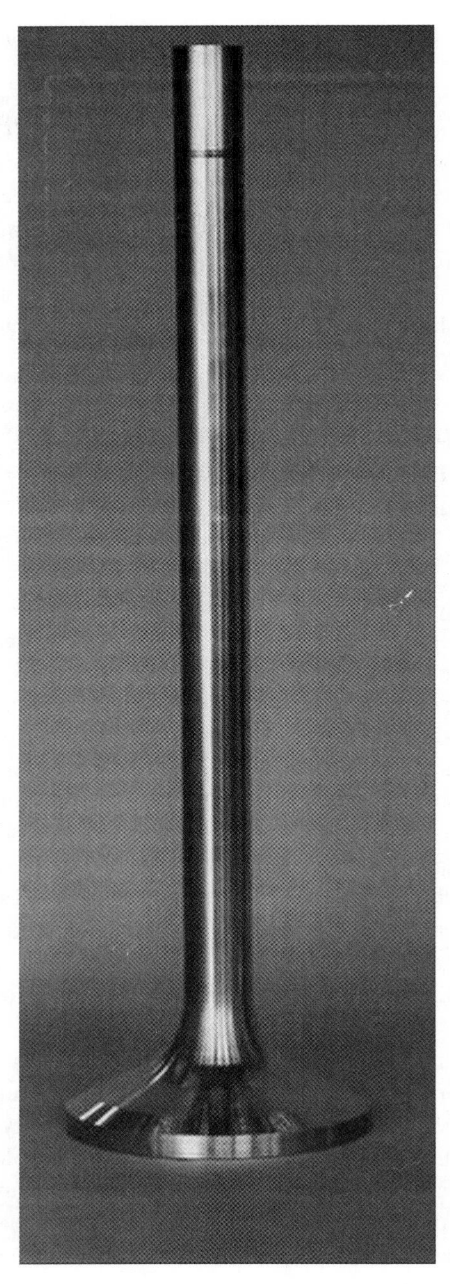

12.1 Typical forged and machined Nimonic 80A ship diesel valve (courtesy of Böhler Schmiedetechnik GmbH & Co KG, Austria).

12.2.2 Titanium alloys

One of the major drivers for the fast development of titanium alloys in the last few decades has been the aerospace and aircraft industry. Owing to their superior balance of specific weight and mechanical and physical properties, titanium alloys are being used in almost every part of modern aircraft and helicopters (see Fig. 12.2). Typical applications are in high-load-transfer components in the structure and in the landing gear, and in tracks, mountings and pylon parts in the wing area. In rotorcraft, typically the main and rear rotor shafts, the rotor star, and the connections to the blades are made of titanium alloys. In aircraft engines, the fan, compressor discs, blades, blisks (bladed discs) and aerofoils, as well as some parts of the frame, are made of titanium alloys.

In addition to aerospace applications, titanium and titanium alloys are also used in the chemical and power industries, in automotive engineering, in medical applications and in sports equipment. In the traditional areas of the chemical and power industries, titanium is commonly used as a corrosion-resistant material. For example, titanium has become increasingly common in offshore structures in recent years (Schutz and Watkins, 1998).

Another field in which the use of titanium is growing is that of consumer products, especially jewellery and sporting goods. In the field of jewellery, rings and watchcases are often made of titanium. The most common application in the area of sporting goods is in golf club heads (see Fig. 12.3). Other examples are provided by tennis rackets and bicycle frames (Yamada, 1996). The benefits of

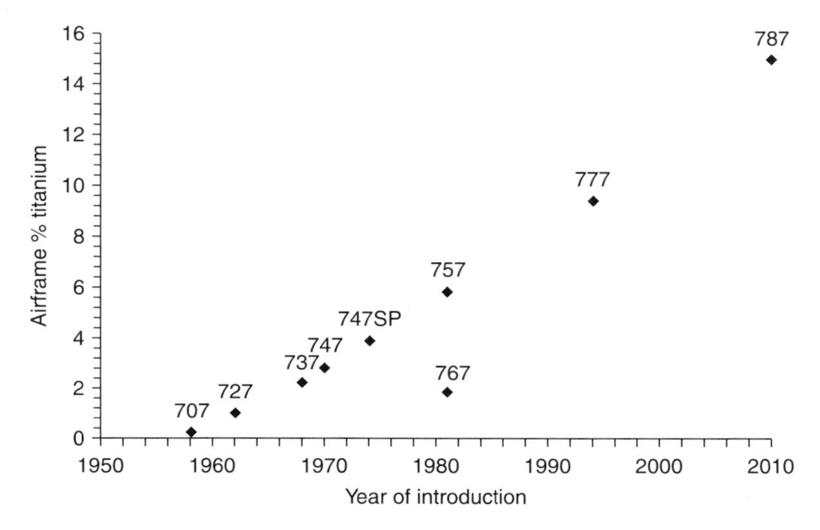

12.2 Titanium usage in Boeing aircraft airframes (Lütjering and Williams, 2007; http://www.boeing.com/commercial/787family/programfacts.html, 2011).

12.3 Golf club head made of titanium alloy (Callaway Diablo Edge Driver 11° GR R).

titanium golf club heads have been widely publicized. The high strength and low density of titanium allow a larger area for striking the ball properly while retaining the club head speed.

A more established application area of titanium is the biomedical field, mainly in the form of CP titanium and Ti–6Al–4V, which have been used in the past as implant materials (see Fig. 12.4). Vanadium-free titanium alloys such as Ti–6Al–7Nb and Ti–5A–2.5Fe were developed because toxicity of vanadium to the human body was suspected. In recent years, efforts have been made to develop β-titanium alloys using non-toxic elements such as Nb, Ta, Zr and Mo as alloying additions (Wang, 1996). The main advantages of these β alloys over the conventional Ti–6Al–4V alloy are a higher fatigue strength, lower modulus of elasticity and improved biocompatibility.

Owing to the high costs, the utilization of titanium alloys in the automobile industry has not gone very far. Potential parts have been identified, for example valves, valve springs and connecting rods in the engine area, and suspension springs, bolts, fasteners and the exhaust system in the car body area. For many years, titanium has been used in high-performance vehicles such as Formula 1 racing cars and off-road racing trucks (Peters and Leyens, 2002), but for application in a family car, the problem of producing low-cost titanium parts has to be solved.

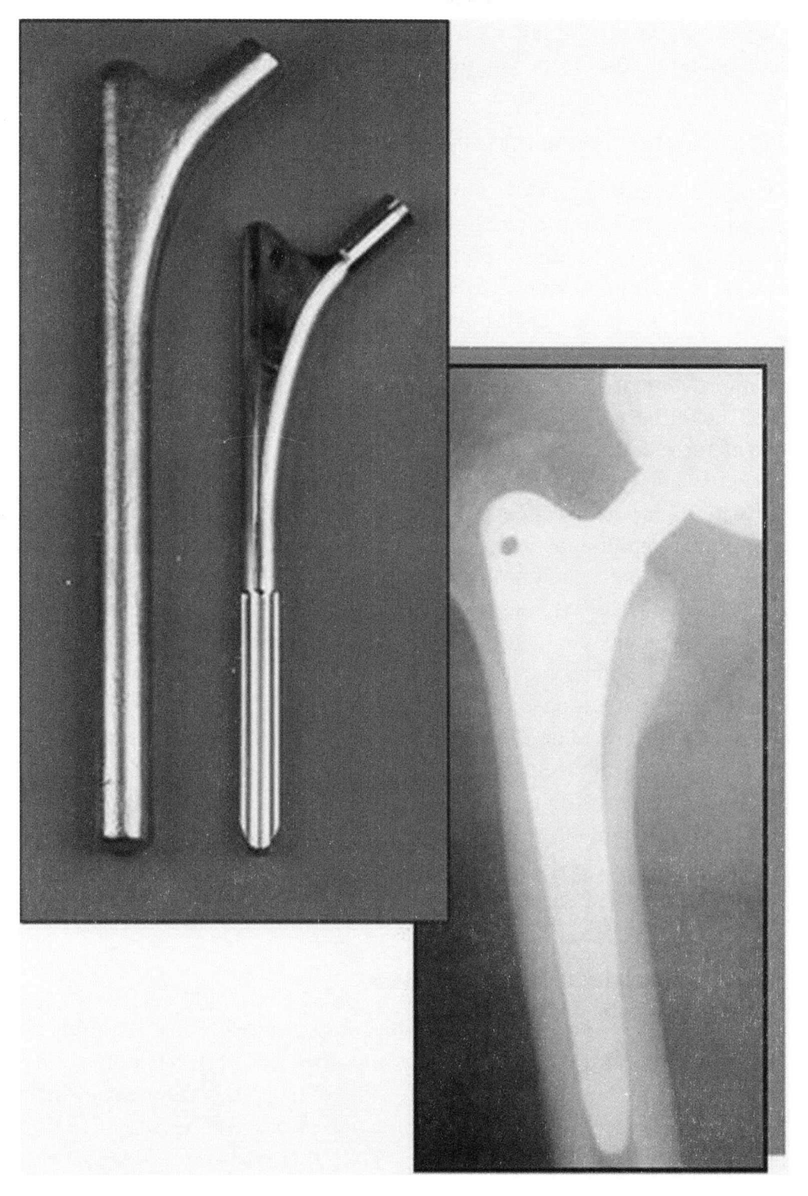

12.4 A typical medical implant (hip joint): upper left image, the part after forging (left) and the final part (right); lower right image, radiograph.

Another, but rare, application area of titanium alloys can be found in architecture, such as in the roof of the Guggenheim Museum in Bilbao, Spain.

12.2.3 Titanium aluminides

Titanium–aluminium alloys have found, besides their application in aeroengines, a small niche application as lightweight high-temperature construction materials for moving parts in high-performance automotive engines, such as turbocharger turbine wheels and exhaust valves. Mitsubishi established the use of investment-cast γ-TiAl turbocharger turbine wheels in the early 1990s in a series-production vehicle, the Lancer Evolution VI. As a result of the reduced weight, this leads to an improvement in the turbocharger response in comparison with the commonly used nickel-based turbine wheels. Another application field of γ-TiAl is in exhaust valves for racing cars to improve fuel economy and engine performance (e.g. by an increase in the allowed rpm). The applications of titanium aluminide are found only in high-performance automotive engines because of the absence of an economical manufacturing process, which would be needed to expand the future applications in the automotive industry (Noda, 1998; Tetsui, 1999; Chandley, 2000; Clemens, 2008; Peters and Clemens, 2010).

Other possible application fields are in connecting rods and piston pins in combustion engines, in gas turbine blades and as a lightweight sheet material (Sauthoff, 1996; Knippscheer and Frommeyer, 2006; MW Racing, 2011).

Various American and European projects have shown the suitability of γ-TiAl for aerospace structures. Sheets for application in hot structures such as thermal-protection systems, divergent flaps and sidewalls have been primarily analysed. γ-TiAl shows high potential for support structures for future supersonic propulsion systems, and as a possible matrix material for intermetallic composites (Loria, 2000; Clemens and Kestler, 2000).

12.3 Production processes

Even though the properties of the three alloy groups discussed in this chapter are distinctly different, the production route does not vary too much, owing to the high quality and purity necessary for these materials. One major difference for these alloys is the price. Comparing the nickel-based superalloy 718 double melt, Ti–6Al–4V and the intermetallic alloy TNB, the cost ratio for 1 kg of semi-finished product is approximately 1:1.5:7 (as of January 2011).

12.3.1 Nickel-based superalloys

Owing to the critical applications of nickel-based superalloys, high purity and homogeneity are a must, and therefore conventional melting technologies such as those used for low-alloy steels are seldom suitable. Thus, various combinations of vacuum induction melting (VIM), electroslag remelting (ESR) and vacuum arc remelting (VAR) are the main techniques used to produce high-quality alloys. Scrap from the production process, recycled material, master alloys and pure metals are used in the VIM furnace to obtain the correct chemistry after melting. The VIM process offers the possibility to limit the oxygen, nitrogen and hydrogen contents in the melt, which is necessary when one wishes to use reactive elements such as aluminium as alloying elements without getting non-metallic inclusions in the final material. After the melting process, the liquid metal is in most cases cast into ladles while still under vacuum. This semi-final product is used for further processing, for instance in the ESR process.

In the ESR process, an electrode produced by VIM or differently is continuously melted under air or gas using heat produced by electrical resistance in a slag. The liquid metal runs through a reactive slag of lower density into a cooled copper crucible. The main advantage of the ESR process is the possibility of influencing the chemical composition of the final ingot by means of reactions with the slag, for example to obtain a reduction in the sulphur content of the material. In addition, the controlled solidification process leads to a very homogeneous material in most of the ingot volume. The VAR process is slightly less complex and therefore easier to control than the ESR process. In the VAR process, an electrode obtained from the VIM, ESR or some other process is melted under vacuum via an electric arc and is again cast into a cooled copper crucible (see Fig. 12.5). The main difference of the VAR from the ESR process is the absence of a slag and therefore no possibility of chemical refinement. Nevertheless, the vacuum influences the content of oxygen and hydrogen and to a limited amount also nitrogen, and the content of high-vapour-pressure elements such as lead or tin in the final ingot. The main purpose of the VAR process is the production of a highly homogeneous material as a result of the controlled melting and solidification process. After these melting processes, a homogenization heat treatment consisting of one or more steps is usually performed to equalize the last small segregations. (For more information, see Sims *et al.* (1987) and Reed (2006).)

Most ingots produced by either double-melt VIM/VAR, VIM/ESR or triple-melt VIM/ESR/VAR are afterwards processed by hot forming. Usually several upsetting and racking processes, implying tight temperature and deformation control, are used to produce the preferred homogeneous microstructure. Hydraulic open die presses and/or rotary forging machines are used to break down the equiaxed structure produced by the remelting process to a fine globular structure, and thus the billet produced shows properties that are less direction-dependent. To prevent detrimental effects of non-metallic inclusions or other defects, the billet material is ultrasonically tested and areas with indications are removed and scrapped. For critical applications such as in aircraft parts, it must be guaranteed that the material of the component can be traced back to its position in the billet and in the ingot, and thus an excellent quality system is necessary for companies producing such materials. (For more information, see Sims *et al.* (1987) and Reed (2006).)

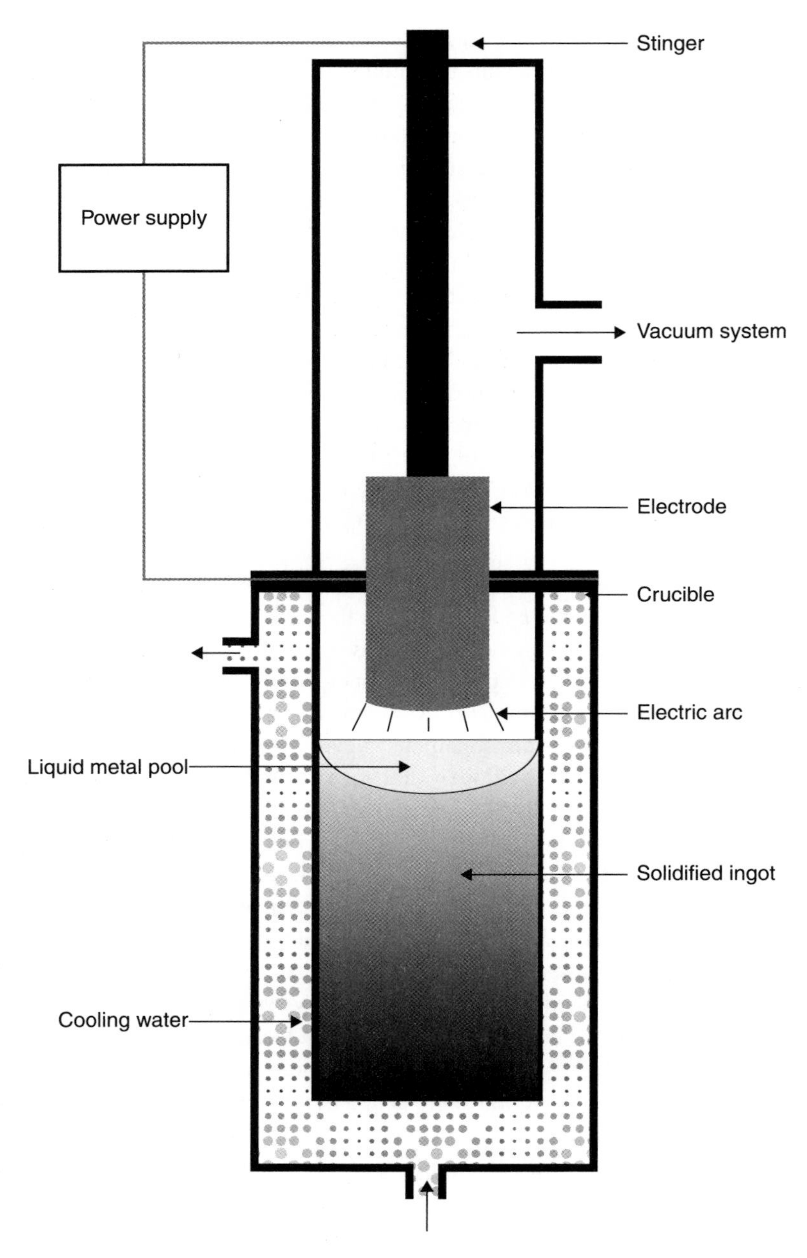

12.5 Sketch of a vacuum remelting furnace (courtesy of Böhler Schmiedetechnik GmbH & Co KG, Austria).

Besides this typical route, powder metallurgy is used to produce nickel-based alloys with properties not achievable by standard methods. Higher amounts of segregating elements, oxide dispersion strengthening and very fine homogeneously distributed grains are achievable using this method. Typically, inert-gas atomization, rotating-electrode processes or centrifugal atomization is used to produce a powder of high purity. After this, the powder is containerized and put into a hot isostatic press, where the application of high temperature and pressure for a specific time leads to a dense material. In some cases, hot extrusion is used additionally to further densify the alloy. (For more information, see Sims *et al.* (1987) and Reed (2006).)

To achieve superior mechanical properties, further processing of the billet material is necessary. For many applications, thermomechanical processing is used to generate the preferred shape and microstructure. Isothermal or classical hydraulic presses, as well as screw presses and hammers, are used to produce parts by closed-die forming processes. Also, rolling into sheets, bars, rods and rings is possible using powerful equipment. Independently of the forming process, the small processing window of these alloys means that the temperature, deformation and handling of the material have to be controlled strictly to achieve the desired properties. After one or more forming steps, various heat treatment procedures (e.g. solution annealing and ageing) are usually necessary to generate the final microstructure and nanostructure. (For more information, see Sims *et al.* (1987) and Reed (2006).)

The extreme gas temperatures of more than 1450°C during the starting phase in the high-pressure turbines of modern engines exceeds the melting point of typical wrought nickel-based superalloys such as Alloy 720 and Waspaloy. Therefore a different production route had to be developed for turbine parts in this environment. Today, the blades and aerofoils of high-pressure turbines are generally produced by investment casting processes in order to generate single-crystal structures, and hollow profiles to allow active cooling. In addition, these superalloys are covered with multiple coatings to prevent them from melting. (For more information, see Sims *et al.* (1987) and Reed (2006).)

The generally superior mechanical properties of nickel-based superalloys are their biggest advantage in service but are detrimental for any machining operation. Typically, boronitrides, silicon–aluminium-based ceramics or in some cases hard metal or titanium-nitride-coated tools are used. In addition, the cutting speed has to be controlled very carefully in order to limit the heat generated during cutting, which limits the tool life. For example, the use of hard metal tools with heat-treatable steel allows cutting speeds ten times higher than for nickel-based alloys.

For several parts, especially the casings of turbine engines, welding must be used to produce their complex structure efficiently. Depending on the composition of the alloy, processes such as shielded metal-arc welding, gas tungsten-arc welding, gas metal-arc welding may be possible, or only resistance, electron beam or laser welding may be possible. Whereas Alloy 718 is described as good weldable superalloy, fourth-generation single-crystal superalloys such as CMSX 4 cannot be welded at all. (For more information, see Sims *et al.* (1987) and Reed (2006).)

12.3.2 Titanium alloys

The high affinity of titanium for oxygen limits the melting methods for the production of pure high-quality titanium alloys. Differently from what was described earlier for nickel-based alloys, the melting of titanium usually starts with a VAR process. Titanium sponge made of titanium ore, alloying elements, and cleaned and sorted scrap are shredded and then compacted in hydraulic presses into briquettes, which are welded together to form the initial VAR electrode. After the first remelting process, the alloy is typically not defect-free and not homogeneous enough, and thus it is necessary to do an additional vacuum arc remelting process. High-purity rotor grades and alloys with high amounts of segregating elements, such as Ti–10V–2Fe–3Al, are usually triple melted to reach sufficient quality. The second melting process used for titanium alloys is the cold hearth melting (CHM) process (Fig. 12.6). In CHM, the electrode is melted into a shallow water-cooled copper hearth using plasma arcs or electron beams as heat sources. Owing to the cooling of the copper crucible, a thin solidified titanium shell is always present and prevents the liquid from reacting with the crucible. A big advantage of the CHM process is that high-density impurities such as tungsten originating from scrap can be removed in three to five cascade-like melt baths because these impurities gravitate down and remain in the mushy zone on the bottom of the crucible. However, even though the control of the CHM process is much better than for VAR processes, a subsequent VAR process has to be performed for rotor grade titanium alloys. (For more information, see Lütjering and Williams (2007).)

The further processing of titanium alloys is very similar to that of nickel-based superalloys. Billet converting, thermomechanical processing and machining are

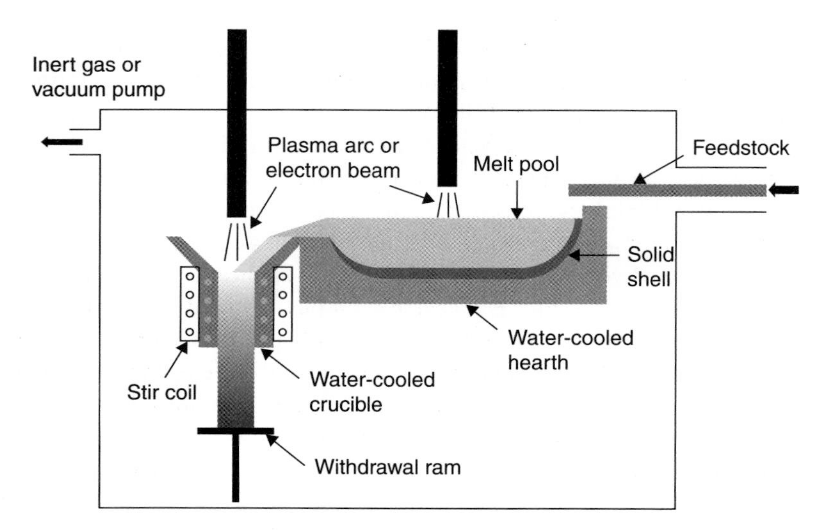

12.6 Schematic diagram of a cold hearth melting facility (courtesy of Böhler Schmiedetechnik GmbH & Co KG, Austria).

typically used to produce final components and have to be controlled tightly to guarantee good mechanical properties (Fig. 12.7). Although investment casting is possible for titanium alloys, the high affinity for oxygen limits this process. For some applications such as in thin sheets, titanium alloys can also be cold formed, but very special equipment is necessary. (For more information, see Lütjering and Williams (2007).)

The joining methods for titanium alloys are also limited owing to the reactivity mentioned above. Most frequently, gas–tungsten arc welding, gas–metal arc welding, electron beam welding, plasma arc welding, laser welding and friction welding are used. In the compressor parts of aircraft engines, in particular, several stages of discs are welded together to form the final drum (Richter *et al.*, 2011).

12.3.3 Titanium aluminides

There are two major methods of production of the pre-material. On the one hand, there is ingot metallurgy. This starts with Ti sponge, Al pellets and additive elements pressed to form a master alloy, followed by a melting process (similar to the production of titanium alloys), resulting in cast products. These products can be either billets or final products (e.g. valves and turbocharger wheels). There are three major casting routes, namely investment casting, counter-gravity investment casting and centrifugal casting. The billets are subsequently hipped to reduce porosity. On the other hand, powder metallurgy is also possible. In this case the powder is hipped to form billet material or final products (e.g. valves) (Kestler and Clemens, 2005).

12.7 Photograph of closed-die forging step for a Ti–6Al–4V aircraft part (courtesy of Böhler Schmiedetechnik GmbH & Co KG, Austria).

Subsequent process steps are possible for both pre-material paths, such as rolling and flattening of sheet material, and extrusion and straightening of bar material. Powder-metallurgical materials need extrusion before forging operations, whereas cast materials can be used directly after hipping. Forging operations can be done isothermally or in a near-conventional forging process.

Both hot-die forging and isothermal forging are unique forging methods initially developed for the aerospace industry. Both of these methods can provide net-shape or near-net-shape forgings. In contrast to conventional forging, the die temperature is held near the stock temperature. Both techniques allow the forging of materials that are difficult or impossible to forge by conventional methods, and have the advantage of lower raw-material input weights, reduced machining time, and uniformity of microstructure and properties.

The die temperatures used in hot-die forging are lower than those in isothermal forging (about 110–220°C below the workpiece temperature). This allows the use of lower-cost die materials. The strain rates are generally low but up to ten times higher than in isothermal presses and have to be controlled carefully to prevent excessive adiabatic heating of the workpiece. Hot-die forging requires a consistent die temperature within the die. Induction, infrared and resistance heating are common methods of die heating.

In isothermal forging, the die and workpiece are maintained at or near the same temperature. Hydraulic presses are primarily used for isothermal forging processes to control strain rates of generally less than $0.5\,s^{-1}$. Mo-based alloys such as TZM are the most widely used die material for this process. These materials retain their strength above 1100°C and are relatively easy to machine. However, this requires the use of vacuum or an inert atmosphere for protection of the die. The enclosure and the atmospheric controls required add high costs to the initial investment (Montero *et al.*, 2005).

12.4 Microstructures and mechanical properties

12.4.1 Nickel-based superalloys

Nickel is a versatile element and will alloy with most metals. The wide solubility ranges between iron, chromium and nickel make many alloy combinations possible. The face-centred cubic structure of the nickel matrix can be strengthened by solid solution hardening, carbide precipitation or precipitation hardening. The excellent mechanical properties of nickel-based superalloys are inherently related to the precipitation of intermetallic phases in the disordered face-centred cubic γ matrix and to their interactions with the grain growth and recrystallization mechanisms. In the following, the focus is on the production of wrought alloys.

In the case of solid solution hardening, cobalt, iron, chromium, molybdenum, tungsten, vanadium, titanium and aluminium are all solid solution hardeners in nickel. At temperatures above $0.6T_m$ (where T_m is the melting temperature in kelvin), which is the range of high-temperature creep, the strengthening is diffusion-dependent, and large, slowly diffusing elements such as molybdenum and tungsten are the most effective hardeners. Carbide strengthening relies mainly on MC, M_6C, M_7C_3 and $M_{23}C_6$ (where M is a metallic carbide-forming element). MC usually occurs in the form of a primary large blocky carbide, random in distribution, and is generally not desired (Fig. 12.8). M_6C carbides are also blocky; when formed in grain boundaries, they can be used to control grain size, but when precipitated in a Widmanstätten pattern throughout the grains, these carbides can impair the ductility and rupture life. M_7C_3 carbides form intergranularly and are beneficial if precipitated as discrete particles. They can cause embrittlement if they agglomerate, forming continuous grain boundary films. This condition will occur over an extended period of time at high temperatures. $M_{23}C_6$ carbides show a propensity for grain boundary precipitation (Fig. 12.9) and are influential in determining the mechanical properties of nickel-based alloys. Discrete grain boundary particles enhance rupture properties. Long-time exposure at 760–980°C will cause precipitation of angular intragranular carbides, and particles along twin bands and twin ends. Heat treatment provides the alloy designer with a means of creating the desired carbide structure and morphology before placing the material in service. The chemistry of the alloy, its prior processing history and the heat treatment given to the material influence carbide precipitation and ultimately the performance of the alloy. Each new alloy must be thoroughly examined to determine its response to heat treatment or high temperature. The topologically close-packed phases are generally undesired, since they are very

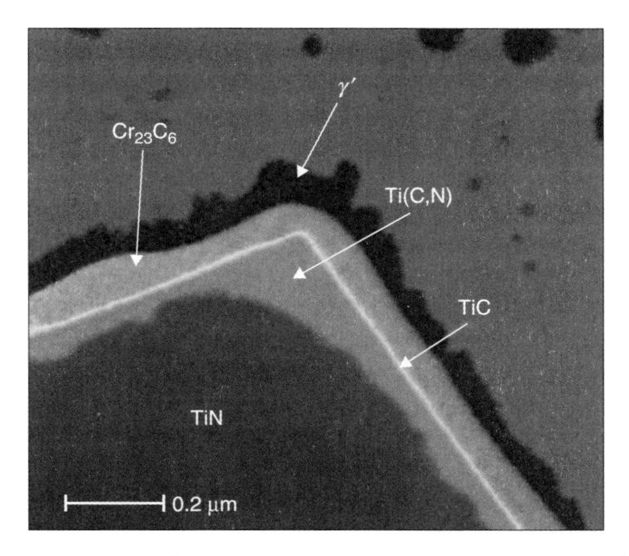

12.8 Energy-filtered transmission electron microscopy (EFTEM) analysis of Ti(C,N) particle with adjacent $Cr_{23}C_6$ and γ' precipitates (Wasle *et al.*, 2003).

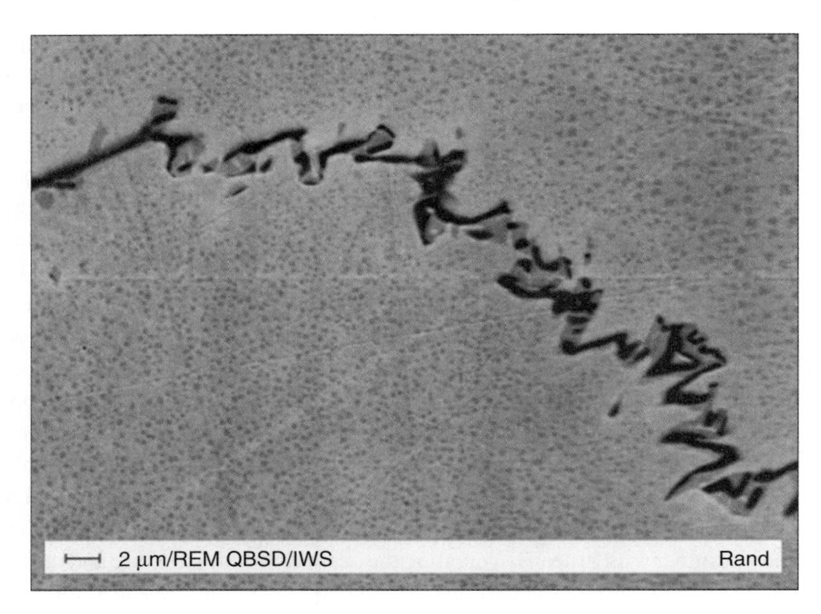

12.9 Grain boundary film of $Cr_{23}C_6$ particles in a γ matrix with γ' particles in Alloy 80A, SEM image (Wasle *et al.*, 2003).

brittle owing to their limited number of slip planes. The most common representatives of this family are the σ, the Laves and the μ phases. Their chemistries and crystallographic structures are rather complex. More information can be found elsewhere (e.g. Durrand-Charre, 1997).

The precipitation of the γ' phase, $Ni_3(Al,Ti,Nb)$, in a nickel matrix provides significant strengthening to the material. This coherent, ordered phase with a type $L1_2$ crystal structure has a lattice constant with a mismatch of 1% or less with the lattice constant of the γ matrix, which allows low surface energy and long-time stability. Precipitation of γ' from a supersaturated matrix yields an increase in strength with increasing precipitation temperature, up to the over-ageing or coarsening temperature. The high-temperature strength of alloys produced by γ' precipitation is a function of the γ' volume fraction, particle size, size distribution and morphology. Depending on the heat treatment and the lattice mismatch, the morphology of these precipitates varies over a wide range, from simple spheres through cubes and octocubes to complex structures, such as octodendrites and dendrites (see Fig. 12.10) (Radis *et al.*, 2009). The amount of γ' formed is a function of the hardener content, i.e. the content of aluminium, niobium and titanium. The effective strengthening produced by γ' decreases above ca. $0.6T_m$ as the particles coarsen. To retard coarsening, the alloy designer can add elements to increase the volume percentage of γ' or add high-partitioning, slow-diffusing elements such as niobium or tantalum to form the desired precipitates. Coarse primary γ' precipitates, located at grain boundaries of the γ matrix, are believed to

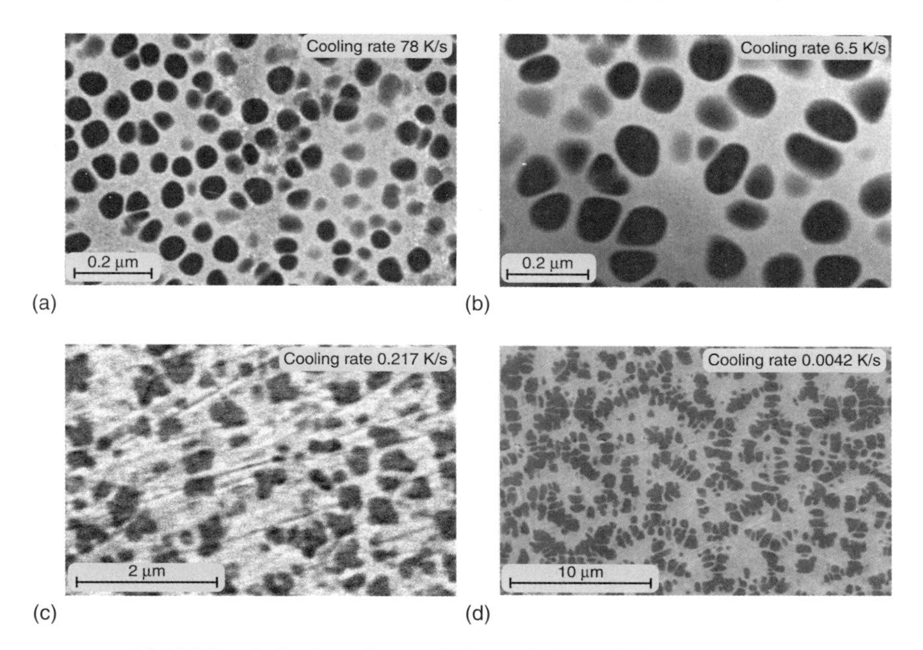

12.10 Morphologies of γ' precipitates in UDIMET 720 Li depending on cooling rate. (a),(b) EFTEM (Cr map) images; (c),(d) SEM (backscattered electron) images (Radis *et al.*, 2009).

be responsible for the inhibition of grain growth at high temperatures, for example in the case of solution annealing of UDIMET 720Li (Jackson and Reed, 1999).

The γ' phase can transform to other (Ni_3X) precipitates if the alloy is supersaturated with titanium or niobium. Titanium-rich metastable γ' can transform to the hexagonal close-packed η phase Ni_3Ti. The formation of η phase can alter the mechanical properties, and the effects of this phase must be determined on an individual-alloy basis. Excess niobium results in metastable η transforming to semi-coherent, lenticular-shaped γ'' (a body-centred tetragonal phase of type $D0_{22}$) and ultimately to the orthorhombic equilibrium δ phase (Ni_3Nb) of type $D0_a$ (Stotter *et al.*, 2008; Radis *et al.*, 2010). Both γ' and γ'' can be present at peak hardness, whereas transformation to the coarse, elongated δ phase results in a decrease in hardness. Since δ phase precipitates preferentially along Nb-enriched grain boundaries, it is also used for controlling grain growth, for example during the thermomechanical processing of the alloy Allvac 718Plus (Zickler *et al.*, 2009). However, the plate-shaped morphology (Stotter *et al.*, 2008) of δ precipitates enhances the notch sensitivity and thus reduces the fatigue strength.

Depending on the application, various mechanical properties can be optimized by tailoring the microstructural features described above. Tables 12.1 and 12.2 give an overview of the chemistry and mechanical properties of selected nickel-based superalloys used for different applications.

Table 12.1 Nominal chemistry of selected nickel-based superalloys

Alloy	Ni	Fe	Cr	Co	Ta	Nb	Ti	Al	Mo	W	C	Others
Alloy 625	61.0	2.5	21.5	0	0	3.6	0.2	0.2	9.0	0	0.05	
Alloy 718	52.5	18.5	19.0	0	0	5.1	0.9	0.5	3.0	0	0.04	
Allvac 718+	51.5	9	19	9	0	5.3	0.75	1.5	2.8	1	0.03	
Nimonic 80A	76.0	0	19.5	0	0	0	2.4	1.4	0	0	0.06	B, Zr
Alloy 720	55.0	0	17.9	14.7	0	0	5.0	2.5	3.0	1.3	0.03	B, Zr
Haynes 282	58	0	19.5	10	0	0	2.1	1.5	8.5	0	0.06	
CMSX 2	66	0	8.0	4.6	5.8	0	0.9	5.6	0.6	7.9	0	
CMSX 4	61	0	6.4	9.5	6.5	0	1	5.7	0.6	6.3	0	2.9 Re

Table 12.2 **Ultimate tensile strength of selected nickel-based superalloys**

	UTS (MPa)		
Alloy	Room temperature	650°C	760°C
Alloy 625	965	835	550
Alloy 718	1435	1228	950
Nimonic 80A	1000	795	600
Alloy 720	1570	1455	1455
Haynes 282	1147	1050	870
CMSX 2	1185		1295

12.4.2 Titanium alloys

Owing to the high variety of different alloying concepts, titanium alloys have been classified into three different categories dependent on the concentration of β-stabilizing elements (see Fig. 12.11). The α alloys, including the so-called CP grades (commercially pure) with different oxygen contents to increase their strength, are mostly used because of their outstanding corrosion resistance and relatively easy processability. Depending on the strength necessary, higher alloying contents may be necessary for typical tube and sheet materials.

The most widely used group of titanium alloys is the α+β class, including the 'workhorse' Ti–6Al–4V. The properties of α+β alloys are influenced on the one

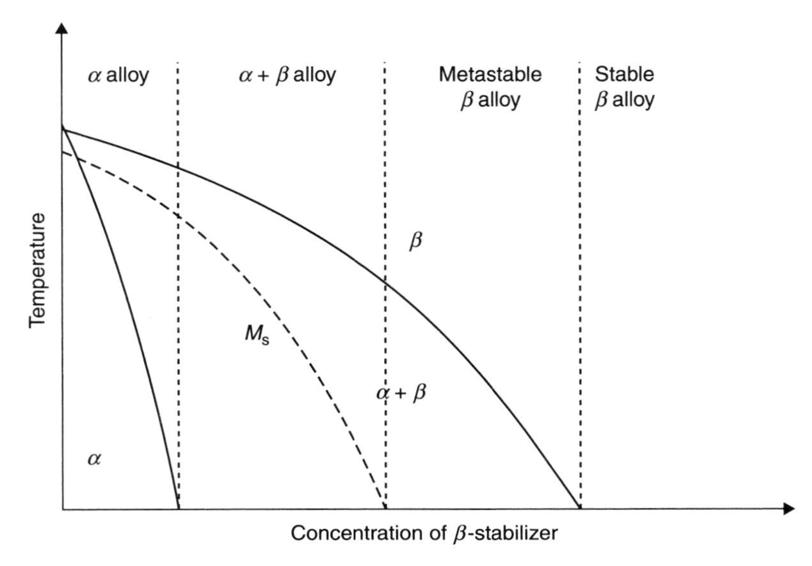

12.11 Schematic pseudo-binary phase diagram of titanium alloys (courtesy of Böhler Schmiedetechnik GmbH & Co KG, Austria).

hand by their alloying composition and on the other hand by the microstructure, which can be varied strongly by different heat treatments. So-called near-α-alloys such as Ti–6Al–2Sn–4Zr–2Mo–0.1Si and Ti–8Al–1V–1Mo, with low contents of β phase at room temperature, are used because of their good high-temperature properties. Working temperatures of more than 500°C are possible for these alloys, and thus they are ideal candidates for high-pressure compressor discs in aircraft engines. One of main reasons why Ti–6Al–4V is used more often than all other titanium alloys put together is the variability of mechanical properties with heat treatment. Figures 12.12–12.15 show examples of how the microstructure of Ti–6Al–4V can be varied with different heat treatments below the β transus temperature (Oberwinkler, 2010). The mechanical properties of bimodal structures can be influenced by the $\alpha+\beta$ processing temperature and therefore the fraction of spherical primary α, and subsequently by the cooling rate. Slower cooling rates such as in Fig. 12.13 result in coarser lamellae (α and β plates); faster cooling results in a fine bimodal structure with higher strength (see Fig. 12.14). Whereas mill annealing at temperatures around 700°C leads to stress relief only, a final heat treatment below the solution temperature of the intermetallic phase Ti_3Al (~550°C for Ti–6Al–4V) leads to precipitation of fine, dispersed particles in the α phase. The fully lamellar structure which is obtained by β-processing (forming or heat treatment above the α solution temperature) is also strongly influenced by the cooling rate. A coarse lamellar structure shows very high fracture toughness

12.12 Microstructure of Ti–6Al–4V $\alpha+\beta$ alloy, forged and mill annealed (Oberwinkler, 2010).

12.13 Microstructure of Ti–6Al–4V $\alpha+\beta$ alloy, forged, solution annealed with slow air cooling and subsequently mill annealed (Oberwinkler, 2010).

12.14 Microstructure of Ti–6Al–4V $\alpha+\beta$ alloy, forged, solution annealed with fast cooling and subsequently mill annealed (Oberwinkler, 2010).

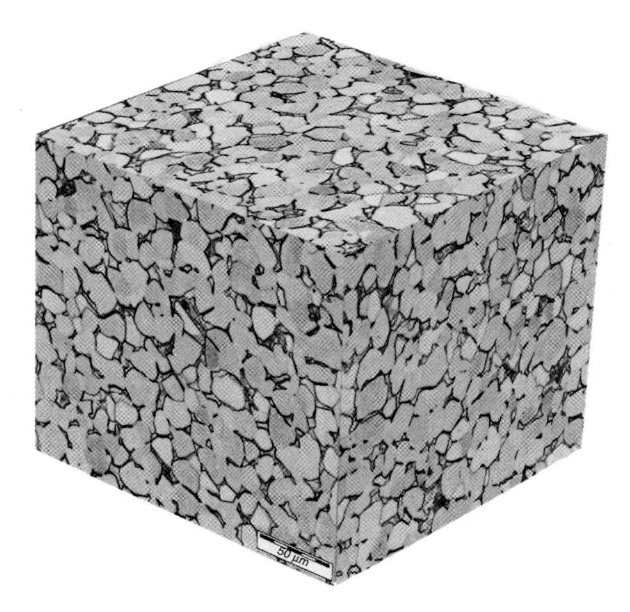

12.15 Microstructure of Ti–6Al–4V $\alpha+\beta$ alloy, forged, and recrystallization annealed (Oberwinkler, 2010).

properties. A fully equiaxed microstructure is usually obtained by recrystallization annealing (see Fig. 12.15), resulting in high ductility and fracture toughness. (See also Lutjering and Williams (2007) and Oberwinkler (2010).)

The last class of titanium alloys is the so-called β alloys such as Ti-17 (Ti–5Al–2Sn–2Zr–4Mo–4Cr) and Ti–10V–2Fe–3Al. An example of the microstructure of a β-processed Ti-17 part is shown in Fig. 12.16. β alloys do not transform martensitically upon quenching to room temperature, resulting in a metastable (seldom stable) β phase. The main characteristic of β alloys is that they can be hardened to much higher yield strength levels than $\alpha+\beta$ alloys. The β alloys are especially good in environments where hydrogen pickup is possible because the β phase has a higher hydrogen tolerance than the α phase. Typically, structural parts such as tracks and landing gear but also some engine parts in low-temperature areas are made from these alloys. (See also Lutjering and Williams (2007) and Warchomicka (2008).)

12.4.3 Titanium aluminides

In any materials selection situation, one must consider which specific properties of a material are crucial for the structural design. One common method is to examine the capabilities of materials by use of property 'cross-plots', as documented by Ashby. The strength properties of TiAl alloys become interesting

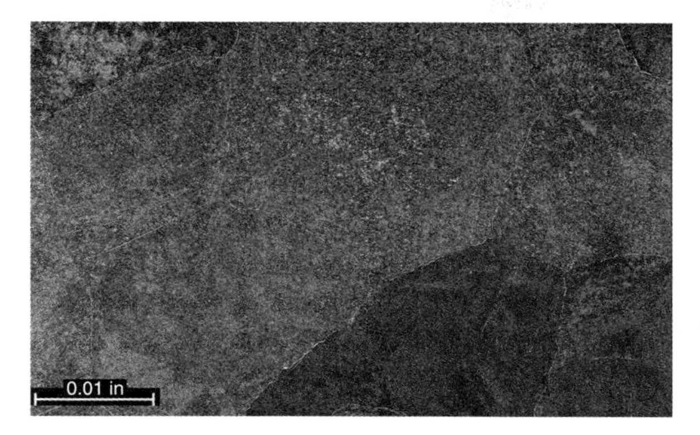

12.16 Microstructure of β-forged Ti17 (courtesy of Böhler Schmiedetechnik GmbH & Co KG, Austria).

relative to structural metals when the density of the material is important (see Fig. 12.17). TiAl alloys are half the density of Ni-based alloys, and their density is even less than that of titanium alloys. Nearly all mechanical design involves selection of materials based on their elastic properties relative to other properties such as temperature, strength or thermal expansion coefficient. When the modulus-versus-yield-strength properties of intermetallics are compared, they are unremarkable relative to steel, titanium and nickel alloys. However, when the modulus relative to density is examined, it can be shown that some intermetallics exhibit a combination of properties which is intermediate between those of common structural alloys and structural ceramics. This is a direct reflection of the change in density and in the characteristics of the atomic bonding (Dimiduk, 1999).

Modern mechanical design calls for control of the fracture properties of structural materials used under both static and cyclic conditions. The deformation and fracture characteristics of TiAl alloys depend sensitively on the composition, the partitioning of ternary and quaternary elements and the microstructure. The nature and scale of the microstructure play an important role in the strengthening of two- and three-phase alloys. Figure 12.18 shows an example of the microstructure of a modern TiAl alloy after heat treatment. Alloys with a fully lamellar microstructure are beneficial for fracture toughness and high-temperature strength. However, ductility at low and ambient temperatures is poor. Duplex structures are conducive to tensile ductility but show poor high-temperature strength and creep resistance.

TiAl alloys are considered to replace higher-density alloys such as nickel-based and iron-based superalloys over certain ranges of temperature and stress. Therefore, their mechanical properties have to be assessed against the high standard set by these alloys (Appel *et al.*, 2000).

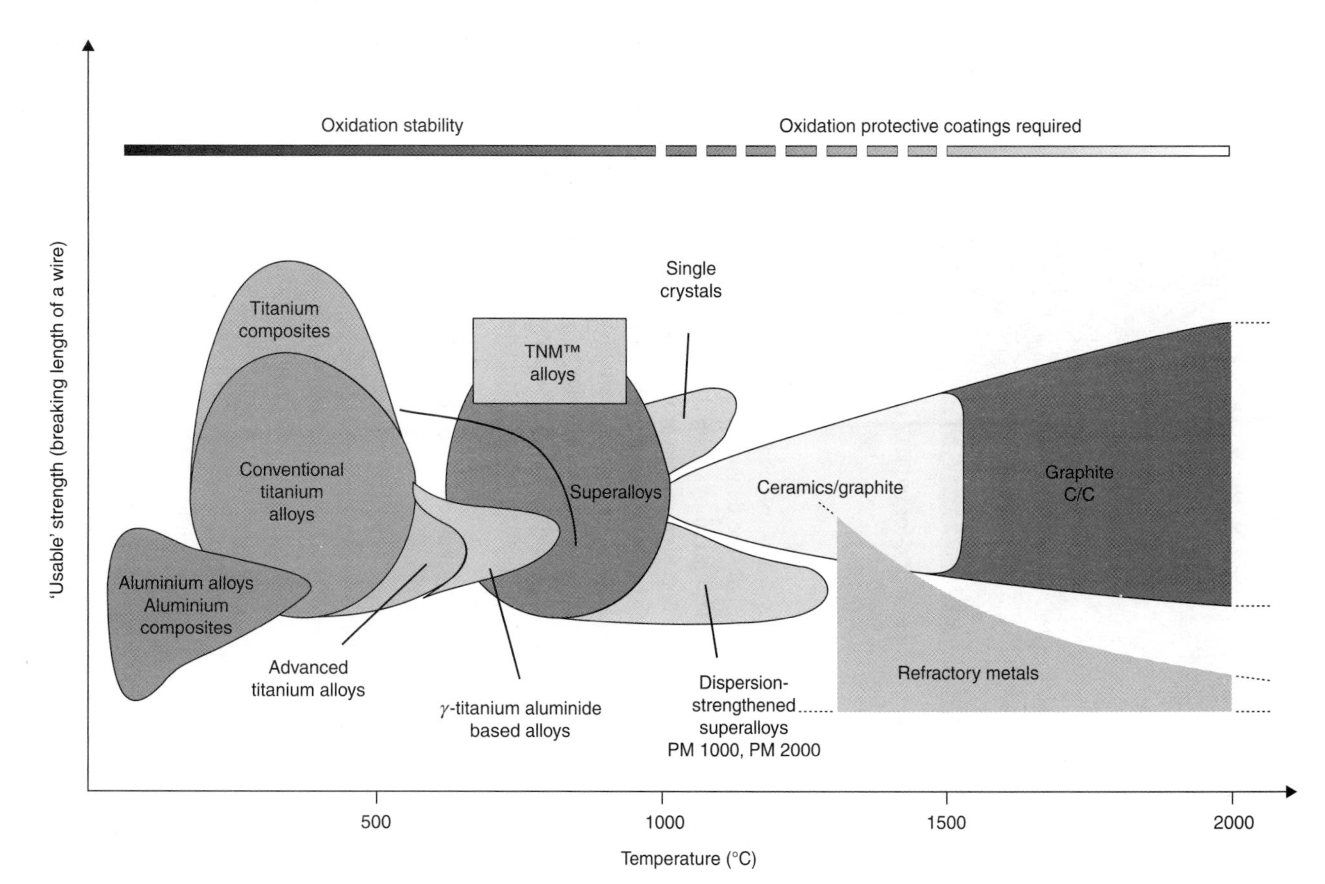

12.17 Ashby plot of strength vs temperature (Clemens, 2010).

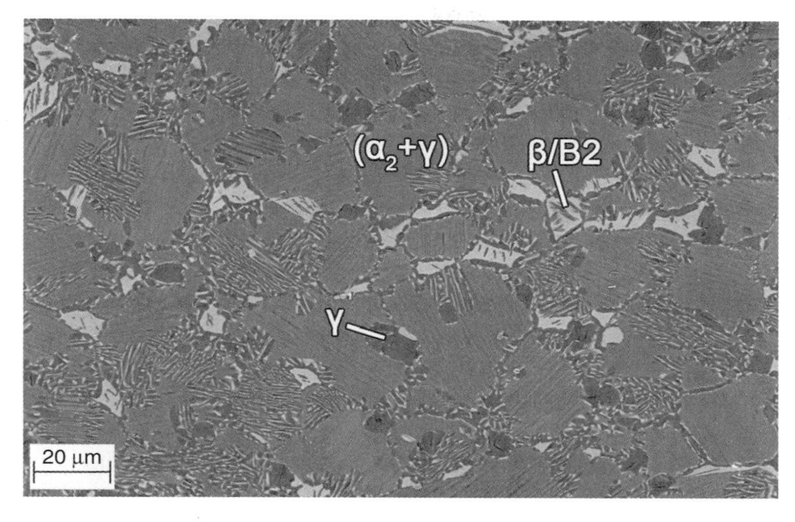

12.18 SEM micrograph of Ti–43.5Al–4Nb–1Mo–0.1B alloy (TNM) after heat treatment (Wallgram *et al.*, 2009).

At room temperature, the fracture toughness versus modulus relationship shows properties at the level of cast irons. Additional comparison with Ti- and Ni-based alloys shows that none of the intermetallics will exhibit toughness levels close to those of titanium alloys and they can only approach the lower range of nickel alloys. However, this situation changes more favourably towards intermetallics as the temperature increases. Nevertheless, the low room temperature fracture toughness and ductility of these alloys require careful handling during the machining and assembly of components such as turbine blades.

Current research on the durability and life-limiting behaviour of intermetallic alloys under fatigue conditions should provide knowledge that will be useful for design approaches. In particular, aluminides exhibit steep crack growth curves as a function of the cyclic stress-intensity range. Therefore, studies have to focus on understanding the microstructure and metallurgical aspects of thresholds for the growth of long cracks (Dimiduk, 1999).

Hence the microstructure of TiAl alloys has to be tailored carefully in order to achieve balanced mechanical properties for specific fields of application.

12.5 Materials modelling and process simulation

Production processes have to be controlled very carefully for all these materials in order to produce parts with the desired microstructure and mechanical properties. One possible way to reduce costs and development time and guarantee stable production processes is the use of modern modelling and simulation technologies.

12.5.1 Nickel-based superalloys

In the context of the thermomechanical processing of superalloys, finite difference methods (FDMs) and finite element methods (FEMs) are being used to investigate the development of temperature and strain during cogging, rolling, forging and heat treatment processes. As a direct result of these simulations, the forces and energies necessary in the forming processes and the heating and holding times for the heat treatment can be evaluated. In order to guarantee good simulation outputs, special focus has to be given to the material data used in the models. Depending on the problem which has to be solved, physical parameters for the elastic, plastic and thermal material behaviour, as well as the properties of interfaces and boundaries, may have to be fed into the simulation software. Even though most of the commercial finite element codes, such as ABAQUS™, FORGE™ and DEFORM™, provide a material database, the quality and limits of validity of the database have to be proven very carefully before use. Especially in the case of superalloys, most companies that simulate thermomechanical processes prefer to develop their own databases, and thus their processes are covered in the most reliable way (Stockinger *et al.*, 2010). For closed-die forging processes, dimensional information such as folding, underfilling of the dies and the shape of the flash can be evaluated directly using finite element simulation. As a result of this, the design of each forming operation can be optimized in such a way as to reduce material input and prevent the occurrence of folding and other dimensional defects. Another effect leading to possible scrap in a very late stage of the production process can be modelled quite accurately with modern FEM codes. In complex geometrical structures such as forgings, sometimes the necessary rapid cooling leads to high residual stresses, which result in distortion during subsequent machining processes. Elastoplastic finite element models allow the prediction of residual stresses after the thermomechanical processing of nickel-based superalloys and offer the possibility to design optimum machining strategies in order to limit geometrical deviations in the final part (Cihak *et al.*, 2006a,b; Krempaszky *et al.*, 2005).

Even though a lot of experience with the forming of these alloys helps in designing the process, the complex relationship between temperature, strain, strain rate and microstructure development in the material cannot be estimated very accurately. Owing to the strong effect of microstructural properties such as the grain size and the size and distribution of precipitates on the mechanical properties, sophisticated models to predict these parameters locally are highly appreciated today.

Very commonly, semi-empirical models, described for instance by Sellars and Whiteman (1979), Cahn and Haasen (1983), Brand *et al.* (1996), Sommitsch (1999), Huang *et al.* (2001), Stockinger and Tockner (2005) and Huber *et al.* (2008), are used to model physical phenomena such as dynamic, post-dynamic and static recrystallization and grain growth as a function of the parameters of thermomechanical processes. A big advantage of such semi-empirical models is that they can be implemented easily in commercial finite element codes, without slowing down the simulation to inefficient computation times. Using these models, it becomes possible to predict the grain size and its distribution across a part after thermomechanical processing. The weakness of such models is the relatively high experimental effort necessary to generate the up to 40 material parameters required, which, furthermore, are valid only for one alloy. Additionally, an accurate prediction of the precipitation and dissolution processes of second phases is not possible. But, in order to predict mechanical properties such as yield strength, the modelling of second phases and also dislocation densities is a must.

For processes with higher symmetry such as rolling, finite difference methods with coupled physical models are possible owing to the faster simulations that are possible (Sommitsch, 1999). For complicated three-dimensional processes such as cogging or closed-die forging processes, a full coupling of such models is generally not efficient enough. Therefore, a different approach is sometimes used to model second phases and dislocation densities, in which the development of the temperature and strain with time at some points is tracked through the process using finite element simulation, and this data is then used for more physical models, such as in the case of thermokinetic simulations using, for instance, MatCalc (Radis *et al.*, 2010).

As a consequence of the development of methods for microstructure simulation, models predicting complex mechanical properties such as fatigue strength and creep resistance have been developed (Stoschka *et al.*, 2010).

12.5.2 Titanium alloys

Given the growing importance of titanium-based materials in the aerospace, automotive, sports and medical sectors, modelling the microstructure and properties of titanium and its alloys is now a vital part of research into the development of new applications. The development of titanium alloys has been an important task for the manufacturing industry. Traditionally, such alloys and processes are designed empirically, i.e. using a mixture of experience and intuition. This is both costly and time-consuming. It is now highly desirable to develop advanced computer-based models for the design of alloy compositions and processing routes. Also, the simulation of microstructure evolution is of high importance owing to its impact on the mechanical properties. The types of phases, the grain size and grain shape, and the morphology and distribution of the fine microstructure determine the properties and therefore the applications of titanium alloys. The final microstructure is formed during thermomechanical processing and heat treatment. It is therefore important to understand the correlation between processing parameters, microstructure and (mechanical) properties.

There are several different approaches to simulating the thermomechanical processing of titanium and titanium alloys. Semi-empirical approaches are usually used to calculate microstructural parameters such as grain size and phase fractions. The calculation time for these relatively simple approaches is in most cases insignificant owing to the advanced computer technology available today. A disadvantage of these approaches is that validation of the models against experimental data is necessary for each material and process data setting.

For example, the kinetics of phase transformations in titanium alloys can be modelled within the framework of the Avrami theory by means of the Johnson–Mehl–Avrami (JMA) equation. This theory has been adapted for different mechanisms of phase nucleation and growth. The JMA kinetic parameters for these transformations for different temperatures, alloys and conditions were obtained by using data from experimental results (Malinov *et al.*, 2001).

Another alternative for the modelling of microstructure in titanium is the cellular automata method. A cellular automaton (CA) is an algorithm that describes the discrete spatial or temporal evolution of a complex system by applying local deterministic or probabilistic transformation rules to the cells of a lattice. The lattice is typically regular, and its dimensions can be arbitrary. In general, CA modelling utilizes a regular lattice that is divided into cells of equal size. Each cell is characterized by several different states. By taking into account the states of the cells in its neighbourhood, the state of the cell can be made to change by time stepping according to transition rules (Sommitsch *et al.*, 2010).

In practice, CA modelling is used to simulate grain-coarsening behaviour in the single- and two-phase areas of titanium alloys. The output of the CA model can be displayed in the form of statistical data on the grain size and the phase distribution, as well as images of virtual micrographs, which offer the possibility of easy comparison with experimental results. On the other hand, the simulation time for an annealing process can be up to several hours.

12.5.3 Titanium aluminides

Because titanium aluminides are one of the youngest material groups used for technological purposes, as well as because of their complex microstructural features and intermetallic properties, just a few modelling techniques have been published so far. Thermodynamic modelling of the stability of different phases shows reasonably good results (see for instance Chladil *et al.* (2005) and Schmoelzer *et al.* (2010)). To optimize heat treatment processes based on thermodynamics, it is necessary to increase the accuracy of the current simulations, and therefore the databases which are used in software packages such as ThermoCalc and MatCalc, have to be adapted.

Modelling the complex microstructure evolution that occurs during the processing of modern titanium–aluminium alloys such as TNM is still a challenge, owing to the multiple phase changes between the solidus temperature and room temperature (see Schmoelzer *et al.*, 2010). Some research activities on phase field models describing the lamellar $\alpha_2 \rightarrow \alpha_2 + \gamma$ transformation have been published (e.g. Wen *et al.*, 2001) but, owing to the high simulation effort, these models are not currently being used to optimize thermomechanical treatments in real production processes.

For massive forming processes such as extrusion and isothermal die forging, FDMs and FEMs are frequently used to describe the strain, temperature and strain rate, as described in Section 12.5.1. Special effort has to be put into the generation and validation of material data such as flow curves and thermo-physical properties up to 1400°C in order to obtain reasonable results in simulations.

12.6 Process and materials optimization: case study

Semi-finished material is produced from nickel-based superalloys by the following process chain: melting (vacuum induction melting); remelting (electroslag remelting and/or vacuum arc remelting); primary forming (cogging/breakdown and coarse rolling); secondary forming (open die forging, radial forging and continuous rolling); and heat treatment and finishing (see Fig. 12.19). In order to fulfil the specifications concerning mechanical properties and grain structure, process optimization needs to be performed by employing numerical modelling and simulation. The development and application of process and structure models for this purpose, and comparison of these models with experiment, are described theoretically and by examples using case studies in the following paragraphs.

The materials examined in these case studies are Alloy 80A and Alloy 718. Alloy 80A is a nickel-based superalloy with a relatively low alloying content, which is being used today in a variety of applications, for example in power plant engineering (primarily for blades), in tool production (e.g. for diamond wheels), in fastening elements (e.g. bolts used in turbine engineering) and in engine construction (mainly in valves and fittings). Alloy 718 is the workhorse of the nickel-based superalloys and was developed for gas turbine applications in the late 1950s. It is currently used in cast, wrought and powder forms (Radavich, 1989). This alloy has excellent strength, ductility and toughness over a wide range of temperatures. A major attribute of Alloy 718 is its process versatility. It can be fabricated over a relatively wide range of temperatures, forging reductions and strain rates to produce microstructures and associated properties for specific requirements. Alloy 718 is unique among nickel-based alloys because of its outstanding weldability and good resistance to strain–age cracking. It possesses excellent corrosion and oxidation resistance in addition to good formability (Klopp, 1995).

The matrix of these alloys is principally of the face-centred cubic lattice type, containing finely dispersed γ' precipitates and carbides. In Alloy 80A, the precipitates are fine and homogeneously distributed and can be of globular or angular shape. They are face-centred cubic and coherent with the matrix, thus exhibiting low surface energy and good long-term stability at elevated temperatures. The carbides are usually of the MC type (with M = Ti, Ta, Nb, W)

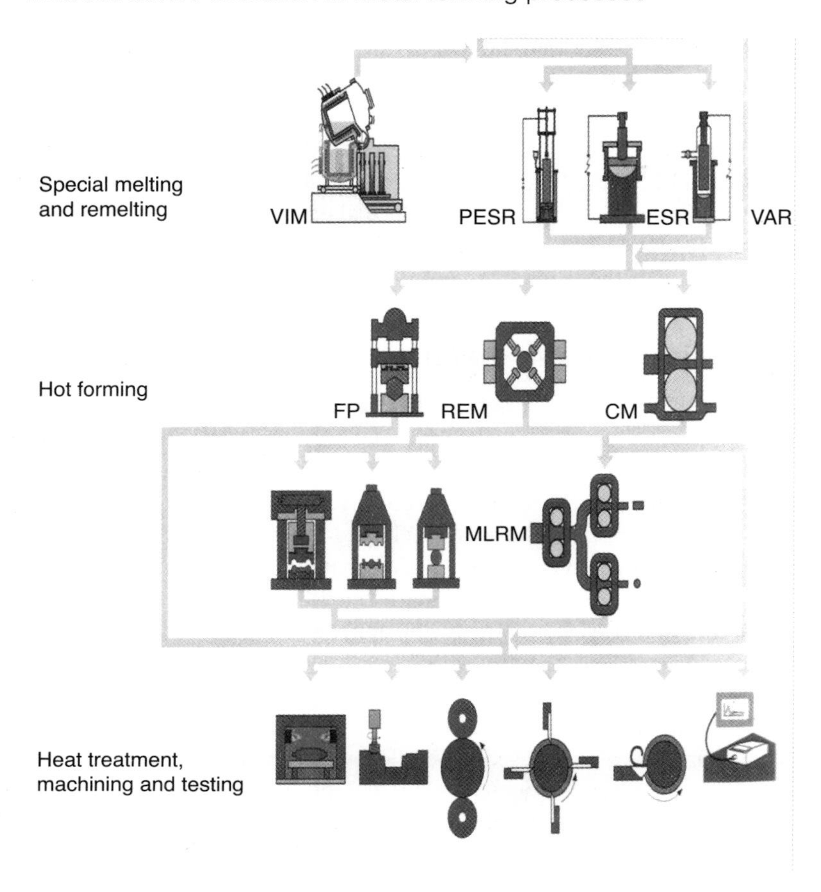

12.19 Production route for semi-finished nickel-based alloys (courtesy of Böhler Edelstahl GmbH & Co KG, Austria). (PESR, pressure electroslag remelting; FP, forging press; RFM, rotary forging machine; CM, cogging mill; MLRM, multi-line rolling mill.)

or the $M_{23}C_6$ type (with M = Cr, Fe, Ni, Co, Mo). It is important to note that these carbides do not form continuous grain boundary films that would affect the material's fracture resistance. Undesirable phases, such as sigma phases and Laves phases, may entail embrittlement, a decrease in strength or other problems.

In the fully heat-treated condition, i.e. solution heat treated and aged, wrought Alloy 718 contains the following precipitate phases. The major strengthening precipitate is the gamma double-prime (γ'') phase (Ni_3Nb). A lower amount of γ' phase may also be present in the alloy, depending on the Nb content and the heat treatment. The γ'' phase is only metastable in Alloy 718. The stable phase is δ, which plays an important role in the control of grain size for wrought Alloy 718 as its presence can inhibit grain coarsening during processing and heat treatment (Sundararaman *et al.*, 1997; Kalluri *et al.*, 1994; Thamboo *et al.*, 2000). Other phases that are present include primary MC phases ((Nb,Ti)(C,N)), especially Nb-rich carbides, carbonitrides and Ti-rich nitrides. A phase that is very detrimental during operation is the Laves phase $(Fe,Ni)_2Nb$, which must be removed by homogenization (Schirra *et al.*, 1991). M_6C and the detrimental alpha-chromium (α-Cr) phase have been found after very long-time exposure at higher temperatures (Radavich, 1989).

An increasing content of alloying elements in these alloys causes the recrystallization temperature to approach the solidus–liquidus range, leading to a further reduction in the already very small hot-forming window, and thus to increasing hot-forming problems. The borderline between nickel-based superalloys that are still forgeable and those that can only be produced by casting methods is usually regarded as corresponding to a γ' factor (= Ti + Al) of 7.5 to 8% (Sims *et al.*, 1987).

Both Alloy 80A and Alloy 718 can be produced via double or triple melting as discussed above. VIM ensures that the raw materials, consisting of virgin material and revert, are mixed together homogeneously in the electrode, but it does not provide a solid product that is suitable for subsequent hot working. Before the remelting procedure, the electrodes are heat-treated to improve their ductility, and thus the effects of the melting cycles can be avoided or minimized in order to reduce the susceptibility to cracking due to internal stresses (Kuchler *et al.*, 2003; Helms *et al.*, 1996).

In ESR, an alternating current is passed through a consumable electrode, a molten slag in which the tip of the electrode is immersed, and a continuously solidifying ingot. The current heats the slag above its melting temperature; the molten slag then melts the electrode into a water-cooled mould. The process takes place in air under a metallurgically active slag, and the energy for melting is provided by resistance heating of the slag. The advantages of the ESR process can be summarized as follows: the molten slag acts as a refining agent to remove oxide clusters (Yue and Dominique, 1989; Tien, 1989; Cordy *et al.*, 1984), the ingots exhibit a low sulphur content and a high cleanness level, and the remaining inclusions are finely dispersed. Hence ingot formation by ESR provides an optimum microstructure with isotropic properties and good hot-forming characteristics. Process automation ensures high reproducibility. The disadvantages of the ESR process are higher gas and trace element contents owing to the absence of a vacuum, the loss of reactive alloying elements, and a tendency towards freckle formation. Freckles are long, discontinuous chain-type segregations of nitrides, carbides and carbonitrides arranged almost parallel to the ingot axis, which occur in the area between the centre and the mid-radius position of the ingot. The tendency towards freckle formation increases with the complexity of the alloy. Freckle formation is promoted by the buoyancy effect of the residual melt (enriched in elements of low specific gravity such as Ti) in the dual-phase range. The more extensive this dual-phase range is, the more intense is the buoyancy effect and the more likely the formation of freckles.

VAR entails the continuous melting of a consumable electrode under vacuum into a water-cooled copper crucible by a direct-current arc to produce an ingot with a structure and chemistry superior to the electrode being melted. The VAR process reduces the levels of volatile tramp elements. Successful application for any particular alloy/ingot diameter combination is believed to be dependent on achieving quasi-steady thermal/solute conditions at the solidification interface. Equilibrium phase relationships dictate the partition of solutes at the intricately shaped dendritic solidification interface, and the local conditions determine the size and geometry of local features and the chemical homogeneity. The local thermal environment is strongly influenced by fluid flows, which in turn are driven by global temperature gradients (convection) and magnetohydrodynamic forces created by the current distribution of the arc. Even though the design and application of VAR have evolved to a sophisticated level, a basic understanding of the metal vapour arc and its coupling with the metallurgy of the process has not yet been achieved. At this point in time, application to larger ingot sizes with more rigorous quality standards is limited by this lack of understanding (Zanner *et al.*, 1989).

In recent years, the performance requirements for nickel-based superalloys have been steadily increasing, resulting in, among other things, a trend towards lower thermomechanical processing temperatures and finer grain sizes. Although improved control of melting processes has reduced the incidence of many melt-related defects, such as freckles, the trend towards finer grain sizes has brought into increased prominence solute-lean defects (lean particularly of Nb and Ti), or white spots, which are generally associated with VAR (Zanner *et al.*, 1989). VAR is less prone to positive segregation than ESR, and is therefore capable of melting larger ingots. However, ESR is more resistant to the formation of white spots or negative segregation (Jackman *et al.*, 1993). Compared with VAR, ESR provides better retention of volatile elements, such as magnesium, which may favourably affect ductility. Conversely, volatile elements detrimental to ductility (such as lead and bismuth) are not removed as effectively in ESR. Triple melting combines the advantages of both ESR and VAR (Moyer *et al.*, 1994). The maximum ingot cross section that can be melted without unacceptable segregation is dependent on the alloy composition. Superalloys highly prone to segregation are limited to diameters of the order of 500–600 mm, whereas some superalloys can be melted to diameters of 1000 mm or larger. A popular ingot size for segregation-prone alloys such as Alloy 718 is 500 mm in diameter and approximately 4.5 tonnes in weight (Forbes Jones and Jackman, 1999).

Process modelling has become a viable tool for optimizing the ESR and VAR processes. A comprehensive modelling approach for the simulation of ingot solidification phenomena in secondary remelting processes is depicted in Fig. 12.20. Based on process- and material-dependent input data, a deterministic

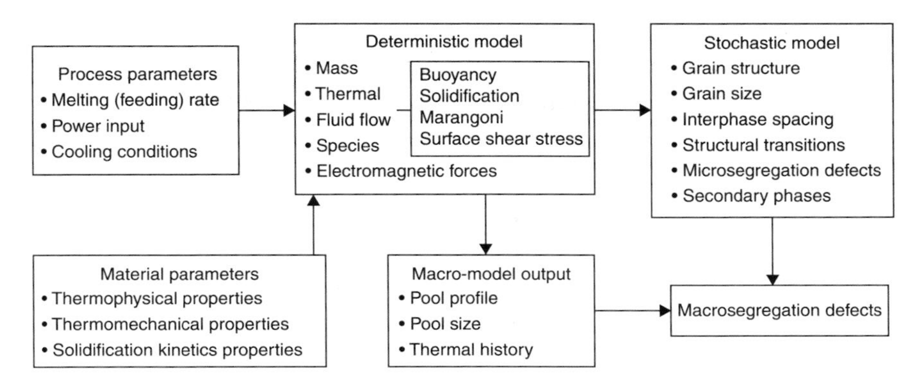

12.20 Flow diagram showing the modelling requirements for secondary remelting processes (Nastac *et al.*, 1998).

model performs macroscopic mass, heat transfer, fluid flow, electromagnetic and species-transport computations to provide temperature, velocity and concentration-field output. From the macroscopic temperature field, the pool profile and size and the shape and size of the mushy region can be determined. In addition, macrosegregation-related defects, such as freckles and tree-ring patterns, can be obtained from the concentration field. A stochastic model could also use the results from a macroscopic model to predict the evolution of the grain structure of solidifying ingots (Nastac *et al.*, 1998). Reiter *et al.* (2003), for example, modelled the VAR process with the help of a FEM-based program that was adapted to include the influence of the magnetohydrodynamic effect on the fluid flow. Furthermore, the heat balance of the system was modelled accurately by subroutines that included the heat fluxes, the formation of shrinkage gaps and the additional heat due to radiation from the arc, and by inverse calculations to obtain the heat flow boundary conditions. Practical experiments were carried out on ingots of Alloy 718 of diameter 410 mm (16 in) and 510 mm (20 in) at different melting rates. The resulting pool profiles were investigated and compared with calculations.

Typically, large ingots in the as-cast condition show serious interdentritic segregation, for example in the case of Alloy 718, which is high in Nb, resulting in the formation of the Nb-rich Laves phase (up to 30% Nb), coarse NbC and δ phase in the interdentritic regions, while the primary dendritic areas are Nb-depleted (Kuchler *et al.*, 2003). Detrimental phases must be reduced by a proprietary heat treatment. If, for example, the Laves phase is present as a continuous or semi-continuous grain boundary network in forged Alloy 718 workpieces, significant reductions in mechanical properties such as ductility and strength can be observed (Schirra *et al.*, 1991). Therefore, the ingots undergo a common multi-step homogenization procedure. In order to avoid cracking during homogenization, heating and cooling cycles are used which preclude the formation of thermal stresses sufficiently large to initiate cracking of the ingot. The larger the ingot is, the more distinct this phenomenon is because of the increased temperature gradients.

The need for uniform microstructures and properties in industrial components requires that careful consideration be given to the choice of the mechanical hot-deformation techniques utilized, during all stages of processing. During ingot breakdown by either rolling or forging, the primary processor must use reduction schedules which are both economical and attempt to transform the inhomogeneous cast ingot structure into a uniform structure. Subsequent working during final processing can also further homogenize the microstructure (Weis *et al.*, 1989).

After the homogenization treatment and conditioning, the ingots are heated for the billet breakdown process with the aim of obtaining a fine, recrystallized grain structure. A single deformation step is not sufficient to refine the grain size in all areas of the billet; hence, in the case of billet forging, a forging practice consisting of multiple upset and drawing stages is commonly applied. As an example, for Alloy 718, the billet forging process consists of high-temperature forging above the δ-solvus to break up the as-cast ingot structure, followed by low-temperature forging in the range of the δ-solvus to obtain a fine-grained microstructure. To find a balance between the promotion of recrystallization and recovery and at the same time avoiding surface cracking, it is important to precisely control the process with regard to the forging temperature, the amount of strain per forging pass and the time required for forging (Dandre *et al.*, 2000a; Guimaraes and Jonas, 1981). In general, owing to the size of the pieces of material being formed, the deformation requirements during primary breakdown are significantly different from those for final component fabrication. For example, the average strains imposed in each deformation cycle during breakdown of large ingots are typically low and non-uniform and result in non-uniform recrystallized microstructures. During working, as the size of the worked product decreases, the potential for imposing higher and more uniform strain fields, which correspondingly produce more uniform recrystallized microstructures, increases.

Investigations of the deformation behaviour of as-cast microstructures have shown that unlike the behaviour during the deformation of pre-cogged structures, dynamic recrystallization seems to have a very small influence on the evolving structure (Weis *et al.*, 1989; Zhao *et al.*, 1997). The mechanisms responsible for softening are primarily dynamic recovery, adiabatic heating and its effect on dynamic recovery and shear band formation (Guimaraes and Jonas, 1981; Dandre *et al.*, 2000b); additionally, static recrystallization occurs during the periods of dwell between successive passes (Mataya and Matlock, 1989; Mataya *et al.*, 1994). Wasle *et al.* (2003) investigated the evolution of the microstructure of a vacuum-arc-remelted as-cast ingot made from Alloy 80A during cogging. The primary breakdown of the structure was simulated by Gleeble compression tests. The very large grains in the cast structure facilitate dynamic recovery processes at the expense of recrystallization (see Fig. 12.21).

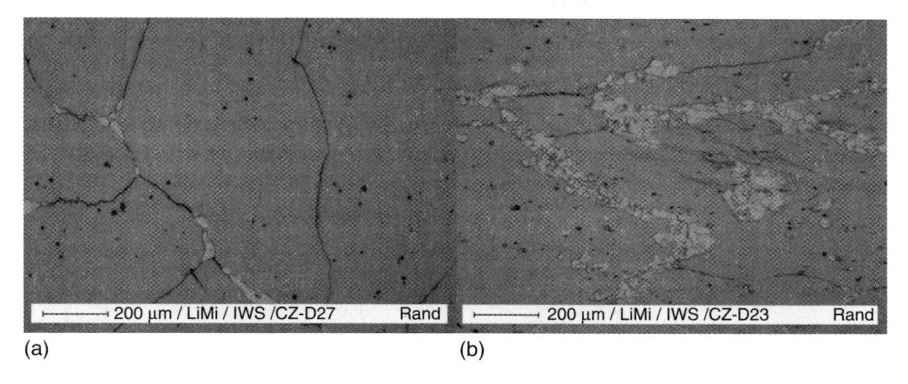

12.21 Recrystallizing grains (light grey) in the shell at 1180°C at a strain rate of 1 s^{-1} and a strain of (a) 0.1 and (b) 0.8 (Wasle *et al.*, 2003).

In creep-resistant nickel-based alloys such as Alloy 80A, a γ'-free mixed crystal occurs only above the γ' dissolution temperature. For formability reasons, hot forming of these materials generally also has to be performed in this temperature range. The microstructure obtained by solution annealing depends not only on the annealing temperature but also on the structural condition prior to annealing, i.e. after heat treatment. The processes occurring during hot forming are determined by the temperature employed, as well as by the ratio and speed of reduction. The processes of formation of new grains and grain growth start in the temperature range where γ' dissolution begins. If the hot forming process fails to result in a uniform, dynamically recrystallized structure, annealing will lead to either grain coarsening or secondary recrystallization. At deformation temperatures below the γ' dissolution temperature, with incipient γ' precipitation, the structure is characterized by strongly deformed grains, which are in part dynamically recovered or recrystallized (see Fig. 12.22). In materials with a low stacking fault energy, such as nickel-based alloys, the dislocation arrangement remains of high energy owing to the fact that the dislocations are linked to the sliding planes, and recrystallization prevails over recovery. A dynamically recrystallized structure is characterized by recrystallization twins with coherent twin boundaries, which are straight and parallel to each other.

A forming temperature slightly above or below the γ' dissolution temperature results in a fully dynamically recrystallized structure with different grain sizes. As schematically illustrated in Fig. 12.23, the surface zone, which cools off too much, undergoes incomplete recrystallization, depending on the temperature and transformation conditions during forming. In order to avoid the undesirable formation of coarse grains, the forming parameters need to be adjusted to obtain the highest possible structural uniformity, with regard to grain size and the distribution of the residual deformation energy, as only this allows a uniform grain structure to be obtained over the entire cross section after solution annealing.

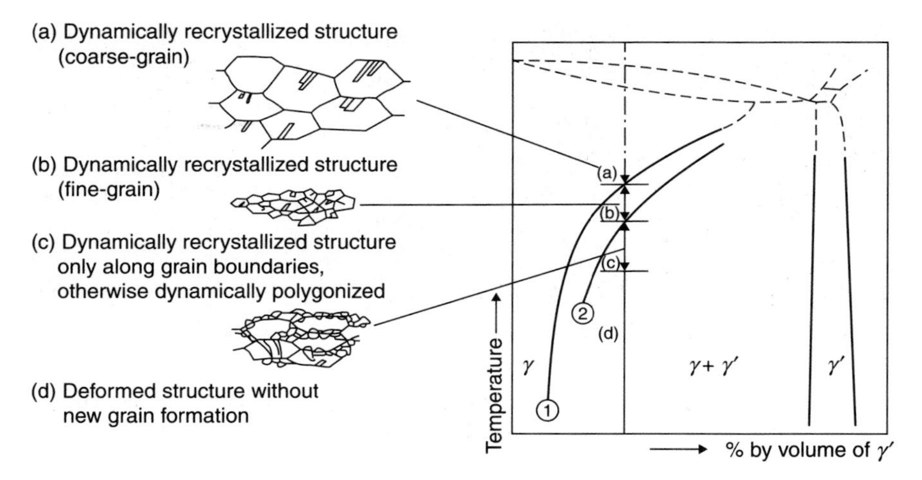

12.22 Grain structure formation as a function of hot-forming temperature (Lehnert *et al.*, 1993).

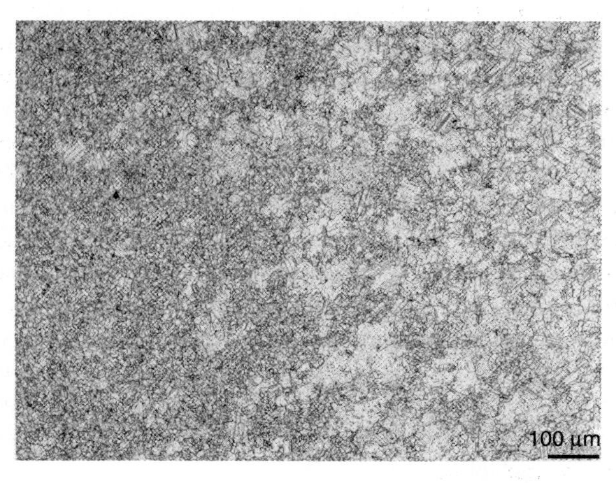

12.23 Unevenly distributed temperature leads to incomplete recrystallization (Stockinger and Tockner, 2005).

In recent years, the constantly expanding capacity of modern computers has allowed the simulation of more complex numerical problems by means of, for example, the finite element method. Thus, the prediction of local plasto-mechanical, thermal or structural conditions is possible. The number of expensive trials can be reduced significantly if program modules are available to compute microstructural characteristics. In this case, optimization of the microstructure by variation of the forming conditions can be attempted (Brand et al., 1996).

It is generally thought that the ability to model and predict, as well as to observe and classify, is a sign of scientific maturity. In this sense, materials science is definitely becoming a mature field (Brechet, 2004). But materials science has a special status owing to the close relation it bears to engineering applications such as metal forming. In that sense, maturity also presupposes the ability to provide guidelines for materials design by choosing appropriate hot forming and heat treatment conditions. Recently, the development of microstructure during the hot forming of nickel-based alloys was modelled using a physical approach by Sommitsch and Mitter (2006), including an evaluation of the thermodynamic equilibrium and of multi-component diffusion, which is necessary for the description of the particle kinetics and for the simulation of the evolution of the grain structure due to recrystallization and grain growth. These authors illustrated a new, physically based continuum-mechanical microstructure model that is expected to lead to a better understanding of microstructure genesis and materials properties, as well as to provide guidance for the optimization of materials processing routes and the resulting structures.

During thermomechanical processing, a dislocation substructure is developed as deformation is imposed. The stored energy can provide the driving force for various restorative processes such as dynamic recovery and recrystallization. The process of recrystallization may be static or dynamic, and involves the nucleation of dislocation-free grains and their subsequent growth. Upon completion of recrystallization, the energy can be reduced further by grain growth, in which the grain boundary area is reduced. The kinetics of recrystallization and grain growth processes are complex. In order to predict the grain size distribution in finished components, a basic understanding of the evolution of the microstructure during complex manufacturing sequences, including the primary working processes (ingot breakdown, rolling and extrusion), final forging or rolling, and heat treatment, must be obtained. As the temperature falls below the liquidus temperatures of second phases, the interaction of particles with recrystallization and grain growth and with mobile dislocations, i.e. the interaction with the flow stress, has also to be accounted for. A precise description of precipitation kinetics is additionally needed for the modelling of heat treatment and ageing processes (see Fig. 12.24).

Especially in the case of nickel-based superalloys such as Alloy 718 and Waspaloy, which are used in critical components such as low- and high-pressure turbine discs for aircraft engines (see Fig. 12.25), tight process control is necessary, together with extensive process modelling. These materials operate in an environment of high creep and fatigue stresses at elevated temperatures. Therefore it is obvious that such components will have challenging mechanical requirements and narrow specifications. To achieve not only a correct, defect-free final geometry but also the required properties, it is necessary to produce a defined microstructure

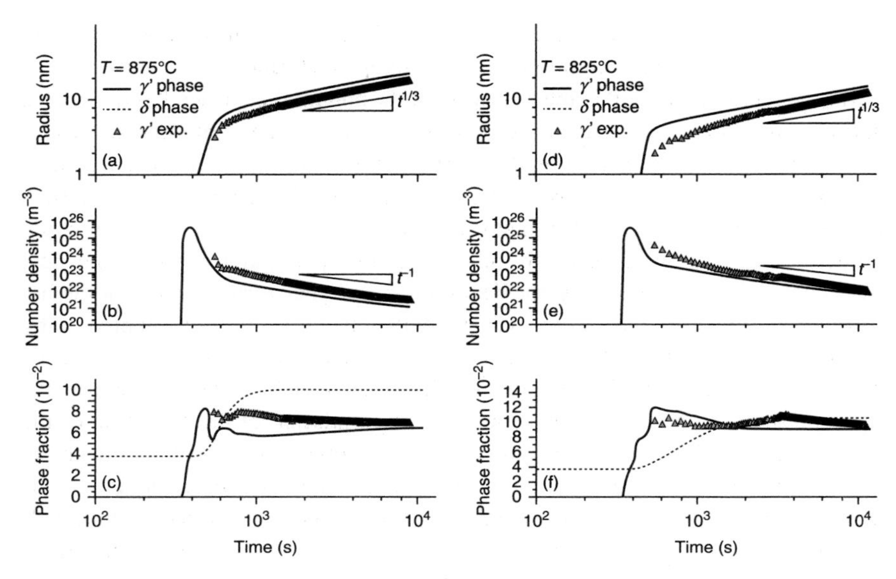

12.24 Radius, number density and volume fraction of γ′ precipitates versus time and comparison with experimental results at temperatures of 825°C and 875°C (Radis *et al.*, 2010).

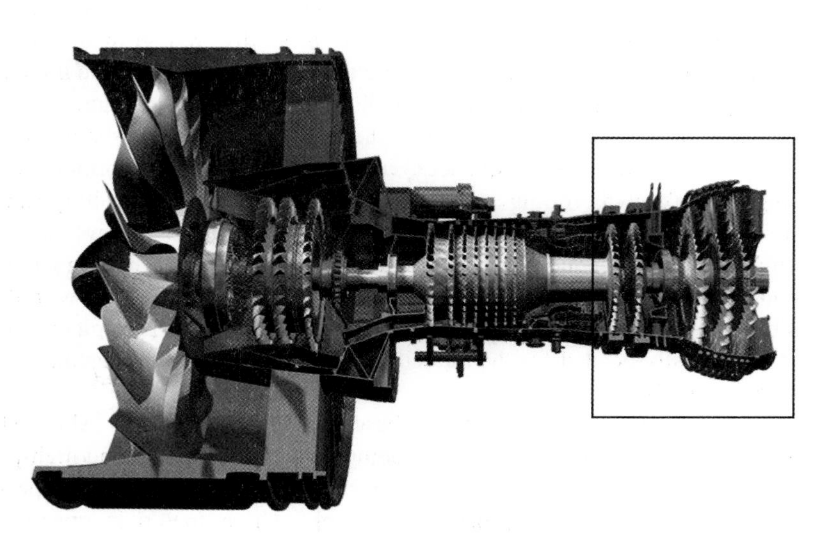

12.25 Sketch of an aircraft engine (courtesy of MTU, Germany).

throughout the whole component. The most frequently used nickel-based alloy for this application today is the ca. 40-year-old Alloy 718 (see also Table 12.1). If Alloy 718 is used in turbine discs, the preforming route described above has to be chosen even more carefully in order to generate a microstructure with a grain size at least as fine as ASTM 5.

The steps after cogging consist in most cases of two or more closed-die forging operations, which are performed to produce geometries as close as possible to the final shape of an engine disc. Owing to the high cost of nickel-based superalloys such as Alloy 718, these forging steps are usually designed without a flash in order to reduce scrap material. Therefore, the design of each forging step has to be done carefully to guarantee complete filling of the dies. Additionally, each forging process has to be optimized with regard to the strain and temperature distribution in the part, so that a fine grain size (often ASTM 10 or finer) and a good distribution of δ-phase can be guaranteed in the final part. Most forging companies use semi-empirical microstructure models coupled to finite element codes to describe the complex influences of strain, strain rate and temperature on recrystallization, grain growth and therefore microstructural parameters such as grain size. In Fig. 12.26, a comparison of simulation results of a forging process performed at the same temperature but with two different die designs is shown. It can be seen

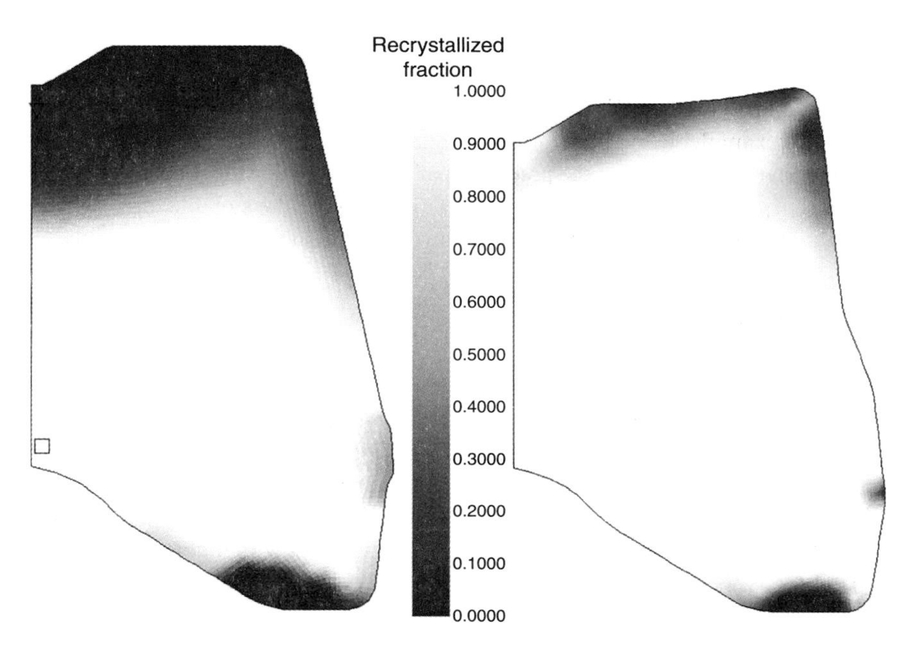

12.26 Comparison of simulations of the recrystallized fraction of an Alloy 718 billet forged with two different blocker die geometries (Stockinger and Tockner, 2005).

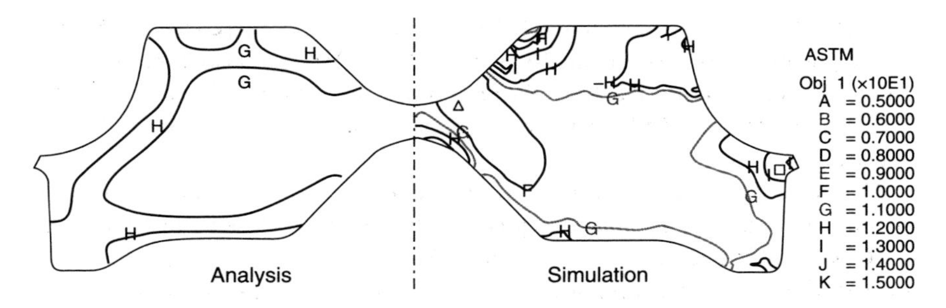

12.27 Comparison of an experimental grain size analysis mapped onto a forged engine disc and a microstructure simulation result using the same process parameters (Stockinger and Tockner, 2005).

that the right-hand design shows much less area with incomplete recrystallization and therefore a higher probability of coarser grains. Although complete recrystallization after the first forging operation will lead to a fine grain size, it is of high importance to introduce significant deformation throughout the whole cross section during final forging too. Areas with strains below the value for the onset of dynamic recrystallization may recrystallize statically during subsequent solution annealing, which results in coarser local structures. Some customers allow direct ageing of Alloy 718, which reduces the risk of local coarse-grain areas and has an additional beneficial effect on some mechanical properties. Figure 12.27 demonstrates the advantage of microstructure models using a final forged disc as an example, using the blocker design on the right-hand side of Fig. 12.26. This comparison shows that the prediction of the grain size using the finite element model is in good agreement with the analysis and is therefore suitable for designing a forging process upfront with regard to the desired final microstructure.

12.7 Future trends

12.7.1 Nickel-based alloys

The environmental effect of stationary and flying gas turbines is currently the major driver for further research and development regarding nickel-based alloys. The demand for higher efficiency combined with lower air pollution due to exhaust of CO and NO_x leads to the need for an increase in the working temperature. This consequently means the development of new alloys with higher-temperature capability, as well as the development of new technologies and modification of old technologies to produce parts out of these alloys. As an example, in the case of large rotors for turbines in the field of energy production, producing ingots for remelting with the same quality as those currently used, and stronger forging equipment, will be two challenges for the near future.

12.7.2 Titanium alloys

Owing to the increase in titanium alloy usage in modern aircraft such as the Boeing 787, the high material costs are a driver for the development of lower-cost materials for applications with less demand on mechanical properties. For smaller parts such as those used in the automotive industry, production technologies such as severe plastic deformation may offer the possibility to use unalloyed titanium as a replacement for more expensive alloys without loss in room temperature properties (Semenova *et al.*, 2008).

Another driver is increasing life expectancy and therefore the growing demand for human bone replacements. In this case, the development process must take account of the compatibility of the material with the human body with regard to chemical elements, as well as physical properties such as Young's modulus.

12.7.3 Titanium aluminides

As in the case of nickel and titanium alloys, a major driver for the future development of titanium aluminides will be the aircraft and aerospace industry and its environmental effects. Higher production volumes will force the development of more efficient processing technologies in order to decrease costs and thus increase efficiency.

12.7.4 Materials simulation

Particularly in the case of titanium aluminides, efficient material models to describe microstructural changes and damage and to predict mechanical properties must be developed in the future in order to optimize the thermomechanical processes used.

With increasing computer speed, the models used for industrial processes will become more and more sophisticated and more accurate, and will be used more frequently too. Huge efforts are currently being made to link simulations along the whole production process. Because these simulations have to be done across several companies or research institutes, interfaces have to be defined very carefully. If the results are to be used for design purposes too, the quality of the simulation results has to be proven statistically, and normal process variations have to be implemented in the simulations. Additionally, standards for data exchange between the partners in the simulation operation must be defined. Finally, the methods, the advantages of these standards, and the risks, have to be carefully documented in order to get these methods approved by responsible authorities such as the Federal Aviation Administration.

12.8 Sources of further information and advice

Appel F, Brossmann U, Christoph U, Eggert S, Janschek P, *et al.* (2000), 'Recent progress in the development of gamma titanium aluminide alloys', *Advanced Engineering Materials 2*, 11, 699–720.

Clemens H and Kestler H (2000), 'Processing and applications of intermetallic γ-TiAl-based alloys', *Advanced Engineering Materials 2*, 9, 551–570.
Kracke A (2010), 'Superalloys, the most successful alloy system of modern times – past, present and future', in Ott E A, Groh J R, Banik A, Dempster I, Gabb T P, *et al.* (eds), *Proceedings of the 7th International Symposium on Superalloy 718 and Derivates*, Warrendale, PA: The Minerals, Metals & Materials Society, 13–50.
Lütjering G and Williams J C (2007), *Titanium*, 2nd edn, Berlin: Springer.
Reed R C (2006), *The Superalloys: Fundamentals and Applications*, New York: Cambridge University Press.
Sha W and Malinov S, 2009, *Titanium Alloys: Modeling of Microstructure, Properties and Applications*, Cambridge: Woodhead.
Sims C, Stoloff N and Hagel W (1987), *Superalloys II: High Temperature Materials for Aerospace and Industrial Power*, New York: Wiley.
Sjöberg G (2010), 'Casting superalloys for structural applications', in Ott E A, Groh J R, Banik A, Dempster I, Gabb T P, *et al.* (eds), *Proceedings of the 7th International Symposium on Superalloy 718 and Derivates*, Warrendale, PA: The Minerals, Metals & Materials Society, 117–130.
Sommitsch C (2006), *Numerical Modelling and Simulation of Production Processes of Semi-Finished Nickel-Based Superalloys*, Habilitation thesis, Graz University of Technology, Austria.

12.9 References

Appel F, Brossmann U, Christoph U, Eggert S, Janschek P, *et al.* (2000), 'Recent progress in the development of gamma titanium aluminide alloys', *Advanced Engineering Materials 2*, 11, 699–720.
Asthana R, Kumar A and Dahotre N B (2006), *Materials Science in Manufacturing*, Burlington, MA: Elsevier.
Brand A J, Karhausen K and Kopp R (1996), 'Microstructural simulation of nickel base alloy Inconel 718 in production of turbine discs', *Materials Science and Technology*, 12, 963–969.
Brechet Y (2004), 'Physically based models for industrial materials: what for?', in Raabe D, Roters F, Barlat F and Chen L Q (eds), *Continuum Scale Simulation of Engineering Materials*, Weinheim: Wiley-VCH, 231.
Cahn R W and Haasen P (1983), *Physical Metallurgy*, Amsterdam: Elsevier.
Chandley D (2000), 'Use of gamma titanium aluminide for automotive engine valves', *Metallurgical Science and Technology*, 18, 8–11.
Cihak U, Staron P, Clemens H, Homeyerc J, Stockinger M and Tockner J (2006a), 'Characterization of residual stresses in turbine discs by neutron and high-energy X-ray diffraction and comparison to finite element modeling', *Materials Science and Engineering A*, 437(1), 75–82.
Cihak U, Staron P, Stockinger M and Clemens H (2006b), 'Characterization of residual stresses in compressor discs for aeroengines', *Advanced Engineering Materials*, 8, 1088–1092.
Chladil H, Clemens H, Leitner H, Bartels A, Gerling R and Marketz W (2005), 'Experimental studies of phase transformations in a carbon containing Ti–45Al–7.5Nb alloy and related thermodynamic simulations', *Advanced Engineering Materials*, 7, 1131–1134.
Clemens H (2008), 'Intermetallische Werkstoffe für Anwendungen in Automobil- und Flugzeugtriebwerken', *BHM Berg- und Hüttenmännische Monatshefte*, 153, 337–341.

Clemens H (2010), 'Entwicklungstrends bei intermetallischen Titanaluminiden', Presentation given at the 8th Werkstoffkongress, Leoben, Austria.

Clemens H and Kestler H (2000), 'Processing and applications of intermetallic γ-TiAl-based alloys', *Advanced Engineering Materials 2*, 9, 551–570.

Cordy J T, Kelley S L and Lherbier L W (1984), 'Chemistry and structure control in remelted superalloy ingots', in Bhat G K and Lherbia L W (eds), *Proceedings of Vacuum Metallurgy Conference on Specialty Metals Melting and Processing*, Warrendale, PA: ISS, 69–74.

Dandre C A, Walsh C A, Evans R W, Reed R C and Roberts S M (2000a), 'Microstructural evolution of nickel-base superalloy forgings during ingot-to-billet conversion', in Green K A, Pollock T M and Kissinger R D (eds), *Proceedings of Superalloys 2000*, Warrendale, PA: The Minerals, Metals & Materials Society, 85–94.

Dandre C A, Roberts S M, Evans R W and Reed R C (2000b), 'Microstructural evolution of Inconel 718 during ingot breakdown: process modelling and validation', *Materials Science and Technology*, 16, 14–25.

Dimiduk D M (1999), 'Gamma titanium aluminide alloys – an assessment within the competition of aerospace structural materials', *Materials Science and Engineering A*, 263, 281–288.

Durrand-Charre M (1997), *The Microstructure of Superalloys*, Amsterdam: Gordon and Breach.

Forbes Jones R M and Jackman L A (1999), 'The structure evolution of superalloy ingots during hot working', *JOM*, 51, 27–31.

Guimaraes A A and Jonas J J (1981), 'Recrystallization and aging effects associated with the high temperature deformation of Waspaloy and Inconel 718', *Metallurgical Transactions*, 12, 1655–1666.

Helms A D, Adasczik C B and Jackmann L A (1996), 'Extending the size limits of cast/wrought Superalloy ingots', in Kissinger R D, Deye D J, Anton D L and Cetel A D (eds), *Proceedings of Superalloys 1996*, Warrendale, PA: The Minerals, Metals & Materials Society, 427–433.

Huang D, Wu W T, Lambert D and Semiatin S L (2001), 'Computer simulation of microstructure evolution during hot forging of Waspaloy and nickel alloy 718', in Srinivasan R and Semiatin S L (eds), *Proceedings of Microstructure Modeling and Prediction During Thermomechanical Processing*, Warrendale, PA: The Minerals, Metals and Materials Society, 137–146.

Huber D, Stotter C, Sommitsch C, Mitsche S, Poelt P, *et al.* (2008), 'Microstructure modeling of the dynamic recrystallization kinetic during turbine disc forging of the nickel based superalloy Allvac 718 Plus™', in Reed R, Green K A, Caron P, Gabb T P, Fahrmann M G *et al.* (eds), *Proceedings of the 11th International Symposium on Superalloys*, Warrendale, PA: The Minerals, Metals and Materials Society, 855–861.

Jackman L A, Maurer G E and Widge S (1993), 'New knowledge about white spots in superalloys', *Advanced Materials and Processes*, 143(5), 18–25.

Jackson M P and Reed R C (1999), 'Heat treatment of UDIMET 720Li: the effect of microstructure and properties', *Materials Science and Engineering*, 259, 85–97.

Kalluri S, Rao K B S, Halford G R and McGaw M A (1994), 'Deformation and damage mechanisms in Inconel 718 superalloy', in Loria E A (ed.), *Proceedings of Superalloys 718, 625, 706 and Various Derivates*, Warrendale, PA: The Minerals, Metals & Materials Society, 593–606.

Kestler H and Clemens H (2005), 'Production, processing and application of (TiAl)-based alloys', in Leyens C and Peters M (eds), *Titanium and Titanium Alloys: Fundamentals and Applications*, Weinheim: Wiley-VCH.

Klopp W D (1995), 'IN 718', in *Aerospace Structural Metals Handbook*, Code 4103, CINDA/USAF CRBA, Handbook Operation, Purdue University, 1–102.

Knippscheer S and Frommeyer G (2006), 'Neu entwickelte TiAl-Basislegierungen für den Leicht-bau von Triebwerks- und Motorkomponenten – Eigenschaften, Herstellung, Anwendung', *Materialwissenschaft und Werkstofftechnik*, 37, 724–730.

Kracke A (2010), 'Superalloys, the most successful alloy system of modern times – past, present and future', in Ott E A, Groh J R, Banik A, Dempster I, Gabb T P, *et al.* (eds), *Proceedings of the 7th International Symposium on Superalloy 718 and Derivates*, Warrendale, PA: The Minerals, Metals & Materials Society, 13–50.

Krempaszky C, Werner E A and Stockinger M (2005), 'Measurement of macroscopic residual stress and resulting distortion during machining', *Materials Science and Technology*, 4, 109–118.

Kuchler H, Wieser V, Bauer R, Stampler F and Bode R (2003), 'Alloy 718 for large forgings', *Proceedings of the 15th International Forgemasters Meeting*, Kobe City, Japan, 492–499.

Lehnert W, Cuong N, Wehage H and Werners R (1993), 'Simulation der Austenitkornfeinung beim Walzen', *Stahl und Eisen*, 113, 103.

Loria E A (2000), 'Gamma titanium aluminides as prospective structural materials', *Intermetallics*, 8, 1339–1345.

Lütjering G and Williams J C (2007), *Titanium*, 2nd edn, Berlin: Springer.

Malinov S, Markovsky P, Sha W and Guo Z (2001), 'Resistivity study and computer modelling of the isothermal transformation kinetics of Ti–6Al–4V and Ti–6Al–2Sn–4Zr–2Mo–0.08Si alloys', *Journal of Alloys and Compounds*, 314, 181–192.

Mataya M C and Matlock D K (1989), 'Effects of multiple reductions on grain refinement during hot working of alloy 718', in Loria E A (ed.), *Proceedings of Superalloy 718: Metallurgy and Applications*, Warrendale, PA: The Minerals, Metals & Materials Society, 155–178.

Mataya M C, Nilsson E R and Krauss G (1994), 'Comparison of single and multiple pass compression tests used to simulate microstructural evolution during hot working of alloys 718 and 304L', in Loria E A (ed.), *Proceedings of Superalloys 718, 625, 706 and Various Derivates*, Warrendale, PA: The Minerals, Metals & Materials Society, 331–343.

Montero R E, Housefield L G and Mace R S (2005), 'Isothermal and hot-die forging', *ASM Handbook*, 14, 183–192.

Moyer J M, Jackman L A, Adasczik C B, Davis R M and Jones R F (1994), 'Advances in triple melting superalloys 718, 706 and 720', in Loria E A (ed.), *Proceedings of Superalloys 718, 625, 706 and Various Derivates*, Warrendale, PA: The Minerals, Metals & Materials Society, 39–48.

MW Racing (2011), *The Next Level of Performance*™. Märkisches Werk GmbH, Haus Heide 21, 58553 Halver, Germany. Available from: http://www.mwracing.eu, [accessed 2 February 2011].

Nastac L, Sundarraj S, Yu K-O and Pang Y (1998), 'The stochastic modeling of solidification structures in alloy 718 remelt ingots', *JOM*, 50, 30–35.

Noda T (1998), 'Application of cast gamma TiAl for automobiles', *Intermetallics*, 6, 709–713.

Oberwinkler B (2010), *Fatigue-Proof and Damage Tolerant Lightweight Design of Ti–6Al–4V Forgings*, PhD thesis, Montanuniversität Leoben Austria.

Peters M and Clemens H (2010), 'Titan, Titanlegierungen und Titanaluminide – Basis für innovative Anwendungen', *BHM*, 155, 402–408.

Peters M and Leyens C (2002), *Titan und Titanlegierungen*, Cologne: DGM.

Radavich J F (1989), 'The physical metallurgy of cast and wrought alloy 718', in Loria E A (ed.), *Proceedings of Superalloy 718: Metallurgy and Applications*, Warrendale, PA: The Minerals, Metals & Materials Society, 229–240.

Radis R, Zickler G, Stockinger M, Sommitsch C and Kozeschnik E (2010), 'Numerical simulation of the simultaneous precipitation of δ and γ' phases in the Ni-base superalloy ATI Allvac718Plus', in Ott E A, Groh J R, Banik A, Dempster I, Gabb T P, *et al.* (eds), *Proceedings of the 7th International Symposium on Superalloy 718 and Derivates*, Warrendale, PA: The Minerals, Metals & Materials Society, 569–578.

Radis R, Schaffer M, Albu M, Kothleitner G, Pölt P and Kozeschnik E (2009), 'Multimodal size distributions of γ' precipitates during continuous cooling of UDIMET 720Li', *Acta Materialia*, 57, 5739–5747.

Reed R C (2006), *The Superalloys: Fundamentals and Applications*, New York: Cambridge University Press.

Reiter G, Maronnier V, Sommitsch C, Gäumann M, Schützenhöfer W and Schneider R (2003), 'Numerical simulation of the VAR process with calcosoft-2D and its validation', *LMPC 2003: Proceedings of the International Symposium on Liquid Metal Processing and Casting*, Nancy, France, 77–86.

Richter K-H, Heins K-U and Specht J-U (2011), 'Qualifizierung des Elektronenstrahlschweißens zum Fügen einer Blisktrommel eines Flugtriebwerkes', MTU Aero Engines, Munich. Available from: http://mtu.de/de/technologies/engineering_news/production/Richter_Qualifizierung_Blisktrommel.pdf, [accessed 11 February 2011].

Sauthoff G (1996), 'State of intermetallics development', *Materials and Corrosion*, 47(7), 589–594.

Schirra J J, Caless R H and Hatala R W (1991), 'The effect of Laves phase on the mechanical properties of wrought and cast + HIP Inconel 718', in Loria E A (ed.), *Proceedings of Superalloys 718, 625, 706 and Various Derivates*, Warrendale, PA: The Minerals, Metals & Materials Society, 375–388.

Schmoelzer T, Liss K D, Zickler G A, Watson I J, Droessler L M, *et al.* (2010), 'Phase fractions, transition and ordering temperatures in TiAl–Nb–Mo alloys: an in- and ex-situ study', *Intermetallics*, 18(8), 1544–1552.

Schutz R W and Watkins H B (1998), 'Recent developments in titanium alloy application in the energy industry', *Materials Science and Engineering A*, 243, 305–315.

Sellars C M and Whiteman J A (1979), 'Recrystallization and grain growth in hot rolling', *Materials Science*, 2, 187–194.

Semenova I P, Korshunov A I, Salimgareeva G Kh, Latysh V V, Yakushina E B and Valiev R Z (2008), 'Mechanical behavior of ultrafine-grained titanium rods obtained using severe plastic deformation', *Physics of Metals and Metallography*, 106, 211–218.

Sims C, Stoloff N and Hagel W (1987), *Superalloys II: High Temperature Materials for Aerospace and Industrial Power*, New York: Wiley.

Sommitsch C (1999), *Theorie und Modell der mikrostrukturellen Entwicklung von Nickel-Basis-Legierungen während dem Warmwalzen*, PhD thesis, University of Technology, Graz.

Sommitsch C and Mitter W (2006), 'On modelling of dynamic recrystallisation of fcc materials with low stacking fault energy', *Acta Materialia*, 54, 357–375.

Sommitsch C, Krumphals A, Candic M, Tian B and Stockinger M (2010), 'Modeling of grain growth in one and two phase materials by 2D cellular automata', *Proceedings of ICPNS 2010*, Guilin City, China.

Stockinger M and Tockner J (2005), 'Optimizing the forging of critical aircraft parts by the use of finite element coupled microstructure modeling', in Loria E A (ed.), *Proceedings of the Sixth International Special Emphasis Symposium on Superalloys 718, 625, 706 & Derivatives*, Warrendale, PA: The Minerals, Metals and Materials Society, 87–95.

Stockinger M, Riedler M and Huber D (2010), 'Effect of process modeling on product quality of superalloy forgings', in Ott E A, Groh J R, Banik A, Dempster I, Gabb T P, *et al.* (eds), *Proceedings of the 7th International Special Emphases Symposium on Superalloys 718 and Derivates*, Warrendale, PA: The Minerals, Metals & Materials Society, 183–197.

Stoschka M, Riedler M, Stockinger M, Maderbacher H and Eichlseder W (2010), 'An integrated approach to relate hot forging process controlled microstructure of in 718 aerospace components to fatigue life', in *Seventh International Special Emphasis Symposium on Superalloys 718, 625, 706 & Derivatives*, Pittsburgh, PA, 751–765.

Stotter C, Sommitsch C, Wagner J, Leitner H, Letofsky-Papst I, *et al.* (2008), 'Characterization of δ-phase in superalloy Allvac718Plus', *International Journal of Materials Research*, 99, 376–380.

Sundararaman M, Mukhopadhyay P and Banerjee S (1997), 'Carbide precipitation in nickel base superalloys 718 and 625 and their effect on mechanical properties', in Loria E A (ed.), *Proceedings of Superalloys 718, 625, 706 and Various Derivates*, Warrendale, PA: The Minerals, Metals & Materials Society, 367–378.

Tetsui T (1999), 'Gamma Ti aluminides for non-aerospace applications', *Current Opinion in Solid State and Materials Science*, 4(3), 243–248.

Thamboo S V, Yang L and Schwant R C (2000), 'Large forgings of alloy 706 and alloy 718 for land based gas turbines', *Proceedings of the 14th International Forgemasters Meeting*, Wiesbaden, Germany, 330–335.

Tien J K (1989), *Superalloys, Supercomposites and Superceramics*, Boston, MA: Academic Press, 49–97.

Wallgram W, Schmoelzer T, Das G, Güther V and Clemens H (2009), 'Technology and mechanical properties of advanced γ-TiAl based alloys', *International Journal of Materials Research and Advanced Techniques (Zeitschrift für Metallkunde)*, 100, 1021–1030.

Wang K (1996), 'The use of titanium for medical applications in the USA', *Materials Science and Engineering A*, 213, 134–137.

Warchomicka F G (2008), *Microstructural Behaviour of Near β-Titanium Alloys during Thermomechanical Processes*, PhD thesis, Vienna University of Technology.

Wasle G, Sommitsch C, Kleber S, Buchmayr B and Clemens H (2003), 'Physikalische Simulation der mehrstufigen Primärumformung von Alloy 80a', *BHM*, 148, 293–298.

Weis M J, Mataya M C, Thompson S W and Matlock D K (1989), 'The hot deformation behavior of an as-cast alloy 718 ingot', in Loria E A (ed.), *Proc. Superalloy 718, E.A.*, Warrendale, PA: The Minerals, Metals & Materials Society, 135–154.

Wen Y H, Chen L Q, Hazzledine P M and Wang Y (2001), 'A three-dimensional phase-field model for computer simulation of lamellar structure formation in γTiAl intermetallic alloys', *Acta Materialia*, 49(12), 2341–2353.

Yamada M (1996), 'An overview on the development of titanium alloys for non-aerospace application in Japan', *Materials Science and Engineering A*, 213, 8–15.

Yue K O and Dominique J A (1989), 'Control of solidification structure in VAR and ESR processed alloy 718 ingots', in Loria E A (ed.), *Proceedings of the International Symposium on the Metallurgy and Applications of Superalloy 718*, Warrendale, PA: The Minerals, Metals & Materials Society, 33–48.

Zanner F, Williamson R L, Harrison R P, Flanders H D, Thompson R D and Szeto W C (1989), 'Vacuum arc remelting of alloy 718', in Loria E A (ed.), *Proceedings of Superalloy 718*, Warrendale, PA: The Minerals, Metals & Materials Society, 17–32.

Zhao D, Guillard S and Male A T (1997), 'High temperature deformation behavior of cast alloy 718', in Paton N, Cabral T, Bowen K, and Tom T (eds), *Superalloys 718, 625, 706 and Various Derivates*, Warrendale, PA: The Minerals, Metals & Materials Society, 193–204.

Zickler G A, Radis R, Hochfellner R, Schweins R, Stockinger M and Leitner H (2009), 'Microstructure and mechanical properties of the superalloys ATI Allvac 718Plus', *Materials Science and Engineering*, 523, 295–303.

Index

AA7055 Al alloy
precipitation behaviour and strength, 290–3
experimental yield strength vs model predictions, 293
precipitates HRTEM and FFT-filtered images, 291
precipitates size and volume fraction evolution, 292
precipitates TEM images, 291
ABAQUS, 325, 362
accelerated cooling, 147
accumulative roll bonding (ARB), 126
additivity rule, 154–5
advanced high-strength steels (AHSS), 13–14
complex-phase steels, 14
dual-phase steels, 14
elongation and tensile strength of steels, 13
martensitic steels, 14
transformation-induced plasticity steels, 14
advanced thermomechanical processing, 238–40
ageing behaviour
age hardening, 276–83
alloying elements and precipitates in Al alloys, 278
heat treatable Al alloys, 277–9
precipitation effect on mechanical properties, 281–3
aluminium alloy processing, 267–94
case studies and applications, 286–93
characterisation and test methods, 284–6
cold working effect on microstructure and properties, 275–6
future trends, 293–4
hot working effect on microstructure and properties, 269–75
evolution of strength of AA7055 Al alloy, 281
HRTEM and FFT-filtered images of AA7055 Al alloy, 280
HRTEM and FFT-filtered images of Al alloy η' phase particles, 283
precipitates in grains and grain boundaries, 282
SAED patterns of AA7055 Al alloy, 280
TEM images of precipitates in AA7055 Al alloy, 279
typical mechanical performance of aluminium alloys, 283
ageing heat treatment, 314
Airbus, 331
aircraft industry, 339, 341
allowable truncation error, 190
Alloy 718, 347, 365–7, 369, 373, 375
Alloy 80A, 365, 367, 370–1
aluminium alloys
ageing behaviour and microstructure evolution, 267–8, 267–94
AA5086 Al alloy EBSD map, 277
ageing behaviour and age hardening, 276–83
cold working effect on microstructure and properties, 275–6
future trends, 293–4
hot working effects on microstructure and properties, 269–75
case studies and applications, 286–93
AA7055alloy precipitation behaviour and strength prediction, 290–3
ageing behaviour and microstructure evolution, 286–90
chemical components, 287
final forgings with different deformations optical microstructure, 288
final forgings with different deformations TEM microstructure, 288
forging alloy micrographs after ageing treatments, 289
original microstructure, 287
room temperature properties, 290
rotating socket sleeve, 287
characterisation and test methods, 284–6
atom probe microscopy lattice rectification, 285
grain deformation and evolution during compression, 286

microstructure characterisation methods, 284
modelling technology, 285–6
performance-testing methods, 284–5
parts from various plastic forming methods, 268
aluminium panels
microstructure control in creep-age forming, 298–333
applications and future trends, 331–3
precipitation control, 301–7
precipitation hardening modelling, 313–31
process and importance, 298–301
stress/strain ageing testing methods, 307–13
approximation-based optimisation (ABO), 41–2
algorithm, 41
flow chart, 43
idea, 41
artificial ageing, 301
artificial neural networks (ANNs), 39–41
feed forward with an n-2-1 structure, 40
single artificial neurone, 39
ASTM 5, 375
atom probe tomography (3D APT), 284
austenisation
austenite grain growth, 215
carbon concentration and distribution profile, 216
austenite homogenisation, 216–17
boron steel constitutive modelling, 218–22
austenite fraction modelling evolution, 218–19
austenite fraction vs temperature, 222
dilatometry strain evaluation, 219
experimental vs computed starting and finishing temperatures, 221
incubation time modelling, 218
strain-temperature relationships, 221
unified constitutive equations, 220
unified material model determination and validation, 220–2
hot stamping and cold die quenching experimental study, 217–18
test programme, 217
mechanism, 214–17
isothermal TT diagram, 215
nucleation, 214–15
austenite
deformation after transformation, 12
deformation during transformation, 12
deformation prior to transformation, 12
autoclave age forming *see* creep-age forming
Avrami coefficient, 155–7
Avrami theory, 364

B-1B' Long Range Combat Aircraft, 331
β-titanium alloys, 342, 358
bainite transformation, 146
cooling process constitutive modelling, 224–34
model calibration, 230–1
nucleation and growth, 225–8
pre-deformed material modelling, 231–4
volume fraction evolution, 228–30
hot stamping and cold die quenching experimental study, 223–4
critical temperatures upon pre-deformation, 224
test programme and temperature profile, 223
mechanism, 222–3
batch severe plastic deformation, 122–6
illustration, 124
three basic options for billet rotation, 125
Brinell hardness, 284
brittle–ductile transition (BDT) temperature, 243

CALPHAD method, 153
carbide strengthening, 351
Cauchy density function, 200
Cauchy stress, 326
cellular automata (CA), 29–30, 162, 164–7, 364
nucleus of the a phase and the surrounding cells in ferrite-austenite state, 165
cellular automata model with finite element (CAFE), 175
chill casting, 115–16
classical plasticity theory, 321
Coble creep mechanism, 252
coefficients, 167–71
cold die quenching
phase transformations modelling in steels, 210–35
cooling, 222–34
future trends, 234–5
heating, 214–22
cold hearth melting, 348
cold working, 269
columnar-grain-structured superalloys, 338
complex-phase steels, 14
computational thermodynamics, 153
Considère criterion, 247, 259
constant-stress creep-ageing test, 308–9
creep-ageing curves for different stress levels, 309
material preparation and test programme, 308
constitutive modelling
bainite transformation during cooling process, 224–34
model calibration, 230–1
boron steel CCT diagram vs predicted starting temperatures, 230
predicted vs experimental bainite fractions, 231
nucleation and growth, 225–8
bainite free energy volume changes, 225

bainite growth rate size vs temperature, 227
free energy change vs nucleus size, 226
pre-deformed material modelling, 231–4
experimental vs predicted boron steel CCT diagrams, 233
pre-deformed steel predicted CCT diagrams, 233
volume fraction evolution, 228–30
bainite nucleation rate vs temperature, 229
constrained Voronoi tessellation method (CVTM), 29
continuous annealing, 149–52, 174–5
dual-phase structure, DP600 characterisation, image map and austenite distribution, 151–2
experimental set-up in a Gleeble 380 simulator for physical modelling, 149
temperature, ferrite recrystallisation kinetics and volume fraction changes, 174
thermal cycle and volume fractions, 150
continuous billets
SPD process, 126–8
classical ECAP and I-ECAP, 128
illustration, 127
continuous-cooling transformation (CCT), 147
continuous dynamic recrystallisation, 273
continuous recrystallisation, 241
crankshaft forging, 55–8
bending-tool displacement, 57
decision variables and accuracy values, 58
second-order response surface, 58
TR forging process, 56
creep-age forming, 294
aluminium panels microstructure control, 298–333
applications and future trends, 331–3
A380 aircraft wing panel, 331
current applications, 331
future trends, 332–3
precipitation control, 301–7
AA7055 microstructure, 305
average precipitate size after ageing, 306
creep-aged 7055 alloy micrograph, 306
precipitation hardening, 301–2
stress-ageing sample fixture, 302
stress effect on age-precipitation hardening, 304
stress/strain age hardening, 302–7
yield strength variation upon different creep-age conditions, 307
precipitation hardening modelling, 313–31
constitutive modelling, 318–20
finite element modelling, 324–5
mechanisms of precipitation hardening, 313–18
modelling verification and discussion, 325–31
physically based unified creep-ageing constitutive equations, 321–4
process and importance, 298–301
process diagram, 299
stress/strain ageing testing methods, 307–13
Al alloy age-forming mechanism, 312
constant-strain stress relaxation ageing test, 309–10
constant-stress creep-ageing test, 308–9
constant-stress strain relaxation ageing curves, 311
constant-stress vs constant-strain ageing, 310–13
creep damage
constitutive equations determination, 205, 207
Al alloy material constants at 150°C, 207
experimental vs computed curves, 205
cubic coarsening law, 319
cyclic extrusion compression (CEC), 122–3, 123, 124

DEFORM, 362
deformation continuous-cooling transformation (DCCT), 147
deformation-induced ferrite transformation (DIFT), 239–40
deformation texture, 276
design-of-experiment (DoE), 38
differential equation model, 157–8
parameters, 158
diffusion, 158–62
calculated changes in the carbon distribution in the austenite, 163–4
carbon distribution in the austenite, 162
solution domain for the two-dimensional problem, 161
diffusion bonding (DB), 133
superplastically formed wide-chord fan blade, 134
digital material representation (DMR) approach, 250, 255
dilatometer, 147
dilatometric tests, 147–9
DIL 805 dilatometer, 148
dilatation curve, 148
dilatometry strain, 219
discontinuous dynamic recrystallisation, 24–8
304L stainless steel application, 27–8
multiple homogeneous equivalent media, 24–5
overall volume changes, 26–7
RX-NR and NR-NR interfaces migration, 26
RX-NR interfaces migration, 25–6
dislocation density, 184
dislocation hardening, 323
dual phase steels, 14, 146–7
dynamic ageing effects, 320
dynamic grain growth, 184
dynamic recovery, 272

dynamic recrystallisation, 68–72, 272–4
chemical composition and austenite grain size influence on peak strain, 71
equations describing the peak strain, 71
grain size dependence on Z for various Nb and Nb-Mo microalloyed steels, 72
stress–strain curve of hot-deformed austenite, 69
thermomechanical process, 104
metadynamic recrystallised grain size evolution, 105
simulation of torsion tests, 105
V, Mo, Ti and Nb effect on solute retardation parameter (SRP), 70

elastic domain, 181
electron backscattered diffraction (EBSD), 247, 250, 259, 284
electroslag remelting, 345, 367–8
energy factor, 231–2
equal-channel angular pressing (ECAP), 122–3, 126–8
ESH constant-stress creep machine, 308
Euler method, 185–7
Eulerian formulations, 17
Eurotherm heat generator, 310
evolutionary algorithms, 46–9, 192

fast evolutionary programming (FEP), 200–1
constants for constitutive equations, 201
material constants determination results, 202
fast Fourier transform (FFT), 279
ferritic transformation, 146, 158
fibrous texture, 6
Fick's second law, 159, 214
finite difference methods (FDMs), 362
finite element (FE), 121, 145
finite element method (FEM), 239, 286, 326–8, 362
boundary and loading conditions, 327
loading process, 327
flow stress, 20–1, 21–4
focused ion beam (FIB) method, 250
FORGE, 362
FORGE2 code, 54
front-tracking method, 160–1

gamma double-prime phase, 366
Gaussian mutation, 200
general objective function, 197–8
genetic algorithm, 43–6, 199
description basic terms, 44
genetic operators, 45
geometrical dynamic recrystallisation, 273
Gleeble 3800 simulator, 204
global sensitivity factors, 169
gradient-based method (GBM), 199–200
grain boundary, 114–15
migration, 20

grain enhancement
severe plastic deformation for refinement of properties, 114–35
overview, 114–17
processes, 122–8
severe thermomechanical treatment, 117–22
ultrafine-grained metals applications, 131–4
ultrafine-grained metals properties produced by SPD, 129–31
grain growth, 5
after recrystallisation, 76–7
coefficients of Eq. 4.13 for various steel grades, 77
grain refinement, 68–76
metal forming, 104–9
modelling methods, 88–103
physical models, 96–103
semi-empirical models, 88–96
precipitation interactions, 77–88
recrystallisation in hot working of steels, 67–109
grain interface, 30–2
grain refinement, 68–76
dynamic recrystallisation, 68–72
recrystallised grain size, 76
severe plastic deformation for enhancement of properties, 114–35
overview, 114–17
processes, 122–8
severe thermomechanical treatment, 117–22
ultrafine-grained metals applications, 131–4
ultrafine-grained metals properties produced by SPD, 129–31
static and metadynamic recrystallisation, 72–5
types of heavy deformation processing, 10
grain size, 114–15
control, 237–8
grain structure
refining mechanism by SPD, 117–18
grain/subgrain size distributions in AA1070, 119
skewed chessboard-like microstructure in Al99.992, 118
refining new approaches, 116–17
refining traditional methods, 115–16
Guinier-Preston zones, 305
Gulfstream G-IV aircraft, 331

Hall–Petch model, 122
Hall–Petch relation, 248, 251–3
Hawk aircraft, 331
heterogeneous deformation, 242
high-pressure torsion (HPT), 122–3
high resolution transmission electron microscopy (HRTEM), 279

homogeneous equivalent media (HEM), 24–5
hot-die forging, 350
hot forming
  unified constitutive equations determination for modellling of steel, 180–207
    case studies, 201, 203–5
    equation form, 181–5
    integration methods, 185–91
    objective functions for optimisation, 191–8
    optimisation methods for materials constants determination, 198–201
hot-rolled dual phase steels, 147
hot stamping
  phase transformations modelling in steels, 210–35
    cooling, 222–34
    future trends, 234–5
    heating, 214–22

incremental ECAP (I-ECAP), 128
incubation time, 218
inert gas condensation, 116
ingot metallurgy, 349
inoculants, 115
interfacial energy, 226
invariant-planestrain deformation, 146
inverse analysis, 58–62, 167–8
  flow chart, 60
  measured and calculated loads using an FE model, 63
  metamodel-based inverse analysis flow chart, 60
isothermal forging, 350

Jacobian matrix, 187
Johnson–Mehl models, 8
Johnson–Mehl–Avrami–Kolmogorov models, 19

Khan–Huang–Liang (KHL) flow stress model, 254
kinematic hardening, 183

Lagrangian formulations, 17
Laguerre–Voronoi tessellation method (LVTM), 29
laminar cooling, 171, 173–4
  cost function for a required volume fraction of martensite, 173
Lancer Evolution VI, 344
Laves phase, 367
least squares method, 191–4
  errors definition and k and n determination, 191
  experimental vs computed curves and sums of errors squares, 193
level set method (LSM), 30–2
lever rule, 148
logarithmic error *see* true error

MARC simulation, 326
martensite
  strengthening, 12
  transformation, 146
martensitic steels, 14
MatCalc, 363, 364
Mayes 100 kN testing machine, 308
mechanical alloying, 116–17
medical implants, 132–3
  plates made of nanostructured titanium, 133
metadynamic recrystallisation, 72–5
  dependence of $t_{0.5}$ on applied strain, 75
  equations to determine the time $t_{0.5srx}$, 74
metal forming, 104–9
  microstructure evolution, 3–15
    advanced high-strength steels (AHSS), 13–14
    carbon steels predicting models, 6–9
    control techniques, 12–13
    future trends, 14–15
    mechanism, 4–6
    overview, 3–4
    strengthening mechanism and mechanical properties, 10–12
  modelling microstructure techniques, 17–32
    coupling between homogenous microstructure description and constitutive law, 20–4
    discontinuous dynamic recrystallisation, 24–8
    future trends, 32
    models general features based on state variables, 18–20
    prediction importance, 17–18
    recrystallisation modelling at microscopic scale, 28–32
  optimising modelling techniques, 35–64
    future trends, 62–4
    metamodelling and optimisation strategy application, 52–62
    nature-inspired optimisation techniques, 42–52
    optimisation strategy, 36–42
  process problem, 59, 61
  thermomechanical process based on dynamic recrystallisation, 104
  thin-slab direct rolling, 104, 106–9
metal processing industry, 339
metallurgical modelling, 17
metamodel-driven optimisation (MDO), 36–8
  flow charts, 37
metamodelling, 52–62
  application to inverse analysis, 58–62
  general classification of optimisation problems in metal forming process, 52
micro-manufacturing, 133–4
  dome component made by micro-bulging, 135

microalloyed steels
deformation-induced grain refinement principles, 240–3
effect on mechanical properties, 243–5
flow stress multiscale modelling, 249–58
DMR approach microstructure designs, 255
equivalent plastic stress distributions, 256
sample results, 256
uniform elongation determination examples, 257
microstructure evolution and work hardening modelling, 237–59
applications, results and discussion, 245–9
future trends, 258–9
thermomechanical and severe plastic deformation processing, 239–40
microalloying elements, 73
microstructure control
nickel, titanium and special alloys processing, 337–77
future trends, 376–7
materials modelling and process simulation, 361–5
microstructures and mechanical properties, 350–61
production processes, 344–50
special alloys application, 339–44
process and materials optimisation, 365–76
aircraft engine sketch, 374
Alloy 718 billet recrystallised fraction simulations, 375
experimental vs simulated grain size analysis, 376
grain structure formation vs hot-forming temperature, 372
incomplete recrystallisation from uneven temperature, 372
precipitation kinetics illustration, 374
recrystallising grains in shell, 371
secondary remelting process diagram, 369
semi-finished nickel-based alloys production route, 366
microstructure coupling
homogenous description and constitutive law, 20–4
modelling strain-hardening stages and flow stress, 21–4
strain hardening, recovery and flow stress, 20–1
microstructure evolution
advanced high-strength steels (AHSS), 13–14
aluminium alloy processing, 267–94
age hardening, 276–83
case studies and applications, 286–93
characterisation and test methods, 284–6
future trends, 286–93
application, results and discussion, 245–9
multiaxial compression tests schedules, 246
steels chemical composition, 245
ultra fine-grained steels microstructures, 247
Y-MA and IF steels flow stress and stress-strain curves, 247
Y-MA steel electron microscopy results, 248
carbon steels predicting models, 6–9
equation of recrystallisation and grain growth, 8
integrated model conceptual scheme, 7
Johnson–Mehl-type transformation models, 9
cold working effect on microstructure and properties, 275–6
deformation texture, 276
grain shape changes, 276
substructure, 276
control techniques, 12–13
deformation after austenite transformation, 12
deformation during austenite transformation, 12
deformation prior to austenite transformation, 12
rolling, 12–13
coupling between homogenous description and constitutive law, 20–4
discontinuous dynamic recrystallisation, 24–8
future trends, 14–15, 32
gradient functional components, 15
hot working effects on microstructure and properties, 269–75
cast structure and properties, 269–70
coarse grain structures in Al alloys, 270–1
dynamic recovery and crystallisation, 271–4
fibrous structure formation, 274–5
mechanism, 4–6
metal forming, 3–15
modelling in microalloyed steels, 237–59
deformation-induced grain refinement principles, 240–3
effect on mechanical properties, 243–5
flow stress multiscale modelling, 249–58
future trends, 258–9
thermomechanical and severe plastic deformation processing, 239–40
modelling techniques in metal forming processes, 17–32
models general features based on state variables, 18–20
overview, 3–4
concept and goal of prediction and control technology, 4
precipitation-hardening mechanism, 314–16
7000 series Al alloys microstructure, 316
AA7xxx at different ageing stages, 315
precipitate radius evolution during ageing, 317

prediction importance, 17–18
recrystallisation modelling at microscopic scale, 28–32
strengthening mechanism and mechanical properties, 10–12
grain refinement, 10
martensite strengthening, 12
precipitation hardening, 10–11
solid resolution hardening, 11
work hardening, 11
model parameters, 100–3
particle number density, volume fracture and size evolution, 102
prediction of softening comparison, 103
summary of recovery, recrystallisation and precipitation modules, 101
molecular dynamic modelling, 285
molecular-dynamics computer simulation, 122
Monte Carlo method, 29–30, 285
multiaxial compression tests, 246
multiaxial constitutive equations, 325–6
multiaxial forging (MF), 123, 125
multiple-step fitting process, 324

nature-inspired optimisation, 42–52
evolutionary algorithms, 46–9
genetic algorithms, 43–6
particle swarm optimisation, 49–50
simulated annealing algorithm, 50–2
near-α-titanium alloys, 356
new-phase particle, 228
Newton's linearisation method, 188
nickel alloys
future trends, 376
materials modelling and process simulation, 362–3
microstructure control in processing, 337–77
microstructures and mechanical properties, 350–61
$Cr_{23}C_6$ particles grain boundary film, 352
nickel-based superalloys, 350–5
Ti particle EFTEM analysis, 351
UDIMET 720 Li precipitates morphologies, 353
production processes, 344–50
nickel-based superalloys, 344–7
vacuum remelting furnace sketch, 346
special alloys application, 339–44
nickel-based superalloys, 339
Nimonic 80A ship diesel valve, 340
nickel-based superalloys
materials modelling and process simulation, 362–3
microstructures and mechanical properties, 350–5
nominal chemistry, 354
ultimate tensile strength, 355
production process, 344–7
special alloys application, 339

niobium (Nb), 77–9
non-crystallisation temperature(T_{nr})
definition, 79–80
various thermomechanical processes, 80
parameters, 84–8
changes with the addition of Mo to Nb-bearing steels, 87
pass strain influence for Nb and Nb-Ti steels, 86
three different ranges of the effect of interpass time, 85
physical metallurgical principles, 80–4
interactions between deformation, recovery, recrystallisation and precipitation, 81
non-equilibrium grain boundary, 130
normalised precipitate size, 322
NR grains, 24
NR-NR interfaces
RX-NR migration, 26
nucleation, 20, 226

objective functions, 194–8
general objective function, 197–8
shortest-distance method, 194–5
errors definition, 195
universal multi-objective function, 195–7
definition of errors, 196
optimisation
future trends, 62–4
metamodelling and strategy application, 52–62
application to inverse analysis, 58–62
crankshaft forging process, 55–8
general classification of optimisation problems in metal forming process, 52
tool shape, 53–5
modelling techniques in metal forming process, 35–64
nature-inspired optimisation techniques, 42–52
strategy, 36–42
approximation-based optimisation strategy, 41–2
artificial neural networks, 39–41
classical procedure flow chart for real material processing, 36
metamodel-driven optimisation strategy, 36–8
response surface methodology, 38–9
ordinary differential equations, 180
Orowan mechanism, 249

particle swarm optimisation (PSO), 49–50
peak ageing size, 316
pearlitic transformation, 146
phase field method, 30–1, 286
phase transformation
application in rolling and annealing in dual phase steels, 171–5

continuous annealing, 174–5
laminar cooling, 171, 173–4
cooling, 222–34
bainite transformation constitutive modelling, 224–34
bainite transformation experimental study, 223–4
bainite transformation mechanism, 222–3
experimental techniques, 146–52
continuous annealing, 149–52
dilatometric tests, 147–9
material selection for models validation, 146–7
future trends, 175–6
basic concept of the DMR *vs.* conventional approach, 176
heating, 214–22
austenisation experimental study, 217–18
austenisation mechanism, 214–17
boron steel austenisation constitutive modelling, 218–22
hot stamping and cold die quenching in steels, 210–35
automotive industry applications, 213
future trends, 234–5
HSDQ process diagram, 211
process, 210–12
USIBOR 1500P and DUCTIBOR 500P applications, 213
USIBOR 1500P CCT diagram, 212
USIBOR 1500P chemical composition, 212
modelling, 153–71
CCT diagrams comparison, 172
coefficients determined using inverse analysis, 171
coefficients identification, 167–71
general information, 153–4
identification, testing and validation of models, 171
managing multi-component systems using prediction of thermodynamics, 153
models, 153–67
numerical solution, 158
volume fractions comparison, 172
modelling in steel, 145–76
overview, 145–6
physical models, 96–103
plastic forming technology, 267
plastic processing
effect of hot working on Al alloy microstructure and properties, 269–75
effect on dynamic recovery and recrystallisation, 271–4
Al subgrains after torsional strain, 272
dynamic recovery, 272
dynamic recrystallisation, 272–4
grains from Al alloys, 273
effect on fibrous structure formation, 274–5
7705 Al alloy flow lines, 275
effects on Al alloys coarse grain structures, 270–1
6005A tube transverse section structure, 271
effects on cast structure and properties, 269–70
cast Al-11 mass % Si alloy microstructure, 269
plastometric tests, 62
powder metallurgy, 347
power law equation, 191
precipitate-coarsening process, 318
precipitation, 5–6
recovery interaction, 84
fractional softening and the recrystallised fraction evolution, 85
recrystallisation interaction, 81–2
illustration, 83
recrystallisation stop temperature (RST) and non-crystallisation temperature definition, 82
variations of T_{nr}, 83
precipitation hardening, 10–11
constitutive modelling, 318–20
material strength increment, 319–20
precipitate radius growth, 318–19
finite element modelling, 324–5
mechanisms, 313–18
binary phase diagram and typical temperature profile, 314
contributions to yield strength, 317–18
heat-treatable alloys, 313
heat treatment, 313–14
microstructure evolution, 314–16
solute, age, and dislocation hardening contributions to yield strength, 317
modelling, 313–31
nanosized VC produced by interphase boundary precipitation, 11
physically based unified creep-ageing constitutive equations, 321–4
material constants determination, 324
unified creep-ageing constitutive equations, 321–4
verification modelling and discussion, 325–31
creep strain, precipitate radius evolution and yield strength, 328–31
creep strain distribution, 330
FE model and numerical procedures, 326–8
multiaxial constitutive equations, 325–6
normalised precipitate size variation and evolution, 330
springback prediction, 328
von Mises stress distribution, 329
precipitation module, 98–100
pseudocode, 45, 47, 48, 50, 51

radial basis functions (RBFs), 39
rapid cooling, 313

rapid transformation annealing (RTA), 240
recovery, 4–5
  precipitation interaction, 84
    fractional softening and the recrystallised fraction evolution, 85
  recrystallisation interaction, 82, 84
recovery module, 98
  stress relaxation curves, 99
recryrstallisation-precipitation interactions, 77–88
  non-crystallisation temperature(T_{nr}), 79–80
  parameters influencing T_{nr}, 84–8
  physical metallurgical principles behind T_{nr}, 80–4
  strain-induced precipitation kinetics of Nb, 77–9
    amount of precipitation as a function of temperature, 78
    solubility products of various nitrides and carbides in austenite, 78
    strain-induced precipitates, 79
recrystallisation, 5
  grain growth, after, 76–7
    coefficients of Eq. 4.13 for various steel grades, 77
  grain growth in hot working of steels, 67–109
  grain refinement, 68–76
  grain size, 76
    austenite equations, 76
  metal forming, 104–9
  modelling at microscopic scale, 28–32
    digital representation of microstructure, 28–9
    grain interface modelling, 30–2
    probabilistic methods, 29–30
  modelling methods, 88–103
    physical models, 96–103
    semi-empirical models, 88–96
  precipitation interactions, 77–88
recrystallisation module, 97–8
repetitive corrugation and straightening (RCS), 123
response surface methodology (RSM), 38–9
retained cementite, 216
Rolls Royce, 332
rotary-die equal-channel angular pressing (RD-ECAP), 270
RX grains, 24
RX-NR interfaces
  migration, 25–6
    mobile surface fractions, 25
  NR-NR migration, 26

scanning electron microscopy (SEM), 284
selected-area electron diffraction (SAED), 279
semi-empirical models, 88–96
sensitivity analysis, 168–71
  start and end temperatures of the transformation, 170
  volume fractions of structural components, 170
severe plastic deformation, 238–41
  grain refinement and enhancement of properties, 114–35
  overview, 114–17
    grain size/boundary effects on metallic materials, 114–15
    refining grain structure new approaches, 116–17
    refining grain structure traditional methods, 115–16
    yield stress *vs.* ferrite grain size, 115
  processes, 122–8
    batch, 122–6
    continuous billets, 126–8
  severe thermomechanical treatment, 117–22
    condition effects, 120–1
    grain/subgrain size distribution in Cu, 119
    material-dependent results, 118–20
    modelling, 121–2
    refining grain structure mechanism, 117–18
  ultrafine-grained metals
    applications, 131–4
    properties, 129–31
shortest-distance method, 194–5
sigma phase, 366–7
simulated annealing algorithm, 50–2
single-crystal superalloys, 338
single-pass hot rolling test, 246
small-angle X-ray scattering (SAXS), 284
solid resolution hardening, 11
solution heat treatment (SHT), 313
sputtering targets, 131–2
  illustration, 132
stacking fault energy, 274
stainless steel, 27–8
  initial grain size effect on grain size evolution and recrystallisation kinetics, 28
  temperature effect on the recrystallisation kinetics and flow stress, 27
state variables
  models general features, 18–20
    notion of microstructure, 18–20
    nucleation and grain boundary migration, 20
static grain growth, 184
static recovery, 323
static recrystallisation, 72–5
  dependence of $t_{0.5}$ on applied strain, 75
  equations to determine the time $t_{0.5mdrx}$, 74
  modelling, 88–93
    additivity principle, 89
    experimental evolution mean recrystallised and uncrystallised grain size, 91–2
    experimental mean total grain size, 93
    predicted evolution of grain size, 90

steels
hot forming modelling, 180–207
case studies, 201, 203–7
constitutive equations integration methods, 185–91
unified constitutive equations form, 181–5
phase transformation, 145–76
application in rolling and annealing in dual phase steels, 171–5
experimental techniques, 146–52
future trends, 175–6
modelling methods, 153–71
overview, 145–6
recrystallisation and grain growth, 67–109
grain growth after recrystallisation, 76–7
grain refinement, 68–76
metal forming, 104–9
modelling methods, 88–103
precipitation interactions, 77–88
strain-hardenable alloys, 268
strain hardening, 20–1
modelling stages, 21–4
curves for a strain rate of $0.1\,s^{-1}$, 22
curves for a strain rate of $0.01\,s^{-1}$, 22
curves for a strain rate of $1\,s^{-1}$, 23
curves for a strain rate of $10\,s^{-1}$, 23
experimental compression and modal curve, 24
strain-induced dynamic transformation (SIDT), 240
strain-induced precipitation
kinetics of Nb, 77–9
modelling of Nb(C,N), 93–6
interpass fractional softening *vs.* mean interpass temperature, 95
model structure flow chart, 95
strain recovery, 20–1
stress relaxation, 298
structural elements, 131
submicron-grained steels *see* ultra fine steels
submodelling, 254
superalloys, 338
superplastic forming (SPF), 133
superplastic material model, 193
supersaturated solid solution (SSSS), 278

tailored microstructure distribution, 234
tangent line method *see* Euler method
tensile testing, 284–5
texture, 6
thermo-mechanical control processing (TMCP), 3
ThermoCalc, 364
thin-slab direct rolling, 104, 106–9
austenite grain size distributions, 107
after the second and third insterstand forT_i = 1060°C, 108
after the second and third insterstand forT_i = 1090°C, 108
V and Nb solute drag influence on static recrystallisation, 107
time–temperature–transformation (TTT), 147
titanium alloys
future trends, 377
material simulation, 377
titanium aluminides, 377
materials modelling and process simulation, 363–4
titanium aluminides, 364–5
microstructure control in processing, 337–77
microstructures and mechanical properties, 355–8
β-forged Ti17, 359
fast air cooled Ti–6Al–4V α+β alloy, 357
forged and mill annealed Ti–6Al–4V α+β alloy, 356
forged and recrystallisation annealed Ti–6Al–4V α+β alloy, 358
pseudo-binary phase diagram schematic, 355
slow air cooled Ti–6Al–4V α+β alloy, 357
strength vs temperature Ashby plot, 360
titanium aluminides, 358–61
TNM micrograph after heat treatment, 361
production processes, 348–9
cold hearth melting facility schematic, 348
Ti–6Al–4V aircraft part closed die forging step, 349
titanium aluminides, 349–50
special alloys application, 341–4
Boeing aircraft airframes titanium usage, 341
golf club head, 342
medical implants, 343
titanium aluminides, 344
tolerance *see* allowable truncation error
tool shape, 53–5
cross-section of the approximation surface *g*, 55
final results for the approximation function *g*, 55
two-stage axisymmetric forging process, 53
transformation, 5
dynamic continuous-cooling transformation (CCT) diagrams, 6
transformation-induced plasticity steels, 14
transmission electron microscopy (TEM), 245–6, 248, 259, 284
trapezoid rule, 188
true error, 197
twist extrusion (TE), 123, 126

UDIMET 720Li, 353
ultra fine steels, 237
ultrafine-grained metals
applications, 131–4
medical implants, 132–3
micro-manufacturing, 133–4
sputtering targets, 131–2
structural elements, 131

superplastic forming/diffusion bonding of aerospace components, 133
properties produced by SPD, 129–31
tensile force *vs.* elongation curves, 129
unified constitutive equations, 220
unified theories, 321
uniform elongation, 244
universal multi-objective function, 195–7
USIBOR 1500 P, 211

vacuum arc remelting, 345, 368
vacuum induction melting, 345, 367
vacuum melting, 338
vertex models (VMs), 30–1
Vickers hardness, 284
viscoplastic constitutive equations
case studies, 201, 203–7
creep damage determination, 205, 207
steel and Al alloy viscoplastic-damage determination, 201, 203–5
determination for steel hot forming modelling, 180–207
form for hot metal forming, 181–5
cyclic hardening diagram, 182
deformation mechanisms, 181
uniaxial viscoplastic equations, 183–5
viscoplastic flow, recrystallisation and hardening modelling, 181–3
integration methods, 185–91
error control methods, 190–1
explicit forward Euler method, 186
forward integration, 185–6
implicit numerical integration, 186–8
numerical difficulties, 185
numerical integration flow chart, 189
objective functions for optimisation, 191–8
least squares method, 191–4
objective functions, 194–8
optimisation methods for material constants determination, 198–201
background, 198–9
fast evolutionary programming (FEP), 200–1
gradient-based method (GBM), 199–200
viscoplastic-damage
determination in steel and Al alloy, 201–5
AA6082 computed vs experimental stress-strain relationships, 204
boron steel and AA6802 material constants, 206
boron steel computed vs experimental stress-strain relationships, 204
viscoplasticity, 181
volume changes, 26–7
Voronoi tessellation method (VTM), 29

warm working, 269
Waspaloy, 373
weighted distance, 198
weighting factor, 197
work hardening, 11
modelling in microalloyed steels, 237–59
application, results and discussion, 245–9
deformation-induced grain refinement principles, 240–3
effects on mechanical properties, 243–5
flow stress multiscale modelling, 249–58
future trends, 258–9
thermomechanical and severe plastic deformation processing, 239–40

Zwick 1484 machine, 310